CONTENTS

* Short answers are given in this book, with full worked solutions for all exercises, large data set activities, exam-style questions and extension questions available to teachers by emailing education@harpercollins.co.uk

INTRODUCTION

Mathematics underpins everything. From a few axioms (basic statements accepted automatically), it is possible to derive all the Pure Mathematics contained within this book. When some modelling assumptions are included, techniques in Pure Mathematics can be used to explore Applied Mathematics, such as Statistics and Mechanics. A sound knowledge of Mathematics is necessary to understand how the world and the rest of the universe behave. For example, it is essential to the investigation of the Big Bang, the mapping of the human genome, understanding the environment and the financial markets.

The aim of this book is to help make your study of advanced mathematics interesting and successful. However, the content can be demanding and there is no need to worry if you do not understand it straight away. Discuss ideas with other students, and of course check with your teacher or tutor. Most important of all, keep asking questions.

This book covers the requirements of AS Mathematics and the first year of A-level Mathematics. If you are going on to study to the full A-level, you will also need the Year 2 Student Book which builds on and extends this content.

Good luck and enjoy your study of Mathematics. We hope that this book will encourage you to study Mathematics further after you have completed your course.

Helen Ball

Kath Hipkiss

Michael Kent

Chris Pearce

FEATURES TO HELP YOU LEARN

Real-life context

Each chapter starts with a real life application of the mathematics you are learning in the chapter.

Learning objectives

A summary of the concepts, ideas and techniques that you will meet in the chapter.

Topic links

See how the material in the chapter relates to other chapters in the book.

Prior knowledge

See what mathematics you should know before you start the chapter, with some practice questions to check your understanding.

Key terms and glossary

Important words are written in **bold**. The words are defined in the glossary at the back of the book.

Stop and think

These boxes present you with probing questions and problems to help you to reflect on what you have been learning.

4 COORDINATE GEOMETRY 1: EQUATIONS OF STRAIGHT LINES

It's not easy comparing mobile phone tariffs from different providers. The following graph provides a simple, visual representation of three tariffs, which can easily be used to make comparisons over time.

From this graph you could identify important features like:

› the tariff with the highest upfront cost

› the tariff with the lowest monthly charge.

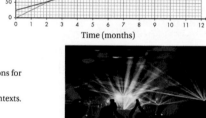

LEARNING OBJECTIVES

You will learn how to:

› write the equation of a straight line in the form $ax + by + c = 0$

› understand and use the gradient conditions for parallel lines

› understand and use the gradient conditions for perpendicular lines

› use straight line models in a variety of contexts.

TOPIC LINKS

You will need your knowledge of the gradient of a straight line to help you understand and solve differentiation problems in **Chapter 8 Differentiation**. The ability to draw and interpret straight line graphs will help you to solve distance, speed and time problems in **Chapter 15 Kinematics**. You will also need to determine and use the equations of straight lines when working with regression lines in **Chapter 12 Data presentation and interpretation** in context.

PRIOR KNOWLEDGE

You should already know how to:

› work with coordinates in all four quadrants

› plot graphs of equations that correspond to straight line graphs in the coordinate plane

› use the form $y = mx + c$ to identify parallel and perpendicular lines

› find the equation of the line through two given points or through one point with a given gradient

Stop and think If you are asked to show that two lines, between two different pairs of coordinates, are parallel do you need to work out the equation of each line or is there a slightly simpler approach?

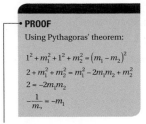

• PROOF

Using Pythagoras' theorem:

$$1^2 + m_1^2 + 1^2 + m_2^2 = (m_1 - m_2)^2$$
$$2 + m_1^2 + m_2^2 = m_1^2 - 2m_1m_2 + m_2^2$$
$$2 = -2m_1m_2$$
$$-\frac{1}{m_2} = -m_1$$

• TECHNOLOGY

Using a graphing software package, try finding the equations of lines passing through other points but that are parallel to the given equation.

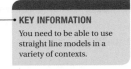

• KEY INFORMATION

You need to be able to use straight line models in a variety of contexts.

Proof

These boxes show proofs or describe what type of and how a proof is being applied.

Technology

These boxes show how you can use your calculator or other technologies such as graphing software to explore the mathematics.

Key information

These boxes highlight information that you need to pay attention to and learn, such as key formulae and learning points.

Explanations and examples

Each section begins with an explanation and one or more worked examples. Some show alternative solutions to get you thinking about different approaches to a problem.

Example 8

Find the equation of the line that is parallel to $3x - y + 7 = 0$ and passes through the point $(1, 8)$.

Write the equation of the line in the form $ax + by + c = 0$, where a, b and c are integers.

Solution

Rearrange the equation into the form $y = mx + c$.

$y = 3x + 7$

$m = 3$ and $(x_1, y_1) = (1, 8)$

Colour-coded questions

The questions in the exercises and the review questions are colour-coded (green, blue and red) to show you how difficult they are. Exercises start with more accessible (green) questions and then progress through intermediate (blue) to more challenging (red) questions. Red questions are particularly appropriate if you are studying for the full A-level.

(CM) **7** In each of the following cases, explain the mathematical manipulation steps required to show whether or not each pair of lines are perpendicular.

a $5x - 3y - 7 = 0$ and $-10y = 6x + 3$

b $9x - 10y - 7 = 0$ and $9y - 10x + 2 = 0$

(PS) **8** A quadrilateral is drawn with vertices at $(0, 6)$, $(2, 10)$, $(4, 4)$ and $(6, 8)$.

(LM) By considering gradients, work out which of the following the quadrilateral could possibly be and state your reasons.

rhombus   kite   parallelogram   square   trapezium   rectangle

Question-type indicators

Many questions in the book require you to think in different ways. Look for these icons:

(CM) Communicating mathematically – this requires an explanation, often in words. This is a key skill for Assessment Objective 2: Reason, interpret and communicate mathematically, and forms part of the overarching themes for AS and A-level Mathematics (OT1: Mathematical argument, language and proof).

(PF) Proof – you need to show how you arrived at the given result using logical steps. This is a key skill for Assessment Objective 2: Reason, interpret and communicate mathematically, and forms part of the overarching themes for AS and A-level Mathematics (OT1: Mathematical argument, language and proof).

(PS) Problem solving – you need to apply a multi-step process, often using concepts from different areas of mathematics. This is a key skill for Assessment Objective 3: Solve problems within mathematics and other contexts, and forms part of the overarching themes for AS and A-level Mathematics (OT2: Mathematical problem solving).

(M) Modelling – you will need to make one or more assumptions in order to proceed. This is a key skill for Assessment Objective 3: Solve problems within mathematics and other contexts, and forms part of the overarching themes for AS and A-level Mathematics (OT3: Mathematical modelling).

Using the large data set activity

These activities in the Statistics chapters can be used in groups or individually to explore your exam board's large data set, becoming familiar with the contexts and practising the statistical skills you've learnt.

> **Using the large data set 14.1**
>
> The large data set contains a lot of raw data. Remember that your data needs to be cleaned before performing any viable calculations.
>
> a Select a random sample of 30 items from a chosen category and find the mean and standard deviation.
>
> b Using the same category, take a systematic sample of 30 items, find the mean and standard deviation.
>
> c Compare the two samples using the mean and standard deviation. Explain ways in which the samples are similar or different.
>
> d Using your knowledge of the large data set, explain why a sample of 30 may not always be possible.

Summary of key points

At the end of each chapter, there is a summary of key formulae and learning points.

---

4 COORDINATE GEOMETRY 1: EQUATIONS OF STRAIGHT LINES

SUMMARY OF KEY POINTS

> The equation of a straight line may be written in the form $ax + by + c = 0$ where a, b and c are integers.

> You can find the equation of a line using the formula $y - y_1 = m(x - x_1)$, where m is the gradient of the line and (x_1, y_1) is a point on the line. (You need to learn this.)

> You can find the gradient m of the line joining two points with coordinates (x_1, y_1) and (x_2, y_2) by using the formula $m = \frac{y_2 - y_1}{x_2 - x_1}$. (You need to learn this.)

> You can find the equation of the line joining two points with coordinates (x_1, y_1) and (x_2, y_2) by using the formula $\frac{y - y_1}{y_2 - y_1} = \frac{x - x_1}{x_2 - x_1}$. (You need to learn this.)

> Two or more lines are parallel if their gradients, m, are the same.

> A line with a gradient of m_1 is perpendicular to a line with a gradient of m_2 when $m_2 = -\frac{1}{m_1}$.

> Two straight lines with gradients m_1 and m_2 are perpendicular when $m_1 m_2 = -1$. (You need to learn this.)

EXAM-STYLE QUESTIONS 4

Answers page 514

1 Line A is parallel to $y = 3x - 5$ and passes through the point $(1, 6)$.

Find the equation of line A. **[3 marks]**

CM 2 **a** Line A passes through $(0, 7)$ and $(2, 3)$ and line B passes through $(6, 4)$ and $(8, 5)$.

Using $m = \frac{y_2 - y_1}{x_2 - x_1}$, find the gradients of lines A and B. **[4 marks]**

b Compare the gradients of lines A and B and decide which line is steeper.

Give a reason for your answer. **[2 marks]**

3 Line A is perpendicular to $x - y - 8 = 0$ and passes through the y-intercept of $y = 9x + 5$.

Find the equation of line A. **[4 marks]**

CM M 4 The graph below shows the journeys of two different runners.

Runner A lives 3 miles away from runner B.

They both go out for a run one morning. After 3 hours, they meet at a point 9 miles away from runner B's home.

a When the runners meet, how far will runner A have run? **[1 mark]**

b If x is time in hours and y is distance in miles, for each runner, find two pairs of x and y values.

Using $\frac{y - y_1}{y_2 - y_1} = \frac{x - x_1}{x_2 - x_1}$ find the equation of the line for each runner in the form $y = mx + c$. **[6 marks]**

c Look at the slopes of the two lines on the graph to decide which runner runs faster.

Give a reason for your answer. **[2 marks]**

124

Exam-style questions

Practise what you have learnt throughout the chapter with questions written in examination style and allocated marks, progressing in order of difficulty.

At the end of the book you will find **Exam-style extension questions**, providing additional practice at A-level standard for each chapter.

1 ALGEBRA AND FUNCTIONS 1: MANIPULATING ALGEBRAIC EXPRESSIONS

What is algebra? Very simply, it is the use of letters and symbols to represent numbers. Numbers are themselves symbols to represent values and quantities. When people first started counting, it would probably have made sense to count in tens because we have ten fingers. However, people in some parts of the world count in base 2 and others count in base 20. In computing, binary code is used to encode portions of data.

At some point in history, people needed to record numbers and started doing so using symbols. The Egyptians used hieroglyphics and the Romans used numerals. The use of symbols to represent numbers evolved over many thousands of years until, over the last several hundred years, the Arabic number system was adopted by most societies.

Algebraic symbols are a good way to write mathematical rules in general terms, which allow them to be manipulated.

LEARNING OBJECTIVES

You will learn how to:

› manipulate polynomials algebraically, including

  › expanding brackets and simplifying by collecting like terms

  › factorisation

  › simple algebraic division

  › use of the factor theorem

› understand and use the binomial expansion of $(a + b)^n$ for positive integer n and the notation $n!$ and ^nC_r

› understand and use the laws of indices for all rational exponents

› use and manipulate surds, including rationalising the denominator.

TOPIC LINKS

The skills and techniques that you learn in this chapter underpin all the other chapters in this book. For example, in **Chapter 3 Algebra and functions 3: Sketching curves** you will need to fully factorise expressions to determine the roots of the associated equations. In both **Chapter 8 Differentiation** and **Chapter 9 Integration** you need to be confident in your manipulation of indices and surds. The ability to form and manipulate quadratic equations will help you to solve distance, speed and time problems in **Chapter 15 Kinematics**. Knowing how to perform binomial expansion will help you to solve problems involving the binomial distribution in **Chapter 13 Probability and statistical distributions**.

PRIOR KNOWLEDGE

You should already know how to:

- ❯ use and interpret algebraic manipulation, including:
  - ❯ ab in place of $a \times b$
  - ❯ $3y$ in place of $y + y + y$ or $3 \times y$
  - ❯ a^2 in place of $a \times a$, a^3 in place of $a \times a \times a$, a^2b in place of $a \times a \times b$
  - ❯ $\dfrac{a}{b}$ in place of $a \div b$
  - ❯ coefficients written as fractions rather than as decimals
  - ❯ brackets
- ❯ use the concepts and vocabulary of expressions, equations, formulae, identities, inequalities, terms and factors
- ❯ simplify and manipulate algebraic expressions (including those involving surds and algebraic fractions) by
  - ❯ collecting like terms
  - ❯ multiplying a single term over a bracket
  - ❯ taking out common factors
  - ❯ expanding products of two or more binomials
  - ❯ factorising quadratic expressions of the form $x^2 + bx + c$, including the difference of two squares
  - ❯ factorising quadratic expressions of the form $ax^2 + bx + c$
  - ❯ simplifying expressions involving sums, products and powers, including the laws of indices
- ❯ calculate with roots, and with integer and fractional indices
- ❯ calculate exactly with fractions, surds and multiples of π
- ❯ simplify surd expressions involving squares (e.g. $\sqrt{12} = \sqrt{4 \times 3} = \sqrt{4} \times \sqrt{3} = 2\sqrt{3}$ and rationalise denominators).

You should be able to complete the following questions correctly:

1   Expand and simplify the following expressions:

    a $3(a + b)$ **b** $c(4 - d)$

    c $e^2(f - g + eh)$ **d** $i(5 + j) + 7(k - 2j)$

    e $l(l + 5) - 6(l + l^2)$ **f** $m(2m - n) - 2(m + n)$

2   Expand and simplify the following expressions:

    a $(x + 1)(x + 2)$ **b** $(x - 3)(x + 4)$

    c $(x - 2)(x - 3)$ **d** $(x - 3)^2$

    e $(2x + 1)(x + 2)$ **f** $(x + 1)(x + 2)(x + 3)$

3 Simplify the following expressions:

a $a^5 \times a^6$ **b** $a^{\frac{1}{2}} \times a^{\frac{1}{2}}$ **c** $\dfrac{b^9}{b^5}$ **d** $(c^{-3})^2$ **e** $(2d)^4$ **f** $\dfrac{\left(e^2 \times f^8\right)}{ef}$

4 Simplify the following expressions:

a $\sqrt{2} \times \sqrt{3}$ **b** $\sqrt{5} \times \sqrt{8} \times \sqrt{3}$ **c** $\sqrt{72}$

d $\dfrac{\sqrt{54}}{\sqrt{3}}$ **e** $\dfrac{1}{\sqrt{5}}$ **f** $\dfrac{\sqrt{8} \times \sqrt{6}}{\sqrt{12}}$

1.1 Manipulating polynomials algebraically

You need to be able to **expand brackets**, collect like **terms** and **simplify**. As part of your GCSE maths course you will have practised these techniques. You need to be able to apply them to more complex **expressions** like the one in Example 1.

Example 1

Expand and simplify $h\left(h^3 - \dfrac{h}{2}\right) + 2h^2(h^4 + h + 5)$.

Solution

Multiply all the terms inside each bracket by the term outside the bracket.

$$h\left(h^3 - \dfrac{h}{2}\right) + 2h^2\left(h^4 + h + 5\right) = h^4 - \dfrac{h^2}{2} + 2h^6 + 2h^3 + 10h^2$$

Reorder with the indices in descending order.

$$= 2h^6 + h^4 + 2h^3 + 10h^2 - \dfrac{h^2}{2}$$

> When an expression contains terms with different indices, the terms tend to be written in descending order of index.

Simplify.

$$= 2h^6 + h^4 + 2h^3 + \dfrac{20h^2}{2} - \dfrac{h^2}{2}$$

$$= 2h^6 + h^4 + 2h^3 + \dfrac{19h^2}{2}$$

> Fractional coefficients can be written as, for example, $\dfrac{19h^2}{2}$ or $\dfrac{19}{2}h^2$.

From your GCSE course, you should be familiar with the word 'binomial', which means an expression with two terms. There are a number of ways to expand two binomials, as shown in Examples 2 and 3 overleaf.

Example 2

Expand and simplify $(a^2 - 3)(a^2 + 4)$.

Solution

Multiply all the terms in the second bracket by those in the first bracket.

$$(a^2 - 3)(a^2 + 4) = a^2(a^2 + 4) - 3(a^2 + 4)$$
$$= a^4 + 4a^2 - 3a^2 - 12$$

Collect like terms.

$$= a^4 + a^2 - 12$$

Alternatively:

Draw a grid.

	a^2	-3
a^2	a^4	$-3a^2$
4	$4a^2$	-12

$$(a^2 - 3)(a^2 + 4) = a^4 + 4a^2 - 3a^2 - 12$$

Collect like terms.

$$= a^4 + a^2 - 12$$

Alternatively:

Use FOIL (First, Outer, Inner, Last).

First
$(a^2 - 3)(a^2 + 4)$ $(a^2 - 3)(a^2 + 4)$
                              Outer

$(a^2 - 3)(a^2 + 4)$ $(a^2 - 3)(a^2 + 4)$ Last
           Inner

$$(a^2 - 3)(a^2 + 4) = a^4 + 4a^2 - 3a^2 - 12$$

Collect like terms.

$$= a^4 + a^2 - 12$$

Example 3

Expand and simplify $(4b^3 - 3)(4b^3 + 3)$.

Note that the two terms are in the form $(a - b)(a + b)$.

$(a - b)(a + b) = a^2 - b^2$ is called 'the difference of two squares'.

Solution

Using your preferred method, multiply all the terms in the second bracket by those in the first bracket.

$$(4b^3 - 3)(4b^3 + 3) = 4b^3(4b^3 + 3) - 3(4b^3 + 3)$$
$$= 16b^6 + 12b^3 - 12b^3 - 9$$

If a bracket is preceded by a negative term then *all* the terms inside the bracket need to be multiplied by the negative term.

Collect like terms.

$$= 16b^6 - 9$$

Stop and think   If you substitute any real value of b into $(4b^3 - 3)(4b^3 + 3)$, do you get the same result when you substitute the same value of b into $16b^6 - 9$?

You can extend the same techniques to expand three simple **binomials**.

Example 4

Show that $(2a^2 - 3b^2)(2a^2 + 3b^2)(b - a) = 4a^4b - 4a^5 - 9b^5 + 9ab^4$.

Solution

Using your preferred method, expand $(2a^2 - 3b^2)(2a^2 + 3b^2)$.

$$(2a^2 - 3b^2)(2a^2 + 3b^2) = 4a^4 - 9b^4$$

Now you have

$$(2a^2 - 3b^2)(2a^2 + 3b^2)(b - a) = (4a^4 - 9b^4)(b - a)$$

Multiply all the terms in the second bracket by those in the first bracket.

$$(4a^4 - 9b^4)(b - a) = 4a^4(b - a) - 9b^4(b - a)$$
$$= 4a^4b - 4a^5 - 9b^5 + 9b^4a$$
$$= -4a^5 + 4a^4b - 9b^5 + 9ab^4$$

PROOF

When you are asked to 'show that …' you are being asked to demonstrate a proof. If you have to show that an expression is equivalent to another expression, as in this case, you need to ensure that you either manipulate the left-hand side (LHS) or right-hand side (RHS) of the equation to be the same as the other side of the equation. In this case it is easier to expand the brackets on the LHS rather than factorising the RHS. See **Chapter 11 Proof** for more about different types of proof.

Exercise 1.1A

Answers page 487

1 Expand and simplify the following expressions:

   a $4x(x^2 + 3x - 2)$

   b $x^2(x + y - xy)$

2 Expand and simplify the following expressions:

   a $(x^2 - 1)(x + y)$

   b $(2 - x)(x^2 + 3x)$

3 Expand and simplify the following expressions:

   a $(x - 1)(x + 2)^2$

   b $(2x + 3)(2x + 2)(1 - x)$

(PF) **4** Show that $(2x - 1)(x + 2) - (1 - x)(3x^2 + 6) = 3x^3 - x^2 + 9x - 8$.

(PS) (PF) **5** Show that the sum of the squares of two consecutive even numbers is also an even number. (Hint: use $2n$ and $2n + 2$ or $2n - 2$ and $2n$ to represent two consecutive even numbers.)

6 Identify and correct the mistakes in the following expansion.

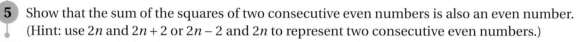

$$(x + 3)(2x^2 + x - 2) - (x + 1)(3 - 2x - x^2)$$

$$= x(2x^2 + x - 2) + 3(2x^2 + x - 2) - x(3 - 2x - x^2) + 1(3 - 2x - x^2)$$

$$= 2x^2 + x^2 - 2x + 3x^2 + 3x - 6 - 3x - 2x^2 + x^3 + 3 - 2x - x^2$$

$$= 3x^2 - 6x + 3$$

(PS) (PF) **7** If p is odd then show that p^2 cannot be even. (Hint: use $2n + 1$ to represent an odd number.)

(PS) (PF) **8** Show that the product of two odd numbers and one even number is always an even number.

1.2 Expanding multiple binomials

Pascal's triangle

You need to be familiar with and know how to use **Pascal's triangle**. It is named after a French mathematician who lived during the 17th century, but it was actually studied many centuries before by mathematicians around the world. If you want to expand $(x + 1)^7$, for example, then some of the numbers in Pascal's triangle can be used for the **coefficients** (the numeric part of a term, but not an index) rather than having to expand and simplify all of the brackets. In **Book 2**, this technique will be used in the context of sequences and series of terms.

> **KEY INFORMATION**
>
> A coefficient is the numeric part of a term. For example, 2 is the coefficient in $2x^3$.

> **KEY INFORMATION**
>
> Pascal's triangle is a triangular pattern of binomial coefficients.

Consider expansions of the following expressions.

$$(a + b)^0 = 1$$

$$(a + b)^1 = 1a^1 + 1b^1$$

$$(a + b)^2 = 1a^2 + 2a^1b^1 + 1b^2$$

$$(a + b)^3 = 1a^3 + 3a^2b^1 + 3a^1b^2 + b^3$$

$$(a + b)^4 = 1a^4 + 4a^3b^1 + 6a^2b^2 + 4a^1b^3 + 1b^4$$

> In these expansions, '1' has been written as a coefficient and as an index, when it wouldn't normally be, to aid the explanation.

And so on. What do you notice?

Firstly, look at the binomial coefficients.

Index of expansion	Pascal's triangle								
0					1				
1				1		1			
2			1		2		1		
3		1		3		3		1	
4	1		4		6		4		1

> This is the sum of 1 and 1 above.

> This is the sum of 3 and 1 above.

The next row of coefficients is generated by starting and ending with a 1, then by adding two coefficients from the previous row.

To generate the coefficients for an expansion with an index of 5, start with 1, then add the two numbers above: $1 + 4 = 5$. Add the two numbers above: $4 + 6 = 10$. Add the two numbers above: $6 + 4 = 10$. Add the two numbers above: $4 + 1 = 5$. End with a 1. So Pascal's triangle now looks like this:

Index of expansion	Pascal's triangle										
0						1					
1					1		1				
2				1		2		1			
3			1		3		3		1		
4		1		4		6		4		1	
5	1		5		10		10		5		1

Note that the first row of Pascal's triangle is for an expansion with an **index** of 0 (not 1). Note also that Pascal's triangle is symmetrical from the top vertex, 1, down through the middle. This is significant because coefficients are repeated in reverse order. Pascal's triangle goes on ad infinitum.

Next, look at the sums of the indices.

In all of the terms in the following expansion, the indices sum to the index in the original expression, i.e. 3. Note that a^0 and b^0 have been added here to complete the picture.

$$(a+b)^3 = 1a^3b^0 + 3a^2b^1 + 3a^1b^2 + a^0b^3$$

Likewise, in all of the terms in the following expansion, the indices add up to the index in the original expression, i.e. 4.

$$(a+b)^4 = 1a^4b^0 + 4a^3b^1 + 6a^2b^2 + 4a^1b^3 + 1a^0b^4$$

> Remember the value of any term with an index of 0 is 1.

KEY INFORMATION

The sum of the indices in each term always totals the index in the original expression.

Stop and think How could you check that your expansion using Pascal's triangle is correct?

Example 5

Use Pascal's triangle to expand:

a $(2x+y)^4$ **b** $(3-x)^3$

Solution

a The index is 4, so from Pascal's triangle the coefficients will be 1, 4, 6, 4 and 1.

In this case, $a = 2x$ and $b = y$.

$$(2x+y)^4 = (1)(2x)^4(y)^0 + (4)(2x)^3(y)^1 + (6)(2x)^2(y)^2 + \\ (4)(2x)^1(y)^3 + (1)(2x)^0(y)^4$$

Simplify.

$$= 16x^4 + 32x^3y + 24x^2y^2 + 8xy^3 + y^4$$

b The index is 3, so from Pascal's triangle the coefficients will be 1, 3, 3 and 1.

In this case $a = 3$ and $b = -x$.

$$(3-x)^3 = (1)(3)^3(-x)^0 + (3)(3)^2(-x)^1 + (3)(3)^1(-x)^2 + \\ (1)(3)^0(-x)^3$$

Simplify.

$$= 27 - 27x + 9x^2 - 3x^3$$

Exercise 1.2A   **Answers page 487**

 1 Use Pascal's triangle to expand the following:

a $(x+2)^3$

b $(2x+1)^4$

2 Use Pascal's triangle to expand the following:

    a $(2-x)^5$ **b** $\left(\dfrac{x}{2}+3\right)^4$

(PS) **3** Expand the expression $(3x+2)(x-1)^4$.

(PS) **4** Find the coefficient of x^2 in the expansion of $(3-2x)^5$.

(PS) **5** Expand the expression $\left(x^3-2x^2+\dfrac{1}{2}\right)^3$. (Hint: let $a=x^3-2x^2$.)

(PS) **6** The coefficient of x^2 in the expansion of $(2x+5)(2+cx)^3$ is 198. Find the value of c.

Factorials

In **Section 1.3**, you will be shown a formula to perform any binomial expansion. To use the formula you need to be familiar with particular notation, starting with **factorials**.

In how many ways can the names of four people, Alisha, Brian, Catriona and Divi, be listed?

The list of possibilities, with Alisha first, is ABCD, ABDC, ACBD, ACDB, ADBC and ADCB.

As there are 6 arrangements with Alisha first, there must also be 6 arrangements with Brian first. Similarly, there will be 6 arrangements with Catriona first and with Divi first. Hence, there are 24 arrangements altogether.

Another way of thinking about this is to say we can choose:

› the first person in 4 ways (as there are 4 people to choose from)

› the second person in 3 ways (as there are now only 3 people to choose from)

› the third person in 2 ways (as there are now only 2 people to choose from)

› the last person in 1 way (as there is now only 1 person to choose from).

So the number of arrangements is $4\times3\times2\times1=24$ ways.

Another way of expressing this is 4! (we say this as '4 factorial').

    $4!=4\times3\times2\times1$

In general, the number of ways of arranging n distinct (different) objects is $n!$, which we call 'n factorial'.

> **KEY INFORMATION**
>
> $n!=n\times(n-1)\times(n-2)\dots\times 2\times1$

> $n!$ is n factorial.

Combinations

Instead of listing all four of the people in the previous scenario, you might just want to list three of their names, in *any* order. How many possibilities are there this time?

The list of possibilities, with Alisha first, is ABC, ABD, ACD.

> **KEY INFORMATION**
>
> $0!=1$

How many other possibilities are there? There is only one: BCD.

We can represent this by $^4C_3$ or $\begin{pmatrix} 4 \\ 3 \end{pmatrix} = \frac{4!}{1!\,3!} = \frac{4 \times 3 \times 2 \times 1}{3 \times 2 \times 1} = 4$

Generally, suppose that you wish to choose r objects from a total of n objects, but the order in which the objects are arranged does not matter. Such a choice is called a **combination**.

The number of combinations of r objects from n distinct objects can be written as

$$^nC_r = \begin{pmatrix} n \\ r \end{pmatrix} = \frac{n!}{(n-r)!\,r!}$$

TECHNOLOGY

Most calculators have a factorial button, which may be labelled as $n!$ or $x!$. To work out the factorial of a number, type in the number then press the factorial button.

Example 6

How many different ways could 7 people standing in a line be arranged?

Solution

The number of ways of arranging 7 people standing in a line is

$$7! = 7 \times 6 \times 5 \times 4 \times 3 \times 2 \times 1 = 5040$$

KEY INFORMATION

In a combination, the order of the items does not matter.

Stop and think When calculating $^nC_r = \begin{pmatrix} n \\ r \end{pmatrix} = \frac{n!}{(n-r)!\,r!}$, what do you notice about the numerator and the sum of the numbers in the product in the denominator?

Example 7

You have 8 pairs of jeans but you want to select 3 of them to take on holiday. In how many ways can this be done?

Solution

The order in which the jeans are picked does not matter. For instance, picking jeans A, B and C will give exactly the same combination as picking jeans B, C and A.

That is, 3 items are being chosen from 8.

$$^8C_3 \text{ or } \begin{pmatrix} 8 \\ 3 \end{pmatrix} = \frac{8!}{(8-3)!\,3!} = \frac{8!}{(5)!\,3!} = 56 \text{ ways (combinations)}$$

TECHNOLOGY

Most calculators have a combinations button, which may be labelled as ^nC_r. To work out the number of combinations, type in the total number of items in the group, n, press the combinations button and then type in the number of items chosen from the group, r.

Answers page 488

1 **a** You like 5 universities but can only put 3 of them down on your application form. In how many ways can this be done?

   b You have 6 friends but can only take 2 of them to the cinema. In how many ways can this be done?

   c You have 8 books but only space for 7 of them on your book shelf. In how many ways can this be done?

2 Calculate the values of the following. What do you notice?

   a $^3C_0$ **b** $^3C_1$

   c $^3C_2$ **d** $^3C_3$

3 Write using ^nC_r the values in the 6th line (for an expression with an index of 5) of Pascal's triangle.

4 Why is using ^nC_r for expansions with a high index preferable to using Pascal's triangle?

(PS) 5 You are going on holiday and want to take some T-shirts with you to wear. You have 16 T-shirts but only have room in your bag for 9 of them. Calculate, without using a calculator, in how many ways this can be done. You must choose the most efficient calculation and show all of your working out.

1.3 The binomial expansion

Instead of using Pascal's triangle, ^nC_r can be used to generate the numeric coefficients of the terms in any binomial expansion. In general terms:

$$(a+b)^n = a^n + \binom{n}{1}a^{n-1}b + \binom{n}{2}a^{n-2}b^2 + \binom{n}{3}a^{n-3}b^3 + \ldots + \binom{n}{r}a^{n-r}b^r \ldots + b^n$$

You do *not* need to remember this formula.

Alternatively:

$$(a+b)^n = a^n + {}^nC_1 a^{n-1}b + {}^nC_2 a^{n-2}b^2 + {}^nC_3 a^{n-3}b^3 + \ldots + {}^nC_r a^{n-r}b^r + b^n$$

> In **Chapter 13 Probability and statistical distributions** you will use the binomial expansion in the context of the binomial distribution, which allows you to work out the probability of getting a particular outcome from a trial.

Example 8

Use the binomial expansion to find the expansion of the expression $(2+3x)^4$.

Solution

State the formula.

$$(a+b)^n = a^n + {}^nC_1 a^{n-1}b + {}^nC_2 a^{n-2}b^2 + {}^nC_3 a^{n-3}b^3 + \ldots + {}^nC_r a^{n-r}b^r + b^n$$

In this case $a = 2$, $b = 3x$ and $n = 4$.

Substitute the values into the formula. Check that in each term the indices sum to the index in the original expression.

$(2 + 3x)^4 = {}^4C_0 2^4 + {}^4C_1 2^3 (3x)^1 + {}^4C_2 2^2 (3x)^2 + {}^4C_3 2^1 (3x)^3 + {}^4C_4 (3x)^4$

Work out the values of the combinations and expand the brackets.

$$= \frac{4!}{4!\,0!}(16) + \frac{4!}{3!\,1!}(8)(3x) + \frac{4!}{2!\,2!}(4)(9x^2) + \frac{4!}{1!\,3!}(2)(27x^3) +$$

$$\frac{4!}{0!\,4!}(81x^4)$$

Simplify.

$$= 16 + (4)(8)(3x) + (6)(4)(9x^2) + (4)(2)(27x^3) + (81x^4)$$

$$= 16 + 96x + 216x^2 + 216x^3 + 81x^4$$

Example 9

Find the coefficient of the x^3 term in the expansion of the expression $(3 - 4x)^5$.

Solution

State the formula.

$(a + b)^n = a^n + {}^nC_1 a^{n-1}b + {}^nC_2 a^{n-2}b^2 + {}^nC_3 a^{n-3}b^3 + \ldots + {}^nC_r a^{n-r}b^r + b^n$

In this case $a = 3$, $b = -4x$ and $n = 5$.

The x^3 term will come from ${}^nC_3 a^{n-3}(bx)^3$.

Substitute the values into this part of the formula.

$${}^nC_3 a^{n-3}(bx)^3 = {}^5C_3 3^2 (-4x)^3$$

Work out the values of the combinations and expand the brackets.

$$= \frac{5!}{2!\,3!}(9)(-64x^3)$$

Simplify.

$$= (01)(9)(-64x^3)$$

$$= -5760x^3$$

So the coefficient of the x^3 term is -5760.

If the expression to be expanded has either $a = 1$ or is manipulated to have $a = 1$ then the formula becomes:

$$(1 + x)^n = 1 + nx + \frac{n(n-1)}{2!}x^2 + \frac{n(n-1)(n-2)}{3!}x^3 + \frac{n(n-1)(n-2)(n-3)}{4!}x^4 + \ldots$$

This is an alternative version of the previously noted formula. Or if x has a coefficient then it becomes:

$$(1 + bx)^n = 1 + n(bx) + \frac{n(n-1)}{2!}(bx)^2 + \frac{n(n-1)(n-2)}{3!}(bx)^3 + \frac{n(n-1)(n-2)(n-3)}{4!}(bx)^4 + \ldots$$

Example 10

Use the formula for $(1 + bx)^n$ to find the expansion of the expression $(2 + 3x)^4$. Use your expansion to find the exact value of 1.97^4.

Solution

You need to write $(2 + 3x)^4$ in the form $(1 + x)^n$.

$$(2 + 3x)^4 = \left[2\left(1 + \frac{3x}{2}\right)\right]^4$$

$$= 2^4\left(1 + \frac{3x}{2}\right)^4$$

State the formula.

$$(1 + bx)^n = 1 + n(bx) + \frac{n(n-1)}{2!}(bx)^2 + \frac{n(n-1)(n-2)}{3!}(bx)^3 +$$

$$\frac{n(n-1)(n-2)(n-3)}{4!}(bx)^4 + \ldots$$

In this case, $bx = \frac{3x}{2}$ and $n = 4$.

Substitute the values into the formula.

$$(2 + 3x)^4 = 2^4\left(1 + 4\left(\frac{3x}{2}\right) + \frac{(4)(3)}{2!}\left(\frac{3x}{2}\right)^2 + \frac{(4)(3)(2)}{3!}\left(\frac{3x}{2}\right)^3 + \right.$$

$$\left. \frac{(4)(3)(2)(1)}{4!}\left(\frac{3x}{2}\right)^4 \right)$$

Expand the brackets and simplify.

$$= 2^4\left(1 + 6x + (6)\left(\frac{9x^2}{4}\right) + (4)\left(\frac{27x^3}{8}\right) + \left(\frac{81x^4}{16}\right)\right)$$

$$= 2^4\left(1 + 6x + \frac{27x^2}{2} + \frac{27x^3}{2} + \frac{81x^4}{16}\right)$$

$$= 16\left(1 + 6x + \frac{27x^2}{2} + \frac{27x^3}{2} + \frac{81x^4}{16}\right)$$

$$= 16 + 96x + 216x^2 + 216x^3 + 81x^4$$

This is the same expression as in Example 8. Sometimes you will be directed as to which formula to use. Usually, though, it is your choice.

To use the expansion to find the exact value of 1.97^4 you need to work out a value for x that will subsequently be substituted.

Let $2 + 3x = 1.97$.

Rearrange.

$\qquad x = -0.01$

State the expansion.

$(2 + 3x)^4 = 16 + 96x + 216x^2 + 216x^3 + 81x^4$

Substitute $x = -0.01$ into the expansion.

$\quad 1.97^4 = 16 + (96)(-0.01) + (216)(-0.01)^2 + (216)(\ 0.01)^3 +$
$\qquad\qquad (81)(-0.01)^4$

$\qquad = 16 - (96)(0.01) + (216)(0.01)^2 - (216)(0.01)^3 + (81)(0.01)^4$

$\qquad = 15.061\,384\,81$

Notice the sign change on each alternate term.

Exercise 1.3A

Answers page 488

1 Using the binomial expansion in the form $(a + b)^n$, write down the expansions of:

  a $(2 + x)^4$

  b $(3 + 5x)^3$

2 Using the binomial expansion in the form $(1 + x)^n$, find the x^3 term in the expansions of:

  a $(1 + 3x)^5$

  b $(2 + 4x)^6$

3 Using the binomial expansion in the form $(a + b)^n$, write down the expansions of:

  a $\left(2 + \dfrac{x}{2}\right)^3$

  b $(3x - 6y)^4$

4 Using the binomial expansion in the form $(1 + x)^n$, find the x^3 term in the expansions of:

  a $(3 + x)^5$

  b $(2 + 3x)^6$

(PS) **5** When $(2 + bx)^7$ is expanded, the coefficient of x^3 is $70\,000$. Find the value of b.

(PS) **6** Using the binomial expansion in the form $(a + b)^n$ find the expansion of $(3 - x)\left(3 - \dfrac{2x}{5}\right)^4$.

(PS) **7** Using the binomial expansion in the form $(1 + x)^n$ find the x^3 term in the expansion of $(2x + 5)\left(7 - \dfrac{x}{2}\right)^4$.

(M) (PS) **8** Using the binomial expansion in a form of your choice expand $\left(3 - \dfrac{5x}{2}\right)^6$ in ascending powers of x up to and including x^3. Use your expansion to find an approximation for 2.75^6.

1.4 Factorisation

Factorising is the opposite of expanding brackets. You need to be able to factorise different sorts of expressions, including quadratic expressions that are in the form $ax^2 + bx + c$, where $a \neq 0$ but b and c could be equal to zero.

Example 11

Factorise the following expressions:

a $7x^2 - 35x$ **b** $6x^2 + 11x - 35$ **c** $16x^2 - 81y^2$

Solution

a Identify the highest common numeric factor and the highest common algebraic factor.
$$7x^2 - 35x = 7x(x - 5)$$

b $c \neq 0$ so the expression will just have two brackets.
$a = 6$ and $b = 11$.
$ac = -210$ and $b = 21 - 10 = 11$ (where 21 and -10 are factors of -210).

$$6x^2 + 11x - 35 = 6x^2 + 21x - 10x - 35$$
$$= 2x(3x - 5) + 7(3x - 5)$$
$$= (3x - 5)(2x + 7)$$

c $c \neq 0$ so the expression will just have two brackets. This is an example of a difference of two squares, as $16x^2 = (4x)^2$ and $81y^2 = (9y)^2$.

$$16x^2 - 81y^2 = (4x)^2 - (9y)^2$$
$$= (4x - 9y)(4x + 9y)$$

> If the expression does not contain a numeric value without an algebraic component, then the factorised expression will usually have just one bracket.

> **KEY INFORMATION**
>
> $a^2 - b^2$ is known as 'the difference of two squares' and is factorised as $(a - b)(a + b)$.

Exercise 1.4A

Answers page 488

1 Factorise the following expressions:

    a $2x + 6$

    b $3x^3 + 9$

    c $8x - 4x^4$

    d $xy + 2xy - xz + 3yz$

2 Factorise the following quadratic expressions:

    a $x^2 - 7x + 12$

    b $2x^2 + 5x + 2$

    c $x^2 + x - 12$

    d $x^2 - 49$

3 Identify and correct the mistakes in the following expansion.

$$14x^2 - 56x = 7x(2x^2 - 8)$$

4 Factorise the following expressions:

a $4xy - 8xy^2$

b $6x^2 - 13x + 6$

c $x^2y^2 + 2xy^3$

d $18x^2 - 12x$

e $81x^2y^2 - 49z^2$

f $2x^3 - 7x^2 + 10x$

(M)(PS) **5** The area of a triangle is given by the expression $4x^2 + 6x - 4$. By factorising the quadratic expression find two possible linear expressions for the base and height of the triangle.

6 Factorise $14x^2 - 17xy - 6y^2$.

(M)(PS) **7** The area of a rectangle is given by the expression $6x^2 + bx + 15$ for b > 0. Given that this expression factorises, find all the possible combinations of dimensions of the rectangle in terms of x.

(PS) **8** Make p the subject of the expression $p(q - r^2) = 2r(3p - q)$.

(PS) **9** The equation of a circle is $x^2 + y^2 - 6x + 10y = -18$ and $r = 4$. Write the equation of this circle in the form $(x - a)^2 + (y - b)^2 = r^2$.

1.5 Algebraic division

You need to be able to manipulate **algebraic fractions** so that they are written in their simplest form. The process is the same as writing a numeric fraction in its simplest form: find the **highest common factor** in the **numerator** and **denominator**, and then cancel.

Example 12

Simplify $\dfrac{3x^3 - 5x}{3x}$.

Solution

Fully factorise the numerator.

$$\frac{3x^3 - 5x}{3x} = \frac{x(3x^2 - 5)}{3x}$$

$\dfrac{x}{x} = 1$ so the fraction can be simplified.

$$= \frac{3x^2 - 5}{3}$$

Algebraic fractions can be written as improper fractions.

The fraction can be left in this format or alternatively

$$= \frac{3x^2}{3} - \frac{5}{3}$$

$\frac{3}{3} = 1$ so the first fraction can be simplified.

$$= x^2 - \frac{5}{3}$$

Or alternatively algebraic fractions can be written as mixed numbers.

Example 13

Simplify $\frac{x^2+5x+6}{x^2+6x+8}$ by factorising.

Solution

Factorise the numerator and the denominator.

$$\frac{x^2+5x+6}{x^2+6x+8} = \frac{(x+2)(x+3)}{(x+2)(x+4)}$$

The numerator and denominator have a common factor of $(x+2)$.

$\frac{(x+2)}{(x+2)} = 1$ so the fraction can be simplified.

$$= \frac{(x+3)}{(x+4)}$$

The factor theorem

Sometimes it isn't easy to spot common **factors**, particularly in high order **polynomials** such as $x^5 + x^4 + x^3 + x^2 + x + 1$. However, if you substitute a value, a, for x into an expression and the value of the expression is zero then $(x - a)$ is a factor of the expression. This result is called the **factor theorem**.

KEY INFORMATION

If $f(a) = 0$ then $(x - a)$ is a factor of $f(x)$.

Example 14

Given the expression $f(x) = x^3 + 6x^2 + 11x + 6$, show that $(x + 2)$ is a factor.

Solution

If $(x + 2)$ is a factor of $f(x)$ then $f(-2) = 0$.

Substitute $x = -2$ into $f(x)$.

$$f(-2) = (-2)^3 + 6(-2)^2 + 11(-2) + 6$$

$$= -8 + 24 - 22 + 6$$

$$= 0$$

$f(-2) = 0$ so $(x + 2)$ is a factor.

Manipulating algebraic fractions

The method for dividing a polynomial by an expression in the form $(ax \pm b)$ where $a \geqslant 1$ is just the same as the method for numeric long division.

Example 15

Divide $x^3 + 6x^2 + 11x + 6$ by $(x + 2)$ and then fully factorise the expression.

Solution

Divide $x^3 + 6x^2 + 11x + 6$ by $(x + 2)$.

$$x + 2 \overline{\smash{)}\ x^3 + 6x^2 + 11x + 6}$$

Divide x^3 by x.

$$\begin{array}{r} x^2 \\ x + 2 \overline{\smash{)}\ x^3 + 6x^2 + 11x + 6} \end{array}$$

Multiply x^2 by $(x + 2)$.

$$\begin{array}{r} x^2 \\ x + 2 \overline{\smash{)}\ x^3 + 6x^2 + 11x + 6} \\ x^3 + 2x^2 \end{array}$$

Work out the remainder by subtracting $x^3 + 2x^2$ from $x^3 + 6x^2$. Carry down the next part of the expression.

$$\begin{array}{r} x^2 \\ x + 2 \overline{\smash{)}\ x^3 + 6x^2 + 11x + 6} \\ x^3 + 2x^2 \\ \hline 0 + 4x^2 + 11x \end{array}$$

Divide $4x^2$ by x.

$$\begin{array}{r} x^2 + 4x \\ x + 2 \overline{\smash{)}\ x^3 + 6x^2 + 11x + 6} \\ x^3 + 2x^2 \\ \hline 0 + 4x^2 + 11x \end{array}$$

It is critical that the division is laid out correctly with the corresponding terms with the same index directly below one another, to avoid errors.

Multiply $4x$ by $(x + 2)$.

$$\begin{array}{r} x^2 + 4x \\ x + 2 \overline{\smash{)}\ x^3 + 6x^2 + 11x + 6} \\ x^3 + 2x^2 \\ \hline 0 + 4x^2 + 11x \\ 4x^2 + 8x \end{array}$$

Work out the remainder by subtracting $4x^2 + 8x$ from $4x^2 + 11x$. Carry down the next part of the expression.

$$\begin{array}{r} x^2 + 4x \\ x + 2 \overline{\smash{)}\ x^3 + 6x^2 + 11x + 6} \\ x^3 + 2x^2 \\ \hline 0 + 4x^2 + 11x \\ 4x^2 + 8x \\ \hline 3x + 6 \end{array}$$

Divide $3x$ by x.

$$\begin{array}{r} x^2 + 4x\ + 3 \\ x + 2 \overline{\smash{\big)}\, x^3 + 6x^2 + 11x + 6} \\ \underline{x^3 + 2x^2} \\ 0\ + 4x^2 + 11x \\ \underline{4x^2 +\ 8x} \\ 3x + 6 \end{array}$$

Multiply 3 by $(x + 2)$. Work out the remainder by subtracting $3x + 6$ from $3x + 6$. Remainder 0.

$$\begin{array}{r} x^2 + 4x\ + 3 \\ x + 2 \overline{\smash{\big)}\, x^3 + 6x^2 + 11x + 6} \\ \underline{x^3 + 2x^2} \\ 0\ + 4x^2 + 11x \\ \underline{4x^2 +\ 8x} \\ 3x + 6 \\ \underline{3x + 6} \\ 0 \end{array}$$

So $x^3 + 6x^2 + 11x + 6 = (x + 2)(x^2 + 4x + 3)$.

Factorise the quadratic.

So $x^3 + 6x^2 + 11x + 6 = (x + 2)(x + 1)(x + 3)$.

KEY INFORMATION

The expression found after the division is known as the **quotient**.

Example 16

Divide $x^3 + 6x^2 + 11x + 6$ by $(x - 3)$ and find the remainder.

Solution

Divide $x^3 + 6x^2 + 11x + 6$ by $(x - 3)$.

$$\begin{array}{r} x^2 + 9x\ + 38 \\ x - 3 \overline{\smash{\big)}\, x^3 + 6x^2 + 11x + 6} \\ \underline{x^3 - 3x^2} \\ 0\ + 9x^2 + 11x \\ \underline{9x^2 - 27x} \\ 38x +\ \ 6 \\ \underline{38x - 114} \\ 120 \end{array}$$

The remainder is 120.

You could now express the original expression as follows:

$$x^3 + 6x^2 + 11x + 6 \equiv (x^2 + 9x + 38)(x - 3) + 120$$

Example 17

Divide $2x^3 - 3x^2 + 1$ by $(2x + 1)$ and then fully factorise the expression.

Solution

Divide $2x^3 - 3x^2 + 1$ by $(2x + 1)$. The cubic is missing an expression in x. We need to add $0x$ to the polynomial to ensure that the division is correct.

$$
\begin{array}{r}
x^2 - 2x + 1 \\
2x + 1\,\overline{\big)\,2x^3 - 3x^2 + 0x + 1} \\
\underline{2x^3 + x^2} \\
0 \quad -4x^2 + 0x \\
\underline{-4x^2 - 2x} \\
2x + 1 \\
\underline{2x + 1} \\
0
\end{array}
$$

So $2x^3 - 3x^2 + 1 = (2x + 1)(x^2 - 2x + 1)$.

Factorise the quadratic.

$$= (2x + 1)\,(x - 1)^2$$

Stop and think

How does the factor theorem assist in the process to fully factorise high order polynomials, for example $x^5 - 5x^4 + 5x^3 + 5x^2 - 6x$?

Exercise 1.5A

Answers page 488

1 Simplify the following expressions:

a $\dfrac{2x + 6}{x + 3}$

b $\dfrac{(x + 1)(x + 2)}{(x - 1)(x + 3)}$

2 By substituting $x = 1$, show that $(x - 1)$ is a factor of $f(x) = x^3 + 4x^2 + x - 6$. Hence, by dividing $f(x)$ by $(x - 1)$, fully factorise $f(x)$.

3 Simplify the following expressions:

a $\dfrac{x^2 + x - 6}{x^2 + 8x + 15}$

b $\dfrac{2x^2 - 5x + 2}{2x^2 + 5x - 3}$

4 Using a combination of the factor theorem and algebraic division, fully factorise the following expressions:

a $3x^3 + 2x^2 - 7x + 2$

b $2x^4 + 3x^3 - 12x^2 - 7x + 6$

c $x^4 - 25x^2 + 144$

5 Express the following improper fractions as mixed numbers:

a $\dfrac{2x^3 + 15x^2 + 31x + 12}{x - 1}$

b $\dfrac{x^3 + 6x^2 + 11x + 6}{x - 2}$

c $\dfrac{x^3 + 4x^2 + x - 6}{x + 4}$

6 Use the factor theorem to find the remainder when:

a $x^4 - 18x + 81$ is divided by $(x - 2)$

b $8x^4 + 2x^3 - 53x^2 + 37x - 6$ is divided by $(x + 1)$

c $2x^5 - 19x^4 + 60x^3 - 65x^2 - 2x + 24$ is divided by $(x + 2)$.

7 Use the factor theorem to express the following improper fractions as mixed numbers:

a $\dfrac{x^4 - 18x + 81}{x - 2}$

b $\dfrac{8x^4 + 2x^3 - 53x^2 + 37x - 6}{x + 1}$

(PS) 8 Fully factorise $x^4 - 7x^3 + 13x^2 + 3x - 18$.

1.6 Laws of indices

In your GCSE course, you met the laws of **indices** (**powers**). You need to be comfortable applying them to expressions including any rational **exponents**.

> See **Section 1.8** for more information about rational and irrational numbers.

❯ $a^m \times a^n = a^{m+n}$

❯ $(a^m)^n = a^{mn}$

❯ $a^m \div a^n = a^{m-n}$

❯ $a^0 = 1$

❯ $a^{-m} = \dfrac{1}{a^m}$

❯ $a^{\frac{m}{n}} = \sqrt[n]{a^m}$

Stop and think If $m = 1$ then what does $a^{\frac{m}{n}}$ mean?

Example 18

Simplify the following expressions:

a $x^{\frac{1}{3}} \times x^1$

b $9x^5 \div 6x^2$

c $(2x^3)^5$

d $(27x^3)^{\frac{1}{3}}$

Solution

a As this is a multiplication involving the same term, add the powers.

$$x^{\frac{1}{3}} \times x^1 = x^{\frac{4}{3}}$$

b As this is a division involving the same term, subtract the powers.

$$9x^5 \div 6x^2 = \frac{9x^5}{6x^2}$$
$$= \frac{3x^3}{2}$$

c As this is a power raised to another power, multiply the powers.

$$(2x^3)^5 = 2^5 \times (x^3)^5$$

Apply the index to the numeric values and multiply the powers on any algebraic terms.

$$= 32x^{15}$$

d As this is a power raised to another power, multiply the powers.

$$(27x^3)^{\frac{1}{3}} = 27^{\frac{1}{3}} \times (x^3)^{\frac{1}{3}}$$

Apply the index to the numeric values and multiply the powers on algebraic terms.

$$= 3 \times x^1$$
$$= 3x$$

Example 19

Evaluate $81^{-\frac{1}{2}}$.

Solution

A negative power means the reciprocal of the expression.

$$81^{-\frac{1}{2}} = \frac{1}{81^{\frac{1}{2}}}$$

A power of $\frac{1}{2}$ means square root.

$$\frac{1}{81^{\frac{1}{2}}} = \pm\frac{1}{9}$$

You must include both the negative and positive roots.

Example 20

Express ±343 as a power of 49.

Solution

What do we know about 343?

$$7^3 = 343$$

We also know that $49^{\frac{1}{2}} = \pm 7$.

Replace 7 in the first statement.

$$\left(49^{\frac{1}{2}}\right)^3 = \pm 343$$

$$49^{\frac{3}{2}} = \pm 343$$

KEY INFORMATION

Laws of indices:

$$a^m \times a^n = a^{m+n}$$

$$(a^m)^n = a^{mn}$$

$$a^m \div a^n = a^{m-n}$$

$$a^0 = 1$$

$$a^{-m} = \frac{1}{a^m}$$

$$a^{\frac{m}{n}} = \sqrt[n]{a^m}$$

Exercise 1.6A

Answers page 488

1 Simplify the following expressions:

  a $a^4 \times 2a^5$

  b $2b^3 \times b$

  c $9c^7 \div 3c^4$

  d $(3d^2)^3$

  e $(e^3)^{-2}$

  f $\left(f^3\right)^{\frac{1}{3}}$

2 Evaluate the following:

  a $64^{\frac{1}{3}}$

  b $25^{\frac{1}{2}}$

  c $27^{-\frac{1}{3}}$

3 Simplify the following expressions:

  a $3a^{-3} \times 2a^7$

  b $4b^2 \div 2b^{-3}$

  c $4c^2 \div 8(c^3)^2$

  d $(d^2)^{\frac{1}{3}} \times d^{\frac{3}{4}}$

  e $5e^2 \times 6e^{-3} \times 3e^4$

  f $2f^2g \div 6f^4g^3$

4 Evaluate the following:

a $64^{-\frac{2}{3}}$ **b** $(-6)^{-2}$ **c** $256^{\frac{1}{4}}$

(PS) **5** Identify and correct the mistakes in the following simplification:

$$(3x^2)^3 \times 2x^{-4} = 3x^6 \times 2x^{-4}$$
$$= 5x^{10}$$

(PS) **6** Express 256 as a power of 8.

(PS) **7** The formula to calculate the area of a circle is πr^2. If the radius of the circle is given by the expression $2x^{-\frac{1}{3}}$, find an expression for the area of the circle in terms of x.

8 Simplify the following expressions:

a $9a^{-2} \times \frac{1}{3}a^{-3} \div \frac{2}{3}a^{-2}$ **b** $\left(6b^{-\frac{1}{3}}\right)^{-2} \div 6b$ **c** $\left(\dfrac{2}{c^{-\frac{1}{2}}}\right)^4 \times 2(c^3)^{\frac{1}{2}}$

(PS) **9** The formula to calculate the volume of a sphere is $\frac{4}{3}\pi r^3$. If the radius of the sphere is given by the expression $\frac{3}{2}x^{\frac{1}{2}}$, find an expression for the volume of the sphere in terms of x.

(PS) **10** If $y = x^3$ and $z = x^5$, show that $z = y^{\frac{5}{3}}$.

(PF)

1.7 Manipulating surds

From your GCSE maths course, you should know that **surds** are roots of rational numbers and are sometimes referred to as exact values. You need to be able to simplify and manipulate algebraic surds using these results:

$$\left(\sqrt{x}\right)^2 = x$$
$$\sqrt{xy} = \sqrt{x}\sqrt{y}$$
$$\left(\sqrt{x} + \sqrt{y}\right)\left(\sqrt{x} - \sqrt{y}\right) = x - y$$

Example 21

Expand and simplify to show that $\left(a + \sqrt{b}\right)^2 = a^2 + 2a\sqrt{b} + b$.

Solution

Expand the brackets.

$$\left(a + \sqrt{b}\right)^2 = \left(a + \sqrt{b}\right)\left(a + \sqrt{b}\right)$$
$$= a^2 + a\sqrt{b} + a\sqrt{b} + \sqrt{b}\sqrt{b}$$

Simplify and collect like terms.

$$= a^2 + 2a\sqrt{b} + b$$

$\sqrt{b}\sqrt{b} = b$

Example 22

Show that:

a $\left(a - \sqrt{b}\right)\left(a + \sqrt{b}\right) = a^2 - b$ **b** $\sqrt{ab}\left(\sqrt{a} - \sqrt{b}\right) = a\sqrt{b} - b\sqrt{a}$

Solution

a Expand the brackets.

$$\left(a - \sqrt{b}\right)\left(a + \sqrt{b}\right) = a^2 + a\sqrt{b} - a\sqrt{b} - \sqrt{b}\sqrt{b}$$

Simplify and collect like terms.

$$= a^2 - b$$

b Expand the brackets.

$$\sqrt{ab}\left(\sqrt{a} - \sqrt{b}\right) = \sqrt{ab}\sqrt{a} - \sqrt{ab}\sqrt{b}$$

Collect like terms.

$$= \sqrt{a^2}\sqrt{b} - \sqrt{a}\sqrt{b^2}$$

Simplify.

$$= a\sqrt{b} - b\sqrt{a}$$

KEY INFORMATION

$$\left(\sqrt{x}\right)^2 = x$$

$$\sqrt{xy} = \sqrt{x}\sqrt{y}$$

$$\left(\sqrt{x} + \sqrt{y}\right)\left(\sqrt{x} - \sqrt{y}\right) = x - y$$

$$\sqrt{ab}\sqrt{a} = \sqrt{a}\sqrt{a}\sqrt{b}$$

Exercise 1.7A

Answers page **489**

1 Simplify the following expressions:

     a $\left(a\sqrt{3}\right)^2$ **b** $\left(\sqrt{5b}\right)^2$

2 Expand and simplify the following expressions:

     a $\sqrt{e}\left(3 + \sqrt{e}\right)$ **b** $2\sqrt{f}\left(4 - 3\sqrt{f}\right)$

3 Simplify the following expressions:

     a $c\left(\sqrt{7c}\right)^2$ **b** $\sqrt{27d^3}$

4 Expand and simplify the following expressions:

     a $\sqrt{gh}\left(\sqrt{g} + h\right)$ **b** $\sqrt{jk}\left(\sqrt{j} - \sqrt{2k}\right)$

5 Expand and simplify $\left(\sqrt{b} - a\right)\left(\sqrt{b} + a\right)$.

(PS) **6** Show that $\left(\sqrt{b} - a\right)^2 = b - 2a\sqrt{b} + a^2$.

(PF)

7 Expand and simplify $\left(\sqrt{a} + \sqrt{b}\right)\left(\sqrt{a} - \sqrt{b}\right)$.

(PS) **8** Show that $\left(\sqrt{ab} + \sqrt{a}\right)^2 = ab + 2a\sqrt{b} + a$.

(PF)

(PS) **9** By finding the difference of two squares and using surds, factorise the expression $a - ab$.

(PS) **10** By finding the difference of two squares and using surds, factorise the expression $b^2 - ab$.

1.8 Rationalising the denominator

A **rational number** is one that can be written as a fraction, whereas an **irrational number** is one that can't. For example, $\sqrt{2}$ is an irrational number and cannot be written as a fraction.

If a fraction has a surd in the denominator then, by default, the denominator is irrational. However, you can manipulate the fraction to eliminate the surd from the denominator to make it a rational number. This process is called rationalising the denominator. Mathematicians generally try not to include surds in the denominator of a fraction.

In general, $\sqrt{\dfrac{x}{y}} = \dfrac{\sqrt{x}}{\sqrt{y}}$.

Rationalising a denominator involves multiplying the fraction by 1 in order not to change the value of the fraction, resulting in the surd becoming a rational number.

In general, to rationalise the denominator of a fraction in the form $\sqrt{\dfrac{b}{a}}$ or $\dfrac{b}{\sqrt{a}}$, multiply the fraction by $\dfrac{\sqrt{a}}{\sqrt{a}}$.

Example 23

Rationalise the denominator of $\dfrac{3}{7 - \sqrt{2}}$.

Solution

Multiply the fraction by 1 in the form $\dfrac{7 + \sqrt{2}}{7 + \sqrt{2}}$

$$\frac{3}{7 - \sqrt{2}} = \frac{3}{7 - \sqrt{2}} \times \frac{7 + \sqrt{2}}{7 + \sqrt{2}}$$

Multiply the numerator.

$$3\left(7 + \sqrt{2}\right) = 21 + 3\sqrt{2}$$

Multiply the denominator.

$$\left(7 - \sqrt{2}\right)\left(7 + \sqrt{2}\right) = 49 + 7\sqrt{2} - 7\sqrt{2} - 2$$
$$= 47$$

After rationalising the denominator:

$$= \frac{21 + 3\sqrt{2}}{47}$$

Here, the denominator is multiplied by the same expression but with the opposite sign. By doing this, the surd (the irrational part) is eliminated.

To rationalise the denominator of a fraction in the form $\dfrac{c}{b \pm \sqrt{a}}$ you multiply the fraction by $\dfrac{b \mp \sqrt{a}}{b \mp \sqrt{a}}$.

Example 24

Express $\dfrac{3}{\sqrt{3}}$ in the form 3^n.

Solution

There are a number of different ways in which to manipulate the given expression.

Option 1: Divide 3 by $\sqrt{3}$.

$$3 \div \sqrt{3} = \sqrt{3} = 3^{\frac{1}{2}}$$

Option 2: Express the values in terms of indices and then simplify.

$$\frac{3}{\sqrt{3}} = 3^1 \div 3^{\frac{1}{2}}$$

Subtracting the indices gives $3^{\frac{1}{2}}$.

Option 3: Rationalise the denominator.

$$\frac{3}{\sqrt{3}} = \frac{3}{\sqrt{3}} \times \frac{\sqrt{3}}{\sqrt{3}} = \frac{3\sqrt{3}}{3}$$
$$= \sqrt{3}$$
$$= 3^{\frac{1}{2}}$$

KEY INFORMATION

$$\sqrt{\frac{x}{y}} = \frac{\sqrt{x}}{\sqrt{y}}$$

To rationalise the denominator of a fraction in the form $\sqrt{\dfrac{b}{a}}$ or $\dfrac{b}{\sqrt{a}}$ you multiply the fraction by $\dfrac{\sqrt{a}}{\sqrt{a}}$.

Exercise 1.8A

Answers page 489

1 Rationalise the denominator in the following expressions:

a $\dfrac{2}{\sqrt{3}}$ **b** $\dfrac{4}{2+\sqrt{3}}$ **c** $\dfrac{5x}{3-\sqrt{2}}$

2 Rationalise the denominator in the following expressions:

a $\dfrac{2\sqrt{2}}{3-\sqrt{2}}$ **b** $\dfrac{4x+5}{3-\sqrt{7}}$ **c** $\dfrac{x\sqrt{2}}{5+\sqrt{2}}$

3 Rationalise the denominator in the following expressions:

a $\dfrac{3-\sqrt{7}}{3+\sqrt{7}}$ **b** $\dfrac{5+\sqrt{3}}{5-\sqrt{3}}$ **c** $\dfrac{y-x}{y+\sqrt{x}}$

4 Express each of the following in the form 2^n:

a $\dfrac{2}{\sqrt{2}}$ **b** $\dfrac{1}{2\sqrt{2}}$ **c** $\dfrac{\sqrt[3]{64}}{2\sqrt{2}}$

5 Rationalise the denominator in the following expressions:

a $\dfrac{x\sqrt{x}}{2-\sqrt{x}}$ **b** $\dfrac{y+\sqrt{x}}{y-\sqrt{x}}$ **c** $\dfrac{\sqrt{x}-2}{-3-\sqrt{y}}$

6 Simplify the following expressions:

a $\dfrac{x}{\sqrt{2}}+\dfrac{y}{3}$ **b** $\dfrac{\sqrt{x}}{1-\sqrt{x}}+\dfrac{\sqrt{y}}{2}$ **c** $\dfrac{2-\sqrt{x}}{-\sqrt{x}-2}-\dfrac{\sqrt{x}+2}{\sqrt{x}}$

SUMMARY OF KEY POINTS

❯ Polynomials can be manipulated algebraically:

  ❯ $a^2 - b^2$ is known as 'the difference of two squares' and is factorised as $(a - b)(a + b)$.

  ❯ If $f(a) = 0$ then $(x - a)$ is a factor of $f(x)$. This result is called the factor theorem.

❯ The method for dividing a polynomial by an expression in the form $(ax \pm b)$ where $a \geqslant 1$ is just the same as the method for numeric long division and is called algebraic division.

❯ The following are useful for the binomial expansion of $(a + bx)^n$ for positive integer n and the notation $n!$ and ^nC_r:

  ❯ $n! = n \times (n - 1) \times (n - 2) \ldots \times 2 \times 1$

  ❯ $0! = 1$

  ❯ $(a + bx)^n = {}^nC_0 a^n + {}^nC_1 a^{n-1}(bx)^1 + {}^nC_2 a^{n-2}(bx)^2 + {}^nC_3 a^{n-3}(bx)^3 + \ldots + {}^nC_n(bx)^n$

  ❯ $(1 + bx)^n = 1 + n(bx) + \dfrac{n(n-1)}{2!}(bx)^2 + \dfrac{n(n-1)(n-2)}{3!}(bx^3 + \dfrac{n(n-1)(n-2)(n-3)}{4!}(bx)^4 + \ldots$

❯ These are the laws of indices for all rational exponents:

  ❯ $a^m \times a^n = a^{m+n}$

  ❯ $(a^m)^n = a^{mn}$

  ❯ $a^m \div a^n = a^{m-n}$

  ❯ $a^0 = 1$

  ❯ $a^{-m} = \dfrac{1}{a^m}$

  ❯ $a^{\frac{m}{n}} = \sqrt[n]{a^m}$

❯ You can use and manipulate surds, including rationalising the denominator:

  ❯ $\left(\sqrt{x}\right)^2 = x$

  ❯ $\sqrt{xy} = \sqrt{x}\sqrt{y}$

  ❯ $\left(\sqrt{x} + \sqrt{y}\right)\left(\sqrt{x} - \sqrt{y}\right) = x - y$

  ❯ $\sqrt{\dfrac{x}{y}} = \dfrac{\sqrt{x}}{\sqrt{y}}$

❯ To rationalise the denominator of a fraction in the form $\sqrt{\dfrac{b}{a}}$ or $\dfrac{b}{\sqrt{a}}$ you multiply the fraction by $\dfrac{\sqrt{a}}{\sqrt{a}}$.

❯ To rationalise the denominator of a fraction in the form $\dfrac{c}{b \pm \sqrt{a}}$ you multiply the fraction by $\dfrac{b \mp \sqrt{a}}{b \mp \sqrt{a}}$.

EXAM-STYLE QUESTIONS 1   **Answers page 489**

1. Simplify the following expressions:

   a $\dfrac{x^2 \times x^5}{x^4}$ [2 marks]

   b $(3x^2)^3 \div 9x^5$ [2 marks]

2. Rationalise the denominator in the following expressions:

   a $\dfrac{y}{\sqrt{x}}$ [2 marks]

   b $\dfrac{3}{2-\sqrt{x}}$ [3 marks]

3. Given that $f(x) = 2x^3 + x^2 - 25x + 12$:

   a use the factor theorem to show that $(x-3)$ is a factor of $f(x)$ [3 marks]

   b use the factor theorem to show that $(x+1)$ is not a factor of $f(x)$.   [3 marks]

4. Express 256 as a power of 2.   [1 mark]

(M) 5. There are 10 different flavours of crisps on the shelf at the supermarket. You have enough money to buy three packets of crisps. In how many ways can you do this? Write down the correct combinations notation before attempting the calculation.   [3 marks]

6. Given that $(x - 7)$ is one factor, use algebraic division to fully factorise $x^3 - 6x^2 - 9x + 14$.   [5 marks]

7. Express $\dfrac{125}{5\sqrt{5}}$ as a power of 5.   [2 marks]

8. Show that $\left(a+\sqrt{b}\right)^2 + \left(a-\sqrt{b}\right)^2$ is rational if a and b are rational.   [3 marks]

9. Rationalise the denominator in the following expressions:

   a $\dfrac{\sqrt{x}-1}{\sqrt{x}+1}$ [3 marks]

   b $\dfrac{2\sqrt{x}}{3-\sqrt{2x}}$ [3 marks]

(PS) 10. Use a combination of the factor theorem and algebraic division to fully factorise $x^4 - 9x^2 - 4x + 12$. Then expand your brackets to show that your factorisation is correct.   [7 marks]

(PS) (CM) 11. Explain what is wrong in each of the following two statements:

   a $\left(4-3x\right)^5 = 4^5\left(1-\dfrac{4x}{3}\right)^4$ [2 marks]

   b In the expansion of $(2-x)^3$, the coefficient of $x > 0$.   [2 marks]

12. Use the factor theorem to express the improper fraction $\dfrac{2x^3 + 15x^2 + 31x + 12}{(x-5)}$ as a mixed number.   [5 marks]

(PS) **13** Find the value of x in the expression $\left(\dfrac{x}{27}\right)^{-\frac{2}{3}} = \dfrac{9}{16}$. **[2 marks]**

(PS) **14** In the expansion of $(a-2x)^4$ the coefficient of x is -216.

    a Find the value of a. **[2 marks]**

    b Use the binomial expansion to expand $(a-2x)^4$ using your value of a from **part a**. **[3 marks]**

    c Using your expansion, calculate $(2.8)^4$. **[2 marks]**

(PS) **15** Simplify $\dfrac{12x^2 - 8x - 15}{42x^2 + 29x - 5}$. **[5 marks]**

(PS) **16** The sum of each row in Pascal's triangle is a power of 2. One of the rows from Pascal's triangle is written erroneously as 1, 6, 16, 20, 15, 6, 1. Identify and correct the mistake. **[2 marks]**

(PS) **17** Show that $\left(a+\sqrt{b}\right) \times \left(a-\sqrt{b}\right)^3$ is irrational if a and b are *not* square numbers. **[3 marks]**

(PS) **18** Simplify the following expressions:

    a $\dfrac{1}{2\sqrt{x}+3} + \dfrac{1}{\sqrt{x}}$ **[4 marks]**

    b $\dfrac{3x}{\sqrt{x}-1} - \dfrac{\sqrt{x}+1}{\sqrt{x}}$ **[4 marks]**

(PS) **19** Fully factorise $4x^4 + 8x^3 - 21x^2 - 18x + 27$. **[10 marks]**

20 Use the factor theorem to express the improper fraction

$$\dfrac{2x^5 - 19x^4 + 60x^3 - 65x^2 - 2x + 24}{x+2}$$

as mixed numbers. **[6 marks]**

(PS) **21** **a** Use the binomial expansion to expand $\left(1+\sqrt{x}\right)^3$. **[3 marks]**

    b Using your expansion from **part a**, write down the expansion of $\left(2-2\sqrt{x}\right)^3$. **[2 marks]**

(M) (PS) **22** A bag contains $(7x-3)$ red balls and $(9-x)$ blue balls. A ball is taken from the bag, and not replaced, and then another ball is taken from the bag. Find a fraction, in terms of x, which has both a quadratic numerator and a quadratic denominator, for the probability that two red balls are taken from the bag. **[6 marks]**

(PS) (CM) **23** **a** Use the binomial expansion to expand $\left(1+\dfrac{x}{2}\right)^6$ up to and including the term in x^3. **[2 marks]**

    b Use your expansion to find an approximation to $(1.05)^6$. **[2 marks]**

    c Explain why your answer to **part b** is only an approximation. **[2 marks]**

(PS) (CM) **24** The height of a cuboid is given by the linear expression $2x+1$. The width and depth of the cuboid can also be expressed as linear expressions in terms of x. The volume of a cuboid is given by the expression $2x^3 + 3x^2 - 11x - 7$, where x is an integer. Is this a possible correct expression for the volume of the cuboid? You must justify your answer. If not, suggest a possible alternative expression and explain its validity. **[6 marks]**

2 ALGEBRA AND FUNCTIONS 2: EQUATIONS AND INEQUALITIES

A footballer kicks a football and the ball follows a path given by the function $f(x) = 4x - x^2$. You should recognise this as a quadratic function. For this function, you could work out the maximum height reached by the ball and the horizontal distance travelled by the ball, from when the footballer first kicks it to when it lands back on the ground. You could also plot the graph of the function on a grid to work out these measurements and you would also see the path that the ball follows. In this chapter you are going to cover a variety of techniques to analyse similar functions.

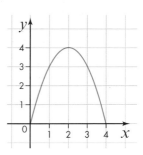

LEARNING OBJECTIVES

You will learn how to:

» work with quadratic functions and their graphs

» calculate and use the discriminant of a quadratic function, including the conditions for real and repeated roots

» complete the square

» solve quadratic equations

» solve simultaneous equations in two variables by elimination and by substitution, including one linear and one quadratic equation

» solve linear and quadratic inequalities in a single variable and interpret such inequalities graphically, including inequalities with brackets and fractions

» express solutions through the correct use of 'and' and 'or', or through set notation.

TOPIC LINKS

You will need the skills and techniques that you learn in this chapter to help you sketch curves in **Chapter 3 Algebra and functions 3: Sketching curves**. The ability to find the solutions to quadratic equations will help you to solve distance, speed and time problems in **Chapter 15 Kinematics**. Knowing how to solve linear and quadratic equations simultaneously will help you to find the points of intersection of circles and straight lines in **Chapter 5 Coordinate geometry 2: Circles**. In **Chapter 6 Trigonometry** and **Chapter 7 Exponentials and logarithms** you will need to solve quadratic equations involving a function of the unknown.

PRIOR KNOWLEDGE

You should already know how to:

- simplify and manipulate algebraic expressions, including:
  - factorising quadratic expressions of the form $x^2 + bx + c$, including the difference of two squares
  - factorising quadratic expressions of the form $ax^2 + bx + c$
- identify and interpret roots, intercepts and turning points of quadratic functions graphically
- deduce roots algebraically and turning points by completing the square
- solve linear equations in one unknown algebraically and find approximate solutions using a graph
- solve quadratic equations algebraically by factorising, by completing the square and by using the quadratic formula
- find approximate solutions to quadratic equations using a graph
- find and use the discriminant of a quadratic equation
- solve two simultaneous equations in two variables algebraically and find approximate solutions using a graph
- translate simple situations or procedures into algebraic expressions or formulae
- derive and solve equations and interpret the solutions
- solve linear inequalities in one or two variables and quadratic inequalities in one variable
- represent the solution set on a number line, using set notation and on a graph.

You should be able to complete the following questions correctly:

1   Solve the quadratic equation $2x^2 - 5x - 3 = 0$ by factorisation.
2   Solve the quadratic equation $x^2 + x - 12 = 0$ by completing the square.
3   By evaluating the discriminant, determine the number of real roots for the quadratic equation $y = 2x^2 - 8x - 13$ and then solve the equation $2x^2 - 8x - 13 = 0$ using the quadratic formula.
4   Solve this pair of simultaneous equations by elimination: $5x - y = 13$ and $2x + y = 1$.
5   Solve this pair of simultaneous equations by substitution: $y = x + 1$ and $y = x^2 - 3x + 2$.
6   Solve the inequality $3x - 5 < 7$ and show the solutions on a number line.
7   On the same grid, shade the regions that represent the following inequalities:
    a $x < 3$ **b** $y < 2$ **c** $2x + y \geqslant 1$

2.1 Quadratic functions

In your GCSE course you plotted the graphs of quadratic **functions** in the form $ax^2 + bx + c$.

You can examine a graph of a **quadratic** curve to identify:

- the axis intercepts

> the roots (solutions) to the **equation**, especially when $y = 0$

> the coordinates of the turning point

> the nature of the **turning point** – whether it is a minimum or a maximum

> the equation of the line of symmetry

> y values for corresponding x values and vice versa.

Example 1

Draw the graph of the equation $y = x^2 - x - 6$ where $-3 \leqslant x \leqslant 4$.

Write down the coordinates of the axis intercepts and the turning point.

Solution

For different values of x, plot the points and join them up with a smooth curve.

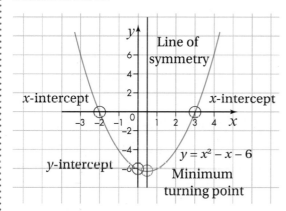

The curve intercepts the x-axis at $(-2, 0)$ and $(3, 0)$ and the y-axis at $(0, -6)$, as highlighted by the red circles.

The line of symmetry, in orange, is at $x = 0.5$. So, substituting $x = 0.5$ into the equation gives $(0.5, -6.25)$ as the coordinates of the minimum turning point (the lowest value of the curve), as highlighted by the green circle.

Example 2

Draw the graph of the equation $y = 2 - 3x - 2x^2$ where $-3 \leqslant x \leqslant 2$.

Use your graph to solve the equation $5 - 3x - 2x^2 = 0$.

Solution

For different values of x, plot the points and join them up with a smooth curve.

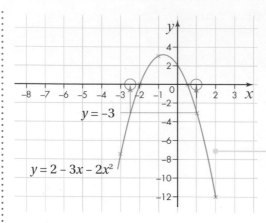

As $a < 0$ the parabola is inverted.

Rearrange the equation so that either the LHS or RHS matches the equation of the drawn curve.

$$5 - 3x - 2x^2 = 0$$

$$2 - 3x - 2x^2 = -3$$

Draw a line at $y = -3$ and read off the corresponding x values, which are -2.5 and 1.

Exercise 2.1A

Answers page 490

1 Draw the graphs of the following equations and write down the coordinates of the axis intercepts and the turning point.

a $y = x^2$ where $-4 \leqslant x \leqslant 4$

b $y = x^2 + 4$ where $-4 \leqslant x \leqslant 4$

c $y = x^2 + 3x - 4$ where $-5 \leqslant x \leqslant 3$

(PS) 2 Find and correct the six mistakes in the following table of values and subsequently draw the graph for the equation.

x	-3	-2	-1	0	1	2	3	4	5
x^2	-9	4	1	0	1	4	9	16	25
$-2x$	6	-4	2	0	-2	-4	-6	-8	-10
-8	-8	-8	-8	-8	-8	-8	-8	8	-8
y	-11	-8	-5	-8	-9	-8	-5	16	7

3 Draw the graphs of the following equations and write down the equation of the line of symmetry and the coordinates of the axis intercepts and the turning point.

a $y = (x + 1)^2$ where $-4 \leqslant x \leqslant 2$

b $y = 2x^2 + 7x + 3$ where $-4 \leqslant x \leqslant 1$

c $y = 4 - 3x - x^2$ where $-5 \leqslant x \leqslant 2$

4 Write down the equation for the following graph. Explain and justify your decision-making process.

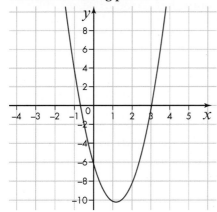

5 Draw the graph of the equation $y = 2 + x - 6x^2$ where $-2 \leqslant x \leqslant 2$.

Use your graph to solve the equation $98 - x - 24x^2 = 0$.

2.2 The discriminant of a quadratic function

Part of the quadratic formula determines how many solutions there are to a quadratic equation.

$$x = \frac{-b \pm \sqrt{b^2 - 4ac}}{2a}$$

You should know from your GCSE course that the expression within the square root, $b^2 - 4ac$, is called the **discriminant**.

If $b^2 - 4ac < 0$ then we say that the quadratic equation doesn't have any real **roots**. When you plot or sketch the curve of a quadratic function where the discriminant is less than zero, it will *not* meet the x-axis.

If $b^2 - 4ac = 0$ then x will have two equal real roots which will both be $-\dfrac{b}{2a}$. When you plot or sketch the curve of a quadratic function where the discriminant is equal to zero, it will meet the x-axis at one point.

If $b^2 - 4ac > 0$ then when the square root of this number is found there will be *two* distinct real roots:

$$x = \frac{-b + \sqrt{b^2 - 4ac}}{2a} \quad \text{and} \quad x = \frac{-b - \sqrt{b^2 - 4ac}}{2a}.$$

When you plot or sketch the curve of a quadratic function where the discriminant is greater than zero, it will meet the x-axis at two different points.

KEY INFORMATION

$b^2 - 4ac$ is called the discriminant.

Although it is possible to find the square root of a negative number and get an imaginary number, you will not learn about them at this level.

Sketching quadratic functions is covered in **Section 3.1**.

KEY INFORMATION

If $b^2 - 4ac < 0$ then the quadratic equation does not have any real roots.

If $b^2 - 4ac = 0$ then the quadratic equation has two equal real roots.

If $b^2 - 4ac > 0$ then the quadratic equation has two distinct real roots.

Stop and think

Work out the value of the discriminant for some quadratic equations that factorise and some that don't. For the equations that do factorise, what is special about the discriminant?

Example 3

Find the discriminant of the equation $x^2 - 7x + 12 = 0$ and hence state the number of solutions to the equation.

Make sure that the equation is written in the form $ax^2 + bx + c = 0$ before you attempt to work out the discriminant.

Solution

$$b^2 - 4ac = (-7)^2 - (4)(1)(12)$$

$$= 49 - 48$$

$$= 1$$

$b^2 - 4ac > 0$ so the equation has two distinct real roots.

You need to ensure that a, b and c have been correctly substituted.

Example 4

Find the quadratic equation which has two equal real roots where $x^2 - kx + 2k = 0$, $k > 0$, and state the value of the two equal roots.

Solution

$$b^2 - 4ac = (-k)^2 - (4)(1)(2k)$$

$$= k^2 - 8k$$

$$k^2 - 8k = 0$$

$$k(k - 8) = 0$$

As $k > 0$, $k = 8$.

So the equation is $x^2 - 8x + 16 = 0$

$(x - 4)^2 = 0$ so the equation has two equal roots at $x = 4$.

To find the roots you need to substitute the value of k back into the original equation and then factorise.

Exercise 2.2A

Answers page 492

1 Work out the discriminant for each of the following equations and hence state the number of solutions to the equation.

   a $x^2 + 2x + 13 = 0$ **b** $2x^2 + 3x + 1 = 0$ **c** $x^2 + 4x + 4 = 0$

(PS) 2 Find and correct the mistakes in the following calculation to work out and evaluate the discriminant for the equation $2x^2 - 5x + 3 = 0$.

$$b^2 - 4ac = (2)^2 - (4)(-5)(3)$$

$$= 4 + 60$$

$$= 64$$

$b^2 - 4ac > 0$ so the equation has two equal real roots.

3 Work out the discriminant for each of the following equations and hence state the number of solutions to the equation.

    a $x^2 - 3x - 5 = 0$ **b** $3x^2 - 2x + 7 = 0$ **c** $x^2 - 12x + 36 = 0$

(PS) **4** Find the two possible quadratic equations which have two equal real roots where $x^2 + (2k + 2)x + (5k - 1) = 0$ and in each case state the value of the two equal roots.

5 Work out the discriminant for each of the following equations and hence state the number of solutions to the equation.

    a $4 + 2x - 3x^2 = 0$ **b** $49 + x^2 - 14x = 0$ **c** $3x^2 + 12 - 5x = 0$

(PS) **6** Find the range of possible values of c for which $2x^2 + 5x - c = 0$ has:

    a no real roots   **b** two equal real roots   **c** two distinct real roots.

2.3 Completing the square

Quadratic expressions can be written in a different format by **completing the square** – a technique that you should remember from your GCSE course.

The expression $x^2 + bx + c$ can be written as $\left(x + \dfrac{b}{2}\right)^2 - \left(\dfrac{b}{2}\right)^2 + c$ by completing the square.

Let's look at how this works.

If you expand $\left(x + \dfrac{b}{2}\right)^2$ you get $x^2 + bx + \dfrac{b^2}{4}$. You should notice that the terms in red are the first part of the original expression whilst $\dfrac{b^2}{4}$ is additional, so it needs to be subtracted.

Working backwards, you have:

$$\left(x + \frac{b}{2}\right)^2 - \left(\frac{b}{2}\right)^2 + c = x^2 + bx + \frac{b^2}{4} - \frac{b^2}{4} + c$$

$$= x^2 + bx + c$$

The expression $ax^2 + bx + c$ factorises to $a\left(x^2 + \frac{bx}{a} + \frac{c}{a}\right)$. By completing the square, this factorised expression can now be written as:

$$a\left(x^2 + \frac{bx}{a} + \frac{c}{a}\right) = a\left[\left(x + \frac{b}{2a}\right)^2 - \frac{b^2}{4a^2} + \frac{c}{a}\right]$$

> **KEY INFORMATION**
>
> $x^2 + bx + c = \left(x + \dfrac{b}{2}\right)^2 - \left(\dfrac{b}{2}\right)^2$
>
> $+ c$ by completing the square.

> **KEY INFORMATION**
>
> $a\left(x^2 + \dfrac{bx}{a} + \dfrac{c}{a}\right) =$
>
> $\left[\left(x + \dfrac{b}{2a}\right)^2 - \dfrac{b^2}{4a^2} + \dfrac{c}{a}\right]$
>
> by completing the square.
>
> Another way of writing this is:
>
> $ax^2 + bx + c =$
>
> $a\left(x + \dfrac{b}{2a}\right)^2 + \left(c - \dfrac{b^2}{4a}\right)$

Example 5

Complete the square for the equation $y = 2x^2 + 3x - 2$ and use the result to work out the coordinates of the turning point.

Solution

You need to factorise the equation so that the coefficient of x^2 is 1.

$$y = 2\left(x^2 + \frac{3x}{2} - 1\right)$$

You need to halve b and complete the square for the first two parts of the expression. You need to remember to subtract any additional values that were not part of the expression.

$$x^2 + \frac{3x}{2} = \left(x + \frac{3}{4}\right)^2 - \frac{9}{16}$$

Check: $\left(x + \frac{3}{4}\right)^2 - \frac{9}{16} = x^2 + \frac{3x}{2}$
$+ \frac{9}{16} - \frac{9}{16}$

Replace $x^2 + \frac{3x}{2}$ in the original expression and collect like terms.

$$y = 2\left[\left(x + \frac{3}{4}\right)^2 - \frac{9}{16} - 1\right]$$

$$= 2\left[\left(x + \frac{3}{4}\right)^2 - \frac{25}{16}\right]$$

As $a > 0$ the curve of this equation will have a minimum turning point. The minimum value of y will be when $x = -\frac{3}{4}$.

$$y = 2\left[\left(-\frac{3}{4} + \frac{3}{4}\right)^2 - \frac{25}{16}\right]$$

$$= 2\left[0 - \frac{25}{16}\right]$$

$$= -\frac{25}{16}$$

Hence the coordinates of the turning point are $\left(-\frac{3}{4}, -\frac{25}{8}\right)$.

Exercise 2.3A

Answers page 492

1. Complete the square for the following expressions:

   a $x^2 + 4x$ **b** $2x^2 - 8x$ **c** $x^2 + 8x + 7$

(PS) 2. Find and correct the mistakes in the following manipulation to complete the square for the expression $x^2 - 10x + 11$.

   $$x^2 - 10x + 11 = (x + 5)^2 - 10 + 11$$
   $$= (x + 5)^2 - 1$$

3. Complete the square for the following expressions:

   a $x^2 - 8x - 5$ **b** $x^2 + 3x - 7$ **c** $2x^2 + 3x + 9$

(M) 4. Complete the square for the equation $y = x^2 - 5x + 7$ and use the result to work out the coordinates of the turning point.

(PS) 5. A quadratic expression has been rewritten as $2\left[(x + 2)^2 - \frac{13}{2}\right]$ by completing the square. What was the original quadratic expression?

6. Complete the square for the following expressions:

   a $5x^2 + \frac{1}{2}x - 3$ **b** $5x^2 - 18$ **c** $5 - 7x - 3x^2$

7 Show that the expression $2x^2 - 3x + 11$ is positive for all values of x. What can you deduce about the value of the discriminant for this expression without calculation?

 8 Show that $15 - 22x - 15x^2 - 2x^3$ can be written as $(1 - 2x)[(x + 4)^2 - 1]$.

2.4 Solving quadratic equations

You have already seen that a quadratic equation may have no real roots, two equal real roots or two distinct real roots. Another word that is used for the root of a quadratic equation is *solution*. Consequently, when you are asked to solve a quadratic equation you are being asked to find the solutions (or roots).

From your GCSE course you will remember that there are a number of ways to solve a quadratic equation:

> factorisation

> using the quadratic formula

> completing the square

> using a calculator.

You need to be comfortable deciding which is the best method to use, depending on the equation that you are being asked to solve.

KEY INFORMATION

The roots of a quadratic equation can also be described as the solutions.

Factorisation

Example 6

Solve the equation $2x^2 - 3x - 9 = 0$ using factorisation.

The factorisation method can only be used if the equation actually factorises.

Solution

The equation is going to be factorised by inspection. You need to examine the factors of 2 and 9 to determine which combination of factors will give the desired coefficients. Once you have selected the numeric values, you need to choose the operators.

$$2x^2 - 3x - 9 = 0$$

$$(2x + 3)(x - 3) = 0$$

Before progressing to solve the equation, expand the brackets to check that the factorisation is correct.

$$2x^2 - 6x + 3x - 9 = 0$$

The factorisation is correct.

Either $2x + 3 = 0$ so $x = -\frac{3}{2}$

or $x - 3 = 0$ so $x = 3$

Using the quadratic formula

Example 7

Solve the equation $2x^2 - 3x - 9 = 0$ using the quadratic formula.

Solution

Before substituting, state the quadratic formula.

$$x = \frac{-b \pm \sqrt{b^2 - 4ac}}{2a}$$

From the equation, $a = 2$, $b = -3$, $c = -9$.

Substitute in the values.

$$x = \frac{3 \pm \sqrt{9 - (4 \times 2 \times -9)}}{4}$$

$$= \frac{3 \pm \sqrt{81}}{4}$$

Either $x = -\frac{3}{2}$ or 3

> The quadratic formula can be used to find the solutions to any quadratic equation with real roots.

> You need to remember this formula.

> If you are not sure whether or not a quadratic equation factorises but you are asked to give the answer to a specified number of decimal places, then it is likely that the equation does *not* factorise.

Completing the square

Example 8

Solve the equation $2x^2 - 3x - 9 = 0$ by completing the square.

Solution

$2x^2 - 3x - 9 = 0$

Factorise the x^2 term.

$$2\left(x^2 - \frac{3x}{2} - \frac{9}{2}\right) = 0$$

Subtract any extra values that completing the square produces, in this case $\frac{9}{16}$.

$$2\left[\left(x - \frac{3}{4}\right)^2 - \frac{9}{16} - \frac{9}{2}\right] = 0$$

Manipulate the equation to make x the subject.

$$2\left[\left(x - \frac{3}{4}\right)^2 - \frac{81}{16}\right] = 0$$

$$\left(x - \frac{3}{4}\right)^2 - \frac{81}{16} = 0$$

$$\left(x - \frac{3}{4}\right)^2 = \frac{81}{16}$$

$$x - \frac{3}{4} = \pm\frac{9}{4}$$

Either $x = -\frac{3}{2}$ or 3

> Completing the square can be used to find the solutions to any quadratic equation with real roots.

Example 9

By completing the square, show that the solutions to

$x^2 + bx + c = 0$ are given by $x = \dfrac{-b \pm \sqrt{b^2 - 4c}}{2}$.

Solution

$x^2 + bx + c = 0$

Subtract any extra values that completing the square produces, in this case $-\dfrac{b^2}{4}$.

$$\left(x + \frac{b}{2}\right)^2 - \frac{b^2}{4} + c = 0$$

Manipulate the equation to make x the subject.

$$\left(x + \frac{b}{2}\right)^2 = \frac{b^2 - 4c}{4}$$

$$x + \frac{b}{2} = \frac{\pm\sqrt{b^2 - 4c}}{2}$$

$$x = \frac{-b \pm \sqrt{b^2 - 4c}}{2}$$

PROOF

You have proved that you can solve quadratic equations in the form $x^2 + bx + c = 0$ using

$x = \dfrac{-b \pm \sqrt{b^2 - 4c}}{2}$

Using a calculator

Example 10

Solve the equation $2x^2 - 3x - 9 = 0$ using a calculator.

You need to ensure that you are familiar with all the functions on your calculator.

Solution

You need to find the equations mode on your calculator then, within this, select the polynomials option. Using the coefficient editor, you then need to enter the coefficients of the equation. When you press the equals button, the calculator should provide the solutions to the equation, which are $x = -\frac{3}{2}$ or 3.

TECHNOLOGY

You can use your calculator to check your solutions to quadratic equations that you have solved by hand.

Exercise 2.4A   **Answers page 493**

1   Solve each of the following quadratic equations by:

  i factorisation   **ii** using the quadratic formula

  iii completing the square   **iv** using your calculator.

  a $x^2 - 9 = 0$ **b** $x^2 + 3x + 2 = 0$

  c $x^2 - 5x - 7 = 0$ **d** $x^2 - 6x = 0$

2 With reference to **Examples 6**, **7**, **8** and **10** above, explain the advantages and disadvantages of each method for solving a quadratic equation. Give examples of quadratic equations that may and may not be solved with each method.

  3 Solve each of the following quadratic equations by choosing the most appropriate method. Provide reason(s) for your choice of method.

    a $(x-4)^2 = 13$ **b** $2x^2 - 5x - 11 = 0$

    c $3x^2 = 5 - 14x$ **d** $7 - 16x + 4x^2 = 0$

 4 **a** Show that $3x^2 + 4x - 11 = 0$ cannot be solved by factorisation.

    b Solve $3x^2 + 4x - 11 = 0$ using an appropriate method.

5 Solve the equation $(x-3)(x-5) = 8$.

6 By completing the square, show that the solutions to $ax^2 + bx + c = 0$ are given by

$$x = \frac{-b \pm \sqrt{b^2 - 4ac}}{2a}$$

2.5 Solving simultaneous equations

In your GCSE course you saw that if you have two equations with two different variables, you can solve the equations **simultaneously** (at the same time) to work out the values of the two variables.

For example, you might visit the local newsagent one week and buy two bars of chocolate and a magazine, at a total cost of £4.99. The next week you might buy the same magazine but just one bar of chocolate, at a total cost of £3.49. Using this information you can form two equations to solve simultaneously.

    $2c + m = 4.99$ ①

    $c + m = 3.49$ ②

There are two main methods you can use to solve simultaneous equations – **elimination** and **substitution**. You need to be comfortable deciding which is the best method to use.

> **Stop and think**   If you were to solve the simultaneous equations above graphically, where would you find the solution on your diagram?

Elimination

Example 11

Solve the simultaneous equations $2x + 3y = -15$ and $3x + 5y = -26$ by elimination.

Solution

Number the equations ① and ②.

$$2x + 3y = -15 \qquad ①$$

$$3x + 5y = -26 \qquad ②$$

Look at the coefficients of each variable. The question asks you to solve by elimination, so you need to find a way to combine the equations to remove one variable.

Multiply ① by 3 and ② by 2.

$$6x + 9y = -45 \qquad ③$$

$$6x + 10y = -52 \qquad ④$$

Subtract ④ from ③.

$$y = -7$$

Substitute y back into ①.

$$2x - 21 = -15$$

Solve.

$$x = 3$$

KEY INFORMATION

When the signs of the coefficients of the variable to be eliminated are different you should add the equations. When the signs are the same you should subtract one equation from the other.

Number the new equations ③ and ④.

Always substitute back into the original equations in case you have made any errors.

Do your solutions work for both equations?

TECHNOLOGY

Using a graphic calculator or graphing software package, plot the graphs of $2x + 3y = -15$ and $3x + 5y = -26$. Is the point of intersection $(3, -7)$?

Exercise 2.5A

Answers page 494

1 Solve the following simultaneous equations by elimination:

    a $2x + y = -4$ and $5x + y = -1$

    b $3x - 2y = 2$ and $5x + y = 25$

    c $x - 2y = 13$ and $-x + 3y = -15$

2 Find and correct the mistakes in the following working.

$$5x + 6y = 6 \qquad ①$$

$$3x - y = 22 \qquad ②$$

Multiply ② by 6.

$$5x + 6y = 6 \qquad ①$$

$$18x - 6y = 132 \qquad ③$$

Subtract ① from ③.

$$13x = 126$$

$$x = \frac{126}{13}$$

Substitute x into ②.

$$\frac{378}{13} - y = 22$$

$$y = \frac{93}{12}$$

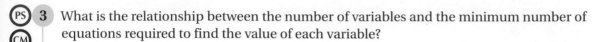 **3** What is the relationship between the number of variables and the minimum number of equations required to find the value of each variable?

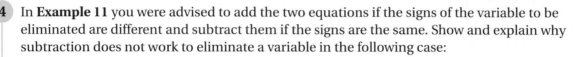 **4** In **Example 11** you were advised to add the two equations if the signs of the variable to be eliminated are different and subtract them if the signs are the same. Show and explain why subtraction does not work to eliminate a variable in the following case:

$$2x + y = -1$$

$$3x - 2y = -12$$

5 Solve the simultaneous equations $3x - 4y = 3$ and $6x + 4y = 3$ by elimination.

6 This year the sum of Helen's age and her daughter's age is 50. A year ago, Helen was 15 times as old as her daughter. By forming a pair of simultaneous equations and solving using elimination, find out how old Helen was when her daughter was born.

7 Solve the simultaneous equations $2x - y = -14$, $3y - 2z = 16$ and $z - x = 3$ by elimination.

An alternative method to elimination for solving simultaneous equations is to use substitution instead. Your choice of method could be dictated by personal preference, complexity of rearrangement for substitution, polynomials of different degrees (one **linear** and one **quadratic**) or a requirement to demonstrate a particular method.

When solving simultaneous equations by substitution you choose one equation and make one of the variables the subject of that equation. You then replace this variable in the untouched equation with your rearrangement.

Substitution

Example 12

Solve the simultaneous equations $2x + y = 7$ and $x - y = 8$ by substitution.

Solution

Number the equations.

$$2x + y = 7 \qquad \qquad ①$$

$$x - y = 8 \qquad \qquad ②$$

You need to choose which equation you want to substitute into the other and which variable is going to be substituted.

Rearrange ②.

$$x = y + 8$$

Substitute for x into ①.

$$2(y + 8) + y = 7$$

Expand and simplify.

$$3y + 16 = 7$$

Solve.

$$y = -3$$

Substitute y into ②.

$$x = 5$$

Equation ② was chosen because the rearrangement is simpler.

Exercise 2.5B

Answers page 494

1 Solve the following simultaneous equations by substitution:

   a $2x + y = -4$ and $5x + y = -1$

   b $3x - 2y = 2$ and $5x + y = 25$

   c $x - 2y = 13$ and $-x + 3y = -15$

2 Find and correct the mistakes in the following working.

$$5x + 6y = 6 \qquad\qquad ①$$
$$3x - y = 22 \qquad\qquad ②$$

Rearrange ②.

$$3x - 22 = y$$

Substitute into ①.

$$5(3y - 22) + 6y = 6$$

Expand and simplify.

$$21y - 110 = 6$$
$$y = \frac{116}{21}$$

Substitute y into ②.

$$3x - 22 = \frac{116}{21}$$

Solve.

$$x = \frac{578}{63}$$

3 Using the substitution method, find where the graphs $4x + 3y = 9$ and $5x - 6y = 60$ intersect.

4 Solve the simultaneous equations $3x - 4y = 3$ and $6x + 4y = 3$ by substitution.

(M) 5 A college student buys two bars of chocolate and seven small bags of sweets and the costs is £1.70. The next day the same student buys three bars of chocolate and eight small bags of sweets. This time the cost is £2.50. The student says that, as he has purchased the same bars of chocolate and bags of sweets and the prices haven't changed, one of the total cost calculations must be wrong. Show, using simultaneous equations, why he is correct.

6 Solve the simultaneous equations $2x - y = -14$, $3y - 2z = 16$ and $z - x = 3$ by substitution.

2.6 Solving linear and quadratic simultaneous equations

Simultaneous equations may not always both be linear. For example, one might be linear and the other quadratic. If this is the case, then usually you will need to use the substitution method – but there are exceptions to this.

Stop and think Can you think of different pairs of equations, one linear and one quadratic, where you could use the elimination method instead of substitution? What is special about these pairs of equations?

Example 13

Solve the simultaneous equations $y - x = 1$ and $2x^2 - 11y = -16$ by substitution.

Solution

Number the equations.

$$y - x = 1 \qquad \qquad ①$$
$$2x^2 - 11y = -16 \qquad ②$$

When you have one linear and one quadratic equation you need to choose which variable is going to be substituted and rearrange the linear equation to make this variable the subject.

Rearrange ①.

$$y = x + 1$$

Substitute for y into ②.

$$2x^2 - 11(x + 1) + 16 = 0$$

Expand and simplify.

$$2x^2 - 11x + 5 = 0$$

Solve.

$$(2x - 1)(x - 5) = 0$$
$$x = \frac{1}{2} \text{ or } x = 5$$

Substitute x into ①.

$$y = \frac{3}{2} \text{ or } y = 6$$

TECHNOLOGY

Using a graphic calculator or graphing software package, plot the graphs of $y - x = 1$ and $2x^2 - 11y = -16$. Do the graphs intersect at two points? What are the coordinates of the points of intersection?

Don't forget to work out the corresponding y values. It is easy to forget!

> **Stop and think**
>
> When you solve two simultaneous equations where one is linear and one is quadratic, how does this relate to the graphs of these two equations?

Exercise 2.6A

Answers page 494

1 Solve the following simultaneous equations:

   a $x + y = 3$ and $2x^2 - y = 25$

   b $2x - y = 20$ and $x^2 + xy = -12$

   c $y = 4x$ and $5 - x^2 = y$

(PS) (CM) **2** For each of the following pairs of simultaneous equations, state which method(s), elimination or substitution, could be used to solve them and why. (You need to justify your answer.)

   a $2x^2 + y = 14$ and $x - 2y = 11$

   b $x^2y - x = 3$ and $x + y = 1$

   c $x - y = 10$ and $xy = 140$

3 Solve the following simultaneous equations:

   a $2x^2 + y = 14$ and $x - 2y = 11$

   b $xy - x = -4$ and $x + y = 1$

   c $x - y = 10$ and $xy = 140$

(M) (PS) **4** Solve the simultaneous equations $2x - 3y = 13$ and $x^2 - y = 7$. What are the coordinates of the points of intersection of this line and this curve?

(M) (PS) **5** The line $y - x = 1$ and the circle $x^2 + y^2 = 64$ intersect at two points. Find the coordinates of the points of intersection, giving your answers to 3 decimal places.

(M) (PS) **6** Do the line $y - x = 10$ and the circle $x^2 + y^2 = 50$ intersect? If so, how many times and what are the coordinates of the points of intersection?

2.7 Solving linear inequalities

An equation is a mathematical statement that shows that two expressions are equal using the = sign. An **inequality** is a mathematical statement that shows that two expressions are not equal (a **strict inequality**), or not entirely equal (an **inclusive inequality**). Inequality statements use these inequality signs:

   < 'less than'

   > 'greater than'

   ≤ 'less than or equal to'

   ≥ 'great than or equal to'

> **KEY INFORMATION**
>
> The inequality symbols are $<, >, \leq$ and $\geq$.

'Less than' and 'greater than' are known as strict inequalities as they do not contain an equals element.

'Less than or equal to' and 'greater than or equal to' are known as inclusive inequalities as they contain an equals element.

To solve an inequality, the manipulations are just the same as if you were solving an equation.

You may need to represent an inequality or a combination of inequalities using **set notation** – this is just a way of listing the **members** of the **set**. For example, the **integer** members of the set $-3 \leqslant x < 2$ are $\{-3, -2, -1, 0, 1\}$.

The notation $\{x: x > a\}$ means all values of x where x is greater than a. For example, $\{x: x > 3\}$ means all values of x where x is greater than 3.

You may need to list the members of two or more combined inequalities. For example, the list of integer members of $1 < x < 5$ *or* $4 \leqslant x < 7$ is the list of all integer members of either set. The integer members of $1 < x < 5$ are $\{2, 3, 4\}$ and the integer members of $4 \leqslant x < 7$ are $\{4, 5, 6\}$. So the integer members of $1 < x < 5$ *or* $4 \leqslant x < 7$ are $\{2, 3, 4, 5, 6\}$. Another way of writing this would be $\{x: 1 < x < 5\} \cup \{x: 4 \leqslant x < 7\}$.

However, the integer members of $1 < x < 5$ *and* $4 \leqslant x < 7$ is the list of all integer members that appear in *both* sets. So the integer member of $1 < x < 5$ *and* $4 \leqslant x < 7$ is $\{4\}$. Another way of writing this would be $\{x: 1 < x < 5\} \cap \{x: 4 \leqslant x < 7\}$.

You may also need to represent inequalities graphically. Use a dotted line to represent a strict inequality on a graph, and use a solid line to represent an inclusive inequality.

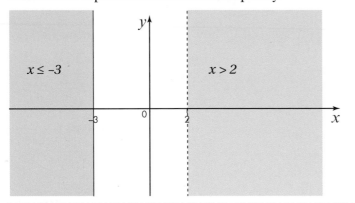

KEY INFORMATION

$<$ and $>$ are strict inequalities.

$\leqslant$ and $\geqslant$ are inclusive inequalities.

$\cup$ = union

$\cap$ = intersection

KEY INFORMATION

$\{x: x > a\}$ means all values of x where x is greater than a.

KEY INFORMATION

The members of $x > a$ or $x < b$ is all the members of $\{x: x > a\}$ and all the members of $\{x: x < b\}$

KEY INFORMATION

The members of $x > a$ and $x < b$ is just the members of $\{x: x > a\}$ that are also in $\{x: x < b\}$

KEY INFORMATION

On graphs, the strict inequalities $<$ and $>$ are represented by dotted lines and the inclusive inequalities $\leqslant$ and $\geqslant$ by solid lines.

Example 14

Solve the linear inequality $5(2x - 1) \leqslant 25$.

Solution

$$5(2x - 1) \leqslant 25$$

Expand the brackets.

$$10x - 5 \leqslant 25$$

Add 5 to both sides of the inequality.

$$10x \leqslant 30$$

Divide both sides by 10.

$$x \leqslant 3$$

Alternatively:

$$5(2x - 1) \leqslant 25$$

Divide both sides of the inequality by 5.

$$2x - 1 \leqslant 5$$

$$2x \leqslant 6$$

$$x \leqslant 3$$

Always ensure that you have included the inequality sign throughout your working and not mistakenly changed it to an equals sign.

Example 15

Solve the linear inequality $\dfrac{4 - x}{3} > 7$.

Solution

$$\frac{4 - x}{3} > 7$$

Multiply both sides of the inequality by 3.

$$4 - x > 21$$

Subtract 4 from both sides.

$$x > 17$$

Add x to both sides of the inequality.

$$0 > 17 + x$$

Subtract 17 from both sides of the inequality.

$$-17 > x$$

Which can be more easily read as $x < -17$.

Alternatively, multiply both sides by -1. This will change the direction of the inequality, as above.

$$x < -17$$

Alternatively:

$$\frac{4 - x}{3} > 7$$

Multiply both sides of the inequality by 3.

$$4 - x > 21$$

Add x to both sides.

$$4 > 21 + x$$

KEY INFORMATION

When you multiply or divide an inequality by a negative number the direction of the inequality reverses.

Subtract 21 from both sides.

$$-17 > x$$

Rewrite.

$$x < -17$$

Example 16

Solve the inequalities $-8 < 3x + 1 \leqslant 10$ and $3 < \frac{4x}{3} < 8$ where $x \in \mathbb{Z}$.

Find $-8 < 3x + 1 \leqslant 10$ or $3 < \frac{4x}{3} < 8$.

Find $-8 < 3x + 1 \leqslant 10$ and $3 < \frac{4x}{3} < 8$.

$\mathbb{Z}$ is the set of all integers, both positive and negative, including zero.

Solution

$$-8 < 3x + 1 \leqslant 10$$

Subtract 1.

$$-9 < 3x \leqslant 9$$

Divide by 3.

$$-3 < x \leqslant 3$$

$\{x: -3 < x \leqslant 3\} = \{-2, -1, 0, 1, 2, 3\}$

$$3 < \frac{4x}{3} < 8$$

Multiply by 3.

$$9 < 4x < 24$$

Divide by 4.

$$\frac{9}{4} < x < 6$$

$\{x : \frac{9}{4} < x < 6\} = \{3, 4, 5\}$

$-8 < 3x + 1 \leqslant 10$ or $3 < \frac{4x}{3} < 8$ is $\{-2, -1, 0, 1, 2, 3, 4, 5\}$

$-8 < 3x + 1 \leqslant 10$ and $3 < \frac{4x}{3} < 8$ is $\{3\}$

Example 17

On the same grid shade the regions that satisfy the inequalities $x + y > -7$, $y \leqslant 2$ and $x < 1$.

Write down the coordinates of any three points in the region that satisfy all three inequalities.

Solution

On a pair of axes draw the line $y = -7 - x$ using a dotted line.

Shade the region $x + y > -7$. (An easier way to think of this is to shade the region $y > -7 - x$.)

Draw the line $y = 2$ using a solid line and shade the region $y \leqslant 2$.

Draw the line $x = 1$ using a dotted line and shade the region $x < 1$.

The region that satisfies all three inequalities is where the shadings overlap.

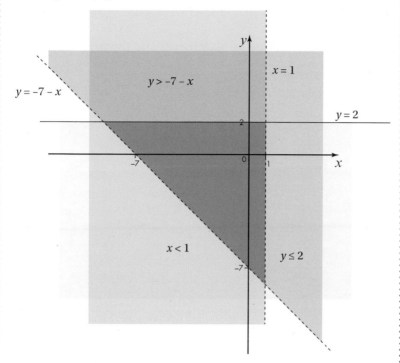

When answering questions on regions defined by inequalities, make sure you clearly state which regions are included and whether the boundaries are included.

Three points that are in the overlap region are, for example, $(-1, 1)$, $(-3, 2)$ and $(0, -5)$.

Exercise 2.7A

Answers page 495

1 Solve the following inequalities:

   a $4x - 3 > 5$ **b** $3x + 7 \leqslant 28$

   c $5x + 6 < x - 2$ **d** $3(x - 2) \geqslant 2 - 4x$

2 On separate pairs of axes shade the regions that satisfy the following inequalities:

   a $4x - 3 > 5$ **b** $3x + 7 \leqslant 28$

   c $5x + 6 < x - 2$ **d** $3(x - 2) \geqslant 2 - 4x$

3 A signal from a mobile phone mast can be picked up by a mobile phone no more than 10 miles away. The circumference of the mobile phone reception has to be at least 55 miles. Find the range of values, to 1 decimal place, for the distance from the mast that a user would have to be in order to be able to receive a signal on their phone.

4 Solve the following inequalities and write down the integers in each set using set notation.

a $4(3x - 2) \geqslant 2x + 2$ and $11 - 2x > \dfrac{x}{5}$

b $7 \leqslant 2x - 1 < 15$ or $-2 < \dfrac{x + 3}{4} < 2$

c $2(x - 3) \leqslant 4x - 12$ and $6x - 5 < -(2 + x)$

d $2 - 9x \geqslant 7x$ or $17 - 3x \geqslant \dfrac{x + 5}{5}$ where $x \geqslant 0$

(PS) 5 Explain what is wrong with each of the following inequality statements and rewrite each one correctly.

a $3 < x < -2$ **b** $-2 > x > 3$ **c** $x < -2$ and $x > 3$

6 The diagram shows a region containing the point (2, 4) bounded by three inequalities. State the three inequalities.

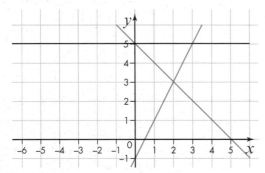

(PS) 7 Match each inequality to its solution.

Inequalities	Solutions
a $9 - x \geqslant 6$	**1** $x \geqslant 3$
b $x - 9 \geqslant -6$	**2** $x \geqslant -3$
c $6 - x \geqslant 9$	**3** $x \leqslant 3$
d $-(9 + x) \leqslant -6$	**4** $x \leqslant -3$

(M) (PS) 8 Marek is going travelling and wants to take as many T-shirts and pairs of jeans as possible. Although the baggage allowance is 23 kg, he doesn't want to take more than 10 kg of belongings. He works out that his toiletries, shoes and duty free will weigh around 3 kg. His T-shirts each weigh 230 g. He found that if he packed 20 T-shirts he would not have weight allowance for more than 5 pairs of jeans. He found that if he packed 25 T-shirts and 3 pairs of jeans he would exceed his self-imposed weight limit. What is the range of values for the weight, to the nearest gram, of a pair of jeans?

2.8 Solving quadratic inequalities

To understand and solve a quadratic inequality you need to solve the associated quadratic equation, sketch the graph of the quadratic function (regardless of whether you are asked to do so or not) and use your sketch to work out the set of values required. You may also need to write the set of values using set notation.

KEY INFORMATION

When solving a quadratic inequality, you should always sketch a graph of the quadratic function.

Example 18

Find the set of values for x for which $x^2 - 3x + 2 < 0$ and show this inequality on a graph.

Solution

Set the quadratic expression equal to 0 and solve the equation.

$$x^2 - 3x + 2 = 0$$

$$(x - 2)(x - 1) = 0$$

$$x = 1 \text{ or } x = 2$$

Sketch the curve of the quadratic equation.

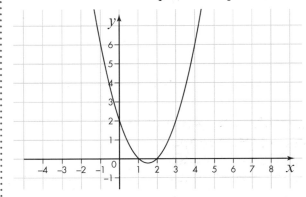

> For more detail on curve sketching see **Chapter 3 Algebra and functions 3: Sketching curves.**

TECHNOLOGY

Using a graphic calculator or graphing software package, plot the graph of $x^2 - 3x + 2 = 0$. What would be the set of values for $x^2 - 3x + 2 > 0$?

So the set of values where the expression is less than 0 is where the curve is below the x-axis: $1 < x < 2$.

Example 19

Find the set of values for x for which $3 + 5x - 2x^2 > x + 3$, show this inequality on a graph and write the set of values for x, where $x \in \mathbb{Z}$, that satisfy this inequality.

Solution

Set the linear expression equal to 0 and solve the equation.

$$x + 3 = 0$$

$$x = -3$$

Find another point on the line $y = x + 3$.

When $x = 0$, $y = 3$.

Set the quadratic expression equal to 0 and solve the equation.

$$3 + 5x - 2x^2 < 0$$

$$(3 - x)(1 + 2x) = 0$$

$$x = 3 \text{ or } x = -\tfrac{1}{2}$$

> You should use the word 'or' here. Do *not* use the word 'and' because mathematically that means where there is an overlap and the two sets do not overlap.

Sketch the straight line of the linear function and curve of the quadratic function on the same pair of axes.

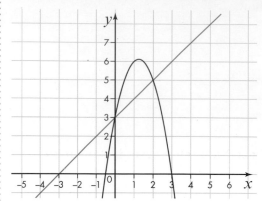

So the set of values where $3 + 5x - 2x^2 > x + 3$ is between $x = 0$ and $x = 2$, $0 < x < 2$.

Using set notation: $\{x: 3 + 5x - 2x^2 > x + 3\} = \{1\}$ where $x \in \mathbb{Z}$.

Alternatively:

$$3 + 5x - 2x^2 > x + 3$$

Rearrange the inequality by subtracting x and 3 from both sides of the inequality.

$$4x - 2x^2 > 0$$

Solve the equation.

$$2x(2 - x) = 0$$

$$x = 0 \text{ or } x = 2$$

Sketch the graph.

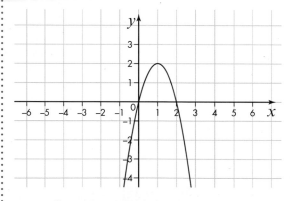

$4x - 2x^2 > 0$ above the x-axis so $0 < x < 2$.

Using set notation: $\{x: 3 + 5x - 2x^2 > x + 3\} = \{1\}$ where $x \in \mathbb{Z}$.

Exercise 2.8A

Answers page 495

1 Solve the following inequalities:

   a $x^2 + 3x - 4 > 0$ **b** $x^2 - 6x + 8 \leqslant 0$

   c $x^2 - 9 \geqslant 0$ **d** $x^2 - 6x < 0$

2 Solve the following inequalities:

   a $2x^2 + 7x < -3$ **b** $-x^2 > 3x - 4$

   c $x^2 > 4$ **d** $3x^2 \leqslant 5x$

(PS) 3 Solve the following inequalities and write down the integers in each set using set notation.

   a $6 - 5x - x^2 > 0$ and $-x + 6 > 0$

   b $2x^2 + 9x - 5 \leqslant 0$ or $2x < 5$

   c $(2x + 1)^2 - 9 \leqslant 0$ and $2x < 6$

   d $4x^2 < 3x$ or $\dfrac{1 - 4x}{4} > 0$

(M) (PS) 4 A signal from a mobile phone mast can be picked up by a mobile phone no more than 10 miles away. The area of the mobile phone reception has to be at least 250 square miles. Find the range of values, to 1 decimal place, for the distance from the mast that a user would have to be in order to be able to use their phone.

5 Find the set of values for x which satisfy the following inequalities and write the set of values for x, where $x \in \mathbb{Z}$, that satisfy each inequality.

   a $x^2 + 4 < 7x - 2$ **b** $2x^2 - 3x - 15 < 2x - 3$

   c $12 + 9x + x^2 > 3(x + 1)$ **d** $x + 1 > 3 - 4x - 3x^2$

6 Solve the following inequalities and write down the integers in each set using set notation.

   a $x^2 + 5x - 6 < 0$ and $x^2 + 3x - 4 < 0$

   b $x^2 + 5x < 6$ or $x^2 + 3x < 4$

   c $x^2 - 6 < -5x$ and $x^2 - 4 > -3x$

   d $x^2 + 5x - 6 > 0$ or $x^2 + 3x - 4 < 0$

SUMMARY OF KEY POINTS

> The discriminant of a quadratic function indicates the number of real roots:
>> > If $b^2 - 4ac < 0$ then the quadratic equation does not have any real roots.
>> > If $b^2 - 4ac = 0$ then the quadratic equation has two equal real roots.
>> > If $b^2 - 4ac > 0$ then the quadratic equation has two distinct real roots.

> Two formulae can be used to complete the square:

>> > $x^2 + bx + c = \left(x + \dfrac{b}{2}\right)^2 - \left(\dfrac{b}{2}\right)^2 + c$ > $a\left(x^2 + \dfrac{bx}{a} + \dfrac{c}{a}\right) = a\left[\left(x + \dfrac{b}{2a}\right)^2 - \dfrac{b^2}{4a^2} + \dfrac{c}{a}\right]$

> $<$ and $>$ are strict inequalities.

> $\leqslant$ and $\geqslant$ are inclusive inequalities.

> $\{x: x > a\}$ means all values of x where x is greater than a.

> The members of $x > a$ or $x < b$ is all the members of $\{x: x > a\}$ with all the members of $\{x: x < b\}$.

> The members of $x > a$ and $x < b$ is just the members of $\{x: x > a\}$ that are also in $\{x: x < b\}$.

> On graphs, the strict inequalities $<$ and $>$ are represented by dotted lines and the inclusive inequalities $\leqslant$ and $\geqslant$ by solid lines.

> When you multiply or divide an inequality by a negative number the direction of the inequality reverses.

EXAM-STYLE QUESTIONS 2

Answers page 495

1   Draw the graph of the equation $y = x^2 + 5x - 6$ where $-7 \leqslant x \leqslant 2$. Write down the coordinates of the axis intercepts and the turning point, and the equation of the line of symmetry. **[5 marks]**

(PF) 2   By evaluating the discriminant, show that $y = 2x^2 - 3x - 7$ has two real roots. **[3 marks]**

3   Solve $3x^2 - 3x - 11 = 0$ using the quadratic formula. Hence solve $3x^2 - 3x - 11 > 0$. **[5 marks]**

4   Solve the simultaneous equations $3x + y = 12$ and $4x - 2y = 26$. **[4 marks]**

5   Solve the inequality $4x - 7 \leqslant 10$ and on a pair of axes shade the region that satisfies this inequality. **[4 marks]**

6   Solve the inequality $x^2 - 9x - 10 \geqslant 0$. **[4 marks]**

(PS) (CM) 7   Write down the equation for the following graph. Explain and justify your decision-making process.

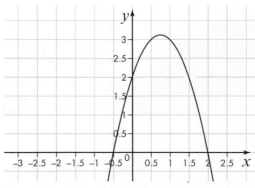

[4 marks]

PS **CM** **8** By evaluating the discriminant for the equation $y = 2x^2 + 3x + 13$, comment on the validity of the statement '$2x^2 + 3x + 13 \geqslant 0$ for all values of x'. **[4 marks]**

9 Complete the square for the equation $y = 4x^2 - 7x - 2$ and use the result to work out the coordinates of the turning point. **[5 marks]**

M **PS** **10** Solve the simultaneous equations $3x + 5y = 2$ and $2x - 6y = 6$. What are the coordinates of the point of intersection of these two lines? **[4 marks]**

M **PS** **11** Solve the simultaneous equations $3x - 2y = 7$ and $x^2 - y^2 = 8$. What are the coordinates of the points of intersection of this line and this curve? **[5 marks]**

12 Solve the following inequalities and write down the integers in each set using set notation.

   a $3(2 - 3x) \leqslant 4x - 5$ or $\dfrac{2x}{5} > 7x + 4$ **[3 marks]**

   b $3(2 - 3x) \geqslant 4x - 5$ and $\dfrac{2x}{5} < 7x + 4$ **[3 marks]**

13 Solve the inequality $6x^2 > 3 - 7x$. **[4 marks]**

14 Solve the following inequalities and write down the integers in each set using set notation.

   a $x^2 + 10x + 21 > 0$ and $3x - 5 > 0$ **[3 marks]**

   b $x^2 + 10x + 21 > 0$ or $3x - 5 > 0$ **[3 marks]**

PS **15** Find the set of values for x for which satisfy the inequality $3x^2 + 5x - 2 > 7x + 15$. **[6 marks]**

PS **16** Find the range of possible values of a $(a \neq 0)$ for which $ax^2 + 5x - 6 = 0$ has:

   a no real roots **[2 marks]**

   b two equal real roots **[2 marks]**

   c two distinct real roots. **[2 marks]**

PS **PF** **17** Show that $a^2x^2 + 3bx + \sqrt{c} = 0$ can be solved using the formula

$$x = \frac{-3b \pm \sqrt{9b^2 - 4a^2\sqrt{c}}}{2a^2}.$$ **[4 marks]**

M **PS** **18** The perimeter of a rectangle is 18 units and the area of the rectangle is 14 units². What are the length and the width of the rectangle? **[7 marks]**

M **PS** **19** The curved surface area of a cone is given by the formula $\pi r l$ where r is the radius of the base of the cone and l is the slant length from the base of the cone to the apex. The slant length of a cone is 13 units. If the curved surface area of the cone cannot exceed 260 units² and the area of the base of the cone must be greater than 60 units², find the range of values for the radius, r, to 1 decimal place. **[7 marks]**

CM **20** Solve the following equations simultaneously. Clearly state the solutions.

$$2x - 3y = -5$$

$$3y + 7z = 2$$

$$7xz = -2$$ **[6 marks]**

3 ALGEBRA AND FUNCTIONS 3: SKETCHING CURVES

The curve created by the cables of the Clifton suspension bridge in Bristol is called a parabola. The forces of gravity, compression and tension act on the suspension bridge to give it this parabolic shape. Parabolic shapes are used widely in structural engineering because they are very good at transferring bending forces into compression and tension across large spans.

The graph of a quadratic function is also a parabola. Consequently, if you plotted the graph of the Clifton suspension bridge you could work out the quadratic equation for the curve and, conversely, if you knew the quadratic equation you could sketch the curve.

LEARNING OBJECTIVES

You will learn how to:

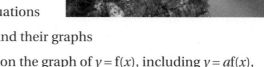

> use and understand the graphs of functions

> sketch curves defined by simple equations including quadratics, cubics, quartics and reciprocals

> interpret the algebraic solution of functions graphically

> use the intersection points of graphs to solve equations

> understand and use proportional relationships and their graphs

> understand the effect of simple transformations on the graph of $y = f(x)$, including $y = af(x)$, $y = f(x) + a$, $y = f(x + a)$ and $y = f(ax)$.

TOPIC LINKS

You will need the skills and techniques that you learn in this chapter to help you understand the results of differentiation in **Chapter 8 Differentiation**. The ability to sketch curves will help you to understand and solve area problems in **Chapter 9 Integration**. The ability to sketch curves will also help you to understand and solve distance, speed and time problems in **Chapter 15 Kinematics**.

PRIOR KNOWLEDGE

You should already know how to:

> expand products of two or more binomials

> factorise quadratic expressions of the form $ax^2 + bx + c$, including completing the square

> work with coordinates in all four quadrants

> identify and interpret roots, intercepts and turning points of quadratic functions graphically

> deduce roots algebraically and deduce turning points by completing the square

> recognise, sketch and interpret graphs of:

> > linear functions

> > quadratic functions

> > simple cubic functions

> > the reciprocal function $y = \dfrac{1}{x}$ with $x \neq 0$

> > the trigonometric functions (with x in degrees) $y = \sin x$, $y = \cos x$ and $y = \tan x$ for angles of any size

> sketch translations and reflections of a given function

> solve problems involving direct and inverse proportion, including graphical and algebraic representations.

You should be able to complete the following questions correctly:

1 Factorise and solve each of the following quadratic equations:

   a $x^2 - 3x - 28 = 0$ **b** $x^2 - 49 = 0$ **c** $x^2 + 9 = 6x$

   d $5x^2 = x$ **e** $9x^2 - 25 = 0$ **f** $6x^2 - 13x = -5$

2 Solve each of the following quadratic equations by completing the square:

   a $x^2 - 8x - 1 = 0$ **b** $x^2 - 2x - 3 = 0$ **c** $x^2 + 6x + 2 = 0$

3 For each of the quadratic functions in **question 2**, work out the coordinates of the turning point and say whether it is a maximum or a minimum.

4 Find the discriminant for each of the following quadratic equations. In each case, explain what the value of the discriminant means.

   a $x^2 - 6x + 9 = 0$ **b** $x^2 - 2x + 15 = 0$ **c** $x^2 - 7x - 1 = 0$

3.1 Sketching curves of quadratic functions

A sketch shows the shape of a curve along with its key features – for example intercepts with the axes, and turning points. It is *not* an accurate plot of a curve. You need to be able to sketch the graphs of **quadratic** functions in the form $y = ax^2 + bx + c$.

From your GCSE course, you should know that the graph of a quadratic function is the shape of a **parabola**. The orientation of the parabola, whether it is a $\cup$ shape or a $\cap$ shape, is determined by the value of a:

> If $a > 0$ then the graph will be a $\cup$ shape.

> If $a < 0$ then the graph will be a $\cap$ shape.

> If $a = 0$ then it is not a quadratic function.

KEY INFORMATION

If $a > 0$ then the graph will be a $\cup$ shape.

If $a < 0$ then the graph will be a $\cap$ shape.

You should also know that if you work out the value of the **discriminant**, $b^2 - 4ac$, you can determine how many real **roots** (non-imaginary solutions) the quadratic equation has.

› If $b^2 - 4ac > 0$ then the equation has two different real roots and consequently the graph will cross the x-axis at two different points.

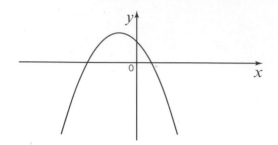

› If $b^2 - 4ac = 0$ then the equation has two equal roots and consequently the graph will touch the x-axis only once.

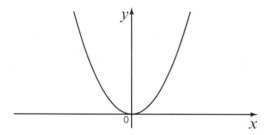

› If $b^2 - 4ac < 0$ then the equation does not have any real roots.

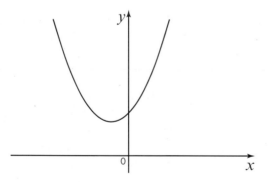

If you know that the discriminant is greater than or equal to zero ($b^2 - 4ac \geqslant 0$), then you can solve the quadratic equation when $y = 0$ to work out the x-intercept(s). Similarly, you can work out the y-intercept by substituting $x = 0$. When you sketch a graph you must clearly label the axis intercepts with their coordinates.

The final point of interest on the graph of a quadratic function is the **turning point**.

› If $a > 0$ then the graph will be a ∪ shape and will have a minimum turning point where the function reaches its lowest value.

› If $a < 0$ then the graph will be a ∩ shape and will have a maximum turning point where the function reaches its highest value.

KEY INFORMATION

If $b^2 - 4ac > 0$ then the equation has two different real roots.

If $b^2 - 4ac = 0$ then the equation has two equal roots.

If $b^2 - 4ac < 0$ then the equation does not have any real roots.

KEY INFORMATION

To work out the x-intercept(s), substitute $y = 0$ into the quadratic equation.

To work out the y-intercept, substitute $x = 0$ into the quadratic equation.

KEY INFORMATION

If $a > 0$ then the graph will have a **minimum** turning point.

If $a < 0$ then the graph will have a **maximum** turning point.

Example 1 covers techniques from your GCSE course, including completing the square to work out the coordinates of turning points.

Example 1

Sketch the graph of $y = x^2 - 4x - 5$.

Solution

$a = 1$, $b = -4$, $c = -5$

$a > 0$ so the graph will be a $\cup$ shape.

$$b^2 - 4ac = 16 + 20 = 36$$

$b^2 - 4ac > 0$ so two different real roots.

Substitute $y = 0$ for the x-intercepts.

$$x^2 - 4x - 5 = 0$$

Factorising gives:

$$(x - 5)(x + 1) = 0$$

$$x = -1 \text{ or } 5$$

So the coordinates of the x-intercepts are $(-1, 0)$ and $(5, 0)$.

Substitute $x = 0$ for the y-intercept.

$$y = -5$$

So the coordinates of the y-intercept are $(0, -5)$.

Completing the square to work out the coordinates of the turning point gives:

$$y = (x - 2)^2 - 9$$

When $x = 2$, y will have a minimum value of -9.

The coordinates of the turning point are $(2, -9)$.

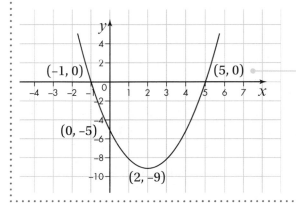

TECHNOLOGY

You can also use a graphic calculator or graphing software package to check the answer. Plot the graph of $y = x^2 - 5x + 6$. Does the sketch look like the accurately plotted graph?

The coordinates of the axis intercepts and the turning point have been clearly labelled on the sketch

Exercise 3.1A

Answers page 497

1 Sketch the graphs of the following equations:

    a $y = x^2 + 6x + 5$ **b** $y = x^2 - 8x + 7$

(PS) 2 Find the quadratic equation for the sketch.

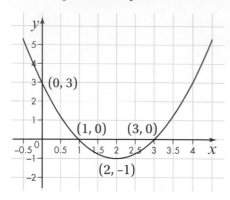

3 Sketch the graphs of the following equations:

    a $y = 3x^2 - 6x + 3$ **b** $y = 2x^2 - 6x$

(PS) (CM) 4 A student sketches the following graph for the equation $y = x^2 - 4x - 21$. Is the sketch correct? If not, point out and explain the mistakes.

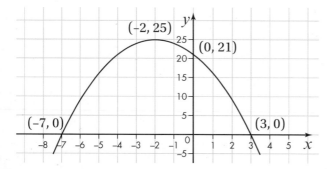

5 Sketch the graphs of the following equations:

    a $y = (4 - x)^2$ **b** $y = 3 - 5x - 2x^2$

(M) (PS) 6 A ball is kicked from a platform that is 3 m above the ground. The ball follows the path of a parabolic curve and lands some distance away from the base of the platform. The graph shows the height, h metres, of the ball over time, t seconds.

    a What is the maximum height the ball reaches above the ground?

    b How many seconds after the ball has been kicked does it hit the ground?

    c Find the equation of the curve.

(M) Modelling   **(PS)** Problem solving   **(PF)** Proof   **(CM)** Communicating mathematically

3.2 Sketching curves of cubic functions

At GCSE you sketched the graphs of **cubic** functions in the form $y = ax^3 + bx^2 + cx + d$. You need to work out how the curve behaves as x tends towards plus and minus infinity, $x \to \pm\infty$, to determine the shape of the curve.

Curves of cubic functions with one repeated root

A cubic function with one repeated real root will not have any turning points. Instead, it will have a point of inflection. At the point of inflection the gradient of the curve is zero and it is not a turning point.

> **KEY INFORMATION**
>
> Consider how the curve behaves when $x \to -\infty$ and $x \to +\infty$.

> **KEY INFORMATION**
>
> A cubic function with one repeated real root will have a **point of inflection**.

Example 2

Sketch the graph of $y = x^3$.

Solution

Substitute $y = 0$ to find the x-intercepts:

$$x^3 = 0$$

$$x = 0$$

So the function has one repeated root, and the coordinates of the point of inflection are (0, 0).

You don't need to substitute $x = 0$ to find the y-intercept, as you have already found $x = 0$ as an x-intercept.

When $x \to +\infty$, a very large, positive number, $y \to +\infty$, and y will be large and positive.

When $x \to -\infty$, a very large, negative number, $y \to -\infty$, and y will be large and negative.

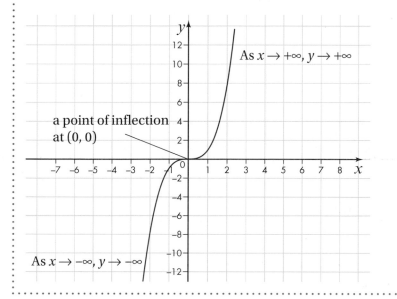

a point of inflection at (0, 0)

As $x \to +\infty$, $y \to +\infty$

As $x \to -\infty$, $y \to -\infty$

> **TECHNOLOGY**
>
> Using a graphic calculator or graphing software package, plot the graph of $y = x^3$ and zoom in on the point of inflection. What happens to the gradient of the curve each side of the point of inflection?

Example 3

Sketch the graph of $y = (x - 1)^3$.

Solution

Substitute $y = 0$ to find the x-intercepts.

$x = 1$ so the equation has one repeated root. The coordinates of the point of inflection are $(1, 0)$.

Substitute $x = 0$ to find the y-intercept. The coordinates of the y-intercept are $(0, -1)$.

When $x \to +\infty$, a very large, positive number, $y \to +\infty$, so y will be large and positive.

When $x \to -\infty$, a very large, negative number, $y \to -\infty$, so y will be large and negative.

TECHNOLOGY

Using a graphic calculator or graphing software package, plot the graphs of $y = (x - 1)^3$ and $y = (2 - x)^3$ on the same pair of axes. What do you notice about how the two curves behave as $x \to \pm\infty$? Why is there a difference?

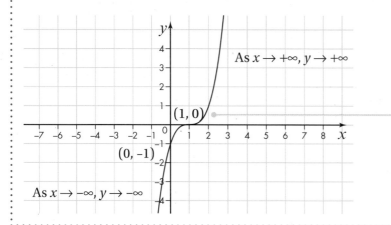

The coordinates of the axis intercepts and point of inflection are included on the graph.

Exercise 3.2A

Answers page 498

1 Sketch the graphs of the following equations:

   a $y = -x^3$ **b** $y = (x - 2)^3$

2 Sketch the graphs of the following equations:

   a $y = (2x - 3)^3$ **b** $y = (3 - x)^3$

PS **CM** **3** A student sketches the following graph for the equation $y = (x + 3)^3$. Is the sketch correct? If not, point out and explain the mistakes.

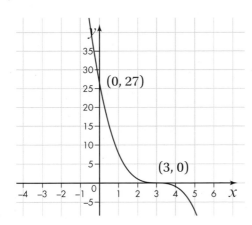

4 Sketch the graphs of $y = -(x - 2)^3$ and $y = -(2 - x)^3$ on the same pair of axes. What do you notice?

PS **5** Which of the following equation(s) match the graph?

A $y = -(3x + 4)^3$ **B** $y = (x + 3)^3$

C $y = (3 - 4x)^3$ **D** $y = -(4x + 3)^3$

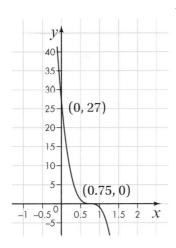

Example 4

Sketch the graph of $y = x^3 - x^2 - 2x$.

Solution

Substitute $y = 0$ to find the x-intercepts.

$$x^3 - x^2 - 2x = 0$$

Factorising gives:

$$x(x + 1)(x - 2) = 0$$

$$x = 0, -1 \text{ or } 2$$

So coordinates of the x-intercepts are $(-1, 0)$, $(0, 0)$ and $(2, 0)$.

You don't need to substitute $x = 0$ for the y-intercept as you have already found $x = 0$ as an x-intercept.

When $x \to +\infty$, a very large, positive number, $y \to +\infty$, so y will be large and positive.

When $x \to -\infty$, a very large, negative number, $y \to -\infty$, so y will be large and negative.

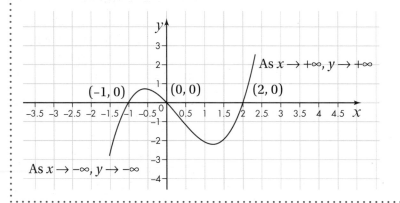

TECHNOLOGY

Using a graphic calculator or graphing software package plot the graph of $y = x^3 - x^2 - 2x$. Can you approximate the coordinates of the two turning points?

Exercise 3.2B

Answers page 498

1 Sketch the graphs of the following equations:

   a $y = (x - 1)(x - 2)(x + 3)$ **b** $y = x(x + 1)(x + 4)$

2 Sketch the graphs of the following equations:

   a $y = (2x - 1)(x + 2)(2x + 3)$ **b** $y = (x - 1)(x - 2)^2$

3 Factorise the following equations and then sketch the graphs:

   a $y = x^3 - 2x^2 - 3x$ **b** $y = x^3 - 7x^2$

4 Sketch the graphs of the following equations:

a $y = (1 - x)(x - 2)(x + 3)$ **b** $y = x^2(1 - x)$

 5 Sketch the graph of $y = x^3 - 2x^2 - x$.

3.3 Sketching curves of quartic functions

You need to be able to sketch the graphs of **quartic** functions in the form $y = ax^4 + bx^3 + cx^2 + dx + e$. You will use the same techniques as you used when sketching the graphs of quadratic and cubic functions.

You can work out the x-intercept(s) by solving the quartic equation when $y = 0$. Similarly, you can work out the y-intercept by substituting $x = 0$. When you sketch a graph you must clearly label the axis intercepts with their coordinates.

Curves of quartic functions with one repeated root

A quartic function with one repeated real root with $a > 0$ has this shape:

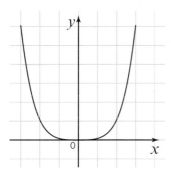

The shape of a quartic function is *not* a parabola – the bottom of a quartic curve appears much flatter than a quadratic. The orientation of the graph is determined by the value of a. If $a > 0$, then $y \rightarrow \infty$ when $x \rightarrow \pm\infty$. If $a < 0$, then the graph will be inverted.

Consequently, a quartic function with one repeated root will have just one turning point.

› If $a > 0$ then the graph will have a minimum turning point where the function reaches its lowest value.

› If $a < 0$ then the graph will have a maximum turning point where the function reaches its highest value.

KEY INFORMATION

A quartic function is one which includes the variable raised to the power 4 and no higher. It can include the variable raised to powers less than 4 and greater than or equal to zero.

KEY INFORMATION

To work out the x-intercept(s), substitute $y = 0$ into the quartic equation.

To work out the y-intercept, substitute $x = 0$ into the quartic equation.

KEY INFORMATION

If $a > 0$ then the quartic curve is a $\cup$ shape. If $a < 0$ then the graph will be inverted.

KEY INFORMATION

If $a > 0$ then the graph will have a minimum turning point.

If $a < 0$ then the graph will have a maximum turning point.

Example 5

Sketch the graph of $y = x^4$.

Solution

$a > 0$ so the graph will not be inverted.

Substitute $y = 0$ to find the x-intercepts. The coordinates of the turning point are $(0, 0)$.

You don't need to substitute $x = 0$ to find the y-intercept as you have already found $x = 0$ as an x-intercept.

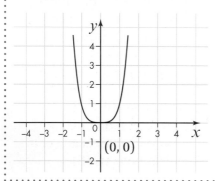

TECHNOLOGY

Using a graphic calculator or graphing software package, plot the graphs of $y = x^4$ and $y = x^2$ on the same pair of axes. How does the quartic curve compare to a quadratic curve?

Example 6

Sketch the graph of $y = (x + 2)^4$.

Solution

$a > 0$ so the graph will not be inverted.

Substitute $y = 0$ to find the x-intercepts. The coordinates of the turning point are $(-2, 0)$.

Substitute $x = 0$ to find the y-intercept. The coordinates of the y-intercept are $(0, 16)$.

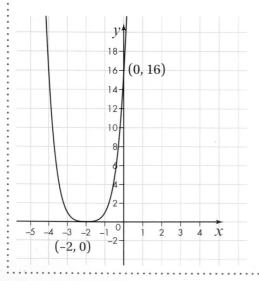

TECHNOLOGY

Using a graphic calculator or graphing software package, plot the graphs of $y = (x + 2)^4$ and $y = (x - 2)^4$ on the same pair of axes. Compare the equations to the coordinates of the turning points. What do you notice?

Example 7

Sketch the graph of $y = -(3-x)^4$.

Solution

$a < 0$ so the graph will be inverted.

Substitute $y = 0$ to find the x-intercepts. The coordinates of the turning point are $(3, 0)$.

Substitute $x = 0$ to find the y-intercept. The coordinates of the y-intercept are $(0, -81)$.

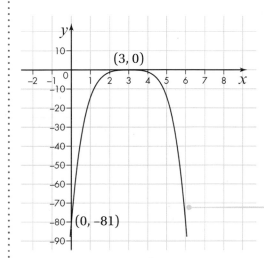

TECHNOLOGY

Using a graphic calculator or graphing software package, plot the graph of $y = -(3-x)^4$. What would be the equation of the quartic equation that was a reflection in the x-axis of the one plotted? Try it!

The graph of the quartic function is inverted because $a < 0$.

Exercise 3.3A

Answers page 500

1 Sketch the graphs of the following equations:

  a $y = -x^4$　　**b** $y = (x-3)^4$

2 Sketch the graphs of the following equations:

  a $y = (5-2x)^4$　　**b** $y = -(x-2)^4$

3 A student sketches the following graph for the equation $y = (3-x)^4$. Is the sketch correct? If not, point out and explain the mistakes.

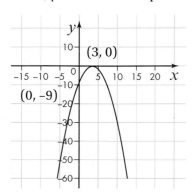

4 Sketch the graphs of $y = -(x + 2)^4$ and $y = (2 - x)^4$ on the same pair of axes. Explain why these two functions do not generate the same curve.

5 Which of the following equation(s) match the graph?

    A $y = -(5x + 4)^4$ **B** $y = (4 - 5x)^4$

    C $y = (5x - 4)^4$ **D** $y = -(4 - 5x)^4$

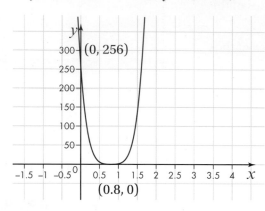

Example 8

Sketch the graph of $y = (x - 1)(x - 2)(x - 3)(x - 4)$.

Solution

Substitute $y = 0$ to find the x-intercepts. The coordinates of the x-intercepts are $(1, 0)$, $(2, 0)$, $(3, 0)$ and $(4, 0)$.

Substitute $x = 0$ to find the y-intercept. The coordinates of the y-intercept are $(0, 24)$.

$a > 0$ so the graph will not be inverted.

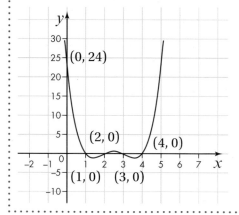

TECHNOLOGY

Using a graphic calculator or graphing software package, plot the graph of $y = (x - 1)(x - 2)(x - 3)(x - 4)$. What do you notice about the distance of each turning point peak/trough from the x-axis?

Example 9

Sketch the graph of $y = (x - 1)^2(x - 2)^2$.

Solution

Substitute $y = 0$ to find the x-intercepts. The coordinates of the x-intercepts are $(1, 0)$ and $(2, 0)$.

Substitute $x = 0$ to find the y-intercept. The coordinates of the y-intercept are $(0, 4)$.

$a > 0$ so the graph will not be inverted.

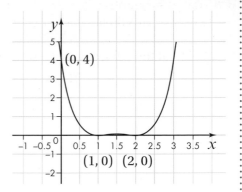

Example 10

Sketch the graph of $y = (x - 1)^3(x + 2)$.

Solution

Substitute $y = 0$ to find the x-intercepts. The coordinates of the x-intercepts are $(-2, 0)$ and $(1, 0)$. The equation has a repeated real root from a cubic function, $x = 1$, and another real root, $x = -2$. At the x-intercept $(1, 0)$, the root from the cubic function, there is a point of inflection.

Substitute $x = 0$ to find the y-intercept. The coordinates of the y-intercept are $(0, -2)$.

$a > 0$ so the graph will not be inverted.

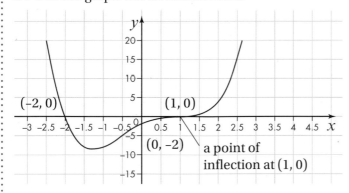

TECHNOLOGY

Using a graphic calculator or graphing software package, plot the graph of $y = (x - 1)^3(x + 2)$. Change the numeric values in the equation. How does this affect the shape of the graph?

Exercise 3.3B
Answers page 500

 1 Sketch the graphs of the following equations:

a $y = (x-1)(x-2)(x+1)(x+2)$

b $y = (x+1)^2(x-4)^2$

2 Sketch the graphs of the following equations:

a $y = (1-x)(2-x)(3-x)(x+1)$

b $y = (x-1)(4x-5)^3$

PS 3 Find the quartic equation for the sketch.

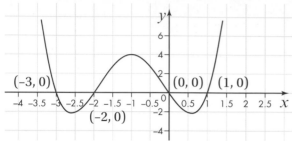

4 Sketch the graphs of the following equations:

a $y = (1-x)(x+2)(x-1)(2-x)$

b $y = x(3-10x)^3$

PS 5 Find the quartic equation for the sketch.

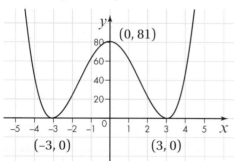

3.4 Sketching curves of reciprocal functions

In your GCSE course you sketched the graph of the reciprocal function $y = \dfrac{1}{x}$.

You now need to be able to sketch the graphs of reciprocals of linear functions, $y = \dfrac{a}{x}$, and quadratic functions, $y = \dfrac{a}{x^2}$.

You need to establish the values of x and y where there are **discontinuities**. A discontinuity is where the value of x or y is not defined. For example, for the function $y = \dfrac{1}{x}$, y is not defined when $x = 0$ and x is not defined when $y = 0$. The straight lines passing through the points of discontinuity, in this case $x = 0$ and $y = 0$, are called **asymptotes**. An asymptote is a straight line that a curve approaches but never meets.

> **KEY INFORMATION**
>
> A discontinuity is where the value of x or y is not defined.

> **KEY INFORMATION**
>
> An asymptote is a straight line that a curve approaches but never meets.

For the function $y = \dfrac{a}{x}$, if $a > 0$ then the graph will be the following shape:

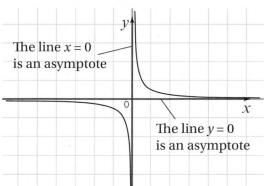

The line $x = 0$ is an asymptote

The line $y = 0$ is an asymptote

If $a < 0$ then the graph will be the following shape:

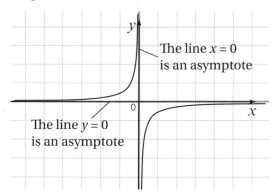

The line $x = 0$ is an asymptote

The line $y = 0$ is an asymptote

Example 11

Sketch the graph of $y = \dfrac{-3}{x+1}$ clearly indicating any asymptotes.

Solution

$a < 0$ so the graph will be a ⌐͞ shape.

When $x = -1$ there is a discontinuity. Therefore, the line $x = -1$ is an asymptote.

When $y = 0$ there is a discontinuity. Therefore, the line $y = 0$ is an asymptote.

As there is a discontinuity at $y = 0$, the graph will never intercept the x-axis.

Substitute $x = 0$ to find the y-intercept. The coordinates of the y-intercept are $(0, -3)$.

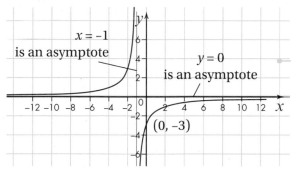

$x = -1$ is an asymptote

$y = 0$ is an asymptote

$(0, -3)$

The asymptotes have been clearly identified on the graph.

TECHNOLOGY

Using a graphic calculator or graphing software package, plot the graphs of $y = \dfrac{-3}{x+1}$ and $y = \dfrac{-3}{x-1}$ on the same pair of axes. What do you notice about the equations of the asymptotes in each case?

Example 12

Sketch the graph of $y = \frac{2}{x-1} + 3$, clearly indicating any asymptotes.

Solution

$a > 0$ so the graph will be a ⌐͞ shape.

When $x = 1$ there is a discontinuity. Therefore, the line $x = 1$ is an asymptote.

When $y = 3$ there is a discontinuity. Therefore, the line $y = 3$ is an asymptote.

Substitute $y = 0$ to find the x-intercepts. The coordinates of the x-intercept are $(\frac{1}{3}, 0)$.

Substitute $x = 0$ to find the y-intercept. The coordinates of the y-intercept are $(0, 1)$.

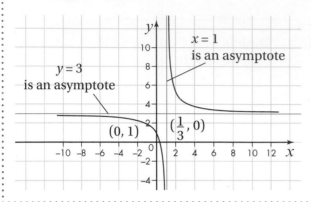

TECHNOLOGY

Using a graphic calculator or graphing software package, plot the graph of $y = \dfrac{2}{x-1} + 3$. Change the value of the '3' in the equation. What happens to the horizontal asymptote? Why do you think this is?

Exercise 3.4A

Answers page 501

(CM) 1 Sketch the graphs of the following equations on the same pair of axes:

  a $y = \dfrac{5}{x}$ **b** $y = \dfrac{10}{x}$

  c Compare the graphs drawn in **parts a** and **b**. What do you notice?

2 For each of the following equations, work out the axis intercepts and the equations of the asymptotes:

  a $y = \dfrac{2}{x+2}$ **b** $y = -\dfrac{3}{x-4}$

(PS) 3 For each of the following y-intercepts and equations of the asymptotes, work out the equation of the graph of a reciprocal linear function:

  a y-intercept at $(0, 1)$, asymptotes at $x = -1$ and $y = 0$

  b y-intercept at $(0, -2)$, asymptotes at $x = -1$ and $y = 0$

(PS) 4 Match the following equations and graphs:

  a $y = -\dfrac{2}{x+2}$ **b** $y = \dfrac{2}{x-2}$

  c $y = \dfrac{2}{-2-x}$ **d** $y = \dfrac{2}{x+2}$

  i

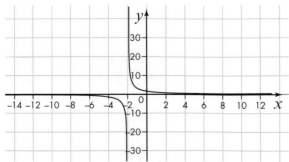

ii

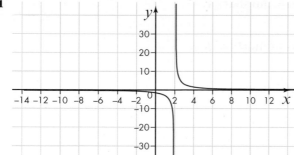

iii

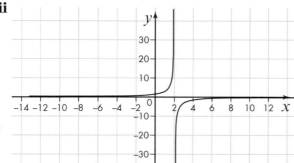

iv

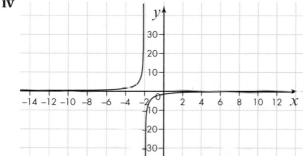

5 Sketch the graphs of the following equations:

a $y = \dfrac{4}{2x-1} + 3$ **b** $y = 5 - \dfrac{2}{x+1}$

6 Is the following a correct graph for the reciprocal function $y = 2 - \dfrac{6}{3x-2}$? Justify your answer.

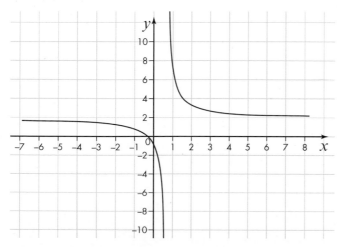

You need to be able to sketch the graphs and know the shapes of reciprocals of quadratic functions, $y = \dfrac{a}{x^2}$.

An easy way to remember their shape is that it is a reflection of itself in the line of the asymptote.

If $a > 0$ then the graph will be the following shape:

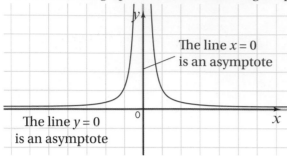

The line $x = 0$ is an asymptote

The line $y = 0$ is an asymptote

If $a < 0$ then the graph will be inverted:

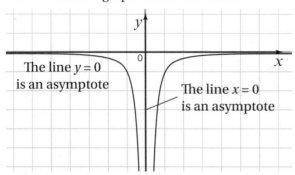

The line $y = 0$ is an asymptote

The line $x = 0$ is an asymptote

Example 13

Sketch the graph of $y = \dfrac{4}{x^2}$, clearly indicating any asymptotes.

Solution

$a > 0$ so the graph will be a ⟋⟍ shape.

When $x = 0$ there is a discontinuity. Therefore, the line $x = 0$ is an asymptote.

When $y = 0$ there is a discontinuity. Therefore, the line $y = 0$ is an asymptote.

Consequently, the graph will never intercept the axes.

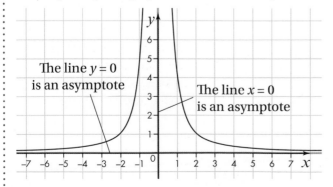

The line $y = 0$ is an asymptote

The line $x = 0$ is an asymptote

TECHNOLOGY

Using a graphic calculator or graphing software package, plot the graph of $y = \dfrac{4}{x^2}$. If you change the value of the numerator in the equation how does this affect the graph?

Example 14

Sketch the graph of $y = -\dfrac{2}{(2x-1)^2}$, clearly indicating any asymptotes.

Solution

$a < 0$ so the graph will be a shape.

When $x = \frac{1}{2}$ there is a discontinuity. Therefore, the line $x = \frac{1}{2}$ is an asymptote.

When $y = 0$ there is a discontinuity. Therefore, the line $y = 0$ is an asymptote.

As there is a discontinuity at $y = 0$, the graph will never intercept the x-axis.

Substitute $x = 0$ to find the y-intercept.
The coordinates of the y-intercept are $(0, -2)$.

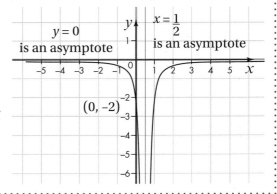

Example 15

Sketch the graph of $y = 5 + \dfrac{1}{(4-x)^2}$, clearly indicating any asymptotes.

Solution

$a > 0$ so the graph will be a 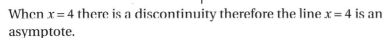 shape.

When $x = 4$ there is a discontinuity therefore the line $x = 4$ is an asymptote.

When $y = 5$ there is a discontinuity therefore the line $y = 5$ is an asymptote.

The graph will never intercept the x-axis because there is an asymptote at $y = 5$.

Substitute $x = 0$ to find the y-intercept. The coordinates of the y-intercept are $(0, \frac{81}{16})$.

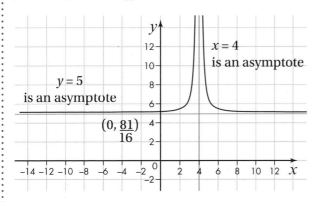

TECHNOLOGY

Using a graphic calculator or graphing software package, plot the graphs of

$y = -\dfrac{2}{(2x-1)^2}$ and $y = \dfrac{2}{(2x-1)^2}$.

What do you notice?

TECHNOLOGY

Using a graphic calculator or graphing software package, plot the graph of

$y = 5 + \dfrac{1}{(4-x)^2}$. If you change the number '5' to a different number, what effect does this have on the curve and asymptotes?

Exercise 3.4B

Answers page 502

 1 Sketch the graphs of the following equations on the same pair of axes:

    a $y = \dfrac{3}{x^2}$ **b** $y = -\dfrac{6}{x^2}$

    c   Compare the graphs drawn in **parts a** and **b**. What do you notice?

2 For each of the following equations work out the axis intercepts and the equations of the asymptotes:

    a $y = \dfrac{1}{(x+2)^2}$ **b** $y = -\dfrac{5}{(x+1)^2}$

 3 For each of the following y-intercepts and equations of the asymptotes, work out the equation of the graph of a reciprocal quadratic function:

    a y-intercept at $(0, 2\frac{1}{9})$, asymptotes at $x = 3$ and $y = 2$

    b y-intercept at $(0, -2\frac{15}{16})$, asymptotes at $x = 4$ and $y = -3$

 4 Match the following equations and graphs. Don't assume that there is a different graph for each equation.

    a $y = -\dfrac{2}{(x+2)^2}$ **b** $y = \dfrac{2}{(x-2)^2}$

    c $y = \dfrac{2}{(-2-x)^2}$ **d** $y = \dfrac{2}{(2-x)^2}$

i

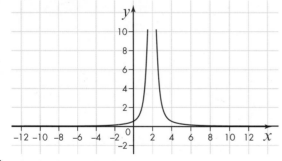

ii

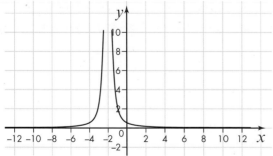

iii

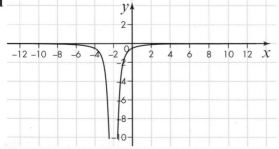

iv

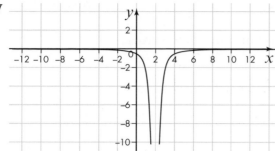

5 Sketch the graphs of the following equations:

a $y = \dfrac{3}{(4x-1)^2} + 2$ **b** $y = 6 - \dfrac{3}{(5x+1)^2}$

(PS) (CM) **6** Is the following a correct graph for the reciprocal function $y = 2 + \dfrac{7}{(2x-3)^2}$? Provide justification for your answer.

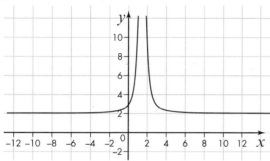

3.5 Intersection points

You need to be able to sketch more than one graph on the same pair of axes to work out the **points of intersection** of the graphs and solve equations.

> **Stop and think**   How will sketching the graphs of two different functions and finding the points of intersection help you to solve an equation which includes those functions?

Example 16

Sketch the graphs of $y = x(x+3)$ and $y = x(x+1)(1-x)$ on the same pair of axes. Find the number of points of intersection.

Solution

$y = x(x+3)$

$a = 1$, $b = 3$, $c = 0$

$a > 0$ so the graph will be a $\cup$ shape.

$b^2 - 4ac = 9$

$b^2 - 4ac > 0$ so two different real roots.

Substitute $y = 0$ to find the x-intercepts. The coordinates of the x-intercepts are $(-3, 0)$ and $(0, 0)$.

You don't need to substitute $x = 0$ for the y-intercept as you have already found $x = 0$ as an x-intercept.

Completing the square to work out the coordinates of the turning point gives

$$y = \left(x + \frac{3}{2}\right)^2 - \frac{9}{4}$$

When $x = -\frac{3}{2}$, y will have a minimum value of $-\frac{9}{4}$.

The coordinates of the turning point are $(-\frac{3}{2}, -\frac{9}{4})$.

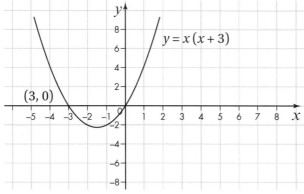

$y = x(x + 1)(1 - x)$

Substitute $y = 0$ to find the x-intercepts. The coordinates of the x-intercepts are $(-1, 0)$, $(0, 0)$ and $(1, 0)$.

The function has at least two different real roots and consequently the curve will have two turning points.

You don't need to substitute $x = 0$ to find the y-intercept as you have already found $x = 0$ as an x-intercept.

When $x \to +\infty$, a very large, positive number, $y \to -\infty$, so y will be large and negative.

When $x \to -\infty$, a very large, negative number, $y \to +\infty$, so y will be large and positive.

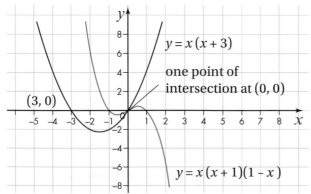

There is one point of intersection at $(0, 0)$.

TECHNOLOGY

Using a graphic calculator or graphing software package, plot the graphs of $y = x(x + 3)$ and $y = x(x + 1)(1 - x)$. How many points of intersection can you see and what are their coordinates?

Example 17

By sketching the graphs of $y = \dfrac{1}{x^2}$ and

$y = -(x+1)(x+2)(x-1)(x-2)$, show that there are four

solutions to the equation $x^2(x+1)(x+2)(x-1)(x-2) = -1$.

Solution

Rearranging $x^2(x+1)(x+2)(x-1)(x-2) = -1$ gives:

$$-(x+1)(x+2)(x-1)(x-2) = \dfrac{1}{x^2}$$

which are the two equations for which the graphs will have been sketched. Consequently, the sketched graphs could show the four solutions to the equation $x^2(x+1)(x+2)(x-1)(x-2) = -1$.

$y = \dfrac{1}{x^2}$

$a > 0$ so the graph will be a ⎯⎯╱╲⎯⎯▶ shape.

When $x = 0$ there is a discontinuity. Therefore, the line $x = 0$ is an asymptote.

When $y = 0$ there is a discontinuity. Therefore, the line $y = 0$ is an asymptote.

Consequently, the graph will never intercept the axes.

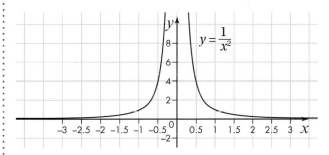

$$y = -(x+1)(x+2)(x-1)(x-2)$$

Substitute $y = 0$ to find the x-intercepts. The coordinates of the x-intercepts are $(-2, 0)$, $(-1, 0)$, $(1, 0)$ and $(2, 0)$.

The function has four different real roots and consequently the curve will have three turning points.

Substitute $x = 0$ to find the y-intercept. The coordinates of the y-intercept are $(0, -4)$.

TECHNOLOGY

Using a graphic calculator or graphing software package, plot the graphs of $y = \dfrac{1}{x^2}$ and $y = -(x+1)(x+2)(x-1)(x-2)$. How could you subtly change the equations so that the graphs do not intersect?

$a < 0$ so the graph will be inverted.

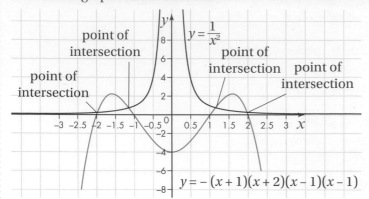

$$y = -(x+1)(x+2)(x-1)(x-1)$$

PROOF

You have proved, using a sketch, that there are four points of intersection between the two curves.

Exercise 3.5A

Answers page 502

1 Sketch the graphs of $y = \dfrac{6}{x}$ and $y = (x-4)(x-8)$ on the same pair of axes. Find the number of points of intersection.

(PS) (CM) 2 Sketch the graphs of $y = -\dfrac{1}{x^2}$ and $y = x^2$ on the same pair of axes. Explain how your graph shows that there are no solutions to the equation $\dfrac{1}{x^2} = -x^2$.

(PS) (CM) 3 Sketch the graphs of $y = x^4$ and $y = x(x-3)(x-4)$ on the same pair of axes. Explain how your graph shows that there are two solutions to the equation of $x^4 = x(x-3)(x-4)$.

(PS) (CM) 4 Sketch the graphs of $y = (x+1)(2x-1)^3$ and $y = (3-x)^2$ on the same pair of axes. Explain how your graph shows that there are two solutions to the equation $\dfrac{(x+1)(2x-1)^3}{(3-x)^2} = 1$.

(PS) (CM) 5 Sketch the graphs of $y = -\dfrac{2}{(x+4)^2}$ and $y = (x+4)^2$ on the same pair of axes. Explain how your graph shows that there are no solutions to the equation $(x+4)^4 = -2$.

(PS) 6 Sketch the graphs of $y = x(x-6)$ and $y = x^2(2-x)$ on the same pair of axes. Find the number of points of intersection, and form and solve an equation to find their coordinates.

7 Sketch the graphs of $y = (x-1)(x+3)(x-2)$ and $y = (x+3)(x+1)(x+2)(x-1)$ on the same pair of axes. Find the number of points of intersection.

8 Sketch the graphs of $y = (3-2x)(x+1)^3$ and $y = \dfrac{4}{(x-3)^2} + 2$ on the same pair of axes. Find the number of points of intersection.

3.6 Proportional relationships

You need to be able to solve problems where two variables are directly proportional to each other. **Direct proportion** doesn't just involve linear relationships but also squares, square roots, cubes and cube roots. You need to be able to express the relationship between the two variables using the proportion symbol, $\propto$, and then find the equation between the two variables involving a **constant of proportionality**. You also need to be able to recognise and use graphs that show proportion.

KEY INFORMATION

The relationship between two variables can be expressed using the proportion symbol, $\propto$.

The constant of proportionality is a multiplier linking two variables.

Stop and think   What constant of proportionality links distance travelled and time if the object is not accelerating or decelerating?

Example 18

The distance, d miles, that a car travels is directly proportional to the time, t hours, of the journey. If $d = 30$ miles when $t = 0.5$ hours, find the constant speed of the car. Sketch the graph of the relationship between d and t and use the graph to work out d when $t = 3$ hours.

Solution

Write down the proportion statement.

$$d \propto t$$

Write down the equation using a constant of proportionality.

$$d = kt$$

Substitute in the known values.

$$30 = 0.5k$$

Rearrange.

$$k = 60$$

The formula is $d = 60t$, so the constant speed is 60 mph.

There is a linear relationship between d and t. When $d = 0$, $t = 0$ and when $d = 30$, $t = 0.5$.

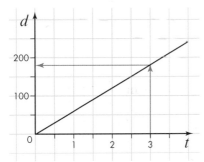

Using the graph, when $t = 3$, $d = 180$ miles.

As the distance, d miles, that a car travels is directly proportional to the time, t hours, for the journey then the graph is of a linear function. Only the positive x and y axes are drawn as it is not possible to have a negative distance or time.

TECHNOLOGY

Using a graphic calculator or graphing software package, plot the graph of $y = 60x$. When $x = 3$, what is the value of y?

Example 19

The total surface area, A, of a hemisphere is directly proportional to the square of its radius, r.

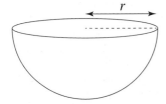

If $A = 48\pi$ cm^2 when $r = 4$ cm, work out the total surface area of a hemisphere with a radius of 9 cm. Sketch the graph of the relationship between A and r.

Solution

Write down the proportion statement.

$$A \propto r^2$$

Write down the equation using a constant of proportionality.

$$A = kr^2$$

Substitute in the known values.

$$48\pi = 16k$$

Rearrange.

$$k = 3\pi$$

So the formula is $A = 3\pi r^2$.

When $r = 9$:

$$A = 243\pi$$

$$a = 3\pi$$

$a > 0$ so the graph will be a $\cup$ shape.

Substitute $A = 0$ to find the x-intercepts.

$$3\pi r^2 = 0$$

$$r = 0$$

So the coordinates of the x-intercept, the y-intercept and the turning point are $(0, 0)$.

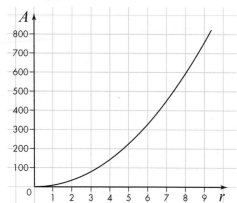

As the total surface area, A, of the hemisphere is directly proportional to the square of its radius, r, then the graph is of a quadratic function. Only the positive x and y axes are drawn as it is not possible to have a negative radius or area.

TECHNOLOGY

Using a graphic calculator or graphing software package, plot the graph of $y = 3\pi x^{-2}$. When $x = 9$ what is the value of y?

Stop and think   If y is inversely proportional to the cube of x, what would the equation of proportionality be and what would the graph of the positive values look like?

Exercise 3.6A

Answers page 503

1 y is directly proportional to x^3. If $y = 16$ when $x = 2$, find:

    a y when $x = 3$ **b** x when $y = 250$

M 2 The distance, d miles, that a car travels is directly proportional to the time, t hours, for the journey. If $d = 15$ miles when $t = 0.25$ hours, find the constant speed of the car. Sketch the graph of the relationship between d and t and use the graph to work out the distance travelled in 4.5 hours.

M 3 The mass, m grams, of a gold ring is directly proportional to its volume, V cm^3. If $m = 9.66$ g when $V = 0.5$ cm^3, work out the mass of a gold ring with a volume of 0.75 cm^3 . Sketch the graph of the relationship between m and V and use the graph to work out the volume of a gold ring with a mass of 15 g.

PS 4 Match the following proportional relationships, graphs and pairs of values. (Note: the same scales have been used on all the graphs.)

    a $y \propto x^3$ **b** $y \propto x^2$ **c** $y \propto x$

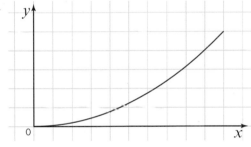

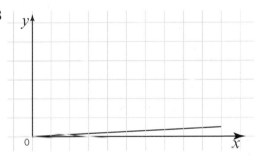

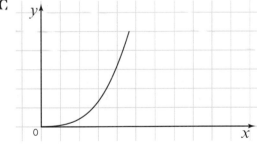

    i $x = 1$, $y = 6$ and $x = 3$, $y = 18$

    ii $x = 1$, $y = 6$ and $x = 3$, $y = 54$

    iii $x = 1$, $y = 3$ and $x = 3$, $y = 81$

M PS 5 The volume, V, of a sphere is directly proportional to the cube of its radius, r. If $V = 36\pi$ cm^3 when $r = 3$ cm, work out the volume of a sphere with a radius of 7 cm. Sketch the graph of the relationship between V and r. Use your graph to find:

    a V when $r = 4$ cm   **b** r when $V = 288\pi$ cm^3

    c Check your answers to **parts a** and **b** by using the equation you found.

PS **6** y is directly proportional to the square root of x. The graph of the relationship between x and y has been sketched.

   a Use the graph to find an equation linking x and y.

   b Find y when $x = 16$.

   c Find x when $y = 50$.

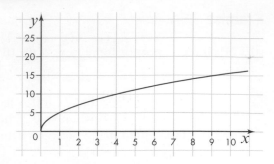

3.7 Translations

In your GCSE course you performed simple **transformations** on the graph of $y = f(x)$ and sketched the resulting graph. The examples below extend these techniques to AS-level content.

> **Stop and think**
>
> If you are given the equation or the graph of a function, how do you perform transformations without drawing the graph from scratch?

Example 20

Sketch the graph of $f(x) = (x - 1)(x - 2)(x - 3)$. On a different pair of axes sketch the graph of $g(x) = x(x + 1)(x - 1)$. Compare the graphs and the equations. What do you notice?

Solution

First you need to sketch the graph of $f(x) = (x - 1)(x - 2)(x - 3)$ using the method described in **Section 3.2**.

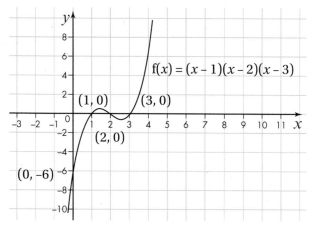

Then sketch the graph of $g(x) = x(x + 1)(x - 1)$.

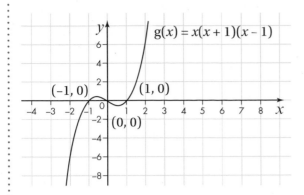

Comparing the graphs, g(x) is a horizontal translation of f(x) of −2 (2 to the left). If you substitute $x + 2$ for x in f(x) you get g(x):

$$f(x + 2) = ((x + 2) - 1)((x + 2) - 2)((x + 2) - 3)$$

So f($x + 2$) = g(x)

KEY INFORMATION

f($x + a$) is a horizontal translation of −a.

Stop and think   What sort of transformation would f($x − a$) be?

Example 21

Given that f(x) = sin x, sketch the graph of f($x − 90°$) for −270° ⩽ x ⩽ 270°, and suggest an equation for f($x − 90°$).

Solution

First you need to sketch the graph of f(x) = sin x, which you should know from your GCSE course.

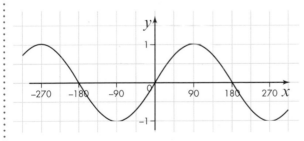

Then you need to sketch the graph of f($x − 90°$) which is a horizontal translation of +90°.

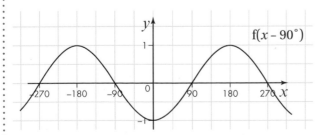

A possible equation for f($x − 90°$) is f($x − 2$) = −cos(x).

Example 22

Sketch the graph of $f(x) = \dfrac{3}{x}$. On a different pair of axes sketch the graph of $g(x) = \dfrac{3}{x} + 4$. State the equations of any asymptotes and the coordinates of any axis intercepts. Compare the graphs and the equations. What do you notice?

Solution

Sketch the graph of $f(x) = \dfrac{3}{x}$ using the method described in **Section 3.4**.

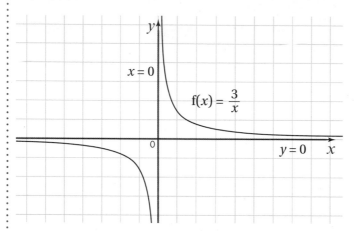

Sketch the graph of $g(x) = \dfrac{3}{x} + 4$ using the method described in **Section 3.4**.

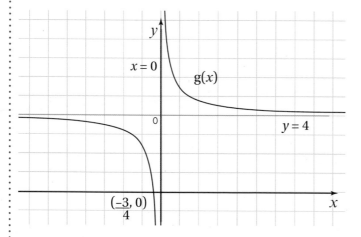

$g(x)$ has a horizontal asymptote at $y = 4$, a vertical asymptote at $x = 0$ and intercepts the x-axis at $\left(-\frac{3}{4}, 0\right)$.

Comparing the graphs, $g(x)$ is a vertical translation of $f(x)$ of 4 (4 up). If you add 4 to $f(x)$ you get $g(x)$.

TECHNOLOGY

Using a graphic calculator or graphing software package, plot the graph of $g(x) = \dfrac{3}{x} + 4$. Change the value of the '4' a few times to be both other positive numbers and negative numbers. What do you notice?

KEY INFORMATION

$f(x) + a$ is a vertical translation of $+a$.

Stop and think What sort of transformation would $f(x) - a$ be?

Example 23

The following graph shows a sketch of the curve f(x). Sketch the graph of f(x) − 2, show the coordinates of P, Q and R and state the number of solutions to the equation f(x) − 2 = 0.

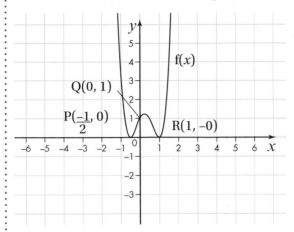

Solution

You need to sketch the graph of f(x) − 2, which is a vertical translation of −2 (2 down).

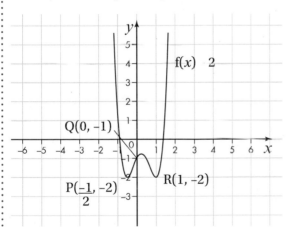

f(x) − 2 = 0 has two solutions (it cuts the x-axis twice).

Exercise 3.7A   **Answers page 504**

1 Given that f(x) = x^2, on different pairs of axes sketch the graphs of:

   a f(x + 3)   **b** f(x) + 3

   c f(x − 4)   **d** f(x + 4)

   For each transformation, state the number of solutions.

2 Given that $f(x) = \dfrac{1}{x}$, on different pairs of axes sketch the graphs of:

a $f(x+2)$ **b** $f(x)+2$

c $f(x-1)$ **d** $f(x)-1$

For each transformation, state the equations of any asymptotes.

PS **3** The following graph shows a sketch of the curve $f(x)$. On different pairs of axes, sketch the graphs of $f(x-2)$ and $f(x)+2$, showing the coordinates of point Q. State the number of solutions to the equations $f(x-2) = 0$ and $f(x)+2 = 0$.

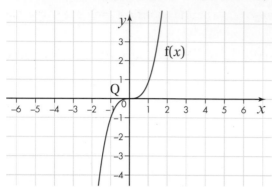

4 Given that $f(x) = \dfrac{1}{x^2}$:

a sketch the graph of $f(x+2)$ and state the equations of any asymptotes

b find the equation of $f(x+2)$

c find the y-intercept of $f(x+2)$.

5 The following graph shows a sketch of the curve $f(x)$. On the same pair of axes, sketch the graphs of $f(x+1)$ and $f(x)+1$, showing any asymptotes and the coordinates of point P in each case. State the number of solutions to the equation $f(x+1) = f(x)+1$.

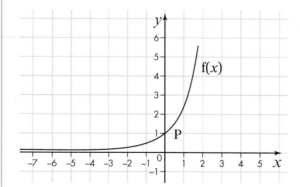

6 Given that $f(x) = \tan x$:

a sketch the graph of $f(x-180°)$ where $-180° \leqslant x \leqslant 180°$

b state the equations of any asymptotes and the coordinates of any axis intercepts

c on a separate pair of axes, sketch the graph of $f(x)+2$

d state the equations of any asymptotes.

7 Given that $f(x) = (x+1)(2x-1)^3$, on different pairs of axes sketch the graphs of:

    a $f(x+2)$ **b** $f(x)+2$

    c $f(x-1)$ **d** $f(x)-1$

    For each transformation, find the equation of the transformed function and state the number of solutions.

 8 Given that $f(x) = 2^{x+1}$, on the same pair of axes sketch the graphs of:

    a $f(x+2)$ **b** $f(x)+2$

    c If originally $f(x) = 2^x$, what transformations would have been required to achieve the same results as in **parts a** and **b**?

    For each transformation, find the equation of the transformed function and state the number of solutions.

 9 Given that $f(x) = x^4 - 5x^3 + 5x^2 + 5x - 6$, show that $f(x-1) = x^4 - 9x^3 + 26x^2 - 24x$.

3.8 Stretches

At GCSE you also saw other simple transformations on the graph of $y = f(x)$ and sketched the resulting graphs. The examples below extend stretch transformations to AS-level content.

Example 24

Sketch the graph of $f(x) = (x+4)(x-3)$. On a different pair of axes sketch the graph of $g(x) = (2x+4)(2x-3)$ and state the solutions to the equation $g(x) = 0$.

Solution

Sketch the graph of $f(x) = (x+4)(x-3)$ using the method described in **Section 3.1**.

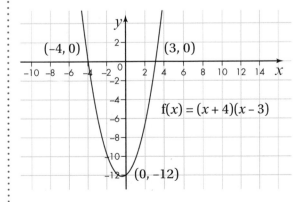

Sketch the graph of $g(x) = (2x + 4)(2x - 3)$.

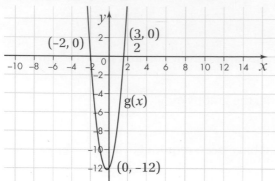

TECHNOLOGY

Using a graphic calculator or graphing software package, plot the graph of $g(x) = (2x + 4)(2x + 3)$. Change the coefficient of x to different values. What do you notice about the relationship between the coefficient of x and the scale factor applied to the stretch?

The solutions of the equation $g(x) = 0$ are $x = -2$ and $x = \frac{3}{2}$ (the x-intercepts).

Comparing the graphs, all the x-coordinates in $f(x)$ have been divided by 2 to get the corresponding coordinates for $g(x)$. The y-coordinates are the same. $g(x)$ is a horizontal stretch of $\frac{1}{2}$.

If you substitute $2x$ for x in $f(x)$ you get $g(x)$:

$$f(2x) = ((2x) + 4)((2x) + 3)$$

So $f(2x) = g(x)$

KEY INFORMATION

$f(ax)$ is a horizontal stretch with a scale factor of $\frac{1}{a}$. Consequently, the x-coordinates will need to be multiplied by $\frac{1}{a}$ and the y-coordinates will remain the same.

Stop and think   What sort of transformation would $f\left(\dfrac{x}{a}\right)$ be?

Example 25

Given that $f(x) = \cos x$, sketch the graph of $f\left(\dfrac{x}{2}\right)$, $-270° \leqslant x \leqslant 270°$, and find the equation of $f\left(\dfrac{x}{2}\right)$.

Solution

First you need to sketch the graph of $f(x) = \cos x$, which you should know from your GCSE course.

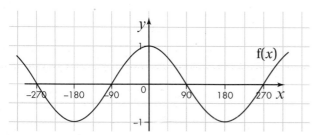

You need to sketch the graph of $f\left(\dfrac{x}{2}\right)$, which is a horizontal stretch with scale factor 2. If you multiply the x-coordinates by 2, the y-coordinates will remain the same.

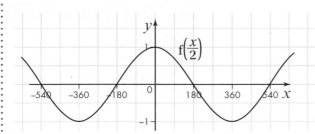

The equation of $f\left(\dfrac{x}{2}\right)$ is $f\left(\dfrac{x}{2}\right) = \cos\left(\dfrac{x}{2}\right)$.

Example 26

Sketch the graph of $f(x) = \dfrac{4}{x^2}$. On a different pair of axes sketch the graph of $g(x) = \dfrac{8}{x^2}$ State the equations of any asymptotes and compare the two graphs. What do you notice?

Solution

Sketch the graph of $f(x) = \dfrac{4}{x^2}$ using the method described in **Section 3.4**.

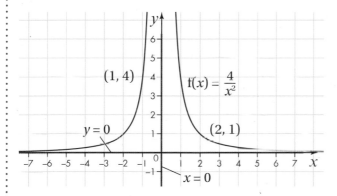

Sketch the graph of $g(x)$.

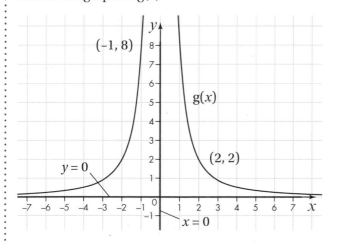

$g(x)$ is a vertical stretch of a scale factor of 2.

Comparing the graphs, all the y-coordinates in f(x) have been multiplied by 2 to get the corresponding coordinates for g(x). The x-coordinates are the same. g(x) is a vertical stretch of 2.

If you multiply f(x) by 2 you get g(x):

$$2f(x) = 2 \times \frac{4}{x^2}$$

So 2f(x) = g(x)

KEY INFORMATION

af(x) is a vertical stretch with a scale factor of a. Consequently, the y-coordinates will need to be multiplied by a and the x-coordinates will remain the same.

Stop and think

What sort of transformation would $\frac{1}{a}$f(x) be?

Stop and think

What sort of transformation would $\frac{b}{a}$f(x) be?

Example 27

The following graph shows a sketch of the curve f(x). Sketch the graph of −3f(x), show the coordinates of P and Q and state the number of solutions to the equation −3f(x) = 0.

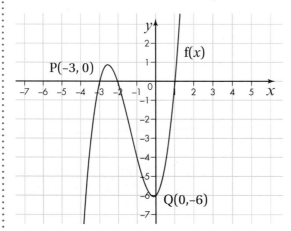

Solution

You need to sketch the graph of −3f(x), which is a vertical stretch with scale factor −3. You need to multiply the y-coordinates by −3 and the x-coordinates will remain the same.

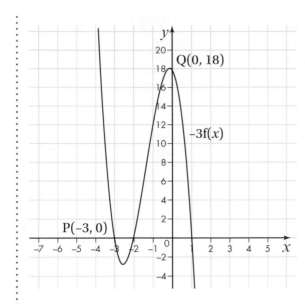

$-3f(x) = 0$ has three solutions (it cuts the x-axis three times).

Exercise 3.8A

Answers page 506

1 Given that $f(x) = x^2$, on different pairs of axes sketch the graphs of:

   a $f(3x)$ **b** $3f(x)$

   c $f(2x)$ **d** $-\dfrac{1}{2}f(x)$

For each transformation, state the number of solutions.

2 Given that $f(x) = \dfrac{1}{x}$, on different pairs of axes sketch the graphs of:

   a $f(2x)$ **b** $2f(x)$

   c $f(-x)$ **d** $-f(x)$

For each transformation, state the equations of any asymptotes.

PS 3 The following graph shows a sketch of the curve $f(x)$. On different pairs of axes, sketch the graphs of $f(-2x)$ and $-2f(x)$, showing the coordinates of points P, Q and R. State the number of solutions to the equations $f(-2x) = 0$ and $-2f(x) = 0$.

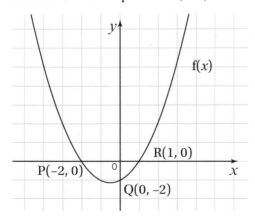

4 Given that $f(x) = \dfrac{1}{(x+1)^2}$:

   a sketch the graph of $2f(x)$ and state the equations of any asymptotes

   b find the equation of $2f(x)$

   c find the y-intercept of $2f(x)$.

5 The following graph shows a sketch of the curve $f(x)$. On the same pair of axes, sketch the graphs of $f(-x)$ and $-f(x)$, showing any asymptotes and the coordinates of point P in each case. State the number of solutions to the equation $f(-x) = -f(x)$.

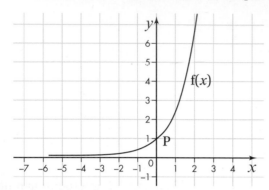

6 Given that $f(x) = \tan x$:

   a sketch the graph of $f(2x)$ where $-180° \leqslant x \leqslant 180°$

   b state the equations of any asymptotes and the coordinates of any axis intercepts

   c on a separate pair of axes, sketch the graph of $2f(x)$

   d state the equations of any asymptotes.

(PS) 7 Given that $f(x) = (x+1)(2x-1)^3$, on different pairs of axes sketch the graphs of:

   a $f(2x)$ **b** $2f(x)$

   c $f\left(\dfrac{x}{2}\right)$ **d** $\dfrac{1}{2}f(x)$

   For each transformation, find the equation of the transformed function and state the number of solutions.

(PS) 8 Given that $f(x) = 2^{x+1}$, on the same pair of axes sketch the graphs of:

   a $f(2x)$ **b** $2f(x)$

   c If originally $f(x) = 2^x$, what transformations would have been required to achieve the same results as in **parts a** and **b**?

   d For each transformation, find the equation of the transformed function and state the number of solutions.

(PS) (PF) 9 Given that $f(x) = x^4 - 5x^3 + 5x^2 + 5x - 6$, show that $f\left(\dfrac{x}{3}\right) = \dfrac{x^4}{81} - \dfrac{5x^3}{27} + \dfrac{5x^{-2}}{9} + \dfrac{5x}{3} - 6$

SUMMARY OF KEY POINTS

- For quadratic functions in the form $y = ax^2 + bx + c$:
  - If $a > 0$ then the graph will be a $\cup$ shape and have a minimum turning point.
  - If $a < 0$ then the graph will be a $\cap$ shape and have a maximum turning point.
  - If $b^2 - 4ac > 0$ then the equation has two different real roots.
  - If $b^2 - 4ac = 0$ then the equation has two equal roots.
  - If $b^2 - 4ac < 0$ then the equation does not have any real roots.
- For cubic functions in the form $y = ax^3 + bx^2 + cx + d$:
  - A cubic equation with one repeated real root will have a point of inflection.
  - A cubic equation with at least two different real roots will have two turning points.
- For quartic functions in the form $y = ax^4 + bx^3 + cx^2 + dx + e$:
  - If the quartic equation has one repeated root and $a > 0$ then the quartic curve is a $\cup$ shape. It will have a minimum turning point. If $a < 0$ then the graph will be inverted and it will have a maximum turning point.
  - If the quartic equation has at least three different real roots or two pairs of different real roots it will have three turning points.
  - If the quartic equation has one real root from a cubic function and another different real root it will have a point of inflection and a turning point.
- For reciprocals of linear functions in the form $y = \dfrac{a}{x}$ and quadratic functions in the form $y = \dfrac{a}{x^2}$:
  - A discontinuity is where the value of x or y is not defined.
  - An asymptote is a straight line that a curve approaches but never meets.
  - To work out the x-intercept(s), substitute $y = 0$ into the equation.
  - To work out the y-intercept, substitute $x = 0$ into the equation.
- For proportional relationships:
  - The relationship between two variables can be expressed using the proportion symbol, $\propto$.
  - The constant of proportionality is a multiplier linking two variables.
- Transformations of functions:
  - $f(x + a)$ is a horizontal translation of $-a$.
  - $f(x) + a$ is a vertical translation of $+a$.
  - $f(ax)$ is a horizontal stretch with a scale factor of $\dfrac{1}{a}$. Consequently, the x-coordinates will need to be multiplied by $\dfrac{1}{a}$ and the y-coordinates will remain the same.
  - $af(x)$ is a vertical stretch with a scale factor of a. Consequently the y-coordinates will need to be multiplied by a and the x-coordinates will remain the same.

EXAM-STYLE QUESTIONS 3

Answers page 508

(PS) 1 The following graph shows a sketch of the curve f(x). The points A, B and C and their coordinates have been marked on the sketch. In each of the following cases state the transformation that has moved A, B and C to the set of new coordinates.

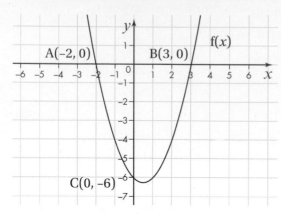

a A(−4, 0), B(1, 0) and C(0, −4) [1 mark]

b A(2, 0), B(−3, 0) and C(0, −6) [1 mark]

c A(−2, 1), B(3, 1) and C(0, −5) [1 mark]

d A(−2, 0), B(3, 0) and C(0, −6) [1 mark]

(M) (PS) 2 A new rollercoaster has been designed for a theme park. Part of its structure will follow the curve of $y = (2x - 1)(x + 2)(x + 3)$ for $-2 \leq x \leq 3$. A supporting cable is to be attached to the rollercoaster in the line of $y = 2x + 3$.

Sketch the graphs of the rollercoaster and supporting cable on the same pair of axes and work out the number of points the cable will touch the rollercoaster within the given interval. Clearly mark these points on your sketch. [5 marks]

(M) 3 The currency of Brazil, r reals, is directly proportional to currency of the United Kingdom, p pounds. If $r = 80$ reals when $t = 16$ pounds, find the exchange rate. Sketch the graph of the relationship between r and p and use the graph to work out:

a 29 pounds in reals [2 marks]

b 250 reals in pounds. [2 marks]

4 Given that f(x) = sin x for −270° ≤ x ≤ 270°, on different pairs of axes sketch the following transformations:

a f(x + 90°) [1 mark]

b −f(x) [1 mark]

c f(x) − 1 [1 mark]

d f$\left(\dfrac{x}{2}\right)$ [1 mark]

5 In January, a breeding pair of rabbits was put on an island where they had no natural predators. Consequently, the rabbits were able to breed freely and the population grew exponentially. The population growth can be modelled using the equation $y = 2^{\frac{x}{2}+1}$ where x is the time in months and y is the population. Sketch the graph of the rabbit population for the first year.

At the same time, a breeding pair of mice was put on a different island where they had no natural predators. Consequently, the mice were able to breed freely and the population grew exponentially. The population growth can be modelled using the equation $y = 2^{x+1}$, where x is the time in months and y is the population. Sketch the graph of the mice population for the first year on the same pair of axes as the rabbit population.

How many times are the populations of the rabbits and the mice the same? **[5 marks]**

6 The area of a shape, $A\,\text{cm}^2$, is directly proportional to the square of one of its length measurements, $l\,\text{cm}$. If $A = 254.504\,\text{cm}^2$ when $l = 9\,\text{cm}$ find the area of a shape with a length of $7\,\text{cm}$. Sketch the graph of the relationship between A and l. **[5 marks]**

7 a Sketch the graphs of $f(x) = (2 - x)(x + 2)^3$ and $g(x) = \dfrac{-1}{(x - 2)^2} + 3$ on the same pair of axes. Clearly highlight any asymptotes or points of intersection. Write down the number of points of intersection. **[3 marks]**

b On a separate pair of axes, sketch the graphs of $-2f(x)$ and $g(x - 2)$. Clearly highlight any asymptotes or points of intersection. Do the transformed graphs intersect? If so, how many times? **[4 marks]**

8 The function $f(x) = y = (x + 3)(2x - 1)^2$ is transformed using single, simple transformations. For each of the following equations, work out the corresponding transformation.

a $y = (x + 3)(2x - 1)^2 + 2$ **[2 marks]**

b $y = (-3 - x)(2x - 1)^2$ **[2 marks]**

c $y = (x + 1)(2x - 5)^2$ **[2 marks]**

d $y = (3 - x)(-2x - 1)^2$ **[2 marks]**

9 The function $y = f(x)$ is shown in the diagram.

On separate diagrams sketch the curves with the following equations clearly showing the coordinates of any axis intercepts and approximate locations of the turning points.

a $f(-x)$ **b** $-\dfrac{1}{2}f(x)$

c $-f(2x)$ **d** $20 - f(x)$ **[8 marks]**

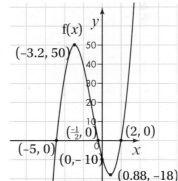

10 a On the same pair of axes, sketch the graphs of $f(x) = x^2 - 2x - 15$ and $x^2 + y^2 = 196$. Show that there are four points of intersection.

b Suggest transformations to satisfy each of the following.

i A vertical translation of $f(x)$ resulting in only two points of intersection.

ii A horizontal translation of $f(x)$ resulting in only two points of intersection. **[8 marks]**

4 COORDINATE GEOMETRY 1: EQUATIONS OF STRAIGHT LINES

It's not easy comparing mobile phone tariffs from different providers. The following graph provides a simple, visual representation of three tariffs, which can easily be used to make comparisons over time.

From this graph you could identify important features like:

> the tariff with the highest upfront cost

> the tariff with the lowest monthly charge.

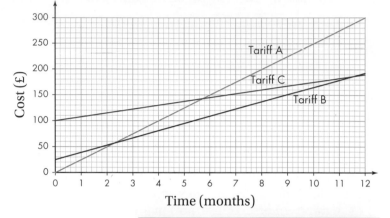

LEARNING OBJECTIVES

You will learn how to:

> write the equation of a straight line in the form $ax + by + c = 0$

> understand and use the gradient conditions for parallel lines

> understand and use the gradient conditions for perpendicular lines

> use straight line models in a variety of contexts.

TOPIC LINKS

You will need your knowledge of the gradient of a straight line to help you understand and solve differentiation problems in **Chapter 8 Differentiation**. The ability to draw and interpret straight line graphs will help you to solve distance, speed and time problems in **Chapter 15 Kinematics**. You will also need to determine and use the equations of straight lines when working with regression lines in **Chapter 12 Data presentation and interpretation** in context.

PRIOR KNOWLEDGE

You should already know how to:

> work with coordinates in all four quadrants

> plot graphs of equations that correspond to straight line graphs in the coordinate plane

> use the form $y = mx + c$ to identify parallel and perpendicular lines

> find the equation of the line through two given points or through one point with a given gradient

> identify and interpret gradients and intercepts of linear functions graphically and algebraically

> recognise, sketch and interpret graphs of linear functions.

You should be able to complete the following questions correctly:

1 Draw the graph of $y = 2x + 3$ for $-3 \leqslant x \leqslant 3$.

   What are the coordinates of the intercepts on the x and y axes?

2 Here are the equations of six lines:

   A $y = 2$ **B** $y = 2x + 3$ **C** $y = -\frac{1}{2}x + 3$

   D $y = 3 - 2x$ **E** $y = 3$ **F** $4x + 2y = 3$

   a Write down the gradient and y-intercept of each line.

   b Identify the lines that are parallel.

   c Identify the lines that are perpendicular.

3 Find the equation of the line through the points $(1, 8)$ and $(2, 0)$ by sketching the graph.

4 Find the equation of the line with a gradient of 2 and passing through the point $(3, 4)$ by sketching the graph.

4.1 Writing the equation of a straight line in the form $ax + by + c = 0$

You should be very familiar with writing the equation of a straight line in the form $y = mx + c$, for example $y = 2x + 3$. You also need to be able to write the equation of a straight line in the form $ax + by + c = 0$ where a, b and c are **integers**. To rewrite an equation that is currently in the form $y = mx + c$ in the form $ax + by + c = 0$, you will need to rearrange the equation.

KEY INFORMATION

The equation of a straight line may be written in the form $ax + by + c = 0$, where a, b and c are integers.

Example 1

Write $y = 2x + 3$ in the form $ax + by + c = 0$, where a, b and c are integers.

Solution

Subtract y from both sides of the equation.

$$0 = 2x + 3 - y$$

Rearrange.

$$2x - y + 3 = 0$$

a, b and c are integers, so this is the final answer.

Stop and think Compare the following example to the one above and work out what is the same and what is different. The equation is still in the form $y = mx + c$ but this time the gradient is both negative and a fraction. How will this affect the method?

Example 2

Write $y = -\frac{1}{2}x + 3$ in the form $ax + by + c = 0$, where a, b and c are integers.

Solution

Add $\frac{1}{2}x$ to both sides of the equation.

$$y + \tfrac{1}{2}x = 3$$

Subtract 3 from both sides of the equation.

$$y + \tfrac{1}{2}x - 3 = 0$$

Multiply both sides of the equation by 2.

$$2y + x - 6 = 0$$

Reorder.

$$x + 2y - 6 = 0$$

a, b and c are integers, so this is the final answer.

Alternatively:

Multiply both sides of the equation by 2.

$$2y = -x + 6$$

Add x to both sides of the equation.

$$2y + x = 6$$

Subtract 6 from both sides of the equation.

$$2y + x - 6 = 0$$

Stop and think As a mathematician, which format of the equation do you find more useful, and why?

Exercise 4.1A   **Answers page 511**

(CM) **1** Which of these equations of lines are not written in the form $ax + by + c = 0$, where a, b and c are integers?

Give reasons for your answers.

a $2x + 3y + 4 = 0$ **b** $-x - 4y - 3 = 0$ **c** $5x + 3y = 1$

d $4x - 3y + 2 = 5$ **e** $x + y - \frac{1}{2} = 0$ **f** $\frac{4}{3}x - \frac{5}{3}y + \frac{7}{3} = 0$

2 Write these equations of lines in the form $ax + by + c = 0$, where a, b and c are integers.

Write down the values of a, b and c in each case.

a $y = 4 + 5x$

b $y = 3 - 2x$

3 Write these equations of lines in the form $ax + by + c = 0$, where a, b and c are integers.

Write down the values of a, b and c in each case.

a $y = \frac{1}{3}x - 7$

b $y = -\frac{2}{5}x + 6$

c $y = \frac{4}{3}x + \frac{7}{2}$

4 Write $8x - 2y + 3 = 0$ in the form $y = mx + c$.

Write down the gradient and the coordinates of the y-intercept.

5 Review and correct this method to write the equation $y = 3 - \frac{5}{2}x$ in the form $ax + by + c = 0$, where a, b and c are integers.

$$y = 2 - \frac{5}{2}x$$

$$2y = 2 - 5x$$

$$5x + 2y = 2$$

So $5x + 2y - 2 = 0$

6 Write $\frac{y}{3} = \frac{x-4}{5}$ in the form $ax + by + c = 0$, where a, b and c are integers.

Write down the values of a, b and c.

(PS) **7** Work out the coordinates of the axis intercepts of the line $-3x - 5y + 2 = 0$.

4.2 Finding the equation of a straight line using the formula $y - y_1 = m(x - x_1)$

If you know the **gradient** m of a line and the coordinates

(x_1, y_1) of a point on the line, then you can use the **formula**

$$y - y_1 = m(x - x_1)$$

(M) **Modelling**   (PS) **Problem solving**   (PF) **Proof**   (CM) **Communicating mathematically**   **103**

to work out the equation of the line. In kinematics in Mechanics, if you know a point on a straight line distance–time graph (for example, that at x hours the object will be y distance from the start) and you know the constant speed (the gradient on a distance–time graph), then you will be able to work out the equation that links the time and the distance travelled for this part of the journey. This method is covered in **Chapter 15**.

Example 3

Find the equation of the line with gradient 2 that passes through the point (3, 7).

Solution

State the formula you are going to use.

$$y - y_1 = m(x - x_1)$$

The values for substitution are $m = 2$ and $(x_1, y_1) = (3, 7)$.

$$y - 7 = 2(x - 3)$$

Expand the bracket.

$$y - 7 = 2x - 6$$

Add 7 to both sides of the equation.

$$y = 2x + 1$$

> The value of m is the gradient stated in the question.

> **TECHNOLOGY**
> You can also use a graphic calculator or graphing software package to check the answer. Plot the graph of $y = 2x + 1$. Does the line have a gradient of 2? Does the line go through the point (3, 7)?

Example 4

A line with a gradient of -1 and the line $y = 2$ meet the y-axis at the same point.

Find the equation of the line in the form $ax + by + c = 0$, where a, b and c are integers.

Solution

The line $y = 2$ meets the y-axis at (0, 2).

State the formula you are going to use.

$$y - y_1 = m(x - x_1)$$

The values for substitution are $m = -1$ and $(x_1, y_1) = (0, 2)$.

$$y - 2 = -1(x - 0)$$

Expand the bracket.

$$y - 2 = -x$$

Add x to both sides of the equation.

$$x + y - 2 = 0$$

> **TECHNOLOGY**
> You can also use a graphic calculator or graphing software package to check the answer. Plot the graph of $y = 2 - x$ (the equation in the form $y = mx + c$). Does the line have a gradient of -1? Does the line go through the point (0, 2)?

> a, b and c are integers, so this is the final answer.

Exercise 4.2A

Answers page 511

1 Find the equation of the line with the given gradient and passing through the given point.

   a $m = 2$ and $(x_1, y_1) = (3, 0)$

   b $m = 3$ and $(x_1, y_1) = (0, 3)$

   c $m = 2$ and $(x_1, y_1) = (3, 4)$

   d $m = -5$ and $(x_1, y_1) = (2, 3)$

2 Find the equation of the line with the given gradient and passing through the given point.

   a $m = -4$ and $(x_1, y_1) = (-2, -5)$

   b $m = -1$ and $(x_1, y_1) = (2, -2)$

3 A line with a gradient of -2 and the line $y = 3$ meet the y-axis at the same point.

Find the equation of the line in the form $ax + by + c = 0$.

4 A line with a gradient of 3 and the line $x = -1$ meet the x-axis at the same point.

Find the equation of the line in the form $ax + by + c = 0$.

(PS) 5 The lines $y = 2x + 4$ and $y = 7 - x$ intersect at the point P.

Find the equation of the line with gradient 3 that passes through the point P.

(PS) 6 A line with a gradient of 3 which passes through the point (1, 1) intersects with another line with a gradient of -1 which passes through (4, 6).

Work out the point of intersection of the two lines.

(PS) 7 A line with a gradient of 5 passes through the point (2, 3).

Does the line passing through the same point with a gradient half the size go through the **origin**?

4.3 Finding the gradient of the straight line between two points

From GCSE mathematics, we know that to find the gradient of a line joining two points, we calculate 'rise divided by run'. Another way of saying this is you divide the difference in the y-coordinates by the difference in the x-coordinates. In more formal terms, if you know or are given the coordinates of two points, (x_1, y_1) and (x_2, y_2), then you can use the formula $m = \dfrac{y_2 - y_1}{x_2 - x_1}$ to work out the gradient m of the line joining these points.

KEY INFORMATION

You can find the gradient m of the line joining two points with coordinates (x_1, y_1) and (x_2, y_2) using the formula

$m = \dfrac{y_2 - y_1}{x_2 - x_1}.$

You need to remember this formula.

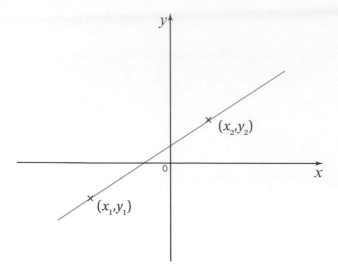

In kinematics in Mechanics, if you know two points on a straight line speed–time graph (that is, if you know that between x_1 and x_2 hours the object will be travelling between speeds y_1 and y_2) then you will be able to calculate the gradient between the two points. On a speed–time graph, the gradient is the acceleration of the object.

Example 5

Work out the gradient of the line joining the points $(-1, 2)$ and $(3, 8)$.

Solution

State the formula you are going to use.

$$m = \frac{y_2 - y_1}{x_2 - x_1}$$

Write down what you know.

Let $(x_1, y_1) = (-1, 2)$

Let $(x_2, y_2) = (3, 8)$

Substitute the values into the formula.

$$m = \frac{8 - 2}{3 - -1}$$

Do the calculations.

$$m = \frac{6}{4}$$

Simplify.

$$m = \frac{3}{2}$$

Alternatively:

State the formula you are going to use.

$$m = \frac{y_2 - y_1}{x_2 - x_1}$$

TECHNOLOGY

You can also use a graphic calculator or graphing software package to check the answer. Plot the points $(-1, 2)$ and $(3, 8)$. Then draw a line between the two points and determine the equation of the line. Is the gradient of the line (the coefficient of x) $\frac{3}{2}$ or an equivalent value?

m is a fraction in its simplest form.

Write down what you know.

Let $(x_1, y_1) = (3, 8)$.

Let $(x_2, y_2) = (-1, 2)$.

Substitute the values into the formula.

$$m = \frac{2 - 8}{-1 - 3}$$

Do the calculations.

$$m = \frac{-6}{-4}$$

Simplify.

$$m = \frac{3}{2}$$

m is a fraction in its simplest form.

Stop and think   Compare the next example to the one above and work out what is the same and what is different. This time the gradient and the coordinates of one point are given. Some information is given about the second point. How will this affect the method?

Example 6

A line with a gradient of 2 passes through the point (5, 6).

What are the coordinates of the x-intercept of the line?

Solution

Write down what you know.

$$m = 2$$

$$x_1 = ?, \ y_1 = 0 \ (x\text{-intercept})$$

$$x_2 = 5, \ y_2 = 6$$

State the formula you are going to use.

$$m = \frac{y_2 - y_1}{x_2 - x_1}$$

Substitute the values into the formula.

$$2 = \frac{6 - 0}{5 - x}$$

Multiply both sides of the equation by $(5 - x)$.

$$2(5 - x) = 6$$

Expand the bracket.

$$10 - 2x = 6$$

You might find that making a quick sketch of this scenario will aid your understanding.

Add $2x$ to both sides of the equation.

$$10 = 6 + 2x$$

Subtract 6 from both sides of the equation.

$$4 = 2x$$

Divide both sides of the equation by 2.

$$\frac{4}{2} = x$$

Simplifying.

$$x = 2$$

The coordinates of the x-intercept are (2, 0).

TECHNOLOGY

Using a graphic calculator or graphing software package, investigate the location of the x-intercept as the gradient varies. Use both positive and negative and integer and fractional values.

Exercise 4.3A   Answers page 511

1 Find the gradient of the line passing through the given points.

  a (2, 3) and (7, 8)

  b (5, 9) and (3, 3)

  c (1, −3) and (3, −9)

2 Find the gradient of the line passing through the given points.

  a ($8a$, $5a$) and ($3a$, $3a$)

  b (a, a) and ($3a$, $-5a$)

3 Review and correct this method to find the gradient of the line between the points (1, 2) and (5, −8).

$$m = \frac{y_1 - y_2}{x_1 - x_2}$$

Let $(x_1, y_1) = (1, 2)$

Let $(x_2, y_2) = (5, 8)$

$$m = \frac{1-5}{8-2}$$
$$m = \frac{-4}{6}$$
So $m = -\frac{2}{3}$

PS **4** Which of the following pairs of points lie on a straight line with a gradient of −3?

  a (1, 1) and (4, −8)

  b (1, 3) and (4, 9)

  c (1, −7) and (4, −16)

  d (1, 7) and (3, 1)

PS **5** Show that the points $(2, 2)$, $(5, \frac{1}{2})$ and $(11, -\frac{5}{2})$ lie on a straight line.

PF

 6 A cyclist climbs to the top of a hill then descends the other side.

The profile of his ascent and descent are plotted on a pair of axes.

The cyclist starts the climb at $(-10, 0)$ and reaches the summit at $(0, \frac{1}{2})$.

The cyclist then descends from the summit to finish at $(2, \frac{3}{8})$.

Which is steepest part of the ride – the ascent or the descent?

7 Find the gradient of the line passing through $(\frac{1}{2}, \frac{1}{3})$ and $(\frac{3}{4}, -\frac{2}{3})$.

4.4 Finding the equation of a straight line using the formula $\frac{y - y_1}{y_2 - y_1} = \frac{x - x_1}{x_2 - x_1}$

If you know or are given the coordinates of two points, (x_1, y_1) and (x_2, y_2), then you can use the formula $\frac{y - y_1}{y_2 - y_1} = \frac{x - x_1}{x_2 - x_1}$ to work out the equation of the line joining these points.

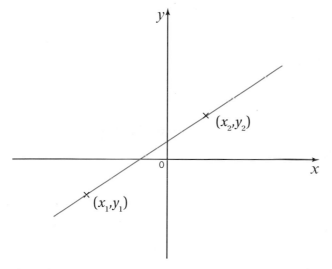

KEY INFORMATION

You can find the equation of the line joining two points with coordinates (x_1, y_1) and (x_2, y_2) using the formula $\frac{y - y_1}{y_2 - y_1} = \frac{x - x_1}{x_2 - x_1}$.

You need to remember this formula.

If you have a straight line conversion graph between different units (for example, temperatures, currencies, units of measure), you can read off two pairs of coordinates to work out the equation of the line joining the points. This allows you to convert any quantity measured in one of the units to the other, for example to convert a temperature in degrees Fahrenheit to Celsius.

PROOF

Show that $\frac{y - y_1}{y_2 - y_1} = \frac{x - x_1}{x_2 - x_1}$.

State the formulae you are going to use.

$$y - y_1 = m(x - x_1)$$

$$m = \frac{y_2 - y_1}{x_2 - x_1}$$

Replace m in the first equation by the value from the second equation.

$$y - y_1 = \frac{y_2 - y_1}{x_2 - x_1}(x - x_1)$$

Rearrange.

$$\frac{y - y_1}{y_2 - y_1} = \frac{x - x_1}{x_2 - x_1}$$

PROOF

By direct proof you have shown that $\dfrac{y - y_1}{y_2 - y_1} = \dfrac{x - x_1}{x_2 - x_1}$.

Example 7

Find the equation of the line between the y-intercept of $y = 3x - 5$ and the point $(9, -8)$.

Write the equation of the line in the form $ax + by + c = 0$, where a, b and c are integers.

You might find it aids your understanding to make a quick sketch of this scenario.

Solution

State the formula you are going to use.

$$\frac{y - y_1}{y_2 - y_1} = \frac{x - x_1}{x_2 - x_1}$$

Write down what you know.

Let $(x_1, y_1) = (0, -5)$.

Let $(x_2, y_2) = (9, -8)$.

Substitute the values into the formula.

$$\frac{y - -5}{-8 - -5} = \frac{x - 0}{9 - 0}$$

Simplify the equation by manipulating the numeric values.

$$\frac{y + 5}{-8 + 5} = \frac{x}{9}$$

$$\frac{y + 5}{-3} = \frac{x}{9}$$

Multiply both sides of the equation by 9.

$$\frac{9(y + 5)}{-3} = x$$

Simplify the left-hand side.

$$-3(y + 5) = x$$

Expand the bracket.

$$-3y - 15 = x$$

Add $3y$ and 15 to both sides of the equation.

$$0 = x + 3y + 15$$

Reorder.

$$x + 3y + 15 = 0$$

TECHNOLOGY

You can also use a graphic calculator or graphing software package to check the answer. Plot the points $(0, -5)$ and $(9, -8)$. Draw a line between the two points and determine the equation of the line.

Exercise 4.4A Answers page 512

1 Find the equation of the line passing through the given points.

  a (2, 3) and (7, 8)

  b (3, 3) and (5, 9)

2 Find the equation of the line passing through the given points.

  a (1, –3) and (3, –9)

  b (–3, –4) and (–7, 6)

3 Review and correct this method to find the equation of the line between the points (1, 2) and (5, –8).

$$\frac{y - y_1}{y_2 - y_1} = \frac{x - x_1}{x_2 - x_1}$$

Let $(x_1, y_1) = (1, 2)$

Let $(x_2, y_2) = (5, -8)$

$$\frac{y - 2}{2 - -8} = \frac{x - 1}{1 - 5}$$

$$\frac{y - 2}{10} = \frac{x - 1}{-4}$$

$$4(y - 2) = 10(x - 1)$$

$$4y - 8 = 10x - 10$$

$$4y = 10x - 18$$

So $\quad y = \dfrac{10x - 18}{4}$

4 Find the equation of the line between the y-intercept of $y = 6 - 2x$ and the point (–1, –2).

Write the equation of the line in the form $ax + by + c = 0$, where a, b and c are integers.

5 Find the equation of the line between the x-intercept of $2x + y - 4 = 0$ and the point (3, –7).

Write the equation of the line in the form $y = mx + c$.

What are the gradient and the coordinates of the y-intercept of this line?

(CM) 6 Line A passes through (2, 7) and (5, 6).

Line B passes through (5, –4) and (7, –6).

Find the equations of lines A and B in the form $y = mx + c$ and state which line is steeper.

Give a reason for your answer.

7 Find the equation of the line passing through the given points.

  a $(\frac{1}{2}, \frac{1}{3})$ and $(\frac{3}{4}, -\frac{2}{3})$

  b $(\frac{2}{7}, -\frac{1}{5})$ and $(-\frac{1}{3}, -\frac{1}{2})$

8 Find the equation of the line between the y-intercept of $2x - 5y + 7 = 0$ and the x-intercept of $y = -3x + 5$ in the form $ax + by + c = 0$, where a, b and c are integers.

 9 Line A passes through $(3, 5)$ and $(4, 9)$.

Line B passes through $(1, -3)$ and $(5, -31)$.

Find the equations of lines A and B and find the point of intersection of lines A and B.

 10 The lines $y = x - 9$ and $y = 5 - x$ intersect at point A.

The lines $y = 7 - 3x$ and $y = 2x + 8$ intersect at point B.

Find the equation of the line between points A and B.

Does this line pass through the point $(1, 2)$?

Justify your answer.

4.5 Parallel and perpendicular lines

Gradient conditions for parallel lines

In your GCSE mathematics course you should have learnt that if two straight lines are **parallel** then the gradients of both lines will be the same. More formally, if you know the equations or gradients of two or more straight lines, and if the value of m in each case is the same, then the lines are parallel.

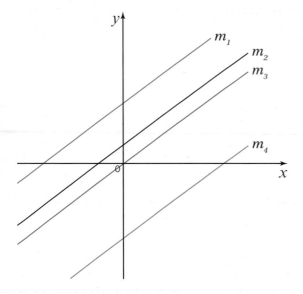

Line 1 has a gradient of m_1.

Line 2 has a gradient of m_2.

Line 3 has a gradient of m_3.

Line 4 has a gradient of m_4.

If $m_1 = m_2 = m_3 = m_4$ then all the lines are parallel.

KEY INFORMATION

Two or more lines are parallel if their gradients, m, are the same.

You need to learn the condition for parallel lines.

If you know that two straight lines are parallel then you can use this information to identify different shapes. For example, a quadrilateral with two pairs of parallel sides must be a square, rectangle, rhombus or parallelogram. If you also know that a pair of lines are perpendicular to one another, then the shape must be either a square or a rectangle.

Example 8

Find the equation of the line that is parallel to $3x - y + 7 = 0$ and passes through the point $(1, 8)$.

Write the equation of the line in the form $ax + by + c = 0$, where a, b and c are integers.

Solution

Rearrange the equation into the form $y = mx + c$.

$$y = 3x + 7$$

Write down what you know.

$$m = 3 \text{ and } (x_1, y_1) = (1, 8)$$

State the formula you are going to use.

$$y - y_1 = m(x - x_1)$$

Substitute the values into the formula.

$$y - 8 = 3(x - 1)$$

Expand the bracket.

$$y - 8 = 3x - 3$$

Add 8 to both sides of the equation.

$$y = 3x + 5$$

Subtract y from both sides of the equation.

$$0 = 3x - y + 5$$

Rearrange.

$$3x - y + 5 = 0$$

> **TECHNOLOGY**
>
> Using a graphic calculator or graphing software package, try finding the equations of lines passing through other points but that are parallel to the given equation.

> The equation is not yet in the required form.

Stop and think   If you are asked to show that two lines, between two different pairs of points, are parallel do you need to work out the equation of each line, or is there a slightly simpler approach?

Example 9

Line A passes through (1, 5) and (3, 11).

Line B passes through (4, 5) and (5, 8).

Show that lines A and B are parallel.

Solution

State the formula you are going to use.

$$m = \frac{y_2 - y_1}{x_2 - x_1}$$

Write down what you know for line A.

Let $(x_1, y_1) = (1, 5)$.

Let $(x_2, y_2) = (3, 11)$.

Substitute the values into the formula.

$$m = \frac{11 - 5}{3 - 1}$$

Do the calculations.

$$m = \frac{6}{2}$$

Simplify.

$$m = 3 \text{ for line A.}$$

Write down what you know for line B.

Let $(x_1, y_1) = (4, 5)$.

Let $(x_2, y_2) = (5, 8)$.

Substitute the values into the formula.

$$m = \frac{8 - 5}{5 - 4}$$

Simplify the equation by manipulating the numeric values.

$$m = 3 \text{ for line B.}$$

The gradients of lines A and B are the same therefore lines A and B are parallel.

It is easier (because there are fewer manipulations so it is less prone to error) to determine the gradient of the line joining two points than to find the equation of the line passing through two points.

TECHNOLOGY

You can also use a graphic calculator or graphing software package to check the answer. Plot the points (1, 5) and (3, 11). Draw a line between the two points and determine the equation of the line. On the same graph, plot the points (4, 5) and (5, 8). Use the software to draw a line between the two points and determine the equation of the line. Do the lines look parallel? Is the value of m the same in both equations?

Exercise 4.5A
Answers page 512

1 Work out whether each pair of lines are parallel.

  a $y = 2x - 6$ and $y = 2x + 9$

  b $y = -3x - 5$ and $y = 5 - 3x$

(PF) **2** Line A passes through $(1, 7)$ and $(3, 11)$.

Line B passes through $(2, -3)$ and $(5, 3)$.

Show that lines A and B are parallel.

(PS) **3** Find the equation of the line that is parallel to $y = 7x - 2$ and passes through the x-intercept of $y = x - 3$.

4 Work out whether each pair of lines are parallel.

  a $4x - y + 2 = 0$ and $-y = 4x + 3$

  b $6x - 2y + 2 = 0$ and $y = 3x - 3$

(PF) **5** Line A passes through $(-1, 3)$ and $(3, -1)$.

Line B passes through $(-2, 2)$ and $(-3, 3)$.

Show that lines A and B are parallel.

(PS) **6** Find the equation of the line that is parallel to $-4x + y + 7 = 0$ and passes through the y-intercept of $2x - y + 3 = 0$.

7 Work out whether each pair of lines are parallel.

  a $5x - 3y - 7 = 0$ and $6y = 10x + 3$

  b $8x - 3y - 7 = 0$ and $6y - 8x + 2 = 0$

(PS) (PF) **8** The four sides of a rectangle have the equations $2x - y + 6 = 0$, $2x + 4y - 44 = 0$, $2x - y - 4 = 0$ and $2x + 4y - 24 = 0$.

Show which pairs of lines form parallel sides.

Gradient conditions for perpendicular lines

In your GCSE mathematics course you should have learnt that if two straight lines are **perpendicular** (intersect at right angles), then the gradient of one line is the negative **reciprocal** of the other. For example, if the gradient of a line is 2 then the gradient of a line perpendicular to it will be $-\frac{1}{2}$. The product of a number and its negative reciprocal is -1.

$-\frac{1}{2}$ is the negative reciprocal of 2 and vice versa.

More formally you can say:

❭ a line with a gradient of m_1 is perpendicular to a line with a gradient of m_2 when $m_2 = -\dfrac{1}{m_1}$

❭ two straight lines with gradients m_1 and m_2 are perpendicular when $m_1 m_2 = -1$.

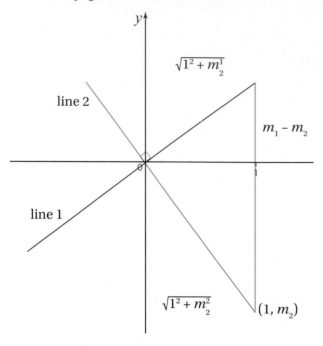

Line 1 has a gradient of m_1.

Line 2 has a gradient of m_2.

If $m_1 m_2 = -1$, then the lines are perpendicular.

PROOF

Using Pythagoras' theorem:

$$1^2 + m_1^2 + 1^2 + m_2^2 = (m_1 - m_2)^2$$

$$2 + m_1^2 + m_2^2 = m_1^2 - 2m_1 m_2 + m_2^2$$

$$2 = -2 m_1 m_2$$

$$\frac{1}{m_2} = -m_1$$

Example 10

Find the equation of the line that is perpendicular to $4x - y - 9 = 0$ and passes through the point $(3, 5)$.

Write the equation of the line in the form $ax + by + c = 0$, where a, b and c are integers.

Solution

Rearrange the equation into the form $y = mx + c$.

$$y = 4x - 9$$

$$m_1 = 4$$

Find the gradient of the line perpendicular to it.

$$m_1 m_2 = -1$$

$$4 m_2 = -1$$

$$m_2 = -\tfrac{1}{4}$$

Let $(x_1, y_1) = (3, 5)$.

State the formula you are going to use.

$$y - y_1 = m(x - x_1)$$

Substitute the values into the formula.

$$y - 5 = -\tfrac{1}{4}(x - 3)$$

Multiply both sides of the equation by -4.

$$-4(y - 5) = x - 3$$

Expand the bracket.

$$-4y + 20 = x - 3$$

Rearrange the equation into the form $ax + by + c = 0$.

$$x + 4y - 23 = 0$$

Example 11

Show that the lines $2x + y - 5 = 0$ and $x - 2y + 4 = 0$ are perpendicular.

Solution

Rearrange the first equation into the form $y = mx + c$.

$$y - 5 = -2x$$

$$y = -2x + 5$$

$$m_1 = -2$$

Rearrange the second equation into the form $y = mx + c$.

$$2y = x + 4$$

$$y = \tfrac{1}{2}x + 2$$

$$m_2 = \tfrac{1}{2}$$

Show that $m_1m_2 = -1$

$$m_1m_2 = (-2)\left(\tfrac{1}{2}\right) = -1$$

So the lines $2x + y - 5 = 0$ and $x - 2y + 4 = 0$ are perpendicular.

TECHNOLOGY

Using a graphic calculator or graphing software package, try plotting lines that intersect but are not perpendicular to each other. What do you notice about the products of their gradients?

Exercise 4.5B   **Answers page 513**

1 Work out whether each pair of lines are perpendicular.

   a $y = 2x + 6$ and $y = \tfrac{1}{2}x + 9$

   b $y = -3x - 5$ and $y = \tfrac{1}{3}x + 5$

2 Review and correct this method to find the equation of the line that is perpendicular to $y = -\frac{1}{2}x + 3$ and passes through the point $(1, 0)$.

$$m = -\frac{1}{2}$$

Let $(x_1, y_1) = (1, 0)$.

$$y - y_1 = m(x - x_1)$$
$$y - 1 = \frac{1}{2}x - 0$$

So $\qquad y = \frac{1}{2}x - 1$

(CM) 3 Are the following pairs of lines perpendicular? You must provide justification for your answers.

a $3x - y + 2 = 0$ and $-y = \frac{1}{3}x + 3$

b $12x - 4y + 4 = 0$ and $y = 12x - 12$

(PS) (PF) 4 Line A passes through $(1, 1)$ and $(3, 9)$.

Line B passes through $(4, 0)$ and $(8, -1)$.

Show that lines A and B are perpendicular.

(PS) (M) 5 The four sides of a rectangle have the equations $2x - y + 6 = 0$, $2x + 4y - 44 = 0$, $2x - y - 4 = 0$ and $2x + 4y - 24 = 0$.

Show which lines are perpendicular.

(PS) 6 Find the equation of the line that is perpendicular to $8x - 2y - 7 = 0$ and passes through the y-intercept of $4x - 2y + 5 = 0$.

(CM) 7 In each of the following cases, explain the mathematical manipulation steps required to show whether or not the lines are perpendicular.

a $5x - 3y - 7 = 0$ and $-10y = 6x + 3$

b $9x - 10y - 7 = 0$ and $9y - 10x + 2 = 0$

(PS) (CM) 8 A quadrilateral is drawn with vertices at $(0, 6)$, $(2, 10)$, $(4, 4)$ and $(6, 8)$.

By considering gradients, work out which of the following shapes the quadrilateral could possibly be, and state your reasons.

rhombus kite parallelogram square trapezium rectangle

4.6 Straight line models

You need to be able to apply everything that you have learnt about straight line models in this chapter in a variety of different contexts, including real-world scenarios.

KEY INFORMATION

You need to be able to use straight line models in a variety of contexts.

Example 12

On an old mercury thermometer, a temperature of 0 °C reads as 32 °F. Similarly, a temperature of 20 °C reads as 68 °F. The relationship between degrees Celsius (°C) and degrees Fahrenheit (°F) can be modelled as a straight line.

If the temperature in degrees Celsius is taken as x and the temperature in degrees Fahrenheit as y, use

$$\frac{y - y_1}{y_2 - y_1} = \frac{x - x_1}{x_2 - x_1},$$

to find the equation of the line in the form $ax + by + c = 0$, where a, b and c are integers.

Solution

State the formula you are going to use.

$$\frac{y - y_1}{y_2 - y_1} = \frac{x - x_1}{x_2 - x_1}$$

Write down what you know.

Let $(x_1, y_1) = (0, 32)$

Let $(x_2, y_2) = (20, 68)$

Substitute the values into the formula.

$$\frac{y - 32}{68 - 32} = \frac{x - 0}{20 - 0}$$

Simplify the equation by manipulating the numeric values.

$$\frac{y - 32}{36} = \frac{x}{20}$$

Multiply both sides of the equation by 36.

$$y - 32 = \frac{36x}{20}$$

Simplify.

$$y - 32 = \frac{9x}{5}$$

Multiply both sides of the equation by 5.

$$5(y - 32) = 9x$$

Expand the bracket.

$$5y - 160 = 9x$$

Subtract $5y$ and add 160 to both sides of the equation.

$$0 = 9x - 5y + 160$$

Rearrange.

$$9x - 5y + 160 = 0$$

TECHNOLOGY

You can also use a graphic calculator or graphing software package to check the answer. Plot the points (0, 32) and (20, 68). Draw a line between the two points and determine the equation of the line.

Example 13

Anna has a very old mobile phone and is not very pleased with the service that she is receiving from her current mobile phone provider.

The number of different phones and tariffs available are bewildering but she manages to narrow her choice down to three options.

She represents the three different tariffs on a graph.

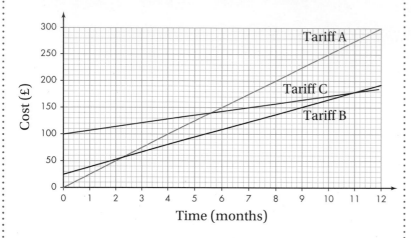

The relationship between time (months) and cost (£) can be modelled as a straight line.

By finding the equations for options A and C, work out at what time option C becomes cheaper than option A.

Solution

You are being asked to find the point of intersection, specifically the *x*-coordinate of the point of intersection, of two straight lines.

How are you going to do this?

You need to break down the problem into mathematical steps. It might be easier to work backwards.

> State the *x*-coordinate of the point of intersection.

> Find the point of intersection of line A and line C.

> Find the equation of line C.

> Find the equation of line A.

> Find two pairs of coordinates on line C.

> Find two pairs of coordinates on line A.

So you need to work through this list of steps from the bottom to the top.

State the formula you are going to use.

$$\frac{y-y_1}{y_2-y_1}=\frac{x-x_1}{x_2-x_1}$$

Write down what you know for line A.

Let $(x_1, y_1) = (0, 0)$.

Let $(x_2, y_2) = (8, 200)$.

Substitute the values into the formula.

$$\frac{y-0}{200-0}=\frac{x-0}{8-0}$$

Simplify the equation by manipulating the numeric values.

$$\frac{y}{200}=\frac{x}{8}$$

Multiply both sides of the equation by 200 and divide by 8.

$$y = 25x$$

Write down what you know for line C.

Let $(x_1, y_1) = (0, 100)$.

Let $(x_2, y_2) = (7, 150)$.

Substitute the values into the formula.

$$\frac{y-100}{150-100}=\frac{x-0}{7-0}$$

$$\frac{y-100}{50}=\frac{x}{7}$$

$$y-100=\frac{50x}{7}$$

$$y=\frac{50}{7}x+100$$

Set the equations of the two straight lines equal to each other and solve this equation to find the x-coordinate of the point of intersection.

$$25x=\frac{50}{7}x+100$$

Multiply both sides of the equation by 7.

$$175x = 50x + 700$$

Subtract $50x$ from both sides of the equation.

$$125x = 700$$

Divide both sides of the equation by 125.

$$x = 5.6$$

Option C becomes cheaper than option A 5.6 months after the start of both contracts.

Stop and think How else could you have found out at what point in time option C becomes cheaper than option A?

Exercise 4.6A

Answers page 513

 1 At a post office a holiday maker who is travelling to the USA exchanges £500 for $800.

Another holiday maker exchanges £100 for $160.

The relationship between pounds (£) and US dollars ($) can be modelled as a straight line.

If the number of pounds is taken as x and the number of dollars as y, use

$$\frac{y - y_1}{y_2 - y_1} = \frac{x - x_1}{x_2 - x_1}$$

to find the equation of the line in the form $y = mx + c$.

 2 On a distance–time graph the equations for two different sections of the journey are
 $y = 2x + 6$, $0 \leqslant x \leqslant 1$ and $y = -2x + 12$, $4 \leqslant x \leqslant 5$, where x is time in hours and y is distance travelled.

Decide which of the following statements are true and which are false.

Give a reason for each of your decisions.

a The lines for the two equations are perpendicular to each other.

b The two lines cross the y-axis at the same point.

c The line for $y = -2x + 12$ is travelling away from the start.

d The speed is the same in the two sections of the journey.

e The lines for the two equations are parallel to each other.

f The line for $y = 2x + 6$ is travelling back to the start.

 3 At the start of Year 12, two students, Anna and Bhavini, buy new printers.

Anna pays £50 for her new printer and estimates that within five years her printer will have cost £950 to buy and run.

Bhavini pays £67.50 for her new printer and estimates that within five years her printer will have cost £667.50 to buy and run.

For each printer, find the equation of the costs in the form $y = mx + c$, where x is time in years and y is the overall cost.

For each printer, interpret the values of m and c.

If Anna and Bhavini both use four ink cartridges per year, what is the cost of one ink cartridge in each case?

What is significant about the x-intercept in each case?

4 A group of friends are going to a party and need to hire a taxi to take them there. They call three different taxi firms to the firm with the cheapest fare.

Taxi firm A	No callout charge.
	Charge per mile.
	Example charge for 10 miles would be £50.
Taxi firm B	Callout charge of £10.
	Additionally, charge per mile.
	Example charge for 10 miles would be £40.
Taxi firm C	Callout charge of £15.
	Additionally, charge per mile.
	Example charge for 10 miles would be £35.

The relationship between miles travelled and overall cost can be modelled as a straight line.

If x is the number of miles travelled and y is the cost, state, for each taxi firm, two pairs of x and y values.

Using $\dfrac{y - y_1}{y_2 - y_1} = \dfrac{x - x_1}{x_2 - x_1}$, find the equation of the line for each taxi firm in the form $y = mx + c$.

Hence work out which taxi firm is cheapest for a journey of 4 miles, and for a journey of 7 miles.

 5 A litre is roughly equivalent to 1.75 pints, and 1 pint is roughly equivalent to 570 millilitres.

The relationship between pints and litres can be modelled as a straight line.

If the number of litres is taken as x and the number of pints as y, find the equation of the line in the form $y = mx + c$.

SUMMARY OF KEY POINTS

› The equation of a straight line may be written in the form $ax + by + c = 0$ where a, b and c are integers.

› You can find the equation of a line using the formula $y - y_1 = m(x - x_1)$, where m is the gradient of the line and (x_1, y_1) is a point on the line. (You need to learn this.)

› You can find the gradient m of the line joining two points with coordinates (x_1, y_1) and (x_2, y_2) by using the formula $m = \dfrac{y_2 - y_1}{x_2 - x_1}$. (You need to learn this.)

› You can find the equation of the line joining two points with coordinates (x_1, y_1) and (x_2, y_2) by using the formula $\dfrac{y - y_1}{y_2 - y_1} = \dfrac{x - x_1}{x_2 - x_1}$. (You need to learn this.)

› Two or more lines are parallel if their gradients, m, are the same.

› A line with a gradient of m_1 is perpendicular to a line with a gradient of m_2 when $m_2 = -\dfrac{1}{m_1}$.

› Two straight lines with gradients m_1 and m_2 are perpendicular when $m_1 m_2 = -1$. (You need to learn this.)

EXAM-STYLE QUESTIONS 4
Answers page 514

1 Line A is parallel to $y = 3x - 5$ and passes through the point $(1, 6)$.

Find the equation of line A. **[3 marks]**

CM 2 **a** Line A passes through $(0, 7)$ and $(2, 3)$ and line B passes through $(6, 4)$ and $(8, 5)$.

Using $m = \dfrac{y_2 - y_1}{x_2 - x_1}$, find the gradients of lines A and B. **[4 marks]**

b Compare the gradients of lines A and B and decide which line is steeper.

Give a reason for your answer. **[2 marks]**

3 Line A is perpendicular to $x - y - 8 = 0$ and passes through the y-intercept of $y = 9x + 5$.

Find the equation of line A. **[4 marks]**

CM 4 The graph below shows the journeys of two different runners.

M Runner A lives 3 miles away from runner B.

They both go out for a run one morning. After 3 hours, they meet at a point 9 miles away from runner B's home.

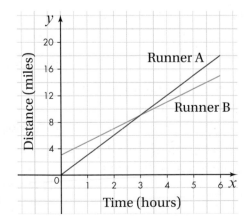

a When the runners meet, how far will runner A have run? **[1 mark]**

b If x is time in hours and y is distance in miles, for each runner, find two pairs of x and y values.

Using $\dfrac{y - y_1}{y_2 - y_1} = \dfrac{x - x_1}{x_2 - x_1}$ find the equation of the line for each runner in the form $y = mx + c$. **[6 marks]**

c Look at the slopes of the two lines on the graph to decide which runner runs faster.

Give a reason for your answer. **[2 marks]**

PS **5** Line A is perpendicular to $y = 2x$ and passes through the point (12, 4).

M
    a Find the equation of line A. **[3 marks]**

    b Find the coordinates of the axis intercepts. **[2 marks]**

    c Find the area of the triangle with vertices at the origin, the x-intercept of line A and the y-intercept of line A. **[2 marks]**

PS **6** The points A(4, 5), B(–2, 8) and C(–10, –8) are the vertices of a triangle.

PF Show that ABC is a right-angled triangle. **[5 marks]**

PS **7** At a fun fair you pay an entrance fee of £5 and then can either pay £2 for each ride or buy a book of tickets for six rides costing £10.

M
The relationship between the number of rides and the overall cost (including the entrance fee) can be modelled as a straight line.

If x is the number of rides and y is the cost, state, for the first payment option, two pairs of x and y values.

Using $\dfrac{y - y_1}{y_2 - y_1} = \dfrac{x - x_1}{x_2 - x_1}$, find the equation of the line for the first option in the form $y = mx + c$.

Hence work out which payment option is cheaper for 4 rides and for 17 rides. **[10 marks]**

PS **8** Line A has a gradient of $\frac{1}{2}$ and passes through the points (8, 5) and (–12, a).

Find the equation of line B, which is perpendicular to line A and passes through (0, a). **[5 marks]**

PS **9** Find the equation of the line, in the form $y = mx + c$, that is perpendicular to the line which passes through the points $(2\sqrt{2}, \sqrt{2})$ and $(\sqrt{2}, 2\sqrt{3})$, and which meets the y-axis at (0, 3). **[5 marks]**

PS **10** The lines $x + 4y - 8 = 0$ and $3x + 5y + 15 = 0$ intersect at point A.

Find the equation of the line that passes through the point of intersection and is perpendicular to $3x + 5y + 15 = 0$. **[7 marks]**

PS **11** A straight line A passes through the points (0, 32) and (5, 41).

CM A straight line B passes through the points (32, 0) and (41, 5).

    a Find the equations of lines A and B in the form $y = mx + c$.

       What do you notice? **[4 marks]**

    b Find the point of intersection of lines A and B. **[3 marks]**

    c On the same pair of axes sketch the graphs of lines A and B, including the point of intersection.

       What transformation links lines A and B? **[3 marks]**

    d If you are told that x in the equation for line A is degrees Celsius and y is degrees Fahrenheit, and vice versa for line B, what can you deduce about the point of intersection? **[1 mark]**

5 COORDINATE GEOMETRY 2: CIRCLES

Ever since humans first looked into the sky and thought about the shape of the Sun and the Moon, they have used circles to solve problems. Do you need to construct the perpendicular bisector of the line segment between two points? Draw two overlapping circles. Do you want your cart to travel smoothly along the road rather than bumping up and down? Try a circular wheel. Coins, lenses, gears, pulleys (see **Chapter 16 Forces**): wherever there has been human technological advancement, a circle is not too far away.

Straight lines and circles are the basis of most sports: a netball court, a football pitch, a racetrack, a snooker table. But why is the circle so ubiquitous? The circle occupies a unique position amongst two-dimensional shapes in that each point on the perimeter is equidistant from the centre of the shape. In netball, in order to shoot you have to be within the semicircle, so the maximum distance is the same the whole way round. This means that there is no possible unfair tactical advantage or disadvantage available.

LEARNING OBJECTIVES

You will learn how to:

> find the centre of the circle and the radius from the equation of a circle

> write the equation of a circle

> find the equation of a circle given different information such as the end points of a diameter

> find the equation of a perpendicular bisector

> find the equation and length of a tangent

> use circle theorems to solve problems.

TOPIC LINKS

In this chapter you will apply the techniques from **Chapter 2 Algebra and functions 2: Equations and inequalities** and **Chapter 4 Coordinate geometry 1: Equations of straight lines**, including finding a gradient, to solve problems involving tangents and radii of circles and points of intersection of circles with straight lines.

PRIOR KNOWLEDGE

You should already know how to:

> write the equation of a circle with a centre at the origin (0, 0) and a radius of r as $x^2 + y^2 = r^2$

> apply circle theorems to find angles

> find the gradient of a line segment

> find a perpendicular gradient

> construct a perpendicular bisector

> find the equation of a line using $y - y_1 = m(x - x_1)$

> write the equation of a line in the form $y = mx + c$

> write the equation of a line in the form $ax + by + c = 0$, where a, b and c are integers

> complete the square on a quadratic function

> use the discriminant $b^2 - 4ac$ to check how many roots a quadratic equation has

> find the point of intersection of two straight lines

> find the points of intersection of a linear graph and a quadratic graph.

You should be able to complete the following questions correctly:

1   Find the gradient of the line segment AB given A(−2, 9) and B(8, 5).

2   Find the equation of the straight line that is perpendicular to the line $y = 10 - 3x$ and which passes through the point (5, 8), giving the answer in the form $ax + by + c = 0$, where a, b and c are integers.

3   Find the coordinates of the point where the lines $y = \frac{2}{3}x - 2$ and $x + 2y = 17$ intersect.

4   Write $x^2 - 14x + 23$ in the form $(x + p)^2 + q$.

5.1 Equations of circles

The equation of a **circle** with its **centre** at the origin (0, 0) and **radius** r has the equation $x^2 + y^2 = r^2$.

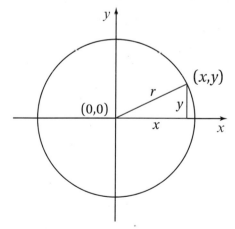

For any point of the circle (x, y) joined to the origin at the centre of the circle:

> x is the horizontal side

> y is the vertical side

> r, the radius, is the hypotenuse.

Hence every single point on the **circumference** of the circle satisfies Pythagoras' theorem.

For example, the circle with equation $x^2 + y^2 = 16$ has its centre at $(0, 0)$ and a radius of $\sqrt{16} = 4$.

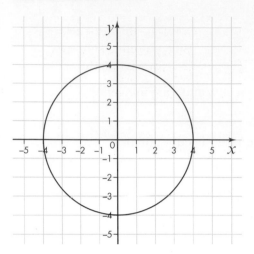

TECHNOLOGY

Draw $x^2 + y^2 = 25$ and $x^2 + y^2 = 36$ using graphing software. You can see how the centre of the circle is always $(0, 0)$ but the radius is the square root of the number on the right-hand side. Predict what $x^2 + y^2 = 81$ will look like, then check.

This can be extended to a circle with any other centre as follows.

Consider a circle with centre (a, b) and radius r.

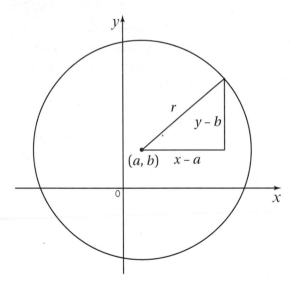

For any point (x, y) on the circumference, a right-angled triangle can be drawn in the same way. This time:

> $(x - a)$ is the horizontal side

> $(y - b)$ is the vertical side

> r, the radius, is the hypotenuse.

Again, every single point on the circumference of the circle satisfies Pythagoras' theorem, so the circle has the equation:

$$(x - a)^2 + (y - b)^2 = r^2$$

KEY INFORMATION

The general equation of a circle is given by
$(x - a)^2 + (y - b)^2 = r^2$.

This circle has a centre at (a, b) and a radius of r.

KEY INFORMATION

A circle's position and size are uniquely defined by its centre and radius.

For example, this circle has a centre at (2, 3) and a radius of 2:

Its equation is $(x - 2)^2 + (y - 3)^2 = 4$.

Note that if one of the coordinates of the centre is negative, such as for a circle with a centre of (38, −16), then the corresponding bracket in the equation will contain a + sign. So, a circle with a centre of (38, −16) and a radius of 24 has the equation $(x - 38)^2 + (y + 16)^2 = 576$.

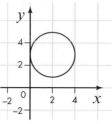

TECHNOLOGY

Draw $(x - 6)^2 + (y + 2)^2 = 9$ using graphing software. You can see that the centre of the circle is (6, −2) and the radius is 3. Predict what $(x + 3)^2 + (y - 5)^2 = 49$ and $(x + 1)^2 + y^2 = 100$ would look like.

Stop and think How does finding the centre of the circle from the equation $(x - a)^2 + (y - b)^2 = r^2$ relate to solving a quadratic equation or to translating a function using f(x) notation such as f(x − 5)? Why is this a requirement of these three situations?

Finding the equation of a circle from its centre and a point on the circumference

Given the centre of the circle and a point on the circumference, you can find the radius by using Pythagoras' theorem.

For coordinates (x_1, y_1) and (x_2, y_2), the length of the line segment between them is given by the formula:

$$d^2 = (x_1 - x_2)^2 + (y_1 - y_2)^2$$

KEY INFORMATION

For coordinates (x_1, y_1) and (x_2, y_2), the length d is given by $d^2 = (x_1 - x_2)^2 + (y_1 - y_2)^2$. This is equivalent to $c^2 = a^2 + b^2$, where the vertices at the ends of the hypotenuse are given as coordinates.

Example 1

The circle C has centre (−7, 13).
The point (5, 4) lies on the circumference of the circle.

Find the equation of the circle.

Solution

Let $(-7, 13) = (x_1, y_1)$ and $(5, 4) = (x_2, y_2)$.

The radius can be found using the formula:

$$d^2 = (x_1 - x_2)^2 + (y_1 - y_2)^2$$
$$r^2 = (-7 - 5)^2 + (13 - 4)^2$$
$$= (-12)^2 + (9)^2$$
$$= 144 + 81$$
$$= 225$$
$$r = \sqrt{225} = 15$$

Note, however, that you only need $r^2 = 225$ for the equation.

TECHNOLOGY

Plot the line segment between the points (1, 5) and (9, 11) using graphing software. This is the hypotenuse of a right-angled triangle. Draw in $y = 5$ and $x = 9$ to complete the triangle. You can therefore use Pythagoras' theorem to find the hypotenuse.

Substitute the centre $(-7, 13)$ and $r^2 = 225$ into the general equation.

$(x + 7)^2 + (y - 13)^2 = 225$

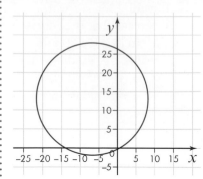

You can also find the centre and radius by completing the square.

Consider the circle with the equation $(x + 11)^2 + (y - 5)^2 = 64$.

Expand the brackets.

$$x^2 + 22x + 121 + y^2 - 10y + 25 = 64$$

Rearrange.

$$x^2 + y^2 + 22x - 10y + 121 + 25 - 64 = 0$$

Simplify.

$$x^2 + y^2 + 22x - 10y + 82 = 0$$

The equation has now been written in the form

$$x^2 + y^2 + 2fx + 2gy + c = 0$$

with $f = 11$, $g = -5$ and $c = 82$.

Note that $f = -a$ and $g = -b$ and $c = a^2 + b^2 - r^2$ and that the coefficients of x^2 and y^2 are the same.

If you are given the equation of a circle in this form, then you can find the centre and radius by rewriting it in the form

$$(x - a)^2 + (y - b)^2 = r^2$$

This can be achieved by completing the square as shown in **Example 2**.

Example 2

Find the centre and radius of the circle which has the equation $x^2 + y^2 - 18x + 10y + 102 = 0$.

Solution

Completing the square on x: $x^2 - 18x = (x - 9)^2 - 81$

Completing the square on y: $y^2 + 10y = (y + 5)^2 - 25$

KEY INFORMATION

When a circle equation is given in the form $x^2 + y^2 + 2fx + 2gy + c = 0$, find the centre and radius by completing the square on x and y.

Hence $x^2 + y^2 - 18x + 10y + 102 = 0$ can be rewritten as

$$(x - 9)^2 - 81 + (y + 5)^2 - 25 + 102 = 0$$

Rearrange.

$$(x - 9)^2 + (y + 5)^2 - 81 - 25 + 102 = 0$$

Simplify.

$$(x - 9)^2 + (y + 5)^2 = 4$$

Comparing the equation with the general equation,
$a = 9$, $b = -5$ and $r = 2$.

The centre of the circle is (9, −5) and the radius is 2.

Exercise 5.1A

Answers page 515

1 State the centre and radius of each circle. Give answers in surds where appropriate.

   a $(x + 5)^2 + (y - 8)^2 = 36$ **b** $(x - 19)^2 + (y - 33)^2 = 400$

   c $x^2 + (y + 4)^2 = 45$ **d** $(x + 3)^2 + (y + 10)^2 = 28$

2 Write the equation of each circle in these two forms:

   $(x - a)^2 + (y - b)^2 = r^2$ and $x^2 + y^2 + 2fx + 2gy + c = 0$, where f, g and c are integers.

   a centre (−5, 9) and radius 7   **b** centre (−11, −1) and radius 13

   c centre (3, 0) and radius $4\sqrt{3}$ **d** centre (14, 6) and radius $2\sqrt{11}$

3 Find the equation of each circle:

   a   **b** 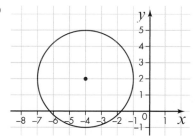   **c**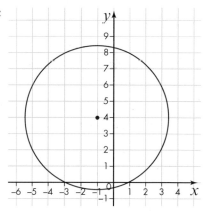

4 Find the centre and radius of each circle:

   a $x^2 + y^2 - 18x + 14y = 14$ **b** $x^2 + y^2 + 8x = 9$

   c $2x^2 + 2y^2 + 20x + 36y + 158 = 0$ **d** $4x^2 + 4y^2 + 120x + 855 = 24y$

5 A circle has centre (−4, 7).

   The point (1, 3) lies on the circumference of the circle.

   Find the equation of the circle in the form $(x - a)^2 + (y - b)^2 = r^2$.

6 Find the equation of the circle with centre (9, 2) that touches the y-axis.

(M) 7 A map is drawn on a coordinate grid, with 1 km divisions on both axes. A youth hostel is located on the map at (1, 3) and a theatre is located at (4, 7). The distance between these locations is given as d km. Find the coordinates of the two locations on the x-axis which are distance d km from the youth hostel.

(PS) 8 A circle has the equation $x^2 + y^2 + px + 6y = 96$.

Given that the circle has a radius of 11 and that p is a positive constant, find the distance of the centre of the circle from the origin.

Finding the equation from the ends of a diameter

If you are given two points on the circumference which are the ends of a **diameter** of a circle, then the centre of the circle can be found by determining the **midpoint** of the line segment between the two points.

The midpoint of (x_1, y_1) and (x_2, y_2) is given by $\left(\dfrac{x_1 + x_2}{2}, \dfrac{y_1 + y_2}{2} \right)$.

This is equivalent to finding the mean of the x-values and the mean of the y-values.

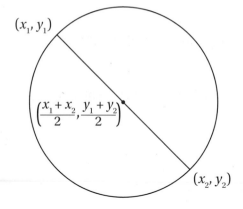

> **TECHNOLOGY**
>
> Plot the line segment between the points (1, 5) and (9, 11) using graphing software. Where is the midpoint? Plot the points (4, 7) and (−2, 13). Predict where the midpoint will be.

> **KEY INFORMATION**
>
> The midpoint of a line segment is given by
> $$\left(\frac{x_1 + x_2}{2}, \frac{y_1 + y_2}{2} \right).$$

Example 3

a Find the centre and radius of the circle C which has a diameter joining the points A(43, 11) and B(13, 27).

b Hence state the equation of C.

Solution

a The centre is the midpoint of the line segment between the points A and B.

Centre $= \left(\dfrac{43 + 13}{2}, \dfrac{11 + 27}{2} \right) = \left(\dfrac{56}{2}, \dfrac{38}{2} \right) = (28, 19)$

The radius is the distance between the centre and a point on the circumference.

Let the centre $(28, 19) = (x_1, y_1)$ and A$(43, 11) = (x_2, y_2)$.

$$d^2 = (x_1 - x_2)^2 + (y_1 - y_2)^2$$
$$r^2 = (28 - 43)^2 + (19 - 11)^2$$
$$= (-15)^2 + (8)^2$$
$$= 225 + 64$$
$$= 289$$
$$r = \sqrt{289} = 17$$

b The circle has the equation $(x - 28)^2 + (y - 19)^2 = 289$.

Exercise 5.1B

Answers page 515

1 The points A$(-17, 25)$ and B$(27, -8)$ are the two ends of the diameter AB of a circle C. Find the length of the diameter.

2 A circle C has diameter DE where D and E are given by $(2, 9)$ and $(14, -7)$.

a Find the centre of C.   **b** Find the radius of C.

c Find the equation of C, giving the answer in the form $(x - a)^2 + (y - b)^2 = r^2$.

3 Find the equation of the circle which has a diameter joining the points $(-3, 8)$ and $(21, -2)$.

CM **PS** **4** A student answered the following question:

'A circle C has diameter AB where A is $(17, 11)$ and B is $(-18, 23)$. Find the equation of the circle.'

Her solution is given below:

$$d^2 = (x_1 - x_2)^2 + (y_1 - y_2)^2$$
$$(17 - -18)^2 + (11 - 23)^2 = 1369$$
$$\text{Centre} = \left(\frac{17-18}{2}, \frac{11+23}{2} \right) = (-0.5, 17)$$
$$\text{Equation is } (x + 0.5)^2 + (y - 17)^2 = 1369$$

a What mistake did the student make?

b What is the correct answer to the question?

PF **5** A circle C has diameter FG and centre X.

Given that F has coordinates $(9, 5)$ and X has coordinates $(5, -3)$, find:

a the coordinates of G

b the radius of C in the form $a\sqrt{5}$

c the equation of C.

d Verify that the point H$(-3, 1)$ lies on C.

6 The line segment joining the points S(−9, 4) and T(5, 10) is a diameter of the circle C.

The line segment joining the points U(1, 14) and V(p, q) is also a diameter of the circle C.

   a Find the values of p and q.

   b Find the equation of the circle.

(PF) 7 Line L has the equation $y = 2x + 14$.

Circle C has the equation $x^2 + y^2 + 6x - 16y = 52$.

L intersects C at points M and N.

Prove that MN is a diameter of the circle.

5.2 Angles in a semicircle

Consider three points A, B and P, on the circumference of a circle C with AB as a diameter, as shown in the diagram. The centre of the circle is O.

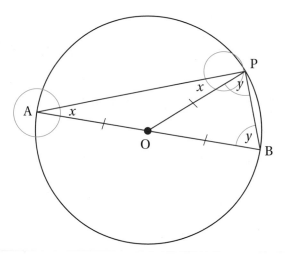

AOP and BOP are both isosceles triangles with OA = OP = OB because they are all radii of C.

Hence ∠OAP and ∠OPA in triangle AOP are the same as each other. Label each one x.

Also ∠OBP and ∠OPB in triangle BOP are the same as each other. Label each one y.

So ∠APB is given by $(x + y)$.

The sum of the interior angles in the triangle APB is given by $(x + x + y + y) = 2(x + y)$.

But the sum of the interior angles in any triangle is 180°.

Hence $2(x + y) = 180°$.

From which $(x + y) = 90°$.

PROOF

This is an example of a geometric proof. By using the fact that the radii are always the same length, you can show that the two triangles are both isosceles and hence that there are pairs of angles that are equal to each other. You can then deduce that ∠APB = 90°.

KEY INFORMATION

For three points on the circumference, if two points are on a diameter, then the triangle is right-angled.

KEY INFORMATION

To show a triangle is right-angled, show that the sides satisfy Pythagoras' theorem or show that the gradients of two sides are perpendicular.

KEY INFORMATION

The rule for perpendicular gradients, m_1 and m_2, is given by $m_1 \times m_2 = -1$.

Therefore $\angle APB = 90°$. In general, any triangle drawn in a circle with two points on a diameter and the third point somewhere on the circumference will be a right-angled triangle.

Since a diameter divides a circle into two equal semicircles, this fact is often stated as 'The angle in a semicircle is a right angle'.

The converse is also true. If you are told that $\angle APB$ is a right angle, then AB is a diameter of a circle with P on the circumference.

Given three points, there are two ways of showing that they are the vertices of a right-angled triangle. The first way is to show that the three sides of the triangle satisfy Pythagoras' theorem, as in **Example 4**. The second way is to show that two of the gradients of the sides satisfy the rule for perpendicular gradients (i.e. $m_1 \times m_2 = -1$), as in **Example 5**.

> **TECHNOLOGY**
>
> If you have two lines drawn accurately using graphing software, you can count squares to check if they are perpendicular. For example, 2 squares right and 1 square down is perpendicular to 2 squares down and 1 square left.

Example 4

Points P, Q and R are given by P(−5, −2), Q(7, −8) and R(3, −16).

a Find:

   i PQ^2 **ii** QR^2 **iii** PR^2

b Hence explain why PR is a diameter of a circle which also has Q on the circumference.

Solution

a Use $d^2 = (x_1 - x_2)^2 + (y_1 - y_2)^2$

   i PQ^2: $d^2 = (-5 - 7)^2 + (-2 - (-8))^2$
              $= (-12)^2 + (6)^2 = 180$

   ii QR^2: $d^2 = (7 - 3)^2 + (-8 - (-16))^2$
              $= (4)^2 + (8)^2 = 80$

   iii PR^2: $d^2 = (-5 - 3)^2 + (-2 - (-16))^2$
               $= (-8)^2 + (14)^2 = 260$

b Since $PQ^2 + QR^2 = PR^2$ $(180 + 80 = 260)$, PQR obeys Pythagoras' theorem and is a right-angled triangle.

   PR is the longest side so it is the hypotenuse of the triangle.

   Since the angle in a semicircle is a right angle, PR is a diameter of a circle with Q on the circumference.

Alternatively:

   Use graphing software to join the points P(−5, −2), Q(7, −8) and R(3, −16) to make a triangle and draw the circle $(x + 1)^2 + (y + 9)^2 = 65$, as shown in the diagram.

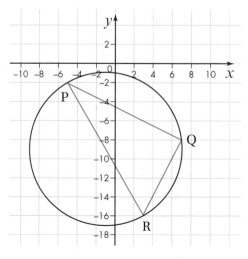

Example 5

Points S, T and U are given by S(7, 8), T(33, −10) and
U(15, −16).

a By considering the gradients of ST, TU and SU, show that
STU is a right-angled triangle.

b Find the equation of the circle C that passes through S, T
and U.

c Given that the point P on the circumference of C is given by
(x, y), show that $(x − 7)(x − 33) + (y − 8)(y + 10) = 0$.

> In practice, only two of the gradients are required to demonstrate the triangle is right-angled.

Solution

a Use gradient $= \dfrac{y_1 - y_2}{x_1 - x_2}$

Gradient of ST $= \dfrac{8 - (-10)}{7 - 33} = \dfrac{18}{-26} = -\dfrac{9}{13}$

Gradient of TU $= \dfrac{(-10) - (-16)}{33 - 15} = \dfrac{6}{18} = \dfrac{1}{3}$

Gradient of SU $= \dfrac{8 - (-16)}{7 - 15} = \dfrac{24}{-8} = -3$

Since the gradients of TU and SU satisfy $m_1 \times m_2 = -1$,
TU and SU are perpendicular and STU is a right-angled
triangle.

b Since TU and SU are perpendicular, ST is the diameter of
the circle.

The centre of the circle is the midpoint of ST.

Midpoint $= \left(\dfrac{7 + 33}{2}, \dfrac{8 - 10}{2} \right) = (20, -1)$.

The radius of a circle is the distance between the centre and
a point on the circumference. Any point will do, such as S.

$$d^2 = (x_1 - x_2)^2 + (y_1 - y_2)^2$$
$$r^2 = (7 - 20)^2 + (8 - (-1))^2$$
$$= (-13)^2 + (9)^2 = 250$$

Hence the centre of the circle is (20, −1) and $r^2 = 250$.

The equation of the circle is given by $(x - 20)^2 + (y + 1)^2 = 250$.

c For any point P(x, y) on the circumference, other than S or
T, PST will be a right-angled triangle and PS and PT will be
perpendicular.

Gradient of PS $= \dfrac{y - 8}{x - 7}$

Gradient of PT $= \dfrac{y - (-10)}{x - 33} = \dfrac{y + 10}{x - 33}$

Use the rule for perpendicular gradients, $m_1 \times m_2 = -1$.

$$\frac{y-8}{x-7} \times \frac{y+10}{x-33} = -1$$

Multiply both sides by $(x-7)(x-33)$.

$$(y-8)(y+10) = -(x-7)(x-33)$$

Add $(x-7)(x-33)$ to both sides.

$(y-8)(y+10) + (x-7)(x-33) = 0$, as required.

This leads to another general equation for a circle based upon the ends of one of its diameters.

For a circle with points (a, b) and (c, d) as the ends of a diameter, its equation can be written as:

$$(x-a)(x-c) + (y-b)(y-d) = 0$$

Exercise 5.2A

Answers page 516

(PF) **1** Points D(–1, 11), E(1, 5) and F(13, 9) lie on the circumference of a circle.

    a Show that $DE^2 + EF^2 = DF^2$.

    b By considering the gradients of DE and EF, show that DE and EF are perpendicular.

    c Find the equation of the circle.

(M) **2** Points P, Q and R are given by (–5, –4), (11, 8) and (13, 2) respectively.

(PF)   **a** Prove that the triangle PQR is right-angled.

A company's logo is designed such that P, Q and R are points on the circumference of a circle C with the triangle PQR cut from C as shown. The logo is to be painted on the side of the company's headquarters and each square is 1 m across. The shaded area will be green and the triangle will be white.

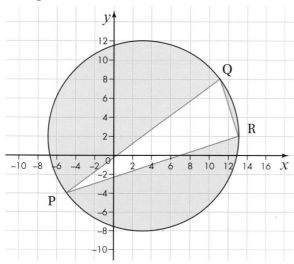

b Find the exact area of the green region.

Given that the green paint costs £4 for a tin that will coat $5\,\text{m}^2$:

c Find the total cost of the green paint required to paint the logo.

3 Points A, B and C lie on the circumference of a circle such that AB is a diameter.

AB $= (2x + 5)$ cm, AC $= (4x - 9)$ cm and BC $= (14 - x)$ cm.

Find the lengths of both possible radii.

PF 4 Points A$(-5, 2)$, B$(1, 0)$ and C$(-3, q)$ lie on the circle $(x + 2)^2 + (y - 1)2 = r^2$, where $q > 0$.

a Find the value of r^2.

b Find the value of q.

c Find the gradients of AB, AC and BC.

d Hence prove that AB is a diameter of the circle.

PF 5 Points S$(5, 6)$ and T$(-10, 1)$ lie on the straight line L_1.

Points U$(9, 4)$ and V$(11, -2)$ lie on the straight line L_2.

a Show that L_1 and L_2 are perpendicular.

L_1 and L_2 intersect at point W.

S, V and W lie on the circumference of a circle C.

b Find the equation of C in the form $x^2 + y^2 + 2fx + 2gy + c = 0$, where f, g and c are integers.

PS 6 Points P$(1, (t + 7))$, Q$(3, (t + 11))$ and R$((t + 17), 3)$ lie on the circumference of circle C_1 such that PR is a diameter.

a Find the equation of the circle in the form $(x - a)^2 + (y - b)^2 = r^2$.

Points J$(-11, 6)$, K$(-1, (p - 8))$ and L$(7, 22)$ lie on the circumference of circle C_2 such that JL is a diameter.

b Find both possible values for p.

5.3 Perpendicular from the centre to a chord

A chord of a circle is any straight line connecting two points on the circumference. A diameter is a special type of chord which passes through the centre of the circle; it is longer than any other chord.

Consider two points, A and B, on the circumference of a circle C, with AB as a chord, as shown in the diagram. The centre of the circle is O and the angle bisector of $\angle$AOB intersects the chord at M.

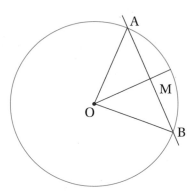

OAM and OBM are congruent triangles:

> OA and OB are the same length because they are both radii of C.

> OM is common to both triangles.

> ∠AOM and ∠BOM are the same because OM is the angle bisector of ∠AOB.

Since two sides and the included angle are common to both triangles, these two triangles are congruent.

As a result, AM and BM are the same length as each other, so M is the midpoint of the chord, and ∠AMO and ∠BMO are the same as each other. Because ∠AMO and ∠BMO are the two angles on a straight line, they sum to 180° and therefore each of them is 90°.

This is often stated as 'the perpendicular from the centre of the circle to a chord bisects the chord'. Essentially, this means that if the midpoint of a chord is joined to the centre of the circle by a straight line L, then the chord and L are perpendicular.

Recall the concept of a perpendicular bisector, which is the locus of all the points equidistant from two points. The perpendicular bisector of a chord will pass through the centre of the circle.

If you are given the ends of a chord then it is possible to find the equation of the straight line that passes through the centre of the circle using this process:

> Find the gradient of the chord.

> Find the gradient of a line perpendicular to the chord.

> Find the midpoint of the chord.

> Substitute the perpendicular gradient, m, and the x- and y-coordinates from the midpoint (x_1, y_1) into the formula $y - y_1 = m(x - x_1)$.

PROOF

This is another example of a geometric proof. Congruent triangles are identical. They have the same sides and the same angles. In this proof, the initial set-up includes two triangles, and by demonstrating that the triangles are congruent you can then make deductions about the point at which the radius intersects the chord.

KEY INFORMATION

The perpendicular from the centre of the circle to a chord bisects the chord.

Example 6

The points G(9, –2) and H(21, 6) are such that GH is a chord of a circle C.

The centre of C has x-coordinate 3.

Find the equation of C.

Solution

Use gradient $= \dfrac{y_1 - y_2}{x_1 - x_2}$.

Gradient of GH $= \dfrac{-2 - 6}{9 - 21} = \dfrac{-8}{-12} = \dfrac{2}{3}$

Gradient of the perpendicular bisector $= -\dfrac{3}{2}$

Midpoint of GH $= \left(\dfrac{9 + 21}{2}, \dfrac{-2 + 6}{2} \right) = (15, 2)$

The equation of the straight line passing through the centre of the circle is given by $y - 2 = -\dfrac{3}{2}(x - 15)$.

Given that the centre of the circle has x-coordinate 3, you can find the y-coordinate by substitution.

When $x = 3$:

$$y - 2 = -\frac{3}{2}(3 - 15)$$

$$= -\frac{3}{2} \times -12$$

$$= 18$$

$$y = 20$$

To find a perpendicular gradient, remember the rule, 'reciprocate and change the sign'. The reciprocal of $\dfrac{a}{b}$ is $\dfrac{b}{a}$. For example, if a line has a gradient of $\frac{5}{3}$, then the perpendicular line has a gradient of $-\frac{3}{5}$. Similarly, if a line has a integer gradient such as 4, then the perpendicular line has a gradient of $-\frac{1}{4}$.

Now you have the centre, (3, 20), you can find r^2. You can use G as the point on the circumference.

$$d^2 = (x_1 - x_2)^2 + (y_1 - y_2)^2$$

$$r^2 = (3 - 9)^2 + (20 - (-2))^2$$

$$= (-6)^2 + (22)^2 = 520$$

Hence the centre of the circle is (3, 20) and $r^2 = 520$.

The equation of the circle is given by $(x - 3)^2 + (y - 20)^2 = 520$.

Example 7

Find the equation of the circle that passes through the points D(9, 9), E(–3, 7) and F(16, 4).

Solution

The centre of the circle will lie on the perpendicular bisector of the two points at the ends of any chord.

Therefore, if you find the equations of two perpendicular bisectors, you can solve them simultaneously to find the centre of the circle.

Given three points, there are three possible perpendicular bisectors to choose from, but you only need two, for example DE and DF.

For DE:

Gradient of DE $= \dfrac{9-7}{9-(-3)} = \dfrac{2}{12} = \dfrac{1}{6}$

Gradient of the perpendicular bisector $= -6$

Midpoint of DE $= \left(\dfrac{9-3}{2}, \dfrac{9+7}{2}\right) = (3, 8)$

The equation of the first straight line passing through the centre of the circle is given by $y - 8 = -6(x - 3)$.

Hence $y = -6(x - 3) + 8$

For DF:

Gradient of DF $= \dfrac{9-4}{9-16} = -\dfrac{5}{7}$

Gradient of the perpendicular bisector $= \dfrac{7}{5}$

Midpoint of DE $= \left(\dfrac{9+16}{2}, \dfrac{9+4}{2}\right) = \left(\dfrac{25}{2}, \dfrac{13}{2}\right)$

The equation of the second straight line passing through the centre of the circle is given by $y - \dfrac{13}{2} = \dfrac{7}{5}\left(x - \dfrac{25}{2}\right)$

Hence $y = \dfrac{7}{5}\left(x - \dfrac{25}{2}\right) + \dfrac{13}{2}$

Put the equations equal.

$$-6(x - 3) + 8 = \dfrac{7}{5}\left(x - \dfrac{25}{2}\right) + \dfrac{13}{2}$$

$$-6x + 18 + 8 = \dfrac{7}{5}x - \dfrac{35}{2} + \dfrac{13}{2}$$

Multiply through by 10.

$$-60x + 180 + 80 = 14x - 175 + 65$$

$$-60x + 260 = 14x - 110$$

$$370 = 74x$$

$$x = 5$$

Substitute for $x = 5$ in $y = -6(x - 3) + 8$.

$$y = -6(5 - 3) + 8$$

$$= -6 \times 2 + 8$$

$$= -4$$

Now you have the centre, $(5, -4)$, you can find r^2. You can use D as the point on the circumference.

$$d^2 = (x_1 - x_2)^2 + (y_1 - y_2)^2$$

$$r^2 = (5 - 9)^2 + (-4 - 9)^2$$

$$= (-4)^2 + (-13)^2$$

$$= 185$$

Hence the centre of the circle is $(5, -4)$ and $r^2 = 185$.

The equation of the circle is given by $(x - 5)^2 + (y + 4)^2 = 185$.

Exercise 5.3A

Answers page 516

1 D(9, 5) and E(7, –3) are the two ends of a chord on circle C.

F is the midpoint of the chord DE.

G is the centre of C. The x-coordinate of G is 4.

a Find the coordinates of F.

b Find the gradient of the chord DE.

c Find the gradient of FG.

d Find the equation of the straight line that passes through F and G.

e Verify that the y-coordinate of G is 2.

f Find the equation of C.

2 JK is a chord of a circle. J has coordinates (3, 7) and K has coordinates (10, 2).

The y-coordinate of the centre is 8.

Find the equation of the circle.

3 A circle has a chord UV for which U has coordinates $(10, 11)$ and the midpoint of the chord, X, has coordinates $(6, 3)$.

   a Find the coordinates of V.

The centre of the circle has x-coordinate 8.

   b Find the equation of the circle.

(M) 4 A ferris wheel is modelled as a circle with centre $(10, 25)$ m and a radius of 17 m. There is a horizontal platform 15 m below the centre for people to board the wheel. This platform extends for 6 m either side of the circumference of the wheel. How long is the platform?

5 A circle has the equation $x^2 + y^2 + 6x + 10y = 151$.

X is the centre of the circle.

   a Find the coordinates of X.

The line $y = 6$ intersects the circle at points P and Q.

   b Find the length of the chord PQ.

   c Find the area of the triangle PQX.

(CM) 6 Points $A(-4, 3)$, $B(2, 7)$ and $C(10, -5)$ lie on the circumference of a circle.

M and N are the midpoints of AB and BC respectively.

   a Find the equation of the perpendicular bisector of AB, giving the answer in the form $ax + by = c$, where a, b and c are positive integers.

   b Find the equation of the perpendicular bisector of BC, giving the answer in the form $ax = by + c$, where a, b and c are positive integers.

The perpendicular bisectors of AB and BC intersect at X.

   c State the size of $\angle MXN$.

   d Hence find the size of $\angle ABC$.

   e What does your answer to **part d** tell you about the line AC?

7 Find the equation of each circle. For each circle, you are given three points on its circumference.

   a $(12, 8)$, $(11, 1)$ and $(20, 4)$

   b $(3, 8)$, $(1, 4)$ and $(-6, 5)$

   c $(19, 4)$, $(17, 0)$ and $(4, -1)$

   d $(-5, -23)$, $(-17, -19)$ and $(-1, -3)$

8 Circle A has the equation $(x + 6)^2 + (y - 6)^2 = 145$.

Circle B has the equation $(x - 8)^2 + (y + 1)^2 = 40$.

   a Find the points of intersection of the two circles.

   b Find the length of the chord shared by the two circles, writing the answer in the form $a\sqrt{5}$.

5.4 Radius perpendicular to the tangent

A tangent is a straight line which just touches a curve at a single point.

Consider the tangent TA where T is the point of contact of the circle C and A is any other point on the tangent, as shown in the diagram. The centre of the circle is O.

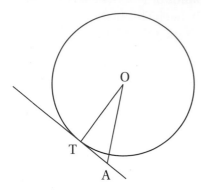

Assume that the perpendicular drawn from O intersects the tangent at A. Hence ∠OAT is a right angle and OT is the hypotenuse of the right-angled triangle OAT, so OT is longer than OA.

But clearly OA is longer than OT because A lies outside the circle. So ∠OAT cannot be a right angle. Since A is any point other than T, then ∠OTA must be a right angle.

You may also observe that since this argument could have been constructed on the other side of T as well, ∠OTA would have to have been acute in both cases, which is a contradiction as the two angles must sum to 180°.

The tangent can also be considered to be the limit of the chord, from the previous proof in which it was shown that the chord and the radius that bisected it were perpendicular.

This theorem is often stated as 'A tangent will always be perpendicular to the radius at the point of contact'.

If you are given the centre, X, of a circle and a point, A, on the circumference then it is possible to find the equation of the tangent at that point using this process:

> Find the gradient of the radius XA.

> Find the gradient of a line perpendicular to XA (this will be the gradient of the tangent).

> Substitute the perpendicular gradient, m, and the x- and y-coordinates from A(x_1, y_1) into the formula $y - y_1 = m(x - x_1)$.

PROOF

This is another example of a geometric proof, this time by *contradiction*, where you start with an initial assumption which you then show cannot be true by a series of logical steps. In this example, you start with an initial assumption that OA is perpendicular to the tangent and then show that this would make OT longer than OA, which is impossible. Hence the initial assumption is incorrect and OA cannot be perpendicular to the tangent, leaving OT as the only possible perpendicular.

KEY INFORMATION

A radius and tangent are perpendicular at the point of contact.

Example 8

A circle C has centre (2, 8). Point T(7, 5) is on the circumference of the circle.

Find the equation of the tangent to the circle that passes through T.

Give the answer in the form $ax + by + c = 0$.

Solution

Gradient of radius $= \dfrac{8-5}{2-7} = -\dfrac{3}{5}$

Gradient of the tangent $= \dfrac{5}{3}$

The equation of the tangent is given by $y - 5 = \dfrac{5}{3}(x - 7)$.

Multiply both sides by 3.

$$3y - 15 = 5(x - 7)$$
$$= 5x - 35$$
$$-5x + 3y + 20 = 0$$

Example 9

Find the length of the tangent from the point P(6, –6) to the circle C with equation $(x + 5)^2 + (y - 4)^2 = 25$.

Solution

Draw a diagram. Let X be the centre of the circle and T be the point where the tangent touches the circle.

From the equation of the circle, X has coordinates (–5, 4) and the radius is 5.

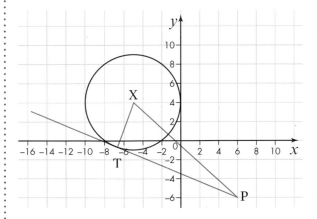

The question has asked for the length of PT. Since the radius and tangent are perpendicular, PXT is a right-angled triangle and you can use Pythagoras' theorem to find PT. XT is the radius and you can find PX since you know the coordinates for both P and X.

KEY INFORMATION

Use Pythagoras' theorem to find the length of a tangent.

For PX, use $d^2 = (x_1 - x_2)^2 + (y_1 - y_2)^2$.

$PX^2 = (6 - (-5))^2 + (-6 - 4)^2$

$\qquad = (11)^2 + (-10)^2 = 221$

Apply Pythagoras' theorem to PXT.

$PT^2 = PX^2 - XT^2$

$\qquad = 221 - 5^2$

$\qquad = 196$

$PT = \sqrt{196} = 14$

The length of the tangent is 14 units.

> **Stop and think**
>
> How could you find the coordinates of the point T where the tangent touches the circle? How many possible pairs of coordinates are there for T? Why is it not necessary to know the position of T to answer the question?

Exercise 5.4A

Answers page 516

1 A circle C has centre (4, 6). The point A(7, 8) lies on the circumference of C.

Find the equation of the tangent to C at point A.

Give the answer in the form $ax + by = c$, where a, b and c are positive integers.

(PF) 2 A circle has the equation $x^2 + (y - 1)^2 = 13$.

a Show that the point T(2, −2) lies on the circle.

b Prove that the tangent at T has the equation $2x = 3y + 10$.

(M) 3 A plaza is modelled on a metre square grid using the equation $(x - 3)^2 + (y - 6)^2 = 64$.

a State the centre of the plaza.

b State the radius of the plaza.

A straight path is to be built from the point (15, 7) to the edge of the plaza such that the path is a tangent of the circle. The path stops when it reaches the plaza.

c Find the length of the path.

(PS) 4 Points A(4, 7) and B(10, 7) are such that AB is a chord of a circle of centre X(7, 11).

Two tangents touch the circle at A and B.

Q is the point of intersection of the two tangents.

a Find the coordinates of Q.

b Find the area of the quadrilateral AXBQ.

5 A circle has the equation $x^2 + y^2 + 8x - 2y - 64 = 0$.

A tangent of the circle passes through the point P(7, 4).

Find the length of the tangent.

6 A circle has the equation $x^2 + y^2 + 354 = 30x + 46y$.

X is the centre of the circle.

A tangent is drawn from the point P(10, q) to the circle at the point T(31, 35).

a Find the value of q.

b Show that $PT^2 + TX^2 = PX^2$.

c Find the area of the triangle PXT.

 a Find the equation of the circle with centre (5, 11) for which the line with equation $y = x + 2$ is a tangent.

b Find the equation of the circle with centre (−8, 23) for which the line with equation $y = 3x + 5$ is a tangent.

If you are trying to find the tangent to a circle with a given gradient, there will be two such tangents, parallel and on opposite ends of a diameter. There are two common methods to find the equations.

Method 1
Given the gradient of the tangent and the equation of the circle, it is possible to find both points where the two tangents would touch the circle:

› Find the perpendicular gradient, which will be the gradient of the diameter.

› Hence find the equation of the diameter (since you know the coordinates of the centre of the circle).

› Solve simultaneously using the equation of the diameter and the equation of the circle.

› Once you know the points where the tangents meet the circle, you can find the equations of the tangents.

Method 2
When a quadratic curve intersects the x-axis at two distinct real roots, the value of $b^2 - 4ac$ is positive. When, however, the curve touches the x-axis without intersecting it, so that the x-axis is a tangent, then the value of $b^2 - 4ac$ is zero. This is a general rule that can be applied to any situation where a line is tangent to a curve. Rather than obtaining two distinct real roots, you get the same real root twice, sometimes called repeated roots.

Hence, to show that a straight line is a tangent to a circle, you can solve their equations simultaneously and show that you obtain repeated roots.

Both methods are demonstrated in **Example 10**.

KEY INFORMATION
$b^2 - 4ac = 0$ for repeated roots such as where a tangent touches a curve.

Example 10

Show that the line $y = 2x + 3$ is a tangent to the circle with equation $(x - 10)^2 + (y - 8)^2 = 45$.

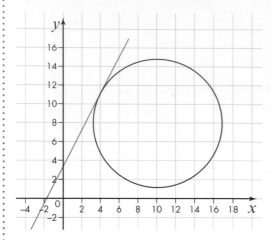

Solution

Method 1

The centre of the circle is (10, 8).

The gradient of the tangent is 2, so the gradient of the diameter is $-\frac{1}{2}$.

The equation of the diameter is given by $y - 8 = -\frac{1}{2}(x - 10)$, which simplifies to $x = 26 - 2y$.

Substitute $x = 26 - 2y$ into the equation of the circle.

$$(x - 10)^2 + (y - 8)^2 = 45$$

$$(26 - 2y - 10)^2 + (y - 8)^2 = 45$$

$$(16 - 2y)^2 + (y - 8)^2 = 45$$

Expand and write in the form $ax^2 + bx + c = 0$.

$$256 - 64y + 4y^2 + y^2 - 16y + 64 = 45$$

$$5y^2 - 80y + 275 = 0$$

Simplify, factorise and solve.

$$y^2 - 16y + 55 = 0$$

$$(y - 5)(y - 11) = 0$$

$$y = 5 \text{ or } 11$$

Substitute for y in $x = 26 - 2y$.

When $y = 5$, $x = 26 - 2 \times 5 = 16$

When $y = 11$, $x = 26 - 2 \times 11 = 4$

The gradient of the tangent is 2.

For Method 1 you use the gradient of the tangent only to find the equation of the line perpendicular to it through the centre. You substitute this equation into the circle equation to find the points of intersection. You then show that one of the two possible y-intercepts is 3 and hence one of the two possible tangents has the equation $y = 2x + 3$.

For the point $(16, 5)$, $y - 5 = 2(x - 16)$ which gives $y = 2x - 27$.

For the point $(4, 11)$, $y - 11 = 2(x - 4)$ which gives $y = 2x + 3$.

Hence $y = 2x + 3$ is a tangent to the circle.

Method 2

Solve simultaneously.

Substitute $y = 2x + 3$ into the equation of the circle.

$$(x - 10)^2 + (y - 8)^2 = 45$$

$$(x - 10)^2 + (2x + 3 - 8)^2 = 45$$

$$(x - 10)^2 + (2x - 5)^2 = 45$$

Expand.

$$x^2 - 20x + 100 + 4x^2 - 20x + 25 = 45$$

Collect like terms and make the right-hand side zero.

$$5x^2 - 40x + 80 = 0$$

Simplify by dividing through by 5.

$$x^2 - 8x + 16 = 0$$

Factorise.

$$(x - 4)^2 = 0$$

Hence the equation has two repeated roots ($x = 4$), which demonstrates that $y = 2x + 3$ is a tangent to the circle.

> For Method 2, you substitute the tangent equation, $y = 2x + 3$, into the circle equation and demonstrate that there is only one solution (due to the repeated roots) and hence only one point of intersection with the circle.

Stop and think   Which method do you think is more efficient? When would each method be preferable to the other?

Example 11

The circle with equation $x^2 + y^2 - 16x + 2y = 39$ has two tangents with a gradient of 5.

Find the equation of each tangent in the form $y = 5x + c$.

Solution

Solve simultaneously by substituting $y = 5x + c$ into the equation of the circle.

$$x^2 + (5x + c)^2 - 16x + 2(5x + c) = 39$$

Expand and write in the form $ax^2 + bx + c = 0$.

$$x^2 + 25x^2 + 10xc + c^2 - 16x + 10x + 2c = 39$$

$$26x^2 + (10c - 6)x + (c^2 + 2c - 39) = 0$$

For $y = 5x + c$ to be a tangent to the circle, $b^2 - 4ac = 0$.

$$b^2 - 4ac = 0$$

$$(10c - 6)^2 - 4 \times 26 \times (c^2 + 2c - 39) = 0$$

$$100c^2 - 120c + 36 - 104c^2 - 208c + 4056 = 0$$

$$-4c^2 - 328c + 4092 = 0$$

$$c^2 + 82c - 1023 = 0$$

$$(c - 11)(c + 93) = 0$$

$$c = 11 \text{ or } -93$$

The tangents are $y = 5x + 11$ and $y = 5x - 93$.

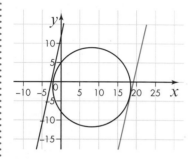

TECHNOLOGY

Plot the circle and tangent using a graphic calculator or graphing software to help you. Note that just demonstrating for yourself that a line appears to be a tangent is *not* a proof, and you need to show a logical and rigorous series of steps leading to a conclusion to actually prove that the line is a tangent to the circle.

Exercise 5.4B

Answers page 517

1 A circle has the equation $x^2 + y^2 = 50$.

Show that each of these straight lines is a tangent to the circle.

a $y = x - 10$ **b** $7y = x + 50$

(PF) 2 Show that the line $y = 4x - 5$ is a tangent to the circle $x^2 + y^2 + 10x - 18y + 38 = 0$.

3 The circle with equation $x^2 + y^2 - 10x - 6y - 6 = 0$ has two tangents with an equation of the form $y = 3x + c$.

Find both possible values of c.

4 The circle with equation $x^2 + y^2 - 12x - 26y + 125 = 0$ has two tangents with a gradient of $-\frac{1}{2}$.

Find the equation of each tangent.

5 The circle with equation $(x - 18)^2 + (y - 10)^2 = 52$ has two tangents with a y-intercept of 24.

Find the equation of each tangent in the form $ax + by + c = 0$, where a, b and c are integers.

SUMMARY OF KEY POINTS

- The general equation of a circle is given by $(x - a)^2 + (y - b)^2 = r^2$:

  - this circle has a centre at (a, b) and a radius of r.

- When a circle equation is given in the form $x^2 + y^2 + 2fx + 2gy + c = 0$, find the centre and radius by completing the square on x and y:

  - For coordinates (x_1, y_1) and (x_2, y_2), the length d is given by $d^2 = (x_1 - x_2)^2 + (y_1 - y_2)^2$. This is equivalent to $c^2 = a^2 + b^2$ where the vertices at the ends of the hypotenuse are given as coordinates.

  - The midpoint of a line segment is given by $\left(\dfrac{x_1 + x_2}{2}, \dfrac{y_1 + y_2}{2} \right)$.

  - The angle in a semicircle is a right angle.

  - Hence, for three points on the circumference, if two points are on a diameter, then the triangle is right-angled.

- To show a triangle is right-angled, show that the sides satisfy Pythagoras' theorem or show that the gradients of two sides are perpendicular. The rule for perpendicular gradients, m_1 and m_2, is given by $m_1 \times m_2 = -1$:

  - The perpendicular from the centre of the circle to a chord bisects the chord.

- A radius and tangent are perpendicular at the point of contact.

- Use Pythagoras' theorem to find the length of a tangent.

- $b^2 - 4ac = 0$ for repeated roots, such as where a tangent touches a curve.

EXAM-STYLE QUESTIONS 5

Answers page 517

 1 A circle has the equation $(x - 9)^2 + (y + 8)^2 = 484$.

    a Gemma says that the centre of the circle is the point $(-9, 8)$.

       Explain why Gemma is incorrect. **[1 mark]**

    b Olivia says that the diameter of the circle is 22.

       Explain why Olivia is incorrect. **[1 mark]**

2 Find the equation of the circle which has a centre of $(7, -3)$ and a diameter of 16.

    Write the answer in the form $x^2 + y^2 + 2fx + 2gy + c = 0$, where f, g and c are integers. **[4 marks]**

3 Identify the coordinates of the four points at which the circle $(x + 5)^2 + (y - 7)^2 = 50$ intersects the x- and y-axes. **[4 marks]**

4 Dilgusha wishes to find the centre and radius of the circle C with the equation $x^2 + y^2 + 6x - 14y + 22 = 0$.

Her working is as follows:

$$x^2 + y^2 + 6x - 14y + 22 = 0$$
$$(x + 6)^2 - 6^2 + (y - 14)^2 + 14^2 + 22 = 0$$
$$(x + 6)^2 + (y - 14)^2 - 36 + 196 + 22 = 0$$
$$(x + 6)^2 + (y - 14)^2 + 182 = 0$$
$$(x + 6)^2 + (y - 14)^2 = -182$$
$$\text{Centre} = (-6, 14)$$
$$\text{Radius} = -\sqrt{182} = -13.5$$

State what is wrong with Dilgusha's solution and give the correct working and answers. **[4 marks]**

5 The motion of an athlete throwing the hammer is modelled using the equation of a circle C on a centimetre square grid. The athlete stands at the centre of C, which is given by the coordinates (60, 130). It is assumed that the athlete remains on this point. The athlete turns on the spot, swinging the hammer which is attached to a metal cord. The hammer is released at the point with coordinates (180, 40). This point lies on the circumference of C. When released, the hammer follows a path which is a tangent to the circle. It is assumed that the hammer's motion is horizontal.

a Calculate the length of the metal cord. **[2 marks]**

b Find the equation of C. **[2 marks]**

c Find the equation of the straight line along which the hammer moves. **[4 marks]**

The hammer lands at the point (5940, 4360).

d Find the distance the athlete has thrown the hammer. **[3 marks]**

e Comment on the assumptions made in this question. **[2 marks]**

6 A circle C has a diameter of AB, where A and B are given by (9, 10) and (13, −2) respectively.

a Find an equation for C. Give the answer in the form $(x - a)^2 + (y - b)^2 = r^2$. **[4 marks]**

b Find the equation of the tangent to C at the point A. Write the answer in the form $ax + by + c = 0$, where a, b and c are integers. **[4 marks]**

7 The equation of circle C is given by $(x + 9)^2 + (y - 4)^2 = 49$.

A tangent is drawn to C from the point P(8, 21).

Find the length of the tangent. **[4 marks]**

8 The circle C has centre X(7, 2) and passes through the point A(31, 12).

The line L is the perpendicular bisector of the line segment between X and A.

a Find an equation for C. **[4 marks]**

b Find an equation for L. [4 marks]

L intersects the circumference at points M and N.

c Find the length of MN, giving your answer in the form $a\sqrt{3}$. [4 marks]

9 The radius of circle C is $5\sqrt{2}$. The centre of the circle is at the point $(-4, 11)$.

a Write down the equation of the circle. [2 marks]

The line L has the equation $x + y = 7$.

The points A and B lie on both C and L. The x-coordinate of A is less than the x-coordinate of B.

b Find the coordinates of A and B. [3 marks]

$P(-3, 18)$ is a third point on the circumference of C. Tangents are drawn to the circle at points A and P.

The tangents intersect at point T.

c Find the coordinates of T. [4 marks]

(PF) (PS) 10 The circle with equation $x^2 + y^2 - 12x - 6y + 36 = 0$ has centre X.

A tangent of equation $y = mx$, where m is non-zero, touches the circle at point T.

The circle touches the x-axis at the point A.

When the lines AX and OT are extended, they meet at the point P.

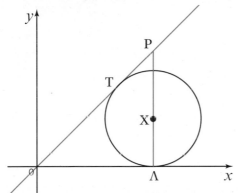

a Prove that the length of XP is 5. [8 marks]

b Find the area of the quadrilateral OAXT. [3 marks]

(PS) 11 The circle C has the equation $(x - 5)^2 + (y - 3)^2 = 2$.

The line $y = mx$ is a tangent to the circle C.

a Show that $23m^2 - 30m + 7 = 0$. [8 marks]

b Find the length of the tangent. [4 marks]

(PS) 12 Circle C_1 has the equation $x^2 + y^2 - 22x - 6y + 65 = 0$.

Circle C_2 has the equation $x^2 + y^2 - 52x - 26y + 585 = 0$.

The circles intersect at P and Q.

Point R is the centre of C_1 and point S is the centre of C_2.

Find the area of the quadrilateral PRQS. [10 marks]

6 TRIGONOMETRY

Trigonometry means 'measuring a triangle'. Mathematicians started to develop this subject thousands of years ago.

In more recent times, trigonometry has been applied to many other areas. Sines and cosines turn out to be useful in many fields you might not have expected. In particular, whenever waves are involved, sines and cosines are the perfect mathematical tool to describe them.

All the electricity that you use in your home or school is delivered by means of an alternating current. This means that the current changes direction in a cycle that repeats 50 times every second. The graph that shows the current looks like this:

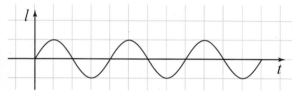

In this chapter you will learn how to find the equation for a graph like this.

All modern communications, such as satellites, mobile phones, computers and televisions, depend on electromagnetic waves. All waves with a repeating cycle can be described using trigonometric functions.

LEARNING OBJECTIVES

You will learn how to:

› define the sine, cosine and tangent for any angle

› use the sine rule, the cosine rule and the formula for the area of a triangle

› understand the graphs of the sine, cosine and tangent functions and interpret their symmetries and periodicity

› use two important formulae connecting $\sin x$, $\cos x$ and $\tan x$

› solve simple trigonometric equations.

TOPIC LINKS

In this chapter you will solve equations and use the algebraic skills that you developed in **Chapter 2 Algebra and functions 2: Equations and inequalities**. You should be familiar with transformations of graphs, which you studied in **Chapter 3 Algebra and functions 3: Sketching curves**. In particular, you should know how the equation changes when a graph is translated, reflected or stretched.

PRIOR KNOWLEDGE

You should already know how to:

› use a calculator to find the sine, cosine or tangent of an angle

› use a calculator to find an angle given its sine, cosine or tangent

› use the equation of a function to produce a graph.

You should be able to complete the following questions correctly:

1 Calculate the length of the shortest side of this triangle.

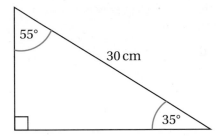

2 Calculate the size of angle C.

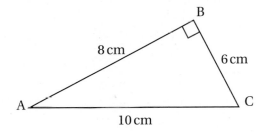

3 Solve the equation $2x^2 + x - 1 = 0$.

6.1 Sine and cosine

When you met the definition of the **sine** and **cosine** of an angle in your GCSE course, it was probably as the ratio of two sides of a right-angled triangle.

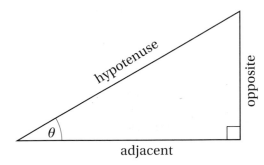

$$\sin\theta = \frac{\text{opposite}}{\text{hypotenuse}}$$

$$\cos\theta = \frac{\text{adjacent}}{\text{hypotenuse}}$$

However, this only works for acute angles. How can you define the sine and cosine of an angle larger than 90°?

You need to change the picture. Imagine a rod OP of length one unit on a coordinate grid. One end, O, is fixed at the origin. The other end, P, starts at (1, 0) and then rotates anticlockwise through angle θ.

θ is the Greek letter 'theta'. We often use Greek letters in mathematics and science to represent unknown variables or constants.

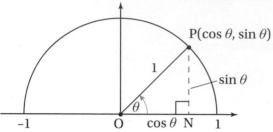

$$PN = OP \times \sin\theta$$
$$= 1 \times \sin\theta = \sin\theta$$
$$ON = OP \times \cos\theta$$
$$= 1 \times \cos\theta = \cos\theta$$

When the angle θ is between 0 and 90°, you should be able to see that the coordinates of P are $(\cos\theta, \sin\theta)$.

We will now use this as the definition of $\sin\theta$ and $\cos\theta$ for any angle. The y-coordinate is $\sin\theta$ and the x-coordinate is $\cos\theta$.

What are the coordinates in this case, when $\theta = 130°$?

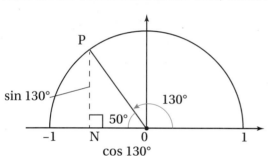

Use the triangle OPN, which has an angle of 50° because $180 - 130 = 50$.

The y-coordinate, $\sin 130° = \sin 50° = 0.766$

The x-coordinate, $\cos 130° = -\cos 50° = -0.643$

This is an example of a general result:

If θ is an obtuse angle,
$$\sin\theta = \sin(180° - \theta) \text{ and } \cos\theta = -\cos(180° - \theta).$$

TECHNOLOGY

Check the values of $\sin 130°$ and $\cos 130°$ with your calculator.

Stop and think   What values does this new definition give for the sine and cosine of 90°? Or for the sine and cosine of 180°?

Example 1

Find all the angles between 0° and 180° that have:

a a sine of 0.5   **b** a cosine of 0.5

Solution

a A calculator gives $\sin^{-1} 0.5 = 30°$ so this is one answer.

Another angle is $180 - 30 = 150°$. There are two possible angles.

b A calculator gives $\cos^{-1} 0.5 = 60°$.

This is the only possible angle because the cosine of an obtuse angle is negative. For example, $180 - 60 = 120$ and $\cos 120° = -0.5$.

Exercise 6.1A

Answers page **518**

1 The sine of 80° is 0.9848.

Write down an obtuse angle which has the same sine.

2 Write down an obtuse angle which has the same sine as:

a 12°   **b** 24°   **c** 36°

d 48°   **e** 72°

3 $\cos 37° = 0.7986$

Use this fact to write down the cosine of an obtuse angle.

(PS) 4 θ is an obtuse angle and $\cos\theta + \cos 81° = 0$.

Work out the value of θ.

5 **a** Solve the equation $\cos x = 0.33$, $0° \leq x \leq 180°$.

     b Solve the equation $\cos x = -0.33$, $0° \leq x \leq 180°$.

(PS) 6 **a** Copy and complete this table.

angle	0°	30°	60°	90°	120°	150°	180°
sine		0.5					
cosine		0.866					

     b Use your table to sketch a graph of $y = \sin x$, $0° \leq x \leq 180°$.

     c Describe the symmetry of your graph.

     d Use your table to sketch a graph of $y = \cos x$, $0° \leq x \leq 180°$.

     e Describe the symmetry of your graph.

     f If $\sin x = \cos x$ find the value of x.

     g If $\sin x + \cos x = 0$ find the value of x.

7 **a** Solve the equation $\sin x = 0.66$, $0° \leq x \leq 180°$.

     b Solve the equation $\sin x = -0.66$, $0° \leq x \leq 180°$.

6.2 The sine rule and the cosine rule

You may remember the sine rule and the cosine rule from your GCSE course.

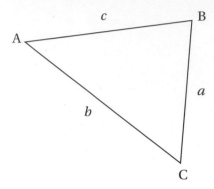

a is the side opposite angle A, *b* is opposite B and *c* is opposite C.

The sine rule

$$\frac{a}{\sin A} = \frac{b}{\sin B} = \frac{c}{\sin C}$$

The cosine rule

$$a^2 = b^2 + c^2 - 2bc \cos A$$

These are true for any triangle.

Stop and think

What happens to the sine rule if $A = 90°$?

What happens to the cosine rule if $A = 90°$?

Proving the sine rule

PROOF

This is a logical deduction using the definition of the sine ratio.

Start with triangle ABC.

Draw BN perpendicular to AC. Use the letter *h* to refer to the length of BN.

From right-angled triangle ABN:

$$\sin A = \frac{h}{c}$$

We can rearrange that as:

$$h = c \sin A \qquad \text{①}$$

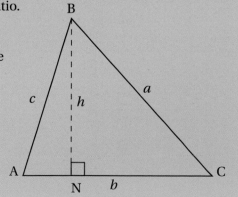

From right-angled triangle CBN:

$$\sin C = \frac{h}{a}$$

Rearrange.

$$h = a \sin C \qquad ②$$

From equations ① and ② we can write:

$$c \sin A = a \sin C$$

Rearrange.

$$\frac{a}{\sin A} = \frac{c}{\sin C}$$

By starting with triangle ABC again and drawing a perpendicular from A to BC we can show that

$$\frac{b}{\sin B} = \frac{c}{\sin C}$$

If we put these results together we get the sine rule.

$$\frac{a}{\sin A} = \frac{b}{\sin B} = \frac{c}{\sin C}$$

Proving the cosine rule

PROOF

This is a logical deduction that uses Pythagoras' theorem.

Start with triangle ABC and draw BN as we did for the sine rule.

Write the length of AN as x, then NC is $b - x$.

Use Pythagoras' rule in triangle ABN.

$$c^2 = h^2 + x^2 \qquad ①$$

And in triangle CBN.

$$a^2 = h^2 + (b - x)^2$$
$$= h^2 + b^2 - 2bx + x^2 \qquad ②$$

Replace $h^2 + x^2$ in equation ② by c^2.

$$a^2 = b^2 + c^2 - 2bx \qquad ③$$

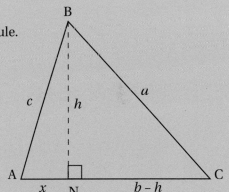

Finally, if you look at the diagram you can see that $x = c \cos A$

When you put this into equation ③ you get $a^2 = b^2 + c^2 - 2bc \cos A$. This is the cosine rule.

The area of a triangle

From your GCSE course, you may also remember this formula for the area of a triangle:

Area = $\frac{1}{2} ab \sin C$

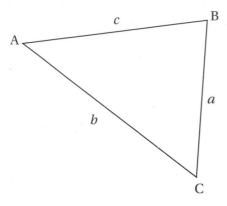

Here is a proof.

PROOF

This is a logical deduction that uses a sine ratio to find an expression for the height of a triangle.

Area = $\frac{1}{2}$ × base × height

$= \frac{1}{2} b \times h = \frac{1}{2} b \times a \sin C$

$= \frac{1}{2} ab \sin C$

This formula is also valid if C is obtuse.

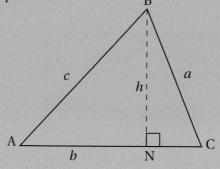

Stop and think   Here is the diagram if C is obtuse.

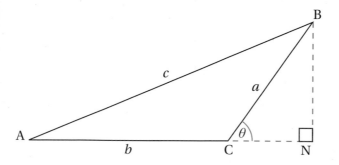

Show that the formula is still correct in this case.

Hint: What can you say about $\sin C$ and $\sin \theta$?

Example 2

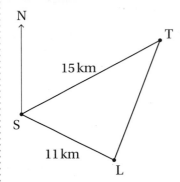

A tower (T) is 15 km from the station (S) on a bearing of 055°.

A lighthouse (L) is 11 km from the station on a bearing of 123°.

Calculate the distance between the tower and the lighthouse.

Solution

The bearing is the angle clockwise from north. The angle S in the triangle is the difference between the two bearings.

$$S = 123 - 55 = 68°$$

You know $S = 68°$, $l = 15$ and $t = 11$.

You can find s using the cosine rule.

$$s^2 = l^2 + t^2 - 2lt \cos S$$
$$= 15^2 + 11^2 - 2 \times 15 \times 11 \cos 68°$$
$$= 222.37\ldots$$
$$s = \sqrt{222.37\ldots} = 14.9$$

The distance is 14.9 km.

See **Chapter 10 Vectors** for more about bearings.

Example 3

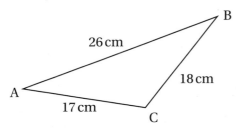

a Find the largest angle of this triangle.

b Find the smallest angle.

Solution

a The largest angle is C because it is opposite the longest side.

To find an angle when you know all the sides, use the cosine rule.

It is easier to use it in this rewritten form:

$$\cos C = \frac{a^2 + c^2 - b^2}{2ac}$$

$$= \frac{18^2 + 17^2 - 26^2}{2 \times 18 \times 17} = \frac{-63}{612}$$

$$= -0.1029\ldots$$

$$C = 95.9°$$

Do not round the numbers on your calculator until you reach the final answer.

b The smallest angle is B because it is opposite the shortest side.

You can use the cosine rule or the sine rule, using your answer to **part a** in the latter case.

$$\frac{\sin B}{b} = \frac{\sin C}{c} \Rightarrow \frac{\sin B}{17} = \frac{\sin 95.9°}{26}$$

$$\Rightarrow \sin B = \frac{\sin 95.9°}{26} \times 17 = 0.6503\ldots$$

$$B = \sin^{-1} 0.6503\ldots$$

$$= 40.6°$$

When you are finding an angle it is convenient to write the fractions this way up.

Sometimes the sine rule can give more than one possible answer when you use it to calculate an angle.

Example 4 demonstrates this.

Example 4

ABC is a triangle.

Angle $A = 50°$, AB = 20 cm and BC = 16 cm.

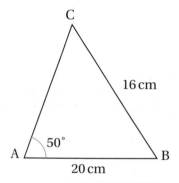

Calculate the size of angle C.

Solution

Use the sine rule.

$$\frac{\sin C}{c} = \frac{\sin A}{a}$$

$$\frac{\sin C}{20} = \frac{\sin 50°}{16}$$

$$\sin C = \frac{20 \times \sin 50°}{16} = 0.9576$$

$\sin^{-1} 0.9576 = 73.2°$

and another possible value is $180 - 73.2 = 106.8°$.

Example 4 gives two possible sizes for angle C.

Why are there two values?

Suppose you try to draw the triangle. You will start with AB 20 cm long. Then draw a line at 50° to this. Point C is on this line somewhere.

Use a pair of compasses open to 16 cm and centred on B to find the position of C.

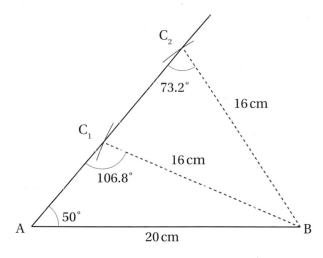

There are two possible positions for C, shown on the diagram as C_1 and C_2. The third angle of the triangle would then be either 73.2° or 106.8°

Stop and think   Suppose that BC in **Example 4** was 20 cm and not 16 cm. Show that in this case there is only one possible answer for the size of angle C.

Exercise 6.2A

Answers page 518

1 **a** Calculate the length of BC.

   b Calculate the area of triangle ABC.

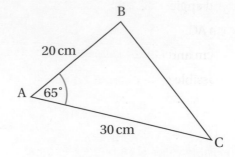

 2 **a** Show that the area of the triangle is
   658 cm² to 3 s.f.

   b Show that the length of BC is 43.4 cm to
   3 s.f.

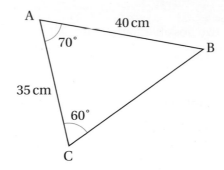

3 Calculate all the angles of this triangle.

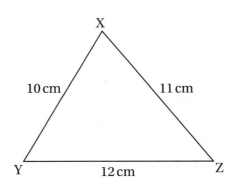

4 **a** Calculate the length of YZ.

   b Calculate the length of XZ.

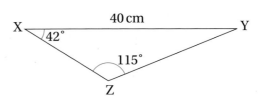

PS 5 ACD is an isosceles triangle.

$A = C = 45°$ and angle ADC = 90°

B is a point on AC.

AD = DC = 12 cm and DB = 9 cm

Find all the possible sizes of angle DBC.

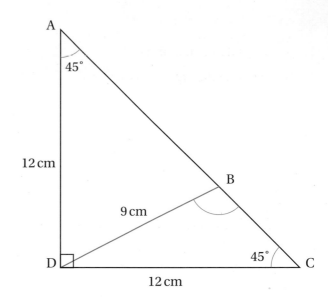

PS 6 ABCD is a quadrilateral.

Calculate the size of angle D.

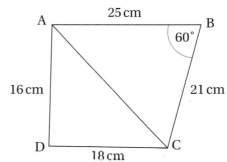

PS 7 A, B and C are three towns.

The distances between them are shown in the diagram.

The bearing of B from A is 100°.

Find the bearing of C from A.

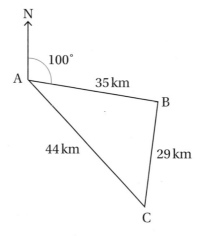

PS 8 A and B are two points on level ground 15 m apart.

ABC is a straight line and CT is a tree.

The angle of elevation of the top of the tree from A is 18° and from B it is 42°.

Calculate the height of the tree.

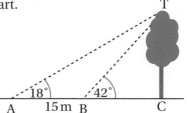

PF **9** XYZ is a triangle and angle X is 30°.

XZ = 40 cm.

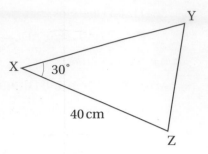

a Find the size of angle Y in the following cases:

i YZ = 32 cm **ii** YZ = 50 cm **iii** YZ = 16 cm

b Prove that the smallest possible value of YZ is 20 cm.

6.3 Trigonometric graphs

In **Section 6.1** we defined the sine and cosine of an obtuse angle using a rotating rod of unit length. We can extend this definition to an angle of any size by rotating the rod further.

In this diagram, angle θ is 220°.

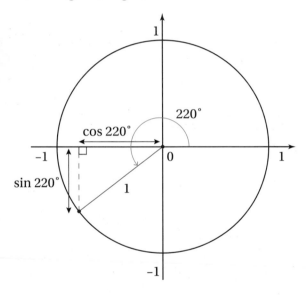

The coordinates of P are both negative so both sin 220° and cos 220° are negative.

The angle between the negative x-axis and OP is 220 − 180 = 40° so

$$\sin 220° = -\sin 40° = -0.6428$$

and $\cos 220° = -\cos 40° = -0.7660$

The following diagram shows when sine and cosine are positive or negative in each of the four **quadrants**.

TECHNOLOGY

Check these values using your calculator.

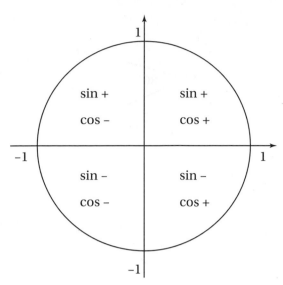

Here are graphs of $y = \sin x$ and $y = \cos x$ for $0° \leqslant x \leqslant 360°$.

$y = \sin x$ $y = \cos x$

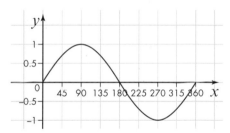   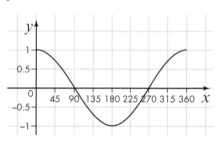

You do not have to stop there. You can rotate OP through more than one revolution, or in the opposite direction, to give the sines and cosines of angles more than 360° or less than 0°. The values will repeat in a cycle every 360°.

The diagram on page 166 shows the position for 220°, and also for 580° (220° + 360°), 940° (220° + 720°), −140° (220° − 360°) and so on.

The following graphs show $y = \sin x$ and $y = \cos x$ for any value of x.

TECHNOLOGY

You can confirm the shapes of these graphs using a graphic calculator or graphing software package.

$y = \sin x$

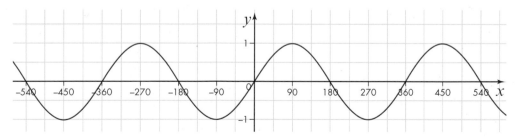

$y = \cos x$

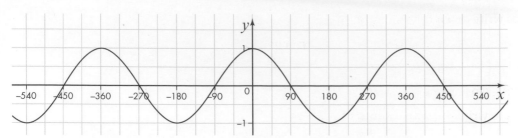

The graphs are periodic with a period of 360°, and extend for ever in both directions.

> **Stop and think**
>
> Describe the symmetries of each of these graphs.
>
> What translations would map the sine graph onto the cosine graph?

Example 5

Solve the equation $\cos x = -0.4$ for $-360° \leqslant x \leqslant 360°$.

Solution

Your calculator will give you one solution: for example, $\cos^{-1}(-0.4) = 113.6°$.

Now look at the graph to find other solutions in the specified range.

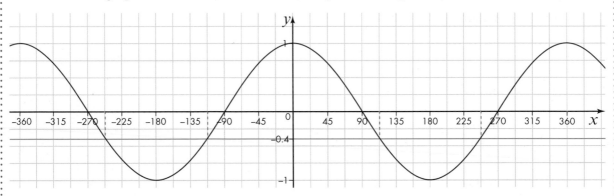

Draw the line $y = -0.4$ and you can see there are four solutions. Use the symmetry of the graph to find them.

They are $x = 113.6°$ or $360 - 113.6 = 246.4°$ or $-113.6°$ or $-246.4°$.

> It is useful to draw a sketch like this.

Exercise 6.3A

Answers page 519

1 $\sin 30° = 0.5$

Use this fact to write down the values of:

a $\sin 150°$ **b** $\sin 210°$ **c** $\sin -30°$

d $\sin -150°$ **e** $\sin -210°$

2 $\cos 15° = 0.966$ to 3 s.f.

Use this fact to write down the values of:

a $\cos 165°$ **b** $\cos 345°$ **c** $\cos -15°$

d $\cos -195°$ **e** $\cos -345°$

3 $\sin 70° = \cos 20° = 0.940$ to 3 s.f.

Use these facts to write down the values of:

a $\cos 380°$ **b** $\sin 430°$ **c** $\sin 790°$

d $\cos -380°$ **e** $\sin -430°$

4 Solve these equations for $0° \leqslant x \leqslant 360°$:

a $\sin x = 0.95$ **b** $\sin x = -0.35$

c $\sin x = -0.812$ **d** $\sin x = 1.5$

5 Solve these equations for $0° \leqslant x \leqslant 360°$:

a $\cos x = -0.25$ **b** $\cos x = 0.1$

c $\cos x = 0$ **d** $\cos x = -1$

6 Solve these equations for $0° \leqslant x \leqslant 720°$:

a $\cos x = 0.22$ **b** $\sin x = -0.22$

c $\sin x = 0$ **d** $\sin x = 1$

7 Solve these equations for $-360° \leqslant x \leqslant 360°$:

a $\cos x = 1$ **b** $\sin x = -0.5$

c $\cos x = 0.7$ **d** $\sin x = -1$

PS 8 Solve the equation $\sin x = \cos x$ for $-360° \leqslant x \leqslant 360°$.

PS 9 One transformation that maps the graph of $y = \sin x$ onto the graph of $y = \cos x$ is a translation of $\begin{bmatrix} -90 \\ 0 \end{bmatrix}$

a Find another translation that maps the graph of $y = \sin x$ onto the graph of $y = \cos x$.

b Find a rotation that maps the graph of $y = \sin x$ onto the graph of $y = \cos x$.

6.4 The tangent function

Look again at this diagram from **Section 6.1**:

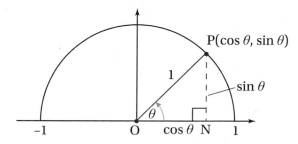

P($\cos \theta$, $\sin \theta$)

$\sin \theta$

1

θ

-1 O $\cos \theta$ N   1

In triangle OPN, the **tangent** of angle $\theta = \dfrac{\text{opposite}}{\text{adjacent}} = \dfrac{\text{PN}}{\text{ON}} = \dfrac{\sin\theta}{\cos\theta}$

We can use this to define the tangent of any angle.

So, for example, $\tan 220° = \dfrac{\sin 220°}{\cos 220°} = \dfrac{-0.6428}{-0.7660} = 0.839$ to 3 s.f.

The tangent of an angle is positive if the sine and cosine have the same sign. It is negative if they have different signs.

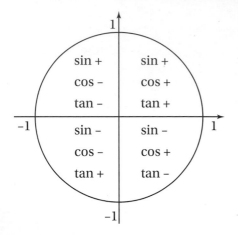

> **KEY INFORMATION**
>
> $\tan\theta = \dfrac{\sin\theta}{\cos\theta}$

> **TECHNOLOGY**
>
> Check that you get this value for $\tan 220°$ on your calculator.

Stop and think   For some values of θ you cannot calculate the tangent. What are those values?

Here is a graph of $y = \tan x$.

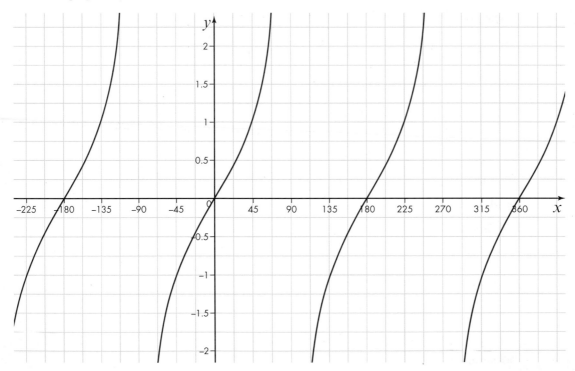

Stop and think   What similarities and differences does this graph have to the graphs of sine and cosine?

This graph has a period of 180°.

There are discontinuities at $x = \pm 90°$, $\pm 270°$, $\pm 450°$ and so on.

> ±90 is a short way of writing 90 or −90. You read it as 'plus or minus 90'.
>
> $\cos x = 0$ for those values of x and so $\tan x$ has no value – you *cannot* divide by 0.
>
> Notice that the tangent function can take any value. This is different from the sine and cosine functions, which must be in the range from −1 to 1.

Example 6

Solve the equation $\tan x = -2$ for $-360° \leqslant x \leqslant 360°$.

Solution

A calculator gives one solution: $\tan^{-1}(-2) = -63.4°$.

The values for the tangent function repeat every 180°.

Other solutions in the specified range are:

$-63.4 - 180 = -243.4°$, $-63.4 + 180 = 116.6°$, $116.6 + 180 = 296.6°$

There are four solutions.

Example 7

Solve the equation $3 \sin x = 2 \cos x$, $0° \leqslant x \leqslant 360°$.

Solution

Divide both sides of the equation by $\cos x$.

$$\frac{3\sin x}{\cos x} = 2$$

But $\frac{\sin x}{\cos x} = \tan x$ so the equation becomes $3 \tan x = 2 \Rightarrow \tan x = \frac{2}{3}$

$$x = \tan^{-1}\left(\frac{2}{3}\right) = 33.7° \text{ or } 180 + 33.7 = 213.7°$$

There are two solutions in the range given.

Exercise 6.4A

Answers page 519

1. $\sin \theta = -0.5$ and $\cos \theta = 0.866$

   Find the value of $\tan \theta$.

2. $\tan \theta = -4.121$ and $\cos \theta = -0.236$

   Find the value of $\sin \theta$.

3. What angles between −180° and 180° have the same tangent as 305°?

4. Solve these equations when $0° \leqslant x \leqslant 360°$.

   a $\tan x = 0.05$ **b** $\tan x = -0.5$

   c $\tan x = 5$ **d** $\tan x = -50$

5. What are the possible values of θ if:

   a $\tan \theta = 0$ **b** $\tan \theta = 1$ **c** $\tan \theta = -1$?

6. Solve the equation $\sin x = 4 \cos x$, $0° \leqslant x \leqslant 360°$.

7 Solve the equation $5 \sin x + 3 \cos x = 0$, $0° \leqslant x \leqslant 360°$.

8 a Describe a translation that will map the graph of $y = \tan x$ onto itself.

b Describe two possible rotations that will map the graph of $y = \tan x$ onto itself.

(PF) 9 a Use the definition of $\tan \theta$ to explain why $\tan \theta > \sin x$ when $0° < \theta < 90°$.

b Illustrate this with a sketch of the graphs of $\sin \theta$ and $\tan \theta$ when $0° < \theta < 90°$.

6.5 Solving trigonometric equations

Here is an equation: $\sin \theta = 0.5$, $0° \leqslant \theta \leqslant 360°$.

The solutions are $\theta = 30°$ or $150°$ and they are shown on this graph.

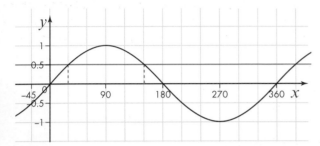

Now consider the equation $\sin 2x = 0.5$, $0° \leqslant x \leqslant 360°$.

This is like the equation above but with $2x$ instead of θ.

That means that $2x = 30°$ or $150°$.

So $x = 15°$ or $75°$.

These are two solutions. However, there are more.

Stop and think   Can you see why there might be more solutions?

If $\sin 2x = 0.5$, then $2x = 30°$ or $150°$ or $390°$ or $510°$ or $750°$ or ...

So $x = 15°$ or $75°$ or $195°$ or $255°$ or $375°$ or ...

The first four of these are in the range $0° \leqslant x \leqslant 360°$ so the equation has four solutions.

They are $x = 15°$ or $75°$ or $195°$ or $255°$.

You can see this on a graph of $y = \sin 2x$.

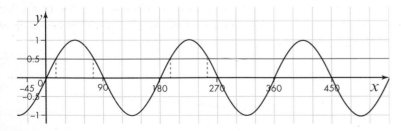

The graph of $y = \sin 2x$ is a stretch of $y = \sin x$ from the y-axis with a scale factor of $\frac{1}{2}$.

The period of the graph is 180°.

You can see that the equation has four solutions in the range $0° \leqslant x \leqslant 360°$.

Example 8

Solve the equation $5\cos(x - 60°) = 4.5$, $0° \leqslant x \leqslant 360°$.

Solution

First divide by 5.

$\cos(x - 60°) = 0.9$

$\qquad x - 60 = 25.8°$ or $-25.8°$ or $334.2°$ or $-334.2°$ or …

Add 60.

$\qquad\qquad x = 85.8°$ or $34.2°$ or $394.2°$ or …

There are two solutions in the range $0° \leqslant x \leqslant 360°$.

$\qquad\qquad x = 34.2°$ or $85.8°$

A graph of $y = 5\cos(x - 60°)$ illustrates the solutions.

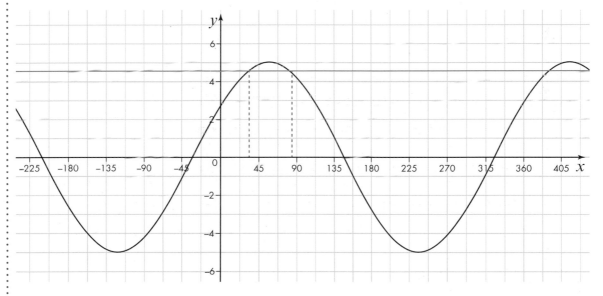

The period of this graph is 360°. The graph of $y = \cos x$ has been stretched from the x-axis with a scale factor of 5 and translated parallel to the x-axis by $\begin{bmatrix} 60° \\ 0 \end{bmatrix}$.

Exercise 6.5A Answers page 519

1. Find the smallest positive solution of each equation:

   a $\sin x = 0.835$ **b** $\sin 2x = 0.835$

   c $\sin(x + 10°) = 0.835$ **d** $\sin(2x + 10°) = 0.835$

2. Solve the equation $\cos 2x = 0.5$, $0° \leqslant x \leqslant 360°$.

3. Solve the equation $\sin(x - 30°) = 0.8$, $0° \leqslant x \leqslant 360°$.

4. Solve the equation $20 \sin 0.5x = 5$, $0° \leqslant x \leqslant 360°$.

5. Solve the equation $\tan 2x = 1.5$, $0° \leqslant x \leqslant 360°$.

6. Solve the equation $100 \sin 10x = 90$, $0° \leqslant x \leqslant 40°$.

7. Solve the equation $5 \sin(2x + 60°) = 3.5$, $0° \leqslant x \leqslant 180°$.

(PF) 8. This is a graph of $y = 20 \sin 50x$:

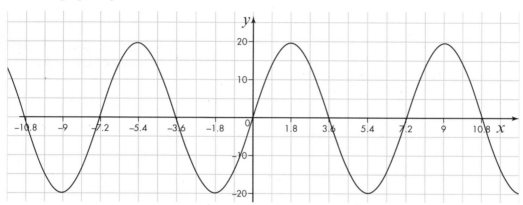

   a Work out the period of the graph.

   b Solve the equation $20 \sin 50x = 10$ when $-5° \leqslant x \leqslant 5°$.

   c Show that there are 100 solutions to the equation $20 \sin 50x = 10$ when $0° \leqslant x \leqslant 360°$.

6.6 A useful formula

Look again at this diagram from **Section 6.1**:

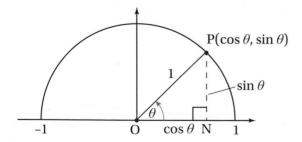

Pythagoras' rule in triangle PON gives $PN^2 + ON^2 = OP^2$.

$PN = \sin\theta$ and $ON = \cos\theta$ and $OP = 1$ so you can write

$$(\sin\theta)^2 + (\cos\theta)^2 = 1$$

We usually write $\sin^2\theta$ instead of $(\sin\theta)^2$ and $\cos^2\theta$ instead of $(\cos\theta)^2$.

The formula becomes

$$\sin^2\theta + \cos^2\theta = 1$$

> This formula is an example of an identity which is true for all values of the variable. We sometimes use the symbol $\equiv$ instead of $=$ to emphasise this. It means 'is identical to'.

Example 9

$\sin x = 0.4$

Find the possible values of $\cos x$.

Solution

As you do not need to find the value of x, you can use the formula $\sin^2 x + \cos^2 x = 1$.

$$0.4^2 + \cos^2 x = 1$$

$$\Rightarrow \cos^2 x = 1 - 0.16 = 0.84$$

$$\Rightarrow \cos x = \pm\sqrt{0.84} \text{ or } \pm 0.917 \text{ to 3 s.f.}$$

> Don't forget to give both possible square roots.

Example 10

Solve the equation $2\sin^2 x = \cos x + 1$, $0° \leqslant x \leqslant 360°$.

Solution

$\sin^2 x + \cos^2 x = 1 \Rightarrow \sin^2 x = 1 - \cos^2 x$

Substitute this in the equation.

$$2(1 - \cos^2 x) = \cos x + 1$$

Rearrange.

$$2 - 2\cos^2 x = \cos x + 1$$

$$2\cos^2 x + \cos x - 1 = 0$$

This is a quadratic equation in $\cos x$.

This factorises.

> Think of it as $2c^2 + c - 1 = 0$ where c is $\cos x$.

$$(2\cos x - 1)(\cos x + 1) = 0$$

Either $\qquad 2\cos x - 1 = 0 \Rightarrow \cos x = \frac{1}{2}$

or $\qquad \cos x + 1 = 0 \Rightarrow \cos x = -1$

The solutions of $\cos x = \frac{1}{2}$ in the given range are $x = 60°$ or $300°$.

The only solution of $\cos x = -1$ in the given range is $x = 180°$

There are three solutions: $x = 60°$, $180°$ or $300°$.

Exercise 6.6A

Answers page 520

1 Use a calculator to confirm numerically that $\sin^2 \theta + \cos^2 \theta = 1$:

    a when $\theta = 75°$ **b** when $\theta = 234°$

2 $\sin \theta = 0.68$

    Find the value of:

    a $\cos \theta$ **b** $\tan \theta$

3 $\cos \theta = 0.44$

    Find the value of:

    a $\sin \theta$ **b** $\tan \theta$

(PF) 4 Here are graphs of $y = \sin^2 x$ (blue) and $y = \cos^2 x$ (red).

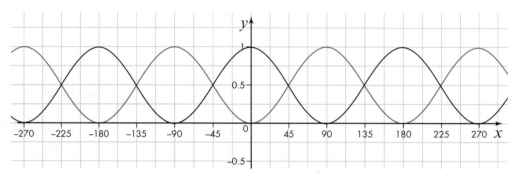

How do the graphs show that $\sin^2 x + \cos^2 x = 1$?

5 Solve the equation $2 \cos x = 3 \cos^2 x$, $0° \leqslant x \leqslant 360°$.

6 Solve the equation $\sin x + 1 = \cos^2 x$, $0° \leqslant x \leqslant 360°$.

7 Solve the equation $4 \cos^2 x = 1$, $0° \leqslant x \leqslant 360°$.

8 Solve the equation $6 \sin^2 x - 5 \sin x + 1 = 0$, $0° \leqslant x \leqslant 360°$.

9 Solve the equation $4 \sin^2 x = 7 \sin x + 2$, $0° \leqslant x \leqslant 360°$.

10 Here are graphs of $y = \cos x$ (red) and $y = \tan x$ (blue).

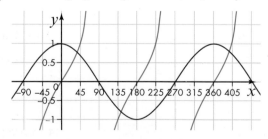

    a How many solutions does the equation $\cos x = \tan x$ have in the range $0° \leqslant x \leqslant 360°$?

    b Calculate the smallest positive solution of the equation.

SUMMARY OF KEY POINTS

> Sine rule:
>
> $\dfrac{a}{\sin A} = \dfrac{b}{\sin B} = \dfrac{c}{\sin C}$

> Cosine rule:
>
> $a^2 = b^2 + c^2 - 2bc \cos A$

> Area of a triangle $= \frac{1}{2} ab \sin C$

> The graphs of the sine and cosine functions are periodic with a period of 360°, and extend for ever in both directions.

> $\tan \theta = \dfrac{\sin \theta}{\cos \theta}$

> The graph of the tangent function has a period of 180°.

> There are discontinuities at $x = \pm 90°$, $\pm 270°$, $\pm 450°$ and so on.

> $\sin^2 \theta + \cos^2 \theta = 1$

EXAM-STYLE QUESTIONS 6

Answers page 520

1 **a** Solve the equation $\sin x = 0.5$, $0° \leqslant x \leqslant 360°$. **[2 marks]**

  b Solve the equation $\sin(2y - 20°) = 0.5$, $0° \leqslant y \leqslant 360°$. **[3 marks]**

(PS) 2 ABCD is a parallelogram.

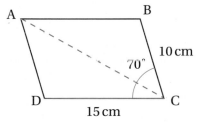

BC = 10 cm, CD = 15 cm and angle BCD = 70°.

Calculate the length of the diagonal AC. **[4 marks]**

3 **a** Sketch the graph of $y = 5 \cos(x + 30°)$, $0° \leqslant x \leqslant 360°$.

   On your sketch, indicate where the graph crosses the coordinate axes. **[2 marks]**

  b Solve the equation $5 \cos(x + 30°) + 4 = 0$, $0° \leqslant x \leqslant 360°$. **[3 marks]**

(PF) 4 The angles of a triangle are 50°, 62° and 68°.

  The longest side of the triangle is 35 cm.

  Show that the length of the shortest side is 28.9 cm to 3 s.f. **[4 marks]**

5 Solve the equation $\sin\theta = 4\cos\theta$, $-180° \leqslant \theta \leqslant 180°$. **[4 marks]**

(PS) (CM) 6 XYZ is a triangle.

XY = 19.5 cm, YZ = 24.9 cm and angle $Z = 50°$.

Carla says, 'Angle X is 78°.'

Larry says, 'Angle X is 102°.'

Who is correct? Justify your answer. **[4 marks]**

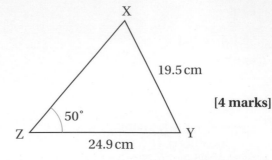

7 Solve the equation $\tan\theta = 2\sin\theta$, $0° \leqslant \theta \leqslant 360°$. **[5 marks]**

8 The sides of a triangle are 15 cm, 18 cm and 21 cm.

Calculate the area of the triangle. **[6 marks]**

9 Solve the equation $6\sin^2 x = 5 + \cos x$, $0° \leqslant x \leqslant 360°$. **[7 marks]**

10 Solve the equation $\sin x° = \sin 2x°$, $0 \leqslant x < 360$. **[3 marks]**

(PS) 11 The banks of a river are parallel.

From point A on one bank, the angles to trees at P and Q on the opposite bank are 41° and 28° respectively, as shown on the diagram.

P and Q are 25 m apart.

Find the width of the river. **[4 marks]**

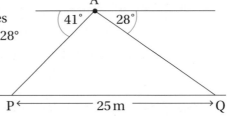

12 $\cos x = -\dfrac{\sqrt{3}}{4}$

Find the possible values of $\tan x$. Write your answers exactly in surd form. **[4 marks]**

13 a Sketch a graph to show that the equation $\tan x = \cos x$ has a solution for $0 < x < 90°$. **[1 mark]**

b Solve the equation $\tan x = \cos x$, $0 < x < 360°$. **[5 marks]**

(M) (PF) 14 The height of the pedal of a bike above the ground after x seconds is

$$0.35 + 0.2\cos(100x + 60°) \text{ m}$$

a Show that initially the pedal is 0.45 m above the ground and that the height decreases at first. **[2 marks]**

b Show that in each revolution of the pedals the height is more than 0.5 m for about 0.8 seconds. **[4 marks]**

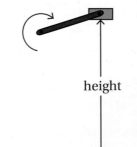

height

7 EXPONENTIALS AND LOGARITHMS

On the right are two common shapes for graphs, particularly if the horizontal axis represents time.

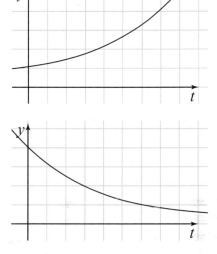

> The first graph could represent the value of savings in a bank, rising prices in a period of inflation or the projected growth in the population of a country.

> The second graph could represent the fall in value of a car as it gets older or the amount of radioactivity from waste from an atomic power station reducing over time.

How can we write the equations of graphs like this? How can we use them? How can we fit a curve like this to real data?

These are questions you will be able to answer after studying this chapter.

LEARNING OBJECTIVES

You will learn how to:

> use and manipulate expressions of the form $y = ka^x$ where the variable x is an index

> use logarithms and understand their connection with indices

> recognise and interpret exponential growth and decay in modelling change.

TOPIC LINKS

In this chapter you will use the laws of indices (see **Chapter 1 Algebra and functions 1: Manipulating algabraic expressions** and further develop your graph-sketching skills (see **Chapter 3 Algebra and functions 3: Sketching curves** and **Chapter 4 Coordinate geometry 1: Equations of straight lines**) to cover new types of graphs. You will draw lines of best fit, which are also explored in **Chapter 12 Data presentation and interpretation**.

PRIOR KNOWLEDGE

You should already know how to:

> manipulate expressions involving powers of numbers

> use the equation of a graph to draw it

> solve linear equations

> find the gradient of a curve at a particular point by drawing a tangent.

You should be able to complete the following questions correctly:

1 Write as a power of 2:

   a 16   **b** $\frac{1}{32}$ **c** $\sqrt{2}$ **d** $\sqrt{8}$ **e** 1

2 Write as a power of n:

   a $n^2 \times n^3$ **b** $n^2 \div n^3$

   c $(n^{-3})^{-2}$ **d** $\sqrt{n} \times \sqrt[3]{n}$

3 Sketch these graphs for $x > 0$:

   a $y = 10 + 0.5x$ **b** $y = 0.5x^2$ **c** $y = 2x^{0.5}$

4 Work out an equation of the straight line that goes through (5, 18) and crosses the y-axis at (0, 20).

5 Estimate the gradient of the graph of $y = x^2$ at the point (1.5, 2.25) by drawing a tangent.

6 The price of a car is £24 000. The price is increased by 4%, then it is increased again by 4%. Work out the new price.

7.1 The function a^x

Suppose you invest some money and it earns 20% per year interest. If you have £10 000 now:

> in 1 year's time you will have £10 000 × 1.2 = £12 000

> in 2 years' time you will have £10 000 × 1.2² = £14 400

> in 3 years' time you will have £10 000 × 1.2³ = £17 280.

More generally, in n years' time you will have £10 000 × 1.2^n.

If the amount you invest is £A, in n years' time you will have £A × 1.2^n.

The important feature is the value of 1.2^n. Here is a graph of $y = 1.2^x$:

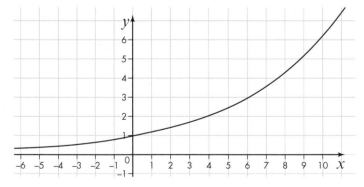

The graph shows how the value of the investment increases over time.

For $x = 1, 2, 3$, and so on, the graph shows the value of the investment in units of £10 000.

For example, after 6 years the value has approximately tripled, to £30 000.

TECHNOLOGY

Make sure you can use graph-plotting software to draw this graph.

The graph shows the value of 1.2^x for any value of x.

Use the graph to estimate 1.2³ and check that your value is approximately 1.2 × 1.2 × 1.2.

$\frac{1}{1.2} = 0.833\,3\ldots$ What power of 1.2 is this? Check that the graph is correct.

Calculate $\sqrt{1.2^5}$ and compare it with the value from the graph.

Check some other values, such as $\sqrt[3]{1.2^{10}}$.

If you had already been saving for a number of years, the graph shows the value in previous years.

For example, 2 years ago the value was £10 000 × 1.2^{-2} = £6694.

Of course, the interest is usually added once a year as a lump sum, so the value would go up in steps, but the graph shows how the value is increasing continually through the year.

Here are some points to notice about the graph of $y = 1.2^x$:

> the value of y is always positive

> the graph passes through the point (0, 1) on the y-axis

> the gradient of the graph is always positive

> as the value of x increases, the gradient of the graph increases.

Example 1

Jasmine buys a car for £15 000. The value of the car decreases by 15% per year.

a Find the value of the car after 3 years.
b Find an expression for the value of the car after n years.
c Show that after 7 years the value of the car is less than £5000.

Solution

a The value after 3 years is £15 000 × 0.85^3 = £9212.
b After n years the value is £15 000 × 0.85^n.
c After 7 years the value is £15 000 × 0.85^7 = £4809, which is less than £5000.

> The multiplier for a decrease of 15% is $1 - 0.15 = 0.85$

Here is a graph of $y = 0.85^x$.

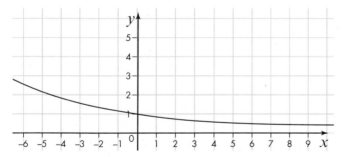

This graph illustrates the falling value of the car in **Example 1**.

Stop and think   Refer back to the four properties of the graph of $y = 1.2^x$, listed above. Which properties are the same for the graph of $y = 0.85^x$ and which are different?

Change the properties which are different so that they apply to the graph of $y = 0.85^x$.

› The increasing value of the investment of £10 000 is an example of **exponential growth**.

› The decreasing value of the value of the car in **Example 1** is an example of **exponential decay**.

Exponential change occurs when the value of a variable increases or decreases by a percentage that remains constant and does not change over time.

People sometimes incorrectly use the word 'exponential' as if it means 'large'. Change can be exponential without being large. Also, change can be large without being exponential.

Look at these graphs of $y = a^x$ for $a = 1.5$, 2 or 3.

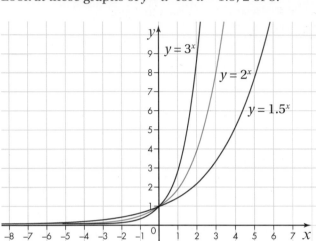

› If $a > 1$, the graph of $y = a^x$ has this shape and represents exponential growth.

Look at these graphs of $y = a^x$ for $a = 0.75$, 0.5 or 0.3:

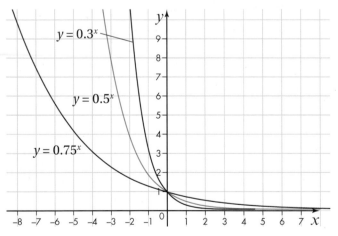

› If $0 < a < 1$, the graph of $y = a^x$ has this shape and represents exponential decay.

KEY INFORMATION

$y = a^x$, $a > 0$ is an example of exponential growth.

$y = a^x$, $0 < a < 1$ is an example of exponential decay.

KEY INFORMATION

The graph of $y = a^x$, where a is positive, passes through $(0, 1)$ and is always above the x-axis. The gradient increases as x increases.

If $0 < a < 1$ the gradient is always negative.

Stop and think   What does the graph of $y = a^x$ look like if $a = 1$?

For a particular value of x, which graph has a larger gradient, $y = 2^x$ or $y = 3^x$?

Exercise 7.1A

Answers page 521

1 **a** Copy and complete this table of values of $y = 1.5^x$. Round the values to 2 d.p.

x	−3	−2	−1	0	1	2	3
$y = 1.5^x$					1.5		

 b Draw a graph of $y = 1.5^x$.

 c Use your graph to find an approximate solution to the equation $1.5^x = 3$.

2 **a** Copy and complete this table of values of $y = 0.6^x$. Round the values to 2 d.p.

x	−3	−2	−1	0	1	2	3
$y = 0.6^x$					0.6		

 b Draw a graph of $y = 0.6^x$.

 c Use your graph to find an approximate solution to the equation $0.6^x = 0.5$.

 d Use your graph to find an approximate solution to the equation $0.6^x = 4$.

3 The population of a town in the year 2020 is 50 000. The population is expected to increase by 8% every decade (10 years).

 a Copy and complete this table.

Year	2020	2030	2040	2050
Predicted population	50 000			

 b Write down a formula for the population (y thousands) x decades after 2020.

 c Sketch a graph of y for $0 \leqslant x \leqslant 3$.

 d Show that, if the model is correct, the population will double by 2110.

 e Explain why the model might not be appropriate for a large number of years.

4 As technology improves, the price of products can go down.

The price of a machine is £500. The price is reducing by 10% per year.

 a Work out the price in 2 years' time.

 b Explain why the price in x years' time will be £500 × 0.9^x.

 c Draw a graph of $y = 500 \times 0.9^x$, $0 \leqslant x \leqslant 5$.

 d Will the price of the machine have halved after 5 years? Give a reason for your answer.

5 **a** On the same axes, sketch the graphs of $y = 1.4^x$ and $y = 1.9^x$.

 b For what values of x is $1.4^x < 1.9^x$?

 c For what values of x is $1.4^x > 1.9^x$?

6 These are graphs of two curves:

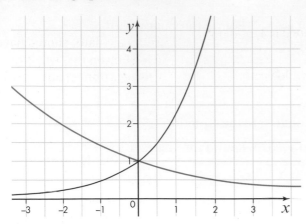

Each graph has an equation of the form $y = a^x$.
Find the value of a for each curve.

7 This graph has an equation of the form $y = Ka^x$:

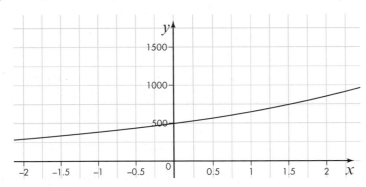

Work out the values of K and a.

PS **8** **a** On the same axes, draw graphs of $y = 2^x$ and $y = \left(\frac{1}{2}\right)^x$.

b Describe a transformation that will map the graph of $y = 2^x$ onto the graph of $y = \left(\frac{1}{2}\right)^x$.

c Generalise the result of **part b**.

9 **a** On the same axes, draw graphs of $y = 2^x$ and $y = 2^{x+1}$.

b Describe a translation that will map the graph of $y = 2^x$ onto the graph of $y = 2^{x+1}$.

c Describe a different transformation that will map the graph of $y = 2^x$ onto the graph of $y = 2^{x+1}$.

7.2 Logarithms

Here are some **powers** of 10:

› $100 = 10^2$

› $10\,000 = 10^4$

› $0.001 = 10^{-3}$

The **logarithm** of a number is the power of 10 which is equal to that number.

So, the logarithm of 100 is 2, the logarithm of 10 000 is 4, and the logarithm of 0.001 is –3.

'log' is the abbreviation for logarithm. So, you would write:

$$\log 1\,000\,000 = 6, \qquad \log 0.000\,01 = -5$$

Every positive number has a logarithm. For example,

$$10^{\frac{1}{2}} = \sqrt{10} = 3.162\,28$$

So $\log 3.162\,28 = 0.5$

> **Stop and think** Why is it possible to find a logarithm only for positive numbers and not for zero or negative numbers?

TECHNOLOGY

Your calculator should have a log button that will give the logarithm of any positive number.

Check the values given here.

If you multiply a number by 10, what happens to its logarithm?

What happens if you multiply by 100, or 1000?

What happens if you divide by 10, or 100, or 1000?

Raising 10 to a power and finding the logarithm are inverse operations.

Here are three laws of logarithms:

$$3 \xrightarrow{\;10^3\;} 1000 \xleftarrow{\;\log 1000\;}$$

1 $\log xy = \log x + \log y$

2 $\log \dfrac{x}{y} = \log x - \log y$

3 $\log x^k = k \log x$

See **Chapter 1 Algebra and functions 1: Manipulating algebraic expressions, Section 1.6**, for more about laws of indices.

These are really just the laws of indices that you already know, written in a different way.

To prove law 1, suppose that $x = 10^a$ and $y = 10^b$.

Then $a = \log x$ and $b = \log y$.

So $\log xy = \log(10^a \times 10^b)$

$\qquad = \log 10^{a+b}$

$\qquad = a + b$

$\qquad = \log x + \log y$

PROOF

Start the proof by writing down what you know.

This is the addition rule for indices.

> **Stop and think** Adapt the proof of law 1 to give a proof of law 2.
>
> The first two lines will be the same.
>
> To prove law 2, suppose that $x = 10^a$ and $y = 10^b$.
>
> Then $a = \log x$ and $b = \log y$.
>
> So $\log \dfrac{a}{b} = \ldots$

To prove law 3, suppose, as before, that $x = 10^a$ and $a = \log x$.

So $\log x^k = \log(10^a)^k$

$\quad\quad\quad = \log 10^{ka}$

$\quad\quad\quad = ka$

$\quad\quad\quad = k \log x$

> This is the multiplication rule for indices

Example 2

$\log x = c$

Write, in terms of c, the logarithms of

a x^3 **b** $\dfrac{10}{x}$ **c** $100\sqrt{x}$

Solution

a $\log x^3 = 3 \log x = 3c$

b $\log \dfrac{10}{x} = \log 10 - \log x$

$\quad\quad\quad = 1 - c$

c $\log 100\sqrt{x} = \log 100 + \log \sqrt{x}$

$\quad\quad\quad\quad = 2 + \log x^{\frac{1}{2}}$

$\quad\quad\quad\quad = 2 + \tfrac{1}{2} \log x$

$\quad\quad\quad\quad = 2 + \tfrac{1}{2} c$

KEY INFORMATION

You need to learn the three laws of logarithms:

- $\log xy = \log x + \log y$
- $\log \dfrac{x}{y} = \log x - \log y$
- $\log x^k = k \log x$

Stop and think   Identify where each of the three laws of logarithms has been used in **Example 2**.

Exercise 7.2A

Answers page 522

1 Without using a calculator, work out the value of:

    a $\log 1000$ **b** $\log 1\,000\,000$

    c $\log 0.001$ **d** $\log 0.000001$

2 Without using a calculator, work out the value of:

    a $\log \sqrt{10}$ **b** $\log \sqrt{100}$

    c $\log \sqrt{1000}$ **d** $\log \dfrac{1}{\sqrt{10}}$

3 $\log 2 = 0.3010$ to 4 d.p.

Without using a calculator, use this fact to work out the value of:

    a $\log 20$ **b** $\log 0.2$

    c $\log 5$ **d** $\log \dfrac{1}{\sqrt{2}}$

4 $\log 8 = 0.9031$ to 4 d.p. and $\log 12 = 1.0792$ to 4 d.p.

Without using a calculator, use these facts to work out the value of:

a $\log 80$ **b** $\log 1.5$

c $\log 9.6$ **d** $\log \left(\frac{2}{3}\right)$

5 $\log x = k$

Write the following in terms of k:

a $\log x^2$ **b** $\log 100x$

c $\log \frac{1}{x}$ **d** $\log \frac{10}{\sqrt{x}}$

6 $\log x = k$ and $\log y = h$

Write the following in terms of k and h:

a $\log xy^2$ **b** $\log \frac{\sqrt{x}}{y}$ **c** $\log \left(\frac{100x^2}{y^3}\right)$ **d** $\log \frac{1}{xy}$

7 Do not use a calculator for this question.

a $\log x = 2$. Work out the value of x.

b $\log 2y = 2$. Work out the value of y.

c $\log(\log z) = 2$. Work out the value of z.

8 Suppose you know the logarithms of 2 and 3.

You could use these to work out the logarithms of other numbers without using a calculator, just using simple arithmetic.

Which of the following can be found easily in this way, without using a calculator?

a $\log 4$ **b** $\log 5$ **c** $\log 6$

d $\log 7$ **e** $\log 8$ **f** $\log 9$

g $\log 11$ **h** $\log 12$

Show how to do so.

Logarithms in other bases

Look at the following expressions:

$$2^5 = 32$$
$$2^{-3} = \frac{1}{8}$$
$$2^{\frac{1}{2}} = 1.41421$$

You can use logarithms to write these expressions, like this:

$$\log_2 32 = 5$$
$$\log_2 \frac{1}{8} = -3$$
$$\log_2 1.41421 = \tfrac{1}{2}$$

In this case $\log_2 32$ means 'log to the base 2 of 32'. The subscript 2 means that you are using powers of 2 in this case.

TECHNOLOGY

The log button on a calculator is *always* base 10. Many calculators also have a button to find the logarithm in any base.

$\log 32 = 1.50514$; if there is no subscript it means a power of 10.

$\log_2 32 = 5$; the subscript means you need a power of 2.

You can choose any positive integer as the **base** of a logarithm.

Here are some equivalent statements:

$$5^3 = 125 \quad \Leftrightarrow \quad \log_5 125 = 3$$

$$8^{\frac{1}{3}} = 2 \quad \Leftrightarrow \quad \log_8 2 = \frac{1}{3}$$

$$\frac{1}{49} = 7^{-2} \quad \Leftrightarrow \quad \log_7 \frac{1}{49} = -2$$

The three laws of logarithms that you saw above for logarithms in base 10 are true for logarithms in any base.

Stop and think	Look through the proofs of the three laws of logarithms (base 10) earlier in this section.
	Check that you can modify the proofs so that they are valid for logarithms in any base.

KEY INFORMATION

If $y = a^x$ then $\log_a y = x$. The base of the logarithm is a where $x \geqslant 0$.

KEY INFORMATION

- $\log_a xy = \log_a x + \log_a y$
- $\log_a \dfrac{x}{y} = \log_a x - \log_a y$
- $\log_a x^k = k \log_a x$

Example 3

Find: **a** $\log_3 243$ **b** $\log_4 32$

Solution

a You have to write 243 as a power of 3.

$243 = 3^5$ so $\log_3 243 = 5$

b $32 = 16 \times 2$

$\quad = 4^2 \times \sqrt{4}$

$\quad = 4^2 \times 4^{\frac{1}{2}}$

$\quad = 4^{2.5}$

$\log_4 32 = 2.5$

TECHNOLOGY

You can use the $\log_a b$ button on your calculator.

Check the values in **Example 3** using the $\log_a b$ button.

However, it is useful to be able to recognise the answer in simple cases when the answer is a small integer.

Exercise 7.2B

Answers page 522

1 Without using a calculator, work out the value of:

   a $\log_2 64$ **b** $\log_5 125$

   c $\log_3 81$ **d** $\log_4 256$

2 Without using a calculator, work out the value of:

   a $\log_9 3$ **b** $\log_9 27$

   c $\log_9 \dfrac{1}{81}$ **d** $\log_9 \sqrt{3}$

3 Match each expression on the left to the corresponding one on the right.

One has been done for you.

$\log_a 2 + \log_a 3$ $\log_a 3$

$2 \log_a 4$ $\log_a 4$

$\log_a 8 - \log_a 2$ $\log_a 6$

$\log_a \dfrac{1}{2} - \log_a \dfrac{1}{6}$ $\log_a 8$

$3 \log_a 2$ $\log_a 16$

4 Do not use a calculator for this question.

Here is a table of values of powers of 2:

x	1	2	3	4	5	6	7	8	9	10
2^x	2	4	8	16	32	64	128	256	512	1024

Use the table to find:

a $\log_2 128$ **b** $\log_2 \dfrac{1}{32}$

c $\log_2 \sqrt{512}$ **d** $\log_2 \sqrt[3]{256}$

5 $\log_3 5 = c$

Without using a calculator, write the following in terms of c:

a $\log_3 15$ **b** $\log_3 0.2$

c $\log_3 125$ **d** $\log_9 5$

6 Find the base for the following logarithms:

a $\log_n 216 = 3$ **b** $\log_n 4 = \dfrac{1}{2}$

c $\log_n 1 = 0$ **d** $\log_n 27 = 1.5$

(PS) 7 $\log_n a = 3.5$ and $\log_n b = 4.5$

Prove that $b = na$.

(PS) 8 **a** Write each of these equations in index form:

     i $x = \log_a b$ **ii** $y = \log_b a$

b Show that $\log_a b = \dfrac{1}{\log_b a}$

7.3 The equation $a^x = b$

Suppose you invest £400 at an interest rate of 6% per year. How long will it take to double the value of your investment?

> The value after n years is £400 × 1.06^n.

To find when the value has doubled you need to solve the equation $400 \times 1.06^n = 800$.

> Divide by 400 and you get $1.06^n = 2$.

How do you solve this equation? The answer is to take logarithms of both sides:

$$\log 1.06^n = \log 2$$

$$n \log 1.06 = \log 2$$

Divide both sides by log 1.06.

$$n = \frac{\log 2}{\log 1.06} = 11.90$$

You do not need to find the values of the logarithms – just use your calculator to do the final division.

In the context of the original question, the answer needs to be a whole number of years.

It takes 12 years before the value of the investment has doubled.

You could have used logarithms to any base to rewrite the equation, but it is simplest to use base 10 because you can find logarithms to base 10 on your calculator.

Example 4

Solve the equation $5^{3x-1} = 12$.

Solution

Take logarithms of both sides.

$$\log 5^{3x-1} = \log 12$$

$$(3x-1) \log 5 = \log 12$$

$$3x - 1 = \frac{\log 12}{\log 5} = 1.5439\ldots$$

$$x = \frac{1.5439 + 1}{3} = 0.848 \text{ to 3 d.p.}$$

A similar method shows that $\log_a b = \dfrac{\log b}{\log a}$

Write $\qquad\qquad c = \log_a b$

Rearrange: $\qquad\qquad b = a^c$

Take logarithms: $\quad \log b = c \log a$

Divide by $\log a$: $\quad \dfrac{\log b}{\log a} = c$

which is the result needed.

PROOF

Start with one side of the formula and try to rearrange until it is equal to the other side. Do not try to change both sides at the same time.

Exercise 7.3A

Answers page 523

1. Solve these equations:

   a $3^x = 11$ **b** $4^x = 175$ **c** $12^x = 6$

2. Solve these equations:

   a $0.5^x = 0.4$ **b** $0.7^x = 0.25$ **c** $0.9^x = 0.55$

3. Solve these equations:

   a $200 \times 1.8^t = 750$ **b** $7000 \times 0.87^t = 4500$ **c** $95 \times 1.04^t = 123$

4 Watson has some shares. Their current value is £12 500. The value of the shares is increasing by 9% a year.

If this rate does not change, work out how long it will take for the value of the shares to:

a increase by 50%   **b** double.

5 Solve these equations:

a $4^{x+2} = 90$ **b** $6^{2x+1} = 35$ **c** $15^{4x-3} = 8$

6 There is $10\,m^2$ of weed on a pond. The area of weed is increasing by 50% each week.

a Find a formula for the area $\left(\dfrac{y}{m^2}\right)$ of weed on the pond in t weeks' time.

b Find the area of weed in three weeks' time.

c The area of the whole pond is $80\,m^2$. How long will it take until the whole pond is covered with weed?

d Increasing the flow of water through the pond decreases the rate of growth to 20% per week. How long would it take to cover the pond now?

e Explain why the model might no longer be appropriate for a large number of weeks.

7 The number of people flying from an airport this year was 1.65 million. The number of travellers is expected to increase by 5% per year.

a How many travellers are expected next year and the year after?

b Write down an expression for the annual number of travellers in t years' time.

c How long will it take until the annual number exceeds 2.5 million?

d Explain why the model might not be correct after a large number of years.

8 Moore's Law in computing says that the number of transistors in an integrated circuit doubles approximately every two years. This is because improvements in technology make it possible to make the transistors smaller and smaller.

a According to Moore's Law, how many years will it take for the number of transistors in an integrated circuit to increase by a factor of 100?

b Sometimes it is said that the time for the number of transistors to double is only 1.5 years. How does this alter your answer to **part a**?

c Some experts believe that Moore's Law will no longer apply in the near future. Can you think of a reason for this?

7.4 Logarithmic graphs

You know how to find the coordinates of points on a line if you know the equation of the line.

This section deals with the inverse question. That is, if you know the coordinates of some points on the line, can you find the equation? For a straight line it is relatively easy.

How do you find the equation of a straight line if you know the coordinates of some points on the line?

For a curve it is more difficult. However, you will usually have some idea of the type of curve you are looking for. Here are two common types.

Curve type 1: $y = ax^n$

In an experiment the length of a pendulum (L metres) and the time (T seconds) it takes to make 10 swings are recorded. This is done for several different lengths of pendulum.

Here are the results in a table:

Length (L m)	Time (T s)
0.6	15.5
0.9	19.0
1.2	21.9
1.5	24.5
1.8	26.8
2.1	29.0

These points are not in a straight line.

You need to find out if an equation of the form $T = aL^n$ will fit these values with a suitable choice of the parameters a and n.

> When you are trying to fit a model to data, the constants are usually called parameters.

Start with $T = aL^n$.

Take logarithms of both sides.

$$\log T = \log aL^n$$
$$= \log a + \log L^n$$
$$= \log a + n \log L$$

> Notice how the laws of logarithms have been used to rearrange the right-hand side.

Look at the final equation. It is in the form $y = c + mx$, with $\log T$ instead of y and $\log L$ instead of x, and with $\log a$ instead of c and n instead of m.

If the equation is correct, a graph of $\log T$ against $\log L$ will be a straight line. The gradient will be n and the intercept on the $\log T$ axis will be $\log a$.

Here are the values of $\log T$ and $\log L$:

Length (L m)	Time (T s)	$\log L$	$\log T$
0.6	15.5	−0.22	1.19
0.9	19.0	−0.05	1.28
1.2	21.9	0.08	1.34
1.5	24.5	0.18	1.39
1.8	26.8	0.26	1.43
2.1	29.0	0.32	1.46

> **TECHNOLOGY**
>
> A spreadsheet software package can calculate the values of the logarithms for you.

Here is the graph:

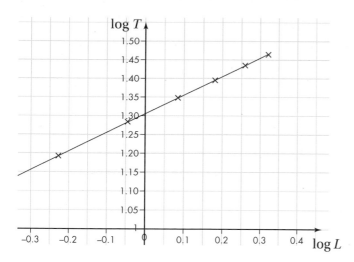

The points have been joined with a straight line.

Drawing a triangle underneath the line will show that the gradient is $\frac{1}{2}$ so $n = 0.5$.

The intercept on the $\log T$ axis is 1.30 so $\log a = 1.30$.

Therefore $a = 10^{1.3} = 20.0$ to 3 s.f.

This means that the equation connecting T and L is $T = 20.0\sqrt{L}$.

> In a real experiment there are often errors in measurements and the points will not be exactly in a straight line. In that case you would use the line of best fit.

Curve type 2: $y = kb^x$

If you have exponential growth or decay, the variable will be the index of a constant value.

A scientist is growing a some bacteria in an experiment. The colony is growing exponentially.

The size of the population, P, after t hours will be given by an equation of the form $P = kb^t$.

The problem is to find the values of the parameters k and b.

Start with $P = kb^t$ and take logarithms of both sides.

$$\log P = \log kb^t$$
$$= \log k + \log b^t$$
$$= \log k + t \log b$$

> This is exactly the same rearrangement as in the previous type of equation but this time the variable (t) is in a different place. The graph will use values of t and not of $\log t$.

Use the values of P and t to draw a graph of $\log P$ against t as shown on the next page.

The gradient is 0.65.

$$\log b = 0.65$$
$$b = 10^{0.65} = 4.47 \text{ to 3 s.f.}$$

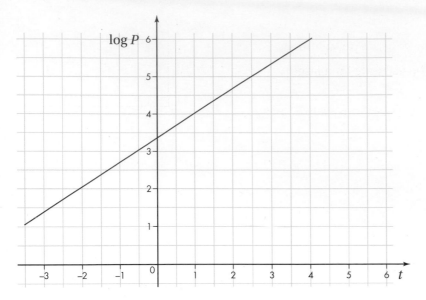

The intercept on the $\log P$ axis is 3.4

$$\log k = 3.4$$

$$k = 10^{3.4} = 2510 \text{ to } 3 \text{ s.f.}$$

The equation, including these parameters, is $P = 2510 \times 4.47^t$

Exercise 7.4A Answers page 523

1 $y = 250x^2$

    a Write a formula for $\log y$ in terms of $\log x$.

    A graph of $\log y$ against $\log x$ will be a straight line.

    b Find the gradient of the line.

    c Find the coordinates of the intercept on the $\log y$ axis.

2 The formula for the volume of a sphere is $V = \frac{4}{3}\pi r^3$.

    Show that a graph of $\log V$ against $\log r$ will be a straight line.

3 In physics, the gravitational force between two masses, m_1 and m_2, a distance r apart, is given by the formula $F = \dfrac{Gm_1m_2}{r^2}$, where G is a constant.

    a Show that a graph of $\log F$ against $\log r$ is a straight line.

    b Find the gradient of the straight line.

    c Find the intercept of the straight line on the $\log F$ axis in terms of the constants m_1, m_2 and G.

 4 The population of a country, P millions, in t years' time is modelled by the formula $P = Ac^t$ where A and c are constants.

    a Show that a graph of $\log P$ against t is a straight line.

    The gradient of the straight line is 0.0128 and the intercept on the $\log P$ axis is $(0, 1.97)$.

b Find the values of A and c.

c Find the annual rate of growth of the population.

d Why might the model not be accurate for a large value of t?

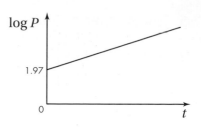

5 The formula connecting two variables, v and t, is $v = ar^n$ where a and n are constants.

Here is a graph of $\log v$ against $\log r$:

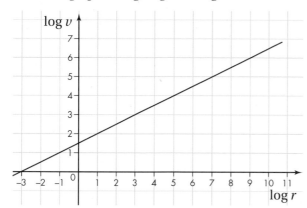

a Find the gradient of the graph.

b Explain why $\log a = 1.5$.

c Find the values of a and n.

d Use the formula to find the value of v when $r = 100$.

6 A scientist suspects that two variables are connected by the formula $x^2y = k$ where k is a constant. The scientist collects some values for x and y and draws a graph of $\log x$ against $\log y$. The points are approximately in a straight line.

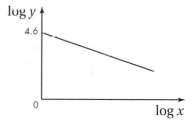

a The scientist says that this shows that the formula is valid. What is the gradient of the line?

b Estimate the value of k to 2 s.f.

 7 The value of a car decreases with age as shown in this table.

Age (t years)	2	4	6	8	10
Value (£y)	18490	13675	10114	7480	5533

A model for the value of the car is the formula $y = Ac^t$ where A and c are parameters.

a Draw a suitable straight line graph to show that this is a suitable model.

b Find the values of A and c.

8 A graph of $\log y$ against $\log x$ is a straight line.

Find a formula for y in terms of x.

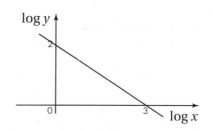

7.5 The number e

Here is the graph of $y = 2^x$.

How can you find the gradient at a particular point with coordinates $(x, 2^x)$?

You can draw the tangent to the curve at any point. The gradient of the tangent at that point is the gradient of the curve at that point.

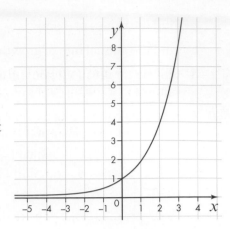

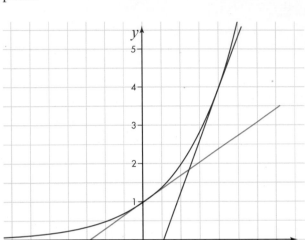

This graph has tangents drawn at (0, 1) and (2, 4).

By drawing a triangle under the tangent, the gradient can be calculated.

Here is a table to show the gradient at different points on the curve $y = 2^x$:

Point	(−3, 0.125)	(−2, 0.25)	(−1, 0.5)	(0, 1)	(1, 2)	(2, 4)	(3, 8)
Gradient	0.087	0.173	0.347	0.693	1.386	2.773	5.545

The gradients are rounded to 3 decimal places.

If you look carefully you will see that the gradient is always 0.6931 × the y-coordinate.

> The value of the multiplier is given to 4 d.p.

This is true for any point on the curve, not just for those in the table.

The gradient of the curve $y = 2^x$ at (x, y) is $0.6931y$.

This method could be used for any curve of the form $y = a^x$ where a is a positive number.

In every case, the gradient is a multiple of the y-coordinate.

Here are expressions for the gradient for different values of a:

y	2^x	3^x	5^x	1.5^x	0.8^x
Gradient	$0.6931y$	$1.0986y$	$1.6094y$	$0.4054y$	$-0.2231y$

Why does the multiplier for $y = 0.8^x$ have a minus sign?

If $y = a^x$ where $a > 0$, the gradient is always a multiplier $\times y$ and the larger a is, the larger the multiplier is. If $a < 1$ then the multiplier is negative.

From the table, it looks as if there should be a value of a between 2 and 3 for which the multiplier is 1.

There is! It is an irrational number (like π) with an infinite decimal expansion that does not recur.

It is denoted by the letter e. Here are the first few terms of the decimal expansion of e:

$$e = 2.718\,281\,828\,459\,045\ldots$$

We now have this important result:

If $y = e^x$ then the gradient at (x, y) is e^x:

The gradient at any point on the curve is equal to the y-coordinate of that point.

Here is a graph of $y = e^x$:

KEY INFORMATION

The gradient at a point on the curve $y = e^x$ is the y-coordinate, or e^x.

TECHNOLOGY

Your calculator should have a button for finding powers of e.

Use it to confirm the value of e shown above.

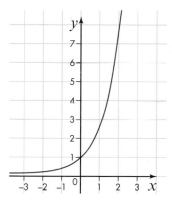

Check by eye that the gradient where $y = 1, 2, 3$ or $\frac{1}{2}$ is indeed $1, 2, 3$ or $\frac{1}{2}$.

For comparison, here are the graphs of $y = 2^x$ (green), $y = e^x$ (red) and $y = 3^x$ (blue):

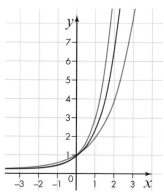

Here are the graphs of $y = e^x$ and $y = 2e^x$:

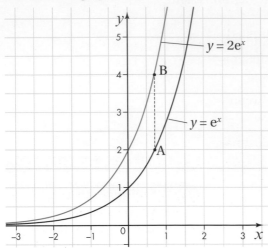

The graph of $y = 2e^x$ is a stretch of $y = e^x$ from the x-axis with a scale factor of 2.

This means that gradients at corresponding points are multiplied by 2.

For example, the gradient at A is 2 and the gradient at B is 4.

The gradient at B is simply the y-coordinate, and this is true for any point on the curve $y = 2e^x$.

If $y = 2e^x$ then the gradient $= 2e^x = y$.

The generalisation is true: if $y = ke^x$ then the gradient $= ke^x = y$ for any value of k.

Stop and think   Sketch graphs to show that this statement is true if $k = 3$.

Sketch graphs to show that this statement is true if $k = \frac{1}{2}$.

Now look at the graph of $y = e^{2x}$.

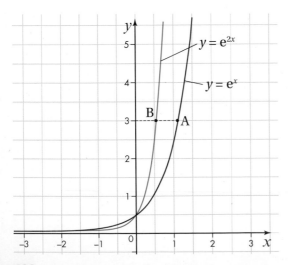

The graph of $y = e^{2x}$ is a stretch of $y = e^x$ from the y-axis with a scale factor of $\frac{1}{2}$.

This means that gradients at corresponding points are multiplied by 2.

For example, the gradient at A is 3 and the gradient at B is 6.

The gradient at B is $2 \times$ the y-coordinate and this is true for any point on the curve $y = e^{2x}$.

If $y = e^{2x}$ then the gradient $= 2e^{2x} = 2y$.

The generalisation is true: if $y = e^{kx}$ then the gradient $= ke^{kx} = ky$ for any value of k.

A graph of $y = ae^{kx}$, $a > 0$, has the following properties:

> it crosses the y-axis at $(0, a)$

> it is always above the x-axis

> if $k > 0$ the gradient is always positive and increases as x increases

> if $k < 0$ the gradient is always negative and gets closer to 0 as x increases.

> **KEY INFORMATION**
>
> If $y = e$ to the power kx then the gradient $= ke$ to the power kx for any value of k.

Example 5

a Sketch the graph of $y = e^{0.5x + 3}$.
b Find the gradient where the curve crosses the y-axis.

Solution

a $y = e^{0.5x + 3}$
$\quad = e^{0.5x} \times e^3 = 20.1e^{0.5x}$
The graph looks like this:

$e^3 = 20.1$ to 3 s.f.

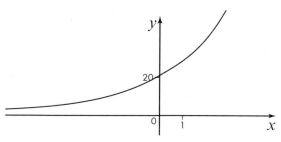

b Gradient $= 0.5 \times 20.1e^{0.5x}$

Where the graph crosses the y-axis, $x = 0$ and gradient $= 10.1$ to 3 s.f.

The graph is steeper than it appears in the sketch – the scales on the axes are not the same.

Exercise 7.5A

Answers page 524

1 Find the value of these expressions to 3 s.f.:

a e^2 **b** $\sqrt{e}$ **c** $\dfrac{1}{e}$

d $\dfrac{e+2}{e-2}$ **e** e^{-3}

CM **2** **a** Find the value of $\left(1+\dfrac{1}{n}\right)^n$ when:

    i $n = 100$ **ii** $n = 10\,000$ **iii** $n = 1\,000\,000$

b How are the answers to **part a** connected to the value of e?

c What do your answers suggest about the value of $\left(1+\dfrac{1}{n}\right)^n$ when n is very large?

3 **a** On the same axes, sketch the curves:

    i $y = e^x$ **ii** $y = e^{-x}$

b Describe the transformation that will map one curve onto the other.

4 Find $f'(x)$ in the following cases:

a $f(x) = 4e^x$ **b** $f(x) = e^{-x}$

c $f(x) = e^{0.5x}$ **d** $f(x) = 10e^{3x}$

5 The equation of a curve is $y = e^x$. Find the gradient of the curve:

a at $(0, 1)$ **b** at the point with an x-coordinate of 2

c at the point with a y-coordinate of 2.

6 Repeat **question 5** for the curve $y = e^{3x}$.

7 The equations of these curves are $y = e^{x+1}$, $y = e^x + 1$ and $y = (e+1)^x$.

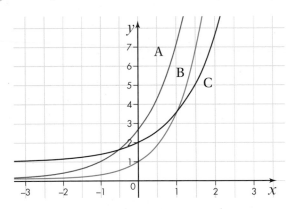

a Match each equation to the appropriate curve on the graph.

b Two of the curves are translations of $y = e^x$. Identify the curves and give the vector of the translation in each case.

c One of the curves is a stretch of $y = e^x$. Identify the curve and describe the stretch.

8 a Find the point where the curve $y = e^{x+2}$ crosses the y-axis.

  b Prove that the curves $y = e^{x+2}$ and $y = e^{x+4}$ are exactly the same shape.

 9 a Work out the value of $1 + \frac{1}{1!} + \frac{1}{2!} + \frac{1}{3!} + \frac{1}{4!} + \frac{1}{5!} + \frac{1}{6!}$ where 3! (3 factorial) = $3 \times 2 \times 1 = 6$ and $4! = 4 \times 3 \times 2 \times 1 = 24$ and so on.

  b Compare your answer to **part a** to the value of e.

  c Add another term to your sum in **part a**. What do you notice about the answer?

  d What does your answer to **part c** suggest? Check your suggestion by adding another term.

 10 a Sketch the graphs of $y = e^{0.5x}$ and $y = e^{2x}$ on the same axes.

  b Describe a stretch that will map $y = e^{2x}$ onto $y = e^{0.5x}$.

11 Find the exact solutions of the equation $2x^2 + ex = e^2$.

7.6 Natural logarithms

In the last section you looked at powers of e. In this section you will look at logarithms to the base e.

$$e^2 = 7.389 \text{ to } 4 \text{ s.f.}$$

You could write that as $\log_e 7.389 = 2$

Logarithms to the base e are sometimes called **natural logarithms** and they are written as ln.

So, you write $\ln 7.389 = 2$

Here are some values of $\ln x$, given to 3 d.p.

x	10	100	1000	0.1	0.01
$\ln x$	2.303	4.605	6.908	−2.303	−4.605

> **KEY INFORMATION**
>
> $\log y$ means logarithm to the base 10 and $\ln y$ means logarithm to the base e, where e / 2.718 28...
>
> If $e^x = y$ then $\ln y = x$.

> **TECHNOLOGY**
>
> You can easily generate values of ln like the ones in the table using a spreadsheet software package.

Example 6

Solve the equation $20e^{2x-4} = 35$.

Solution

Divide both sides of the equation by 20.

$$e^{2x-4} = 1.75$$

Find the ln of each side.

$$2x - 4 = \ln 1.75$$

Rearrange.

$$x = \frac{\ln 1.75 + 4}{2} = 2.280 \text{ to } 4 \text{ s.f.}$$

> **TECHNOLOGY**
>
> Your calculator should have a button for finding ln. It is often the same button as for finding e^x because these are inverse functions.

> **Stop and think**
>
> You could take ln of both sides of the equation immediately.
>
> $\ln(20e^{2x-4}) = \ln 35$
>
> Go on from here to show that this method gives the same answer.
>
> You can use either method. Which method do you prefer?

Here is a graph of $y = \ln x$.

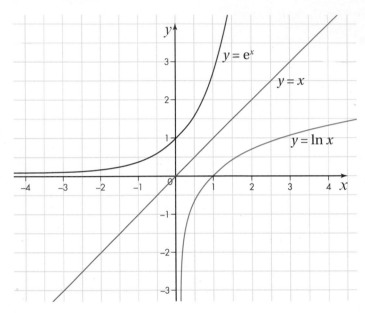

The graph of $y = \ln x$ is a reflection of the graph of $y = e^x$ in the line $y = x$.

This is because e^x and $\ln x$ are inverse functions.

> **Stop and think** Describe the properties of the graph of $y = \ln x$.

Exercise 7.6A

Answers page 524

1 $\ln a = 3$ and $\ln b = 4.5$

Work out the value of:

   a $\ln ab$ **b** $\ln a^4$

   c $\ln \dfrac{b}{a^2}$ **d** $\ln \sqrt{ab}$

2 Solve these equations. Give your answers to 3 s.f.

   a $e^x = 2000$ **b** $e^{-x} = 0.03$

   c $e^{2x-5} = 125$ **d** $e^{-\frac{1}{2}x^2} = 0.5$

3 Solve these equations. Give your answers to 3 s.f.

  a $\ln 0.5x = 4$ **b** $\ln(4x + 2) = 3.5$

  c $4 + \ln 2x = 0$ **d** $\ln\dfrac{120}{x} = 5$

CM 4 What is the connection between $\ln a$ and $\ln\dfrac{1}{a}$ for $a > 0$?

5 a Show that the graph of $y = \ln 4x$ is a translation of the graph of $y = \ln x$.

  Write down the vector that describes the translation.

  b On the same axes sketch the graphs of $y = \ln x$ and $y = \ln 4x$.

6 Solve these equations. Give your answers to 3 d.p.

  a $20e^t = 100$ **b** $40e^{-t} = 35$

  c $250e^{3t} = 8000$ **d** $32.5e^{-0.85t} = 14.8$

7 a Draw a sketch to show that the graphs of $y = 2e^x$ and $y = 6e^{-x}$ cross at one point.

  b Find the coordinates of the point of intersection.

8 The graph of $y = \ln x$ is a reflection of the graph of $y = e^x$ in the line $y = x$.

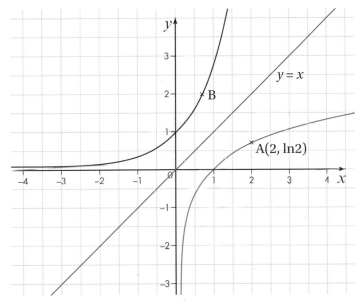

Point A has the coordinates $(2, \ln 2)$ and B is the reflection of A.

  a Write down the coordinates of B.

  b Find the gradient of $y = e^x$ at B.

  c Find the gradient of $y = \ln x$ at A.

  d Generalise your result from **part c**.

9 Here is an equation: $e^{2x} + e^x = 6$.

  a Make the substitution $y = e^x$ and show that you get a quadratic equation in y.

  b Solve the quadratic equation to get two values for y.

  c Use your values for y to solve the original equation.

10 Solve the equation $2e^{2x} - 9e^x + 4 = 0$.

PF **11** Prove that:

    a $\ln(e + e^2) = 1 + \ln(1 + e)$

    b $\ln(e^2 - e^4) = 2 + \ln(1 + e) + \ln(1 - e)$

7.7 Exponential growth and decay

Earlier in this chapter you saw some examples of exponential growth and decay.

It is very common to write equations for exponential growth and decay in terms of powers of e. One reason is that it is then easy to find the rate of change (i.e. the gradient of the curve) at any time.

In **Section 7.1** you looked at an investment of £10 000 earning 20% interest each year.

The value (£y) after t years is given by the formula $y = 10\,000 \times 1.2^t$.

Can 1.2^t be written as a power of e?

Because e^x and $\ln x$ are inverse functions you can write $1.2 = e^{\ln 1.2}$

Then $1.2^t = (e^{\ln 1.2})^t$

$$= e^{(\ln 1.2)t}$$

$$= e^{0.1823t}$$

The formula for the investment is $y = 10\,000e^{0.1823t}$

> **Stop and think**   Show that this formula gives the correct values if $t = 1$ or $t = 2$. You will find those values in **Section 7.1**. There will be small errors. Why?

Of course, the interest is only added once a year so the graph would go up in a sequence of steps. The curve shows how the value would increase if the interest was added continuously – perhaps every day.

You could find the rate of increase of the value at any time.

If $y = 10\,000e^{0.1823t}$ then gradient $= 0.1823 \times 10\,000e^{0.1823t} = 1823e^{0.1823t}$

When $t = 1$, the rate of increase is £2188 per year.

When $t = 2$, the rate of increase is £2625 per year.

These values are the rates of growth at particular moments and they increase as the value of t increases.

This is not the same as the *percentage* growth rate, which is constant at 20% per year.

Example 7

A model for the population of a country (y million people) in t years' time is

$y = 14e^{at}$ where a is a constant.

The population is expected to grow by 5% in the next 10 years.

a Calculate the value of a.

b Estimate the population in 4 years' time.

c Explain why this model might not be appropriate over a long period of time.

Solution

a $y = 14e^{at}$

If $t = 0$, $y = 14e^0 = 14$ ●───────────────

Put $t = 10$ in the formula.

The population in 10 years' time is $14e^{10a}$.

If the percentage growth over the next 10 years is 5%, then

$$\frac{14e^{10a} - 14}{14} = 0.05$$

You can divide every term in the fraction by 14.

$$e^{10a} - 1 = 0.05$$
$$e^{10a} = 1.05$$
$$10a = \ln 1.05$$
$$a = \frac{\ln 1.05}{10} = 0.004\,88 \text{ to 3 s.f.}$$

The population in t years time is $14e^{0.00488t}$ million people.

b If $t = 4$, $y = 14e^{0.00488 \times 4} = 14.28$ million to 2 d.p.

c The exponential model assumes that the rate of growth of the population stays at 5% over any 10-year period. This may not be the case. Changes in the birth rate, in life expectancy, and in levels of immigration and emigration can all affect the rate of population growth.

> Remember that $e^0 = 1$

Carbon-14 is an isotope of carbon that is found in all plants. Carbon-14 is radioactive, which means that over time it decays and changes into a different element.

Carbon-14 dating is used by archaeologists to find the age of artefacts made from wood or other plant material. They can do this by comparing the amount of carbon-14 in the object with the amount that would be present in living plant material. ●───

> Willard F. Libby was awarded the Nobel Prize for Chemistry in 1960 for his work on carbon-14 dating.

Example 8

Suppose there are m milligrams of carbon-14 in a plant when it dies. After t years the amount of carbon-14, in milligrams, will be me^{-kt} where $k = 1.21 \times 10^{-4}$.

a A wooden toy was made 2000 years ago. What percentage of the carbon-14 in the original piece of wood will be left in the toy?

b Calculate the amount of time for half the carbon-14 in a dead tree to decay.

Solution

a Suppose there are initially m mg.

After 2000 years there will be $me^{-1.21\times10^{-4}\times2000} = 0.785m$

The amount of carbon-14 remaining is 78.5% of the original amount.

b When the amount of carbon-14 is reduced to half of what it was,

$$me^{-kt} = \tfrac{1}{2}m$$

Divide by m:

$$e^{-kt} = \tfrac{1}{2}$$
$$-kt = \ln 0.5$$
$$t = \frac{\ln 0.5}{-1.21\times10^{-4}} = 5728$$

It takes 5728 years for half of the carbon-14 to decay.

> This is known as the half-life of carbon-14. You can find the half-lives of many radioactive elements on the internet.

Exercise 7.7A

Answers page 525

1 The value (£v) of an investment after t years is predicted using the formula $v = 5000e^{0.18t}$

  a What is the initial value?

  b Work out the value after:

    i one year   **ii** five years   **iii** 10 years.

CM 2 This graph illustrates exponential growth in the value of a painting. The value is shown in thousands of pounds.

  a What is the current value?

  b How long does it take for the value of the painting to double?

  c Find the percentage increase in value in the first year (from $t = 0$ to $t = 1$).

  d Find the percentage increase in value in the sixth year (from $t = 5$ to $t = 6$).

  e Find the percentage increase in value in any other year.

  f What do you notice about your last three answers?

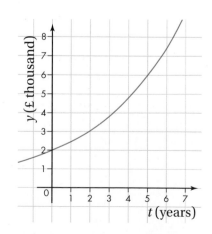

3 The number of starlings in an area of the country is declining. There are currently N birds.

A model for the number (x) t years from now is $x = Ne^{-0.15t}$.

a Show that in 5 years from now the number will have declined to under 50% of the current value.

b What can you say about the population 10 years from now?

4 A model for the population (p million) of a country t years from now is $p = 50e^{0.02t}$.

a Use this model to complete this table of population estimates. Round your answers to 2 d.p.

Number of years from now	0	10	20	30
Population		61.07		

b The table shows the predicted change in population over three 10-year intervals. Show that the *percentage* change in each 10-year interval is the same.

(M) **5** A businessman buys a machine. The value of the machine is expected to fall exponentially and follow this graph.

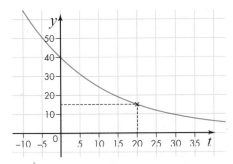

The value after t years is given by the formula $y = Ac^{-kt}$.

Find the values of the parameters A and k.

(M) **6** Radium-223 is a radioactive element.

If you have m mg now, in t days you will have $me^{-0.06t}$ mg.

A scientist has 5.00 mg of radium-223.

a Complete this table to show how the quantity decays. Round the masses to 1 d.p.

Days	0	2	5	10	15
Mass remaining (mg)	5.00		3.70		

b Draw a graph to illustrate your table of values.

c Use your graph to estimate the half-life of radium-233.

d Calculate the half-life using the expression for the mass.

e How long will it take for the mass to be reduced to a quarter of its original value?

7 If you take a drug, the amount in your bloodstream will decay exponentially over time. This means that the drug will have less effect as time goes on.

A patient is taking a drug for a medical condition.

The initial dose is 20 mg.

The amount left in the bloodstream after t hours is $20e^{-0.07t}$ mg.

  a Calculate the amount of the initial dose that is left in the bloodstream after 24 hours.

  b Sketch a graph to show how the amount of the initial dose in the bloodstream declines over 24 hours.

The patient is given a second dose of 20 mg after 12 hours.

  c Find the total amount of the drug in the bloodstream 24 hours after the initial dose.

  d Sketch a graph to illustrate the new situation.

SUMMARY OF KEY POINTS

› A graph of $y = a^x$ with $a > 1$ is an example of exponential growth.

› A graph of $y = a^x$ with $0 < a < 1$ is an example of exponential decay.

› If $y = a^x$ then $\log_a y = x$. The base of the logarithm is a.

› $\log y$ means logarithm base 10 and $\ln y$ means logarithm base e, where e $/ 2.71828$

› The laws of logarithms are:

  › $\log_a xy = \log_a x + \log_a y$

  › $\log_a \dfrac{x}{y} = \log_a x - \log_a y$

  › $\log_a x^k = k \log_a x$.

› Solve equations of the form $y = ax^n$ by taking logarithms of both sides of the equation.

› Straight line logarithmic graphs can be used to estimate parameters (a, b, k and n) in relationships of the form $y = ax^n$ or $y = kb^x$.

› The gradient of $y = e^{ax}$ at (x, y) is ae^{ax} or ay.

› Exponential growth and decay are often described by a relationship of the form $y = ae^{kx}$ where a and k are parameters. For growth, $a > 0$; for decay, $a < 0$.

EXAM-STYLE QUESTIONS 7

Answers page 526

1 Here is a graph of $y = 1.5^x$:

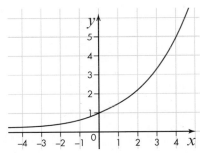

    a Use the graph to find an approximate solution to the equation $1.5^x = 4$.   **[1 mark]**

    b On the same axes sketch the graphs of $y = 1.5^x$ and $y = 1.5^{-x}$.   **[2 marks]**

2 Solve the equation $\log_5 (3t + 8) = 2.1$.

    Round your answer to 3 s.f.   **[3 marks]**

3 The equation of a curve is $y = e^{0.5x + 3}$.

    a Where does the curve cross the y-axis?   **[1 mark]**

    The curve passes through the point $(-2, k)$.

    b Calculate the value of k.   **[2 marks]**

    The curve passes through the point $(h, 100)$.

    c Calculate the value of h.   **[2 marks]**

4 Here is a graph of $y = ce^{-0.5x}$:

a Find the value of c. **[1 mark]**

b Find the gradient at the point where the graph crosses the y-axis. **[2 marks]**

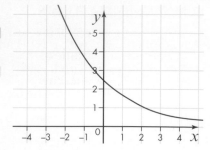

5 $\log_9 x = k$

Write in terms of k:

a $\log_9 3x$ **[2 marks]**

b $\log_3 x$ **[2 marks]**

6 Here is a formula: $r = 0.1e^{0.5t + 1}$

Rearrange the formula to make t the subject. **[3 marks]**

7 The value of a financial investment is modelled by the formula $y = 10e^{0.3t}$ where t is the number of years and y is the value in thousands of pounds.

a Write down the initial value of the investment. **[1 mark]**

b Calculate the annual percentage growth in value. Give your answer to 3 s.f. **[2 marks]**

c Calculate the *monthly* percentage growth in value. Assume that a month is $\frac{1}{12}$ of a year. Give your answer to 3 s.f. **[3 marks]**

8 This table gives estimates of the world population in different years.

Date	1400	1500	1600	1700	1800
Population (millions)	375	460	540	640	945

Sam is trying to model the population by an equation of the form $y = Ae^{kt}$ where y is the population in millions and t is the number of centuries after 1400.

Sam decides to find values of A and k that gives the exact values for 1400 and 1800.

a Show that Sam's value for A is 375. **[1 mark]**

b Find Sam's value for k. **[4 marks]**

c Does Sam's formula give a reasonable value for the population in 1600? Justify your answer. **[2 marks]**

9 A scientist is growing bacteria in a laboratory. The size of the colony of bacteria after t hours is modelled by the formula $y = ka^t$ where k and a are parameters.

Here is a graph of $\log y$ against t:

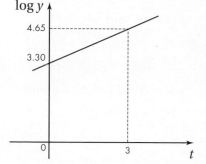

a Calculate the value of k. **[2 marks]**

b Calculate the value of a. **[4 marks]**

c Why might the model stop being appropriate as the number of hours increases? **[1 mark]**

10 The equation of a curve is $y = 20x^n$ where n is a constant.

The point (15, 590) is on the curve.

Calculate the value of n. [5 marks]

11 Solve these equations:

a $e^x = 4e^{-x}$ [3 marks]

b $e^x - 3 = 4e^{-x}$ [5 marks]

(PF) 12 a and b are positive integers.

Prove that $\log_a b = \dfrac{1}{\log_b a}$ [3 marks]

(PF) 13 Prove that $\log_{10}\left(8\sqrt{5}\right) = \dfrac{1}{2}\left(1 + 5\log_{10} 2\right)$ [4 marks]

(M) 14 A scientist records the temperature of a cooling metal at two times.

Time, t (minutes)	0	5
Temperature, y (°C)	200	148

a The scientist wants to fit a formula to this data in the form $y = Ae^{kt}$ where A and k are constants.
Find the values of A and k. [3 marks]

b After 10 minutes the scientist records the temperature again and adds it to the table.

Time, t (minutes)	0	5	10
Temperature, y (°C)	200	148	121

Show that this new reading is not consistent with the model. [1 mark]

c The scientist decides that the first reading ($t = 0$) is probably incorrect and adjusts the model to fit the other two values.

Find the new values for A and k. [4 marks]

d What does the model imply about the temperature after a long period of time? Is this a reasonable result? [1 mark]

15 Two variables, x and y, are related by the formula $y = ax^n$ where a and n are constants.

This table shows three pairs of values of x and y.

x	100	800	5000
y	4000	177	11.3

By drawing a graph, estimate the values of a and n. [4 marks]

16 Solve the equation $2^{2x} + 2^3 = 2^{x+3}$ [4 marks]

8 DIFFERENTIATION

This graph shows the speed of a car over a 40-second interval.

The equation is $y = 20 + 0.0015x(x - 10)(x - 25)(x - 45)$.

The speed is in m s^{-1}. ($20\,\text{m s}^{-1}$ is about $72\,\text{km h}^{-1}$.)

When the speed is increasing the car is accelerating. When the speed is decreasing the car is decelerating. The gradient of the curve at a particular time tells you the acceleration or deceleration at that time. For example, knowing the gradient allows you to estimate the acceleration after 10 seconds, or after 30 seconds.

In this chapter you will learn about differentiation. This is an algebraic method that will help you to find the gradient at any point on the curve. You will see that the gradient can provide information that is useful in real-life situations.

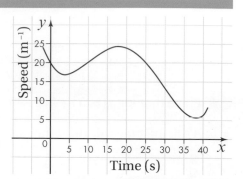

LEARNING OBJECTIVES

You will learn how to:

› find the gradient of the tangent to a curve at any point on the curve

› differentiate expressions containing powers of a variable

› determine geometrical properties of a curve.

TOPIC LINKS

In this chapter you will make use of the algebraic skills you practised in **Chapter 1 Algebra and functions 1: Manipulating algebraic expressions**. You will need to be able to manipulate indices and brackets. You will apply your knowledge of straight lines graphs from **Chapter 4 Coordinate geometry 1: Equations of straight lines**, and methods for finding their equations. The skills learnt in this chapter will be essential when you learn about integration in **Chapter 9 Integration**.

PRIOR KNOWLEDGE

You should already know how to:

› find the gradient of a straight line from its equation

› find the equation of a straight line given two points on the line or one point and the gradient

› manipulate brackets in algebraic expressions

› use and interpret negative and fractional indices.

You should be able to complete the following questions correctly:

1. Here are the equations of some straight lines. Write down the gradient of each one.

   a $y = 7x + 10$ **b** $y = 15 - 0.3x$

   c $y = \frac{1}{2}(x + 4)$ **d** $2x + y = 20$

2. Multiply out the brackets:

   a $2x(x - 4)$ **b** $x^2(x + 3)$

   c $(x + 4)(x - 5)$ **d** $(3x + 1)^2$

3. Write each expression as a power of x.

   a $\dfrac{1}{x^3}$ **b** $\sqrt{x}$ **c** $\dfrac{1}{\sqrt{x}}$

   d $\sqrt[3]{x}$ **e** $\left(\sqrt{x}\right)^3$

8.1 The gradient of a curve

A cyclist is travelling along a straight road and passes a pedestrian.

This graph shows the **distance** of the cyclist from the pedestrian (who is stationary) over the next 40 seconds.

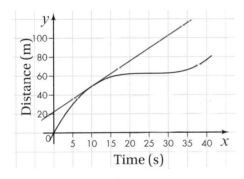

The equation for the curve is $y = 0.004x^3 - 0.3x^2 + 7.5x$. The straight line has the equation $y = 2.7x + 22$ so it passes through (0, 22), (10, 49) and (20, 76).

How can you tell that the speed is changing?

A tangent to the curve has been drawn where $x = 10$.

The speed when $x = 10$ is the **gradient** of the tangent. •

Check that the gradient of the tangent is 2.7. The speed when $x = 10$ is $2.7\,\text{m}\,\text{s}^{-1}$.

> You dealt with the gradient of a straight line graph in your GCSE course.

Example 1

This graph shows the temperature of a liquid over a 7-minute interval.

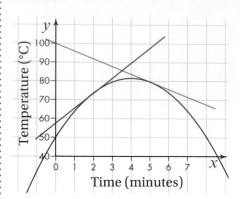

Find the rate of change of the temperature:

a after 2 minutes   **b** after 5 minutes.

Solution

Tangents have been drawn where $x = 2$ and where $x = 5$.

a The gradient of the tangent when $x = 2$ is 8. •────

When $x = 2$ the rate of change of the temperature is $8\,°C$ per minute.

b The gradient of the tangent when $x = 5$ is -4.

When $x = 5$ the rate of change is -4; the temperature is decreasing by $4\,°C$ per minute.

> Check the gradient using a triangle under the tangent.

> **KEY INFORMATION**
>
> The gradient of a curve at any point is equal to the gradient of the tangent at that point.

Exercise 8.1A

Answers page 527

 1 For each of these graphs, work out the gradient and say what it represents.
Put units in your answer.

a The distance of a runner from the start of a race.

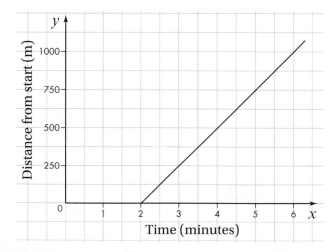

---

214   (M) Modelling   (PS) Problem solving   (PF) Proof   (CM) Communicating mathematically

b The speed of a car.

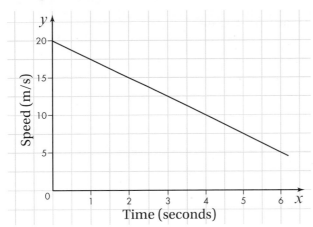

c The amount of petrol in a tank that is being filled.

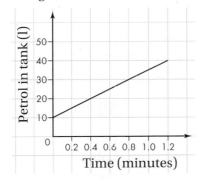

d The volume of air in a balloon as it is inflated.

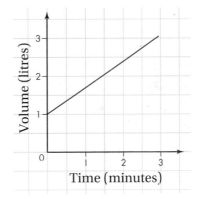

e The amount of charge left on a phone.

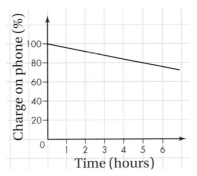

(M) **2** This curve shows the speed of an accelerating racing car.

a A tangent has been drawn at $x = 3$.
Use this to find the acceleration of the car after 3 seconds.

b Estimate the initial acceleration, when $x = 0$.

c Estimate the acceleration after 8 seconds (when $x = 8$).

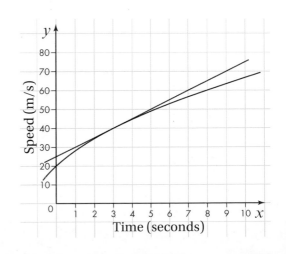

M **3** A bungee jumper drops from a high platform.

This graph shows the distance the jumper has fallen in the first 3 seconds.

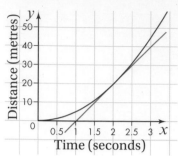

A tangent has been drawn at $x = 2$.

a Find the distance the jumper has fallen after:

    i 1 second   **ii** 2 seconds   **iii** 3 seconds.

b Find the speed at which the jumper is falling after 2 seconds.

c Estimate the jumper's speed after:

    i 1 second   **ii** 3 seconds.

M **4** This graph shows the height of a stone thrown in the air.

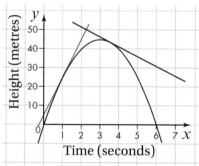

Tangents have been drawn at $x = 1$ and $x = 3.5$.

a Find the speed and the direction of travel after 1 second and after 3.5 seconds.

b The graph has a line of symmetry at $x = 3$. Use this fact to find the speed and the direction of travel after 2.5 seconds and after 5 seconds.

c Estimate the speed after 2 seconds.

M **5** This cone is being filled with water.

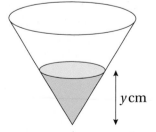

The depth of water (y cm) after x seconds is shown on this graph.

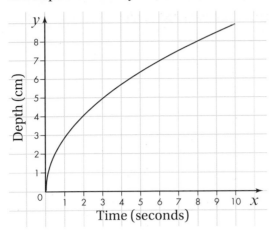

Estimate the rate of change of the height, in cm s^{-1}:

a after 1 second   **b** after 8 seconds.

8.2 The gradient of a quadratic curve

When you know the equation of a curve, it can help you to find the gradient of the curve at any point.

In this section you will look at quadratic curves.

The equation of the simplest quadratic curve is $y = x^2$.

Example 2

a Draw the curve $y = x^2$.

b Draw tangents and use them to find the gradient of the curve at each of these points:

i (1, 1)   **ii** (−2, 4)   **iii** (3, 9)

Solution

a

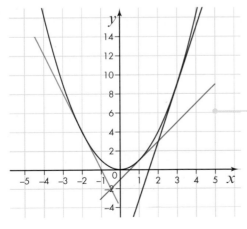

TECHNOLOGY

Use graphical software if it is available. If you draw the curve and the tangents by hand you may not get exactly the same answers as those shown in the solution.

Note that the scales on the axes are not the same.

b Three tangents have been drawn.

   i The gradient at (1, 1) is 2.

   ii The gradient at (−2, 4) is −4.

   iii The gradient at (3, 9) is 6.

Stop and think   You should be able to use your answers to **part b** to write down the gradients at three more points on the curve. Do so.

What is the gradient at (0, 0)?

Here is the graph of $y = x^2$ drawn on a different scale:

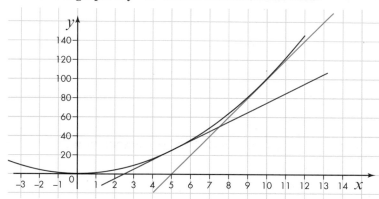

Tangents have been drawn at the points (5, 25) and (10, 100).

What is the gradient of the curve at each of those points?

What is the gradient at the points (−5, 25) and (−10, 100)?

Here is a table showing the gradients at some of the points on the curve $y = x^2$ that you have looked at so far:

Point	(1, 1)	(−2, 4)	(3, 9)	(2, 4)	(−3, 9)	(0, 0)	(−5, 25)	(10, 100)
Gradient	2	−4	6	4	−6	0	−10	20

Stop and think   What is the connection between the x-coordinate of a point and the gradient of the curve at that point?

The gradient at any point on the curve $y = x^2$ is twice the x-coordinate or $2x$.

We write it like this:

> if $y = x^2$ then $\dfrac{dy}{dx} = 2x$.

$\dfrac{dy}{dx}$ is called the **derivative** of y. You say it as 'dee y by dee x'.

The process of finding $\dfrac{dy}{dx}$ is called **differentiation.**

$\dfrac{dy}{dx}$ looks like a fraction but it is not. You should *not* treat the numerator and denominator as separate symbols. In particular, you *cannot* cancel the ds and simplify it to $\dfrac{y}{x}$.

Exercise 8.2A

Answers page 527

1 $y = x^2$

Find the value of y and the value of $\dfrac{dy}{dx}$ when:

a $x = 4$ **b** $x = 2.5$

c $x = -1.2$ **d** $x = 35$

2 Find the gradient of the tangent to the curve $y = x^2$ at the point:

a $(1.5, 2.25)$ **b** $(-0.5, 0.25)$

c $(13, 169)$ **d** $(-3.8, 14.44)$

(PS) 3 There are two points on the curve $y = x^2$ where the y-coordinate is 49. What is the gradient of the curve at each of those points?

4 **a** Draw the curve $y = 0.5x^2$.

    b Draw the tangents to the curve at $(2, 2)$, $(-3, 4.5)$ and $(4, 8)$.

    c Find the gradient of the curve at each of those points.

    d Make a prediction of a formula for $\dfrac{dy}{dx}$ when $y = 0.5x^2$.

    e Choose some other points on the curve, draw the tangents and test your prediction.

5 The points $(10, k)$ and $(-10, k)$ are on the curve $y = 0.5x^2$.

Find the value of k and the gradient at each of these points.

6 **a** Draw the curve $y = 2x^2$.

    b Find an expression for $\dfrac{dy}{dx}$ when $y = 2x^2$. Give some examples to justify your answer.

A general result is that if $y = ax^2$ then $\dfrac{dy}{dx} = 2ax$.

Here are some examples:

› if $y = 3x^2$ then $\dfrac{dy}{dx} = 6x$

› if $y = 5x^2$ then $\dfrac{dy}{dx} = 10x$

› if $y = -2x^2$ then $\dfrac{dy}{dx} = -4x$

› if $y = 0.65x^2$ then $\dfrac{dy}{dx} = 1.3x$.

Stop and think   If $y = bx + c$, where b and c are numbers, what does the graph of y look like? What is $\dfrac{dy}{dx}$?

Here is a general result for any quadratic:

› if $y = ax^2 + bx + c$ then $\dfrac{dy}{dx} = 2ax + b$

The derivative of ax^2 is $2ax$. The derivative of bx is b. The derivative of c is 0.

To differentiate the sum or difference of two terms you just differentiate each term separately. In this case, that means differentiating ax^2, then bx and then c.

There is an alternative notation that is often used.

You can write f(x) instead of y and for the derivative you write f′(x) instead of $\dfrac{dy}{dx}$.

The dash next to the f indicates differentiation.

So you could write $y = 2x^2 - 3x + 5$ and $\dfrac{dy}{dx} = 4x - 3$

or you could write f(x) = $2x^2 - 3x + 5$ and f′(x) = $4x - 3$.

Example 3

f(x) = $4x(5 - x)$

a Work out f′(x).

b Find the gradient of the curve $y = 4x(5 - x)$ at the point where $x = 4$.

c Sketch the graphs of $y = $ f(x) and $y = $ f′(x).

Solution

a f(x) = $4x(5 - x)$

Multiply out the brackets.

f(x) = $20x - 4x^2$

Differentiate.

f′(x) = $20 - 8x$

> Differentiate each term separately.

b f′(4) = $20 - 32 = -12$

The gradient is −12.

> f′(4) means the value of f′(x) when $x = 4$.

c The curve $y = $ f(x) or $y = 4x(5 - x)$ is a quadratic.

Where it crosses the x-axis, $4x(5 - x) = 0$.

Divide by 4.

$x(5 - x) = 0$

So $x = 0$ or 5.

If $x = 2$, say, the value of y is positive so the curve is 'upside down'.

The graph of $y = $ f′(x) or $y = 20 - 8x$ is a straight line with a gradient of −8.

It crosses the x-axis where $20 - 8x = 0 \Rightarrow x = 2.5$.

The graphs of $y = $ f(x) and $y = $ f′(x) are on the right.

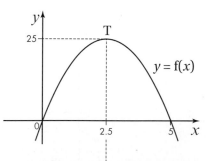

Notice that the point (2.5, 0) on the graph of $y = $ f′(x) corresponds to the point T on the graph of $y = $ f(x) where the gradient is 0.

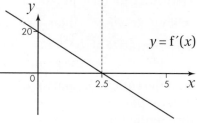

Exercise 8.2B

Answers page 527

1 Work out $\dfrac{dy}{dx}$ for the following curves:

a $y = 3x + 4$ **b** $y = 3x^2 + 4$ **c** $y = 3x^2 + 4x$

2 Work out f′(x) for the following:

a $f(x) = x^2 + 4x - 2$ **b** $f(x) = x^2 - 4x + 2$ **c** $f(x) = 2 + 4x - x^2$

3 Work out $\dfrac{dy}{dx}$ for the following curves. Start by multiplying out the brackets.

a $y = x(x + 6)$ **b** $y = (x - 3)(x + 1)$ **c** $y = (2x + 3)^2$

4 Here is a graph of $y = x^2 + x - 6$:

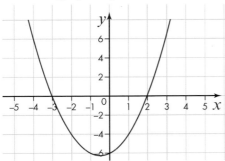

a Work out $\dfrac{dy}{dx}$.

b i Where does the graph cross the y-axis?

ii Work out the gradient of the curve at this point.

c The curve crosses the x-axis at two points. Work out the gradient of the curve at each of these points.

d Sketch a graph of $\dfrac{dy}{dx}$.

5 $f(x) = 6 - x^2$

Here is a graph of $y = f(x)$:

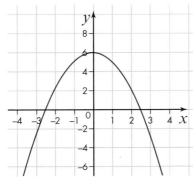

a Work out f′(x).

b Work out the gradient of the curve at (−1, 5) and at (3, −3).

c Sketch a graph of $y = f'(x)$.

6 $f(x) = (x-1)(x-5)$

a Work out $f'(x)$.

b Work out the gradient of the curve $y = f(x)$ at $(1, 0)$ and at $(11, 60)$.

c Sketch graphs of $y = f(x)$ and $y = f'(x)$.

7 The equation of this curve is $y = 15x^2 - 30x - 125$.

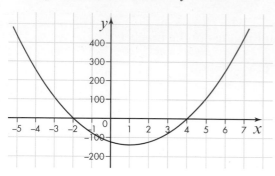

Work out the gradient at:

a $(1, 0)$ **b** $(5, 100)$

8 Look at **Exercise 8.1A question 3**. The equation of the curve is $y = 5x^2$.

Use this fact to check your answers to **part c** of the question.

9 Look at **Exercise 8.1A question 4**. The equation of the curve is $y = 30x - 5x^2$.

Use this fact to check your answer to **part c** of the question.

(PS) 10 Here is a sketch of a curve:

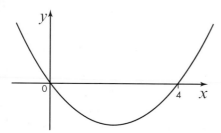

The equation of the curve is $y = kx(x-4)$ where k is a number.

The gradient of the curve at $(4, 0)$ is 2.

Work out the value of k.

8.3 Differentiation of x^2 and x^3

You have been using the result that if $f(x) = x^2$ then $f'(x) = 2x$.

You saw that this was a reasonable result by looking at particular examples. To be certain that it is true you need a proof.

We are looking for the gradient at the point P. The idea is to choose a point near P, find the gradient of the chord joining these two points and then see what happens if the second point is moved closer to P.

Suppose $P(x, f(x))$ is a point on the curve and $Q(x + \delta x, f(x + \delta x))$ is a point close to it, where δx is a small number added to x.

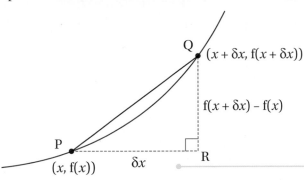

Join P and Q with a straight line. This is a **chord**.

What is the gradient of the chord?

Draw a triangle under PQ, then the gradient of the chord is $\dfrac{QR}{PR}$.

PR is the difference between the x-coordinates and QR is the difference between the y-coordinates.

So the gradient of the chord is $\dfrac{QR}{PR} = \dfrac{f(x+\delta x) - f(x)}{\delta x}$.

Now in this case $\quad f(x) = x^2 \qquad \qquad \textcircled{1}$

and $\qquad \qquad f(x + \delta x) = (x + \delta x)^2 \qquad \textcircled{2}$

The gradient of the chord, $\dfrac{f(x + \delta x) - f(x)}{\delta x} = \dfrac{(x + \delta x)^2 - x^2}{\delta x}$

Multiply out the bracket in the numerator.

$$\dfrac{f(x + \delta x) - f(x)}{\delta x} = \dfrac{x^2 + 2x\,\delta x + \delta x^2 - x^2}{\delta x}$$

$$= \dfrac{2x\,\delta x + \delta x^2}{\delta x}$$

Divide top and bottom by δx:

$$\dfrac{f(x + \delta x) - f(x)}{\delta x} = 2x + \delta x$$

Now look back at the diagram.

Imagine δx getting smaller and closer to 0, so that Q moves down the curve, getting closer and closer to P.

As $\delta x \to 0$, the chord gets closer and closer to the tangent to the curve at P.

As $\delta x \to 0$, the gradient of the chord gets closer to $2x$.

We say that the limit of $\dfrac{f(x + \delta x) - f(x)}{\delta x}$ as $\delta x \to 0$ is $2x$.

We write this more concisely as $\displaystyle\lim_{\delta x \to 0} \dfrac{f(x + \delta x) - f(x)}{\delta x} = 2x$.

PROOF

Read through this proof several times. The first time through, try to get the general idea of where it is going without getting stuck on the details. The second time, try to follow all the steps.

δ is the Greek letter delta and δx is a small change in the value of x. Sometimes h is used as an alternative to δx.

The x^2 terms cancel out.

PROOF

Read through this proof carefully. The questions in **Exercise 8.3A** will help you to understand it more clearly.

You will also see later in the course how the same method can be used for other curves.

This leads to the conclusion that the gradient of the tangent, f'(x), is just 2x.

This completes the proof that if $f(x) = x^2$ then $f'(x) = 2x$.

The derivative of x^2 is $2x$.

> **KEY INFORMATION**
>
> Using the formula
> $$f'(x) = \lim_{\delta x \to 0} \frac{f(x + \delta x) - f(x)}{\delta x}$$ is
> called 'differentiation from first principles'.

Stop and think

Why is a proof like this better than drawing the tangents for different values of x and looking for a pattern in the answers?

We could write $f(x + \delta x) - f(x) = \delta y$, where δy means a small change in the value of y.

Can you think of a reason why we use the notation $\dfrac{dy}{dx}$ for the derivative of y?

Exercise 8.3A
Answers page 528

1 P(3, 9) is a point on the curve $y = x^2$.

   a Work out the gradient of the curve at P.

   Q(3.1, 9.61) is also on the curve $y = x^2$.

   b Work out the gradient of the chord PQ.

   c Work out the gradient of the chord PQ if Q has coordinates $(3.05, 3.05^2)$.

   d Work out the gradient of the chord PQ if Q has coordinates $(3.01, 3.01^2)$.

   e Explain how your answers to **parts b** to **d** confirm your answer to **part a**.

2 P(2, 4) and Q(1.9, 3.61) are two points on the curve $y = x^2$.

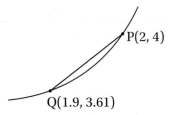

Show that the gradient of the chord PQ is a little smaller than the gradient of the tangent at P.

3 P and Q are two points on the curve $y = 2x^2$.

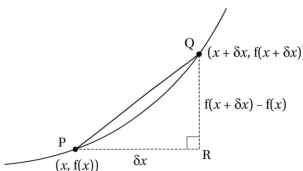

$f(x) = 2x^2$ and $f(x + \delta x) = 2(x + \delta x)^2$

The gradient of the chord is $\dfrac{f(x + \delta x) - f(x)}{\delta x}$.

a Show that $f(x + \delta x) - f(x) = 4x\,\delta x + 2\delta x^2$.

b Show that the gradient of the chord PQ is $4x + 2\delta x$.

c Explain how the result in **part b** shows that if $f(x) = 2x^2$ then $f'(x) = 4x$.

 4 This question is about the function $f(x) = x^3$.

a Sketch the curve $y = x^3$.

P and Q are two points on the curve.

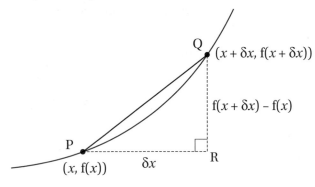

$f(x) = x^3$ and $f(x + \delta x) = (x + \delta x)^3 = x^3 + 3x^2\,\delta x + 3x\,\delta x^2 + \delta x^3$.

b Write down an expression for the length of QR, which is $f(x + \delta x) - f(x)$.

c Show that the gradient of PQ, $\dfrac{f(x + \delta x) - f(x)}{\delta x}$, is $3x^2 + 3x\,\delta x + \delta x^2$.

d Show that as $\delta x \to 0$, $\dfrac{f(x + \delta x) - f(x)}{\delta x} \to 3x^2$.

e If $f(x) = x^3$, what is $f'(x)$?

f Your sketch should show that the gradient of $y = x^3$ is never negative. Is this consistent with your answer to **part d**?

8.4 Differentiation of a polynomial

You now have these results:

> if $y = x^2$ then $\dfrac{dy}{dx} = 2x$

> if $y = x^3$ then $\dfrac{dy}{dx} = 3x^2$.

Or using the function notation:

> if $f(x) = x^2$ then $f'(x) = 2x$

> if $f(x) = x^3$ then $f'(x) = 3x^2$.

These are two examples of a more general result:

> if n is a positive integer and $f(x) = x^n$ then $f'(x) = nx^{n-1}$.

For example, if $f(x) = x^4$ then $f'(x) = 4x^3$ ($n = 4$ and $n - 1 = 3$)

and if $y = x^{10}$ then $\dfrac{dy}{dx} = 10x^9$ ($n = 10$ and $n - 1 = 9$)

> If x^n is multiplied by any constant, just multiply the derivative of x^n by the constant.

> If n is a positive integer and $f(x) = ax^n$ then $f'(x) = nax^{n-1}$.

To differentiate the sum or difference of terms, differentiate each term in turn:

> if $y = f(x) \pm g(x)$ then $\dfrac{dy}{dx} = f'(x) \pm g'(x)$.

Example 4

$f(x) = x^3 - 5x^2 + 2x + 8$

a Show that $(2, 0)$ is on the graph of $y = f(x)$.

b Find the gradient at $(2, 0)$.

Solution

a $f(2) = 2^3 - 5 \times 2^2 + 2 \times 2 + 8 = 8 - 20 + 4 + 8 = 0$

So $(2, 0)$ is on the graph.

b $f(x) = x^3 - 5x^2 + 2x + 8$

Differentiate each term.

$$f'(x) = 3x^2 - 10x + 2$$

At $(2, 0)$, $x = 2$.

$$f'(2) = 3 \times 2^2 - 10 \times 2 + 2 = 12 - 20 + 2 = -6$$

The gradient is -6.

> The derivative of x^3 is $3x^2$.
>
> The derivative of $5x^2$ is $10x$.
>
> The derivative of $2x$ 2.
>
> The derivative of 8 is 0.

Exercise 8.4A

Answers page 529

1 Find $\dfrac{dy}{dx}$ for the following curves:

  a $y = 2x^3$ **b** $y = 0.5x^4$

  c $y = 0.1x^5$ **d** $y = 50x^3$

2 Find $\dfrac{dy}{dx}$ for the following curves:

  a $y = x^3 + 4x^2 - 8x + 1$ **b** $y = 2x^3 - 5x^2 + 6x - 10$

3 Find $f'(x)$ for the following:

  a $f(x) = x^4 + 8x^2$ **b** $f(x) = x^5 - 10x^3 + 2x$

4 This is a graph of $y = x^3 - 2x^2 + 3$.

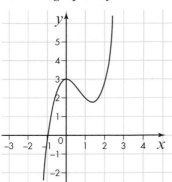

 a Work out $\dfrac{dy}{dx}$.

 b Find the gradient of the curve at:

  i $(2, 3)$ **ii** $(-1, 0)$

 c Look at the graph to check that your answers are reasonable.

5 This is a graph of $f(x) = 0.5x^4 - 2x^2 + x + 3$.

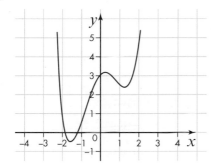

 a Work out $f'(x)$.

 b Find the gradient of the curve at:

  i $(-1, 0.5)$ **ii** $(1, 2.5)$ **iii** $(2, 5)$

 c Look at the graph to check that your answers are reasonable.

6 Find $\dfrac{dy}{dx}$ for the following curves. Start by multiplying out the brackets.

 a $y = x^2(2x + 5)$ **b** $y = x(x - 4)^2$ **c** $y = (x + 1)(x^2 + 1)$

7 This is a graph of $f(x) = x^2(x + 3)$.

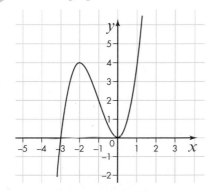

a Work out f′(x).

b Work out the gradient at:

    i (–3, 0)   **ii** (–2, 4)   **iii** (1, 4)

8 This is a graph of $y = x(x - 2)(x - 4)$.

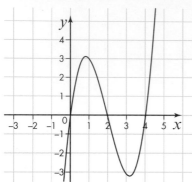

Work out the gradient at each point where the graph crosses the x-axis.

(M) 9 Look at the graph for the cyclist at the start of **Section 8.1**.

The equation of this curve is $y = 0.004x^3 - 0.3x^2 + 7.5x$.

a Show that the value given for the speed after 10 seconds is correct.

b Work out the speed after 20 seconds.

8.5 Differentiation of x^n

You have seen that if $y = x^n$ and n is a positive integer, then $\dfrac{dy}{dx} = nx^{n-1}$.

In fact, this result is true if n is *any* rational number, such as $\frac{5}{2}$ or $-\frac{2}{3}$.

KEY INFORMATION

The derivative of x^n is nx^{n-1} for any value of n.

Example 5

Find $\dfrac{dy}{dx}$ when:

a $y = \dfrac{4}{x}$ **b** $y = 5\sqrt{x}$ **c** $y = x^{\frac{3}{2}}$

Solution

a $y = \dfrac{4}{x} = 4x^{-1}$

Here $n = -1$.

$\dfrac{dy}{dx} = 4 \times (-1)x^{-2}$

$= -4x^{-2}$ or $\dfrac{-4}{x^2}$

b $y = 5\sqrt{x} = 5x^{\frac{1}{2}}$

Here $n = \frac{1}{2}$ and $n - 1 = \frac{1}{2} - 1 = -\frac{1}{2}$.

$\dfrac{dy}{dx} = \dfrac{5}{2} \times \dfrac{1}{\sqrt{x}} = \dfrac{5}{2\sqrt{x}}$

$x^{-\frac{1}{2}} = \dfrac{1}{x^{\frac{1}{2}}} = \dfrac{1}{\sqrt{x}}$

c $y = x^{\frac{3}{2}}$

$$\frac{dy}{dx} = \frac{3}{2}x^{\frac{1}{2}}$$

Example 6

The distance travelled by a runner over a 5-second interval is shown on this graph.

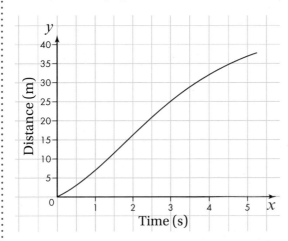

The equation of the curve is $y = 10x^{\frac{3}{2}} - 3x^2$, $0 \leqslant x \leqslant 5$.

Find the speed of the runner after 3 seconds.

Solution

The speed is the gradient of the distance–time graph.

$$\frac{dy}{dx} = 10 \times \frac{3}{2}x^{\frac{1}{2}} - 3 \times 2x$$

$$= 15\sqrt{x} - 6x$$

When $x = 3$, $\frac{dy}{dx} = 15\sqrt{3} - 18 = 8.0$ to 1 d.p.

The speed is $8.0\,\text{m}\,\text{s}^{-1}$.

Exercise 8.5A

Answers page 529

1 Work out $\dfrac{dy}{dx}$ for the following curves:

   a $y = \dfrac{1}{x^2}$ **b** $y = \dfrac{2}{x^3}$

   c $y = \sqrt[3]{x}$ **d** $y = 4x^{\frac{5}{2}}$

2 Differentiate:

   a $y = 10\sqrt{x}$ **b** $y = \dfrac{50}{x} + 10$ **c** $y = 10x^2 - \dfrac{10}{x^2}$

PF **3** This is a graph of $f(x) = \frac{24}{x}, x \neq 0$:

  CM

    a Find $f'(x)$.

    b Find the gradient of the curve at:

      i $(6, 4)$ **ii** $(4, 6)$

      iii $(-2, -12)$ **iv** $(24, 1)$

    c Use your expression for $f'(x)$ to explain why the gradient can never be positive at any point.

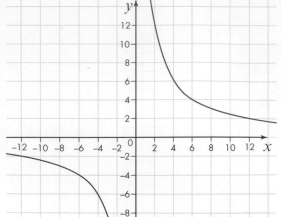

4 This is a graph of $y = \sqrt{x}, x > 0$.

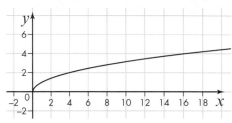

    a Find the gradient of the curve at:

      i $(4, 2)$ **ii** $(9, 3)$ **iii** $(100, 10)$

    b Find the coordinates of the point on the curve where the gradient is $\frac{1}{2}$.

    c Find the coordinates of the point on the curve where the gradient is 1.

5 The equation of a curve is $y = 4\sqrt[3]{x}$. Find the gradient of the curve at $(1, 4)$ and $(8, 8)$.

PF **6** This is a graph of $y = \frac{x}{2} + \frac{2}{x}, x \neq 0$:

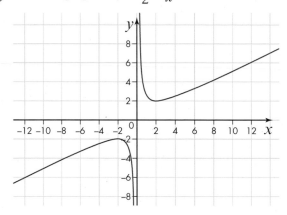

    a Show that the gradient at $(2, 2)$ and at $(-2, -2)$ is 0.

    b Find the gradient at $(0.5, 4.25)$ and at $(4, 2.5)$.

    c Prove that if x is a large number the gradient is approximately 0.5.

7 This is a graph of $f(x) = \frac{2x+5}{x}$, $x \neq 0$:

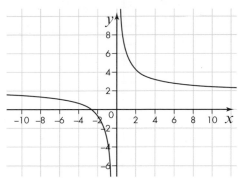

a Find $f'(x)$, $x \neq 0$.

b Find the gradient at:

   i $(2, 4.5)$ **ii** $(10, 2.5)$

c There are two points on the curve where the gradient is -5. Find their coordinates.

It is sometimes convenient to use different letters, other than x and y. You can still differentiate and find the gradient of the graph.

We sometimes call $\frac{dy}{dx}$ the **rate of change** of y with respect to x.

This just means the gradient of the graph of y against x.

Example 7

The area of a circle of radius r is given by the formula $A = \pi r^2$.

a Find the rate of change of A with respect to r.

b Find the value of the rate of change of A when $r = 5$.

Solution

a Differentiate with respect to r. In this case the variable is r. The derivative is the rate of change of A with respect to r.

$$\frac{dA}{dr} = \pi \times 2r = 2\pi r$$

b When $r = 5$, $\frac{dA}{dr} = 10\pi$.

   This means that the gradient of the graph of A against r is 10π or about 31.4 when $r = 5$.

Answers page 529

1 The volume of a sphere of radius r is given by the formula $V = \frac{4}{3}\pi r^3$.

   a Find the rate of change of V with respect to r.

   b Evaluate the rate of change of V with respect to r when $r = 3$.

2 A stone is thrown in the air.

A simple model gives the height, h metres, of the stone after t seconds as $h = 20t - 5t^2$.

   a Find $\dfrac{dh}{dt}$.

   b Find the value of the rate of change of h with respect to t when:

     i $t = 0$ **ii** $t = 2$ **iii** $t = 3$

   c Give an explanation about the movement of the stone for the values in **part b**.

3 $p = 6\sqrt{q}, \quad <2 \quad q \geq 0$

Find the rate of change of p with respect to q when:

   a $q = 4$ **b** $p = 4$

4 The volume of this cylinder is $1000\,\text{cm}^3$.

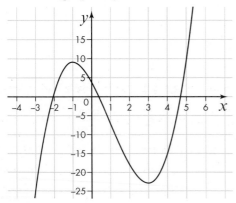

   a Show that $h = \dfrac{1000}{\pi r^2}$.

   b Find the rate of change of h with respect to r.

   c Find the value of the rate of change of h with respect to r when $r = 6$.

8.6 Stationary points and the second derivative

Here is a graph of $y = x^3 - 3x^2 - 9x + 4$:

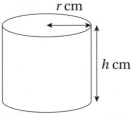

Points where the gradient is zero are called **stationary points**.

You can use differentiation to find stationary points.

If $y = x^3 - 3x^2 - 9x + 4$ then $\dfrac{dy}{dx} = 3x^2 - 6x - 9$.

At a stationary point $\dfrac{dy}{dx} = 0$ and so $3x^2 - 6x - 9 = 0$.

Divide by 3: $x^2 - 2x - 3 = 0$

Factorise: $(x - 3)(x + 1) = 0$

$$x = 3 \text{ or } -1$$

These are the x-coordinates for the stationary points.

Check that the points are A(-1, 9) and B(3, -23).

> A is called a **maximum point** because the y-coordinate is greater at A than at the points on either side of it.

> B is called a **minimum point** because the y-coordinate is less at B than at the points on either side of it.

Notice that A is not the highest point of the curve. It is just higher than points near it.

As you move past A from left to right, the gradient changes from positive to zero to negative. This is true for any maximum point.

As you move past B from left to right, the gradient changes from negative to zero to positive. This is true for any minimum point.

We can summarise this in a table.

Point	$\dfrac{dy}{dx}$	Stationary point	How gradient is changing
A(-1, 9)	0	maximum	$+$ to 0 to $-$
B(3, -23)	0	minimum	$-$ to 0 to $+$

If you did not have a graph of the function, how could you tell whether a stationary point was a maximum or a minimum?

$\dfrac{dy}{dx} = 3x^2 - 6x - 9$ is a function of x. If you differentiate this you get the **second derivative** of y.

Check that second derivative is $6x - 6$.

The symbol for the second derivative of y is $\dfrac{d^2y}{dx^2}$.

$\dfrac{d^2y}{dx^2} = 6x - 6$

Say or read this as 'd two y by d x squared'.

This tells you the rate of change of $\dfrac{dy}{dx}$.

At A, $x = -1$ and $\dfrac{d^2y}{dx^2} = -12$.

This is negative and so the gradient is *decreasing*. This shows that A is a *maximum* point.

At B, $x = 3$ and $\dfrac{d^2y}{dx^2} = +12$.

This is positive and so the gradient is *increasing*. This shows that A is a *minimum* point.

The sign of $\dfrac{d^2y}{dx^2}$ will show you whether a point is a minimum or a maximum.

Here is the table again with an extra column added.

Point	$\dfrac{dy}{dx}$	Stationary point	How gradient is changing	$\dfrac{d^2y}{dx^2}$
A(−1, 9)	0	maximum	+ to 0 to − ⌢	−12 negative
B(3, −23)	0	minimum	− to 0 to + ⌣	12 positive

> **KEY INFORMATION**
> The sign of $\dfrac{d^2y}{dx^2}$ tells you whether the stationary point is a maximum or a minimum.

> **TECHNOLOGY**
> If you have graphical software you can use it to look for stationary points. However, you also need to be able to find them using differentiation.

Example 8

a Find the stationary points of the graph $y = 100 + 300x - x^3$.

b Determine whether each stationary point is a maximum point or a minimum point.

Solution

a
$$y = 100 + 300x - x^3$$

so $\dfrac{dy}{dx} = 300 - 3x^2$.

At a stationary point $\dfrac{dy}{dx} = 0$.

$$300 - 3x^2 = 0$$
$$300 = 3x^2$$
$$x^2 = 100$$
$$x = 10 \text{ or } -10$$

If $x = 10$, $y = 100 + 3000 - 1000 = -2100$

If $x = -10$, $y = 100 - 3000 - 1000 = -1900$

The stationary points are (10, 2100) and (−10, −1900).

b Differentiate $\frac{dy}{dx} = 300 - 3x^2$.

$$\frac{d^2y}{dx^2} = -6x$$

When $x = 10$, $\frac{d^2y}{dx^2} = -60$ which is negative, so $(10, -2100)$ is a maximum point.

When $x = -10$, $\frac{d^2y}{dx^2} = 60$ which is positive, so $(-10, -1900)$ is a minimum point.

> You do not need to find the value of $\frac{d^2y}{dx^2}$; you only need to decide whether it is positive or negative.

Stop and think   *Without* drawing the graph or doing any calculations, what can you say about the stationary points of the graph of the curve $y = x(x - 2)(x - 4)$?

If you use the notation $y = f(x)$ then $\frac{dy}{dx} = f'(x)$ and $\frac{d^2y}{dx^2} = f''(x)$.

You can use the second derivative to help you sketch a curve.

Example 9

Sketch the graph of $f(x) = x^3 - 3x^2 - 9x + 15$ showing the positions of any stationary points.

Solution

$$f(x) = x^3 - 3x^2 - 9x + 15$$

Differentiate.

$$f'(x) - 3x^2 \quad 6x - 9$$

At a stationary point,

$$f'(x) = 3x^2 - 6x - 9 = 0.$$

Divide by 3 and factorise.

$x^2 - 2x - 3 = 0 \Rightarrow (x - 3)(x + 1) = 0$

Either $x - 3 = 0 \Rightarrow x = 3$ or $x + 1 = 0 \Rightarrow x = -1$

$f(3) = 27 - 27 - 27 + 15 = -12$ so $(3, -12)$ is a stationary point.

$f(-1) = -1 - 3 + 9 + 15 = 20$ so $(-1, 20)$ is a stationary point.

Differentiate $f'(x)$ to get $f''(x) = 6x - 6$.

$f''(3) = 18 - 6 = 12 > 0$ so $(3, -12)$ is a minimum point.

$f''(-1) = -6 - 6 = -12 < 0$ so $(-1, 20)$ is a maximum point.

Finally, $f(0) = 15$ so the graph of $f(x)$ crosses the y-axis at $(0, 15)$.

Use all this information to sketch the graph of $f(x)$.

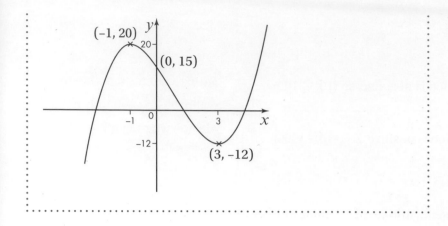

Exercise 8.6A

Answers page 530

(PF) (CM) 1
  a Find the coordinates of the stationary point of the quadratic curve $y = x^2 - 12x - 20$.

  b State whether it is a maximum point or a minimum point. Give a reason for your answer.

  c Sketch the curve.

(PF) (CM) 2
  a Find the coordinates of the stationary point of the graph of $f(x) = 5 + 6x - 2x^2$.

  b State whether it is a maximum point or a minimum point. Give a reason for your answer.

  c Sketch the graph of $f(x)$.

3
  a Sketch the graph of the curve $y = x(6 - x)$.

  b Sketch the graph of $\dfrac{dy}{dx}$.

  c Explain the significance of the point where the graph of $\frac{dy}{dx}$ crosses the x-axis.

  d Sketch the graph of $\dfrac{d^2y}{dx^2}$.

(PF) (CM) 4
  a Find the stationary points of the graph of $f(x) = 2x^3 - 9x^2 + 12x + 8$.

  b State whether each stationary point is a maximum or a minimum. Give a reason for each answer.

  c Make a sketch of the curve, showing the stationary points.

(PF) (CM) 5
  a Find the stationary points on the curve $y = x^3 - 6x^2 - 180x$.

  b State, with a reason, whether each stationary point is a maximum or a minimum.

6 The equation of a curve is $y = x^4 - 2x^2$.

Use differentiation to find the coordinates of the stationary points and determine whether each one is a minimum or a maximum.

7 This is a graph of $y = \frac{10}{x} + \frac{x}{4}$, $x > 0$:

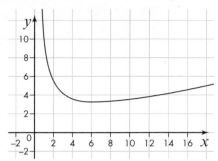

Find the exact value of the x-coordinate of the minimum point.

 8 The speed of a car ($y\,\text{m s}^{-1}$) after x seconds is given by the formula:

$$y = 10 + 6x - 0.5x^2 \quad 0 \leqslant x \leqslant 10$$

Calculate the maximum speed of the car in this 10-second interval.

9 A cricket ball is thrown in the air.

a The height of the ball ($y\,\text{m}$) after x seconds is given by the formula:

$$y = 23x - 5x^2, \qquad 0 \leqslant x \leqslant 4$$

Find the maximum height reached by the ball.

b This simple model does not take account of air resistance. How is that likely to affect the maximum height of the ball?

10 A cuboid has a square cross-section with a side of $x\,\text{cm}$.

The height of the cuboid is $h\,\text{cm}$.

The volume of the cuboid is $1000\,\text{cm}^3$.

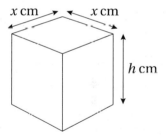

a Show that $h = \dfrac{1000}{x^2}$

b Show that the total surface area of the cuboid is $2x^2 + \dfrac{4000}{x}$

c Find the value of x for which the surface area of the cuboid has a minimum value.

8.7 Tangents and normals

Here is a graph of $y = x - \frac{1}{8}x^2$:

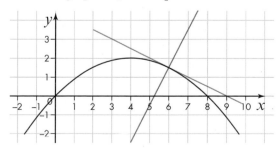

Two lines have been drawn passing through (6, 1.5).

One line (green) is the **tangent** at (6, 1.5).

The other line is **perpendicular** to the tangent. This blue line is called the **normal** at the point (6, 1.5).

You know that if two lines are perpendicular, the product of their gradients is –1.

This is true about the tangent and the normal at any point on a curve.

You can use differentiation to find the equations of the tangent and the normal.

Example 10

Find the equation of the tangent and the normal to the curve $y = x - \frac{1}{8}x^2$ at (6, 1.5).

Solution

If $y = x - \frac{1}{8}x^2$ then $\frac{dy}{dx} = 1 - \frac{1}{4}x$.

If $x = 6$ then $\frac{dy}{dx} = 1 - \frac{1}{4}x = 1 - \frac{3}{2} = -\frac{1}{2}$ or -0.5.

The gradient of the tangent at (6, 1.5) is –0.5.

The equation of the tangent is $y - 1.5 = -0.5(x - 6)$.

$\Rightarrow \quad y - 1.5 = -0.5x + 3 \quad \Rightarrow \quad y = -0.5x + 4.5$

The gradient of the normal at (6, 1.5) is 2 because $2 \times -0.5 = -1$.

The equation of the normal is $y - 1.5 = 2(x - 6)$.

$\Rightarrow \quad y - 1.5 = 2x - 12 \quad \Rightarrow \quad y = 2x - 10.5$

You can check these are correct using a graph plotter.

Stop and think

A student is asked to draw a graph of $y = 2x^2$ and the tangent and normal at (2, 8).

This is what the student draws.

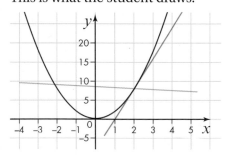

Another student says this must be incorrect. Do you agree? Give a reason for your answer.

Exercise 8.7A

Answers page 531

1 The equation of a curve is $y = (x - 1)(x + 3)$.

   a Sketch the curve.

   b Find $\dfrac{dy}{dx}$.

   c Find the equation of the tangent at $(3, 12)$.

   d Find the equation of the normal at $(3, 12)$.

   e Find the equations of the tangent and the normal at $(-3, 0)$.

2 The equation of a curve is $y = x^3 - 4x^2$.

   a Show that the point $(-1, -5)$ is on the curve.

   b Find $\dfrac{dy}{dx}$.

   c Find the equation of the tangent at $(-1, -5)$.

   d Find the equation of the normal at $(-1, -5)$.

3 The equation of a curve is $y = \dfrac{12}{x^2}$, $x \neq 0$.
   Find the equation of the normal at $(1, 12)$.

4 The equation of a curve is $= 12\sqrt{x}$, $x > 0$.

   a Find the equation of the normal at $(4, 24)$.

   b Find the equation of the tangent at $(100, 120)$.

(PS) 5 The equation of a curve is $y = 10 - x^2$.

   a Find the equation of the tangent at $(2, 6)$.

   b Find the area of the triangle formed by the tangent and the coordinate axes.

(PS) 6 The equation of this curve is $y = x^{\frac{2}{3}}$, $x > 0$.

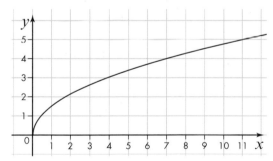

   a Find the equation of the tangent at $(8, 4)$.

   b Find the equation of the tangent at the point where the gradient is $\frac{1}{6}$.

(PS) 7 **a** Sketch the graph of $y = 10x - x^2$.

   b P is the point on the curve with the coordinates $(3, 21)$.

      Find the area of the triangle formed by the tangent at P, the normal at P and the y-axis.

SUMMARY OF KEY POINTS

> If y is a function of x, then the derivative of y is written as $\dfrac{dy}{dx}$ or as $f'(x)$.

> The derivative of ax^n is nax^{n-1} for any values of a and n.

> The value of $\dfrac{dy}{dx}$ at any point on the graph of y is the gradient of the graph at that point:

> > At a turning point $\dfrac{dy}{dx} = 0$

> > A turning point can be a maximum point or a minimum point.

> The second derivative is $\dfrac{d^2y}{dx^2}$ or $f''(x)$:

> > At a maximum point $\dfrac{d^2y}{dx^2}$ is negative and at a minimum point $\dfrac{d^2y}{dx^2}$ is positive.

> The normal at any point on a curve is perpendicular to the tangent at that point.

EXAM-STYLE QUESTIONS 8

Answers page 531

1 $y = 2x + \dfrac{8}{x}, \qquad x \neq 0$

    a Find an expression for $\dfrac{dy}{dx}$.   **[2 marks]**

    b Find the gradient of the graph of $y = 2x + \dfrac{8}{x}$ at the point $(4, 10)$.   **[1 mark]**

2 Here is a graph of $y = f(x)$:

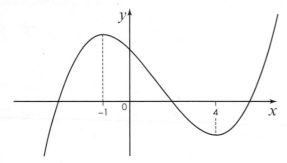

Sketch a graph of $y = f'(x)$ **[3 marks]**

 3 $y = (3x + 2)(x - 1)$

A student writes:

'The derivative of $3x + 2$ is 3 and the derivative of $x - 1$ is 1. So $\dfrac{dy}{dx} = 3 \times 1 = 3$.'

    a Explain what error the student has made.   **[1 mark]**

    b Find the correct expression for $\dfrac{dy}{dx}$.   **[3 marks]**

4 $y = \sqrt{x}\,(x + 4), \qquad x > 0$

Show that $\dfrac{dy}{dx} = 5\frac{1}{6}$ when $x = 9$.   **[5 marks]**

5 $y = x^2(x^2 - 8)$

  a Find $\dfrac{dy}{dx}$.   [2 marks]

  b Find the coordinates of the stationary points of the curve.   [4 marks]

  c Sketch the curve.   [2 marks]

6 The equation of a curve is $y = 2x^3 - 6x^2 - 12x + 4$.

  Find the coordinates of the points on the curve where the gradient is 6.   [6 marks]

7 A drone leaves the ground and flies vertically upwards and then vertically back
   to the ground.

   The time for the whole flight is 20 seconds.

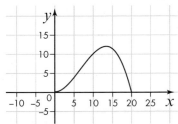

   The height (y m) of the drone after x s is shown in the graph and modelled by
   the equation:

   $y = 0.2x^2 - 0.01x^3$, $0 \leqslant x \leqslant 20$.

  a Show that after 10 seconds the speed is $1\,\text{m s}^{-1}$ upwards.   [3 marks]

  b Show that the ascent takes twice as long as the descent.   [4 marks]

8 $y = x^2 + \dfrac{250}{x}$, $x > 0$.

  a Find $\dfrac{dy}{dx}$.   [2 marks]

  b Show that y has a minimum value and state that value.   [5 marks]

9 This is a sketch of the graph of $y = x + \dfrac{10}{x}$, $x > 0$:

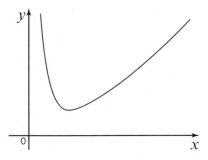

   Find the *exact* values of the coordinates of the minimum point.   [4 marks]

(M) (PF) 10 A rectangular enclosure is being constructed for an outdoor concert.

One side will be a straight wall.

There is 240 metres of fencing available for the other three sides.

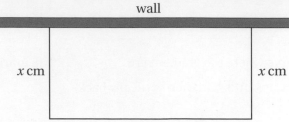

One side of the rectangle is x metres.

a Find an expression for the area of the rectangle, in square metres, in terms of x. **[2 marks]**

The organisers want to make the rectangle as large as possible.

b Prove that the greatest possible area is 7200 m². **[6 marks]**

(PF) 11 $f(x) = 2x^2$

a Show that $f(x + \delta x) - f(x) = 4x\delta x + 2\delta x^2$. **[3 marks]**

b Use the expression in **part a** to prove that $f'(x) = 4x$. **[3 marks]**

12 The equation of a curve is $y = x^2 + 4x - 5$.

Find the equation of the tangent to the curve at $(2, 7)$. **[4 marks]**

(PF) 13 A cone has radius r cm and height h cm.

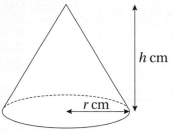

$r + h = 60$

a Show that the volume of the cone is $\frac{1}{3}\pi r^2(60 - r)$. **[2 marks]**

b Show that the cone has a maximum volume when $r = 40$. **[5 marks]**

c Find the maximum volume of the cone. **[1 mark]**

(PS) 14 The equation of a curve is $y = (x + 1)(x - 5)$.

Find the area of the triangle formed by the normal at $(6, 7)$ and the coordinate axes. **[8 marks]**

(PF) 15 The equation of a curve is $y = \dfrac{20}{x}$, $x > 0$.

A tangent is drawn at a point on the curve.

Show that the area of the triangle formed by the tangent and the coordinate axes is always 40, whichever point is chosen. **[5 marks]**

(PF) 16 Differentiate $x^3 - 2x^2$ from first principles. **[4 marks]**

9 INTEGRATION

Engines in petrol and diesel vehicles have pistons moving up and down in cylinders.

This graph illustrates how a two-stroke diesel engine works.

Notice that the axes indicate pressure and volume.

The curve from A to B represents the compression stroke of the piston. As the piston goes in, air in the cylinder is compressed. The pressure increases and the volume decreases.

At B fuel is sprayed in and starts to burn. At C all the fuel has been burnt and the power stroke begins. This is shown by the curve from C to D. The pressure in the cylinder decreases and the volume increases.

From D to A fresh air is blown into the cylinder, expelling the exhaust gases, and the cycle begins again.

The shaded area is important. It measures the work done by the engine to drive the vehicle.

The diagram for a petrol engine is slightly different, but once again there is a cycle on the volume–pressure graph and the area inside the cycle measures the work done.

An engineer designing an engine will want to make this area as large as possible to get the maximum amount of work out of the engine at each stroke.

Calculating the area of a region on a graph is one of the topics in this chapter.

LEARNING OBJECTIVES

You will learn how to:

> use a technique called integration to find the area under a curve

> make use of the connection between differentiation and integration

> find the equation of a curve when you know the gradient and a point on the curve.

TOPIC LINKS

For this chapter, you need to able to manipulate algebraic expressions, in particular those involving indices (**Chapter 1 Algebra and functions 1: Manipulating algebraic expressions**). You will also need to recognise and be able to sketch graphs of simple functions such as quadratics and cubics (**Chapter 3 Algebra and functions 3: Sketching curves**). Integration is strongly linked with differentiation so you need to be familiar with the contents of **Chapter 8 Differentiation**.

PRIOR KNOWLEDGE

You should already know how to:

> differentiate ax^n for any values of a and n, and expressions formed by adding and subtracting terms like this

> manipulate algebraic expressions involving brackets and powers, including fractional and negative powers

> sketch the graphs of quadratic and cubic expressions.

You should be able to complete the following questions correctly:

1 Write $\dfrac{4}{\sqrt{x}}$ in index form.

2 Sketch the graph of $f(x) = -(x - 3)(x + 5)$.

3 Find $\dfrac{dy}{dx}$ for the following $(x > 0)$:

    a $\ y = 4x^3 - 6x + 10$ **b** $\ y = \dfrac{x^2 + 4}{2x}$ **c** $\ y = 6\sqrt[3]{x}$

9.1 Indefinite integrals

For a particular curve, $\dfrac{dy}{dx} = 2x$. This means that the gradient at the point (x, y) is $2x$.

What is the equation of the curve?

One possible equation is $y = x^2$.

However, this is not the only possible answer.

It could also be $y = x^2 + 1$ or $y = x^2 + 10.5$ or $y = x^2 - 3$.

For each of those, $\dfrac{dy}{dx} = 2x$ because the derivative of the final constant is zero.

In fact, $y = x^2 + c$ is a solution, where c is any constant you like.

There is a whole family of curves for which $\dfrac{dy}{dx} = 2x$.

Here are five of them.

> Remember from **Chapter 8 Differentiation**: if $y = x^n$ then $\dfrac{dy}{dx} = x^{n-1}$. In this case $n = 2$.

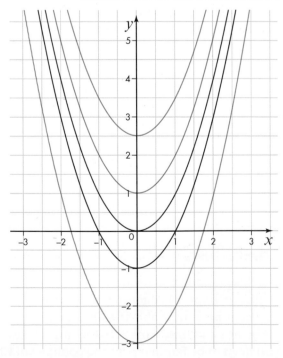

All these curves have the same gradient for a particular value of x.

For example, if $x = 2$ they all have a gradient of 4.

We call $x^2 + c$ the **integral** of $2x$.

The integral of $2x$ is the function we differentiate to get $2x$. There is a special notation to express this.

> It is sometimes called the **indefinite integral** because the value of c is not fixed.

We write $\int 2x \, dx = x^2 + c$, which is read as 'the integral of $2x$ is $x^2 + c$'.

More generally, the derivative of x^{n+1} is $(n+1)x^n$.

That means that the integral of x^n is $\frac{1}{n+1}x^{n+1} + c$, where $n \neq -1$.

To integrate a power of x:

> **KEY INFORMATION**
> $\int x^n \, dx = \frac{1}{n+1}x^{n+1} + c$, where $n \neq -1$.

> increase the index by 1

> multiply by the reciprocal of the new index.

Stop and think   If $n = -1$, what is the value of $n + 1$?

Why does the formula above include the restriction $n \neq -1$?

Example 1

$f'(x) = 6x^2 - \dfrac{2}{x^2}$

Find $f(x)$.

Solution

$$f(x) = \int 6x^2 - \frac{2}{x^2} \, dx$$

$$= \int 6x^2 - 2x^{-2} \, dx$$

$$= \frac{6}{3}x^3 - \frac{2}{-1}x^{-1} + c$$

$$= 2x^3 + 2x^{-1} + c$$

$$= 2x^3 + \frac{2}{x} + c$$

As you can see in **Example 1**, if you have to integrate the sum or difference of two terms, you just integrate each term separately.

Example 2

Find $\displaystyle\int \frac{(x+2)(x-2)}{\sqrt{x}} \, dx$ for values of $x > 0$.

Solution

Multiply out the brackets.

$$\int \frac{(x+2)(x-2)}{\sqrt{x}} \, dx = \int \frac{x^2 - 4}{\sqrt{x}} \, dx$$

Divide by $\sqrt{x}$.

$$\int \frac{(x+2)(x-2)}{\sqrt{x}}\,dx = \int \frac{x^2-4}{\sqrt{x}}\,dx$$
$$= \int x^{\frac{3}{2}} - 4x^{-\frac{1}{2}}\,dx$$
$$= \frac{2}{5}x^{\frac{5}{2}} - \frac{4}{\frac{1}{2}}x^{\frac{1}{2}} + c$$
$$= \frac{2}{5}x^{\frac{5}{2}} - 8x^{\frac{1}{2}} + c$$

Exercise 9.1A   Answers page 533

1 Find:

a $\int 6x\,dx$ **b** $\int 4x+2\,dx$

c $\int 7x-5\,dx$ **d** $\int 3-4x\,dx$

2 Find:

a $\int 2x^4\,dx$ **b** $\int 2x^{-4}\,dx$

c $\int 5x^3\,dx$ **d** $\int \frac{5}{x^3}\,dx$

3 Find:

a $\int 6x^2-4x\,dx$ **b** $\int 2x^4-5x+10\,dx$ **c** $\int 8x-10x^2\,dx$

4 Find:

a $\int \sqrt{x}\,dx$ **b** $\int 2x^{\frac{3}{2}}\,dx$

c $\int 3x^{-\frac{1}{2}}\,dx$ **d** $\int \frac{10}{\sqrt{x}}\,dx$

5 Find:

a $\int x(x-4)\,dx$ **b** $\int x^2(x-8)\,dx$ **c** $\int \frac{10}{x^3}\,dx$

6 Find:

a $\int 4\sqrt{x}\,dx$ **b** $\int \sqrt{4x}\,dx$ **c** $\int \frac{4}{\sqrt{x}}\,dx$

7 $f'(x) = 2x^3 - 2x$

Find an expression for $f(x)$.

(PS) 8 A student writes this incorrect answer to $\int (2x+1)(x-1)\,dx$.

$$\int (2x+1)(x-1)\,dx$$
$$= \int 2x+1\,dx \times \int x-1\,dx$$
$$= (x^2-x)\left(\frac{1}{2}x^2-x\right)+c$$

a Explain the error the student has made.   **b** Find the correct integral.

9 Find:

a $\displaystyle\int \frac{4x^3 - 2}{x^3}\,dx$ **b** $\displaystyle\int \frac{2x + 8}{\sqrt{x}}\,dx$ **c** $\displaystyle\int \frac{4 + 2\sqrt{x}}{x^2}\,dx$

(PS) **10** $f'(x) = g'(x)$

(CM)   What can you say about $f(x)$ and $g(x)$?

Finding the equation of a curve

Suppose you know the derivative, $\dfrac{dy}{dx}$, for a particular curve.

To find the equation of the curve, you need more information.

You also need to know a point on the curve.

Example 3

$y = f(x)$ is the equation of a curve and $f'(x) = 3x^2 + 4x - 3$.

The curve goes through the point $(2, 4)$.

Find the equation of the curve.

Solution

$$y = \int 3x^2 + 4x - 3\,dx$$
$$= x^3 + 2x^2 - 3x + c$$

To find the value of c, put the coordinates $(2, 4)$ into this equation:

$$4 = 8 + 8 - 6 + c$$
$$4 = 10 + c$$
$$c = -6$$

The equation of the curve is $y = x^3 + 2x^2 - 3x - 6$.

Exercise 9.1B   **Answers page 533**

1 A curve passes through $(0, 3)$ and $\dfrac{dy}{dx} = 4x - 2$.

   a Find an expression for y.

   b Find the equation of the curve.

2 For a curve, $\dfrac{dy}{dx} = \sqrt{x}$, $x > 0$.

   a Find an expression for y.

   The point $(9, 25)$ is on the curve.

   b Find the equation of the curve.

3 The gradient of a curve, $\dfrac{dy}{dx}$, is $0.4x + 3$.

   Find the equation of the curve if it passes through:

   a $(0, 0)$ **b** $(0, 5)$ **c** $(5, 0)$

4 The equation of a curve is $y = f(x)$ and $f'(x) = \dfrac{x^2 + 10}{x^2}$.

The curve passes through $(5, 2)$.

Find the equation of the curve.

(PF) 5 The gradient of a curve is given by $\dfrac{dy}{dx} = \dfrac{2}{x^2}$, $x \neq 0$.

a Show that the gradient is always positive,

The curve passes through the point $(2, 4)$.

b Show that the equation of the curve is $y = \dfrac{5x - 2}{x}$.

6 This graph shows a curve with $\dfrac{dy}{dx} = 3x^2 - 3$ that passes through $(0, 2)$.

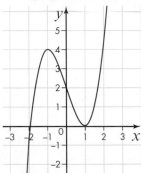

a Find the equation of the curve.

b Show that $(-1, 4)$ and $(1, 0)$ are the only two turning points on the curve.

7 A curve has an equation of the form $y = f(x)$, $x > 0$.

$f'(x) = \dfrac{20}{x^3}$ and the curve passes through the point $(2, 4)$.

a Find the equation of the curve.

b Sketch the graph of the curve.

(M) 8 A straight road passes through points O and A, and $OA = 100$ metres.

Initially, a car is at A. It starts travelling along the road away from O.

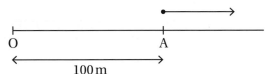

The distance of the car from O after x seconds is y metres.

After x seconds the speed of the car, $\dfrac{dy}{dx}$, is $0.03x^2 (x - 10)^2$, $0 \leqslant x \leqslant 10$.

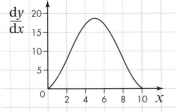

a Find a formula for y in terms of x.

b Show that after 10 seconds the car is $200\,\text{m}$ away from A.

9 A curve passes through the origin, and $\dfrac{dy}{dx} = 3x^2 - 12x + 8$.

Find the coordinates of the points where the curve crosses the x-axis.

(M) **10** A spherical balloon is being inflated.

The radius after x seconds is y centimetres.

$$\dfrac{dy}{dx} = 4x^{-\frac{2}{3}}$$

After 8 seconds the radius is 30 centimetres.

   a Find an equation for y in terms of x.

   b Find the radius after 20 seconds.

9.2 The area under a curve

Here is the graph of $y = 0.3x^2 + 1$:

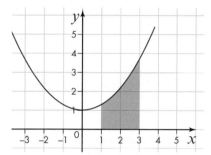

Suppose you want to find the shaded area. This is the region under the curve between $x = 1$ and $x = 3$.

Draw a vertical line at some point x on the x-axis.

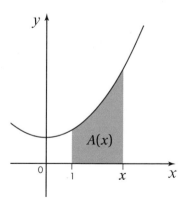

Suppose the area between $x = 1$ and this line is $A(x)$. $A(x)$ is a function of x. As x increases, $A(x)$ increases.

Increase x by a small amount, δx.

> **PROOF**
>
> This proof involves two steps.
>
> First you need to find a general expression for the area from 1 to x.
>
> Then you need to find the particular value for the area when $x = 3$.

> **PROOF**
>
> A similar method, using δx as a small increase in x, was used in proofs in **Chapter 8 Differentiation**.

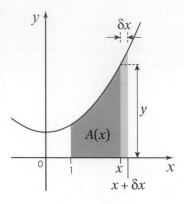

The area shaded yellow, δA, is the difference between $A(x)$ and $A(x + \delta x)$.

This is approximately a rectangle with a height of y and a width of δx.

$$\delta A = A(x + \delta x) - A(x) \approx y\, \delta x$$

Divide by δx.

$$\frac{\delta A}{\delta x} = \frac{A(x + \delta x) - A(x)}{\delta x} \approx y$$

As $\delta x \to 0$, $\dfrac{\delta A}{\delta x} \to y$

$$\lim_{\delta x \to 0} \frac{\delta A}{\delta x} = y$$

This means $\dfrac{dA}{dx} = y$

This shows that $A(x) = \displaystyle\int y\, dx$

$$= \int 0.3x^2 + 1\, dx$$
$$= 0.1x^3 + x + c$$

> **PROOF**
> You have seen this notation for a limit used before, in **Chapter 8 Differentiation**.

So the area from $x = 1$ to any value x is $A(x) = 0.1x^3 + x + c$.

> Do not forget to add the constant c.

What about the **constant of integration**, c?

We are finding the area from 1 to x and so if $x = 1$ the area $A(1) = 0$.

$$0 = 0.1 \times 1^3 + 1 + c$$
$$c = -(0.1 \times 1^3 + 1)$$

> You don't need to work out the exact value of c yet.

The area from 1 to x is $0.1x^3 + x - (0.1 \times 1^3 + 1)$.

In particular, the area from 1 to 3 is

$$[0.1 \times 3^3 + 3] - [0.1 \times 1^3 + 1] = [5.7] - [1.1] = 4.6$$

If you look at the diagram you will see that the area is about 4 or 5 unit squares so this is a reasonable answer.

Look carefully at the last calculation. You can see that it is:

[the value of the integral when $x = 3$]
– [the value of the integral when $x = 1$]

There is a more concise way to write this.

$$\text{Area} = \int_1^3 0.3x^2 + 1\,dx = \left[0.1x^3 + x\right]_1^3$$

Notice the difference between $\int 0.3x^2 + 1\,dx$ and $\int_1^3 0.3x^2 + 1\,dx$.

$\int_1^3 0.3x^2 + 1\,dx$ is called a **definite integral**. It has a particular numerical value.

We said before that $\int 0.3x^2 + 1\,dx$ is an **indefinite integral**. This is a function of x and will include an arbitrary constant, c.

In the example above, a particular expression was used for y and for the limits, but the result is quite general.

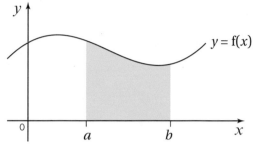

The area under the curve $y = f(x)$ between $x = a$ and $x = b$ is $\int_a^b y\,dx$.

> The limits are put on the integral sign. The integral is put in square brackets with c omitted.

> The words definite and indefinite are sometimes omitted if the type of integral is clear from the context.

> **KEY INFORMATION**
> $\int_a^b y\,dx$ is the area under the curve $y = f(x)$ between $x = a$ and $x = b$.

Example 4

The graph shows a sketch of $y = \dfrac{1}{\sqrt{x}}$, $x > 0$.

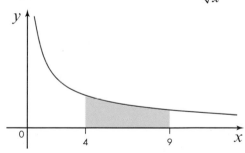

Work out the shaded area under the graph.

Solution

The definite integral $= \displaystyle\int_4^9 \frac{1}{\sqrt{x}}\,dx$

$$= \int_4^9 x^{-\frac{1}{2}}\,dx$$

$$= \left[\frac{1}{\frac{1}{2}}x^{\frac{1}{2}}\right]_4^9$$

$$= \left[2x^{\frac{1}{2}}\right]_4^9$$

$$= [2 \times 3] - [2 \times 2] = 2$$

The area is 2.

So far, all values of *y* have been positive and each curve has been above the *x*-axis.

Suppose the curve is below the *x*-axis. The diagram showing a small increase δ*x* would look like this:

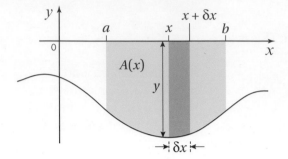

This time *y* is negative. The area shaded green is approximately –*y* δ*x*.

This time you get $\dfrac{\mathrm{d}A}{\mathrm{d}x} = -y$ and $A(x) = -\int y\,\mathrm{d}x$.

In this case, the definite integral will have a negative value.

The area, of course, must be positive, so just take the corresponding positive value.

Example 5

Find the area enclosed by the graph of $y = x(x - 3)$ and the *x*-axis.

Solution

The graph crosses the *x*-axis at 0 and 3.

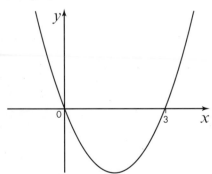

Find the definite integral between 0 and 3.

$$\int_0^3 x(x - 3)\,\mathrm{d}x = \int_0^3 x^2 - 3x\,\mathrm{d}x$$

$$= \left[\frac{1}{3}x^3 - \frac{3}{2}x^2\right]_0^3$$

$$= \left[9 - \frac{27}{2}\right] - [0] = -4\tfrac{1}{2}$$

The integral is negative because the curve is below the *x*-axis. The area is the corresponding positive number.

Area = $4\tfrac{1}{2}$

Look at this graph of $y = 3x^2 - 12$.

$$\int_1^3 3x^2 - 12\,dx = \left[x^3 - 12x \right]_1^3 = [27 - 36] - [1 - 12] = 2$$

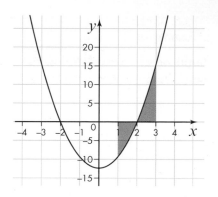

This is the *difference* between the area above the curve and the area below the curve.

If you want to find the total area of the two parts bounded by the curve, $x = 1$ and $x = 3$, you need to find two separate definite integrals.

Stop and think — What are the two separate integrals you need to find? Find them and check that one area is 20 more than the other. What is the total area?

Example 6

The shape of a piece of large metal sheet is the region between the graph of $y = 4 + 0.2x^3$ and the x-axis between the lines $x = -1$ and $x = 2$.

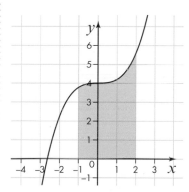

All measurements are in metres.

Calculate the area of the metal sheet.

Solution

The area is given by the definite integral $\int_{-1}^{2} 4 + 0.2x^3\,dx$.

$$\int_{-1}^{2} 4 + 0.2x^3\,dx = \left[4x + 0.05x^4 \right]_{-1}^{2}$$

$$= [8.8] - [-3.95]$$

$$= 12.75$$

The area is $12.75\,\text{m}^2$.

Be careful with the signs when the value of x is negative.

Example 7

The graph shows the curve $y = x^2$ and the line $y = x + 2$.

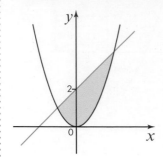

a Find the coordinates of the points where the curve and the straight line cross.

b Find the area between the curve and the straight line.

Solution

a First find the x-coordinates of the points where they cross.

Where they cross:

$$x^2 = x + 2$$

Rearrange.

$$x^2 - x - 2 = 0$$

Factorise.

$$(x + 1)(x - 2) = 0$$

$$x = -1 \text{ or } 2$$

If $x = -1$, $y = -1 + 2 = 1$ and the point is $(-1, 1)$.

If $x = 2$, $y = 2 + 2 = 4$ and the point is $(2, 4)$.

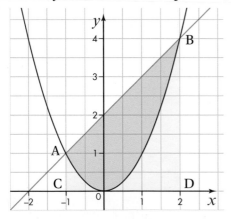

b The area required is the difference between the area of the trapezium ABCD and the area under the curve $y = x^2$ between A and B.

Area under curve $= \int_{-1}^{2} x^2 \, dx = \left[\frac{1}{3} x^3 \right]_{-1}^{2}$

$$= \left[\frac{8}{3} \right] - \left[-\frac{1}{3} \right]$$

$$= 3$$

Area of trapezium $= \dfrac{1+4}{2} \times 3 = 7\frac{1}{2}$

The area between the curve and the line is $7\frac{1}{2} - 3 = 4\frac{1}{2}$.

Exercise 9.2A

Answers page 534

1 Work out:

a $\displaystyle\int_{1}^{2} \frac{1}{3} x^2 \, dx$ **b** $\displaystyle\int_{2}^{4} \frac{1}{3} x^2 \, dx$ **c** $\displaystyle\int_{-1}^{3} \frac{1}{3} x^2 \, dx$

2 Work out:

a $\displaystyle\int_{1}^{2} x^4 \, dx$ **b** $\displaystyle\int_{2}^{4} 10 x^3 \, dx$ **c** $\displaystyle\int_{1}^{5} 10 x^{-2} \, dx$

3 Work out:

a $\displaystyle\int_{3}^{4} 6x^2 - 4x \, dx$ **b** $\displaystyle\int_{0}^{2} 3x^2 + 4x + 3 \, dx$ **c** $\displaystyle\int_{1}^{3} 6x(x-1) \, dx$

4 Find $\displaystyle\int_{1}^{4} f(x) \, dx$ in the following cases:

a $f(x) = 2\sqrt{x}$ **b** $f(x) = \dfrac{20}{\sqrt{x}}$ **c** $f(x) = 8\sqrt[3]{x}$

5 Here is a graph of $y = x^2 - 2x$.

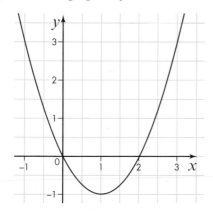

a Find the area under the x-axis enclosed by the curve.

b Find the area bounded by the curve, the x-axis and the line $x = 3$.

c Find the area bounded by the curve, the x-axis and the line $x = -1$.

6 **a** Find the value of:

i $\int_2^3 x^3 \, dx$ **ii** $\int_{-2}^3 x^3 \, dx$ **iii** $\int_{-2}^2 x^3 \, dx$

b Sketch the graph of $y = x^3$ and use it the explain the connection between your answers to **part a**.

7 The shaded shape is cut from the rectangle OABC.

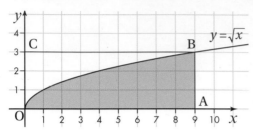

Show that the area of the shape is $\frac{2}{3}$ of the area of the rectangle.

8 This sketch shows the curve $y = 2 + 0.1x^4$.

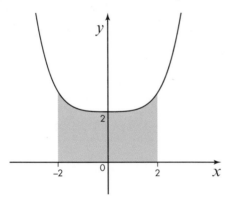

Find the area bounded by the curve, the x-axis, the line $x = 2$ and the line $x = -2$.

9 **a** Find $\int_{-3}^3 9 - x^2 \, dx$.

b Find $\int_0^6 x(6 - x) \, dx$.

c Use sketches to explain the connection between your answers to **parts a** and **b**.

10 **a** Find:

i $\int_0^1 x^2 \, dx$ **ii** $\int_0^1 x^3 \, dx$ **iii** $\int_0^1 x^6 \, dx$

b Write a generalisation to describe all three integrals in **part a**.

11 The speed of a car over a 6-second interval is shown in this graph.

The equation of the curve is $y = x^2(6 - x)$, $0 \leqslant x \leqslant 6$.

Find the distance travelled:

a in the first 3 seconds

b in the fourth second (between $x = 3$ and $x = 4$).

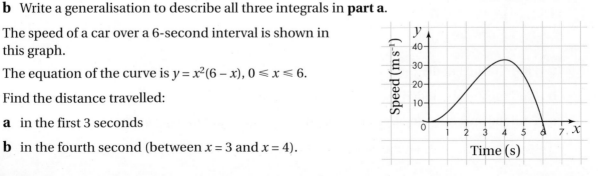

(M) 12 This graph shows the shape of an aircraft wing.

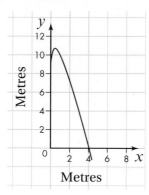

Metres

The equation of the curve is $y = 8 + 8\sqrt{x} - 6x$, $0 \le x \le 4$.

Calculate the area of the wing.

(PS) 13 This is the curve $y = \dfrac{30}{x^2}$, $x \ge 0$.

(CM)

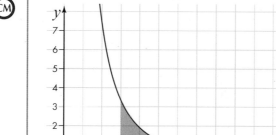

a Calculate the shaded area.

b Calculate the area under the curve between $x = 3$ and $x = 10$.

c Calculate the area under the curve between $x = 3$ and $x = n$.

d Explain why the area under the curve between $x = 3$ and $x = n$ can never be greater than 10.

14 The graph shows the straight line $y = 4 - x$ and the curve $y = x(4 - x)$.

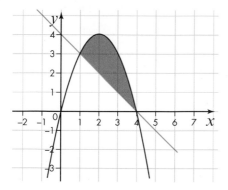

a Find the coordinates of the points where the curve and the straight line cross.

b Find the area between the curve and the straight line.

SUMMARY OF KEY POINTS

› Integration is the reverse process of differentiation.

› The definite integral $\int_a^b f(x)\,dx$ gives the area under the curve $y = f(x)$ between $x = a$ and $x = b$:

  › $\int x^n\,dx = \frac{1}{n+1} x^{n+1}, \qquad n \neq -1$

› To integrate the sum or difference of terms, integrate each term separately.

EXAM-STYLE QUESTIONS 9   Answers page 535

1 Find $\int 4x^2 (2x - 3)\,dx$. [4 marks]

(M) **2** The velocity ($y\,\mathrm{m\,s^{-1}}$) of a cyclist after x seconds is given by the formula $y = 15 - 0.1x^2$.
Find the distance travelled by the cyclist in the first 10 seconds. [4 marks]

3 Find $\int \frac{5 + \sqrt{x}}{x^2}\,dx, x > 0$. [4 marks]

4 A curve passes through $(4, 6)$ and $\frac{dy}{dx} = -\frac{16}{x^2}, x > 0$.
Find the equation of the curve. [4 marks]

(CM) **5** **a** Sketch the graph of $y = \sqrt{x} - 2, 0 \leqslant x \leqslant 9$. [1 mark]

b Calculate $\int_0^9 \sqrt{x} - 2\,dx$. [4 marks]

c Explain precisely what the value of the integral tells you about the graph. [1 mark]

(M) **6** Water is leaking out of a tank. Initially there are 100 litres in the tank. After x hours
(CM) there are y litres of water left in the tank.

The rate at which water is leaking, in litres per hour, is modelled by the formula

$\frac{dy}{dx} = 0.6x - 10, \qquad 0 \leqslant x \leqslant 15$

a Show that after 9 hours the amount of water left in the tank is 34.3 litres. [5 marks]

b Explain why the formula for $\frac{dy}{dx}$ would not be appropriate if $x = 20$. [2 marks]

(PS) **7** A student writes the following answer to an integration problem.
(CM)

$$\int \frac{5x^4 - 8x}{4x^3}\,dx = \frac{\int 5x^4 - 8x\,dx}{\int 4x^3\,dx}$$

$$= \frac{x^5 - 4x^2}{2x^2} + c$$

$$= \frac{1}{2}x^3 + c$$

a The student's answer is incorrect. Explain the error the student has made. [1 mark]

b Work out the correct integral. [4 marks]

(PS) **8** For a particular curve $\frac{dy}{dx} = 2x + 6$. The curve has a minimum point on the x-axis.
Find the equation of the curve. [5 marks]

9 The graph shows the curve $y = kx(4 - x)$.

The area between the curve and the x-axis is 32.

Calculate the value of k. **[6 marks]**

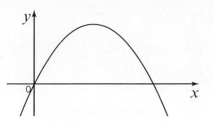

10 This graph shows the curve $y = x^2$.

A is the point with coordinates (a, a^2) and ABCD is a rectangle.

Prove that the area between AB and the curve is $\frac{2}{3}$ of the area of ABCD. **[6 marks]**

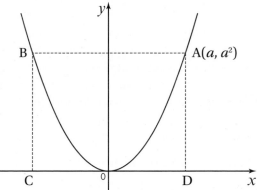

PS 11 The graph shows the lines $y = x^2 - x + 4$ and $y = 2x + 8$.

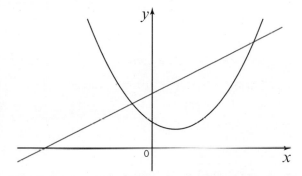

Find the area between the curve and the straight line. **[9 marks]**

12 The graph shows the curves $y = \frac{1}{16}x^3$ and $y = 2\sqrt{x}$, $x \geqslant 0$.

Calculate the area between the two curves. **[5 marks]**

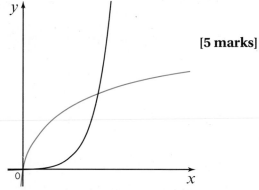

10 VECTORS

The Global Positioning System (GPS) was developed in the US and has been in operation since 1978. It is the most-used satellite navigation (or sat nav) system in the world and was released onto the market in 1994. Whereas in the past you might have had to plot out your route in advance using a road atlas or rely on the navigational skills of your driving companion, all that is required if you have a sat nav is to inform it of your destination. It works out your current location and within seconds computes the most efficient route, from potentially millions of possibilities, using its built-in maps. Sat navs and mobile phones are also able to use the internet to ensure that their maps are up-to-date and accurate, even when roads are changed.

At the heart of this technology is the use of vectors. A vector is a quantity with a size and a direction. Vectors are a key feature in mechanics (such as displacement, velocity, acceleration and forces), but vectors are used in pure mathematics as well. The sat nav calculates the overall distance from starting point to destination – not as the crow flies, but from a huge number of small vectors, each with its own size and direction.

LEARNING OBJECTIVES

You will learn how to:

> write a vector in **i j** notation or as a column vector

> calculate the magnitude and direction of a vector as a bearing in two dimensions

> identify a unit vector and scale it up as appropriate

> find a velocity vector

> apply the triangle law to add vectors

> apply the ratio and midpoint theorems

> add vectors

> multiply a vector by a scalar

> find the position vector of a point

> find the relative displacement between two position vectors.

TOPIC LINKS

This chapter extends the work covered at GCSE on vectors, solving problems in pure mathematics algebraically and geometrically. Applications of vectors from this chapter will then be applied to situations involving mechanics in **Chapter 15 Kinematics** and **Chapter 16 Forces**.

PRIOR KNOWLEDGE

You should already know how to:

> ❭ use Pythagoras' theorem to find the lengths of sides of right-angled triangles

> ❭ use trigonometry (sine, cosine and tangent) to find angles and sides for right-angled triangles

> ❭ use the sine and cosine rules to find angles and sides for non-right-angled triangles

> ❭ simplify surds

> ❭ solve simultaneous linear equations.

You should be able to complete the following questions correctly:

1 Find the length of the line segment AB given the points:

    a A(6, −10) and B(1, 2)

    b A(−5, −4) and B(17, 7) (giving the answer in the form $a\sqrt{5}$).

2 Find the size of the angle H in the right-angled triangle which has a right angle at G and sides GH = 8 cm and GF = 5 cm. Give the answer correct to 1 d.p.

3 Find the length of the side AB in the triangle ABC which has AC = 10 cm, BC = 17 cm and angle ACB = 142°. Give the answer correct to 3 s.f.

10.1 Definition of a vector

Consider two points A(5, 6) and B(−19, 16) on a metre square grid.

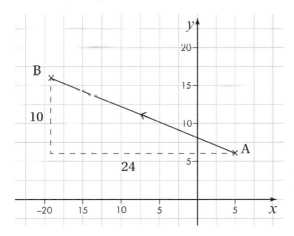

To get from A to B you need to travel 24 m horizontally in the negative x-direction and 10 m vertically in the positive y-direction. The vector $\overrightarrow{AB}$ describes this journey. This is also called the **relative displacement** of B from A.

There are two ways of writing the **vector** $\overrightarrow{AB}$.

The first way of writing the vector is using **i j notation**. **i** is defined as one unit in the positive x-direction (east) and **j** as one unit in the positive y-direction (north). If you have to travel in the negative x- or y-direction, then the coefficient of $\mathbf{i}$ or $\mathbf{j}$ will also be negative. The vector $\overrightarrow{AB}$ would be written as $(-24\mathbf{i} + 10\mathbf{j})$ m.

The second way of writing the vector is to use a **column vector**, as you will have seen when translating shapes in your GCSE course. It has two numbers, one for x and one for y, written vertically in square brackets, and they are the same numbers as the coefficients you used to write the vector in $\mathbf{i}\,\mathbf{j}$ notation.

The vector $\overrightarrow{AB}$ would be written as $\begin{bmatrix} -24 \\ 10 \end{bmatrix}$.

The **magnitude** (or length) of the vector can be found using Pythagoras' theorem. The vector is the hypotenuse of the right-angled triangle for which the $\mathbf{i}$ and $\mathbf{j}$ coefficients are the shorter sides. The magnitude of the vector $\overrightarrow{AB}$ is often written as $|\overrightarrow{AB}|$, with the vector between two vertical lines.

$$c^2 = a^2 + b^2$$
$$c = \sqrt{a^2 + b^2}$$
$$= \sqrt{(-24)^2 + 10^2}$$
$$= \sqrt{576 + 100}$$
$$= \sqrt{676}$$
$$= 26\,\text{m}$$

The magnitude $|\overrightarrow{AB}| = 26\,\text{m}$.

You can also use the formula for finding the distance between two points.

$$d^2 = (x_1 - x_2)^2 + (y_1 - y_2)^2$$

Let $A(5, 6) = (x_1, y_1)$ and $B(-19, 16) = (x_2, y_2)$.

$$d^2 = (5 - (-19))^2 + (6 - 16)^2$$
$$= (24)^2 + (-10)^2$$
$$= 576 + 100$$
$$= 676$$
$$d = \sqrt{676}$$
$$= 26\,\text{m}$$

The direction of the vector can be described as an angle. This is often measured from the positive $\mathbf{i}$ vector. For vectors with a positive $\mathbf{j}$ component, the angle is measured anticlockwise from $\mathbf{i}$ and is positive ($0° < \theta < 180°$). For vectors with a negative $\mathbf{j}$ component, the angle is measured clockwise from $\mathbf{i}$ and is negative ($-180° < \theta < 0°$).

KEY INFORMATION

Vectors can be written in $\mathbf{i}\,\mathbf{j}$ notation or as column vectors.

KEY INFORMATION

Use Pythagoras' theorem to calculate the magnitude (length) of a vector.

KEY INFORMATION

The positive $\mathbf{i}$ vector has the same direction as the positive x-axis.

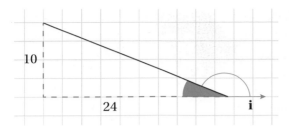

Start by finding the acute angle inside the right-angled triangle.

$$\text{Angle} = \tan^{-1}\left(\frac{10}{24}\right) = 22.6°$$

The angle you need is the obtuse angle measured anticlockwise from **i**.

$$180° - 22.6° = 157.4°$$

Alternatively, you may use a **bearing**. For $\overrightarrow{AB}$, the bearing is a reflex angle. If you calculate the angle between the vector and the x-axis using trigonometry then you can just add it to 270°.

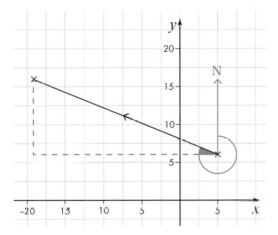

The angle inside the triangle is 22.6° as before.

$$\text{Bearing} = 270° + 22.6° = 292.6°$$

The direction of the vector $\overrightarrow{AB}$ is the bearing 293°, correct to the nearest degree.

A **unit vector** is a vector with a magnitude of 1. The vector $\overrightarrow{AB}$ has a magnitude of 26 m but we can find the unit vector in the direction of $(-24\mathbf{i} + 10\mathbf{j})$ by dividing the vector by 26. Hence the unit vector is $\frac{1}{26}(-24\mathbf{i} + 10\mathbf{j})$ m.

You can use the unit vector to find vectors of other magnitudes in the same direction. For example, if you wished to find a vector with a magnitude of 182 m in the direction of $(-24\mathbf{i} + 10\mathbf{j})$, you could multiply the unit vector by 182.

$$\frac{1}{26}(-24\mathbf{i} + 10\mathbf{j}) \times 182 = 7(-24\mathbf{i} + 10\mathbf{j})\,\text{m}$$

You can also use the unit vector to find a **velocity vector**. The unit vector shows how far you travel if your velocity is $1\,\text{km}\,\text{h}^{-1}$. So, for example, if you were travelling in the direction $(3\mathbf{i} - 4\mathbf{j})$

KEY INFORMATION

Use trigonometry to find the direction of a vector.

KEY INFORMATION

Remember that a bearing should be written with three figures.

KEY INFORMATION

A unit vector has a magnitude of 1.

at $40\,\text{km h}^{-1}$, you would start by finding the unit vector and then multiply this by 40 to find the actual velocity.

$$\sqrt{3^2 + (-4)^2} = 5\,\text{km}$$

The unit vector is therefore $\frac{1}{5}(3\mathbf{i} - 4\mathbf{j})\,\text{km}$

The velocity is $\frac{1}{5}(3\mathbf{i} - 4\mathbf{j}) \times 40 = 8(3\mathbf{i} - 4\mathbf{j}) = (24\mathbf{i} - 32\mathbf{j})\,\text{km h}^{-1}$.

Example 1

Find the magnitude and bearing of the vector $(12\mathbf{i} - 8\mathbf{j})$.

Solution

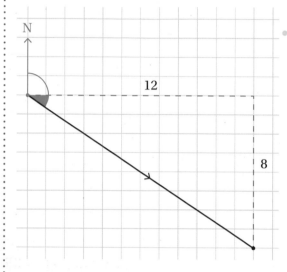

Drawing a diagram may help you visualise the situation in the question.

Use Pythagoras' theorem to find the magnitude of the vector.

$$c^2 = a^2 + b^2$$
$$c = \sqrt{a^2 + b^2}$$
$$= \sqrt{12^2 + (-8)^2}$$
$$= \sqrt{144 + 64}$$
$$= \sqrt{208}$$
$$= 4\sqrt{13}\ \text{units}$$

The magnitude of the vector is $4\sqrt{13}$ units.

Use trigonometry to find the acute angle next to the horizontal.

$$\text{Angle} = \tan^{-1}\left(\frac{8}{12}\right) = 33.7°$$

Add 90° to obtain the bearing.

$$\text{Bearing} = 90° + 33.7° = 123.7°$$

The bearing of the vector is 124°, correct to the nearest degree.

Example 2

A tractor is driven at a speed of $8.45\,\text{km}\,\text{h}^{-1}$ in the direction $(-120\mathbf{i} - 119\mathbf{j})$.

What is the velocity of the tractor?

Solution

The tractor is travelling in the direction $(-120\mathbf{i} - 119\mathbf{j})$.

Start by using Pythagoras' theorem to find the magnitude of the directional vector.

$$\begin{aligned}
\text{Magnitude} &= \sqrt{(-120)^2 + (-119)^2} \\
&= \sqrt{14400 + 14161} \\
&= \sqrt{28561} \\
&= 169\,\text{km}
\end{aligned}$$

Now write the unit vector by dividing by 169.

$$\frac{1}{169}(-120\mathbf{i} - 119\mathbf{j})$$

Finally, multiply the unit vector by 8.45.

$$\begin{aligned}
\frac{1}{169}(-120\mathbf{i} - 119\mathbf{j}) \times 8.45 &= \frac{1}{20}(-120\mathbf{i} - 119\mathbf{j}) \\
&= (-6\mathbf{i} - 5.95\mathbf{j})\,\text{km}\,\text{h}^{-1}
\end{aligned}$$

The velocity of the tractor is $(-6\mathbf{i} - 5.95\mathbf{j})\,\text{km}\,\text{h}^{-1}$.

Exercise 10.1A

Answers page 535

1. Find the angle between each vector and the vector $\mathbf{i}$, giving all answers correct to 1 d.p.

   a $7\mathbf{i} + 4\mathbf{j}$

   b $-10\mathbf{i} + 7\mathbf{j}$

   c $\begin{bmatrix} -6 \\ -5 \end{bmatrix}$

   d $\begin{bmatrix} 4 \\ -9 \end{bmatrix}$

2. For this question, write each bearing correct to the nearest degree.

   a Find the magnitude and bearing of each vector.

   i $24\mathbf{i} + 7\mathbf{j}$

   ii $-12\mathbf{i} - 35\mathbf{j}$

   iii $\begin{bmatrix} 182 \\ -120 \end{bmatrix}$

   b Find the magnitude and bearing of each vector, giving the magnitude in the form $a\sqrt{b}$, where b is prime.

   i $\begin{bmatrix} -16 \\ 8 \end{bmatrix}$

   ii $7\mathbf{i} + \mathbf{j}$

   iii $2\mathbf{i} - 4\mathbf{j}$

(PS) (PF) 3 Rebekah and Henry are discussing a vector of the form $\begin{bmatrix} a \\ a \end{bmatrix}$.

Rebekah says that the vector has a bearing of 045°. Henry says that the vector has a bearing of 225°.

a Who is correct? Explain your answer.

b What form would a vector need to take, to have a bearing of 315°?

(PS) 4 The vector $\begin{bmatrix} a \\ b \end{bmatrix}$ has a magnitude of 130.

If a and b are integers, how many possible pairs of values are there for a and b?

5 a Find the unit vector in the direction:

    i $-4\mathbf{i} + 3\mathbf{j}$ **ii** $143\mathbf{i} - 24\mathbf{j}$

b Find a vector of the given magnitude and in the given direction:

    i magnitude = 52, direction = $-5\mathbf{i} + 12\mathbf{j}$

    ii magnitude = 195, direction = $63\mathbf{i} + 16\mathbf{j}$

(M) 6 A car travels at a speed of $29\,\text{km}\,\text{h}^{-1}$ in the direction $(144\mathbf{i} - 17\mathbf{j})$. What is the velocity of the car?

7 a A vector acts along a bearing of 135° with a magnitude of $7\sqrt{2}$ m. Write the vector as a column vector.

b A vector acts along a bearing of 060° with a magnitude of 2 m. Write the vector as a column vector in surd form.

c A vector acts along a bearing of 210° with a magnitude of 14 m. Write the vector as a column vector in surd form.

(M) (PS) 8 A ship sails 14 km east, then 9 km southeast.

a Calculate the displacement of the ship from its starting point.

b Write the displacement of the ship in the form $(a\mathbf{i} + b\mathbf{j})$ km, where a and b are exact values.

c Calculate the bearing of the ship from its starting point.

> **Stop and think**
>
> The cosine rule could be used to find the displacement in **question 8**.
>
> What other ways are there of finding the displacement?

10.2 Adding and subtracting vectors

To add two vectors in **i j** notation (or written as column vectors), add the **i** terms and the **j** terms separately.

For example, $(3\mathbf{i} + 6\mathbf{j}) + (5\mathbf{i} - 2\mathbf{j}) = (3 + 5)\mathbf{i} + (6 - 2)\mathbf{j} = 8\mathbf{i} + 4\mathbf{j}$

Using column vectors, this would be written as $\begin{bmatrix} 3 \\ 6 \end{bmatrix} + \begin{bmatrix} 5 \\ -2 \end{bmatrix} = \begin{bmatrix} 8 \\ 4 \end{bmatrix}$

Subtraction works in the same way.

For example, $(8\mathbf{i} - 2\mathbf{j}) - (-3\mathbf{i} + 5\mathbf{j}) = (8 - (-3))\mathbf{i} + (-2 - 5)\mathbf{j} = 11\mathbf{i} - 7\mathbf{j}$

Using column vectors, this would be written as $\begin{bmatrix} 8 \\ -2 \end{bmatrix} - \begin{bmatrix} -3 \\ 5 \end{bmatrix} = \begin{bmatrix} 11 \\ -7 \end{bmatrix}$

Types of vectors

Equal vectors	Two vectors are **equal** if they have the same magnitude and the same direction.
Opposite vectors	Two vectors are **opposite** if they have the same magnitude but point in opposite directions.
Parallel vectors	Two vectors are **parallel** if they point in the same direction. They do not need to have the same magnitude.

Parallel vectors have a common factor. For example, $(6\mathbf{i} + 8\mathbf{j})$ and $(9\mathbf{i} + 12\mathbf{j})$ can be written as $2(3\mathbf{i} + 4\mathbf{j})$ and $3(3\mathbf{i} + 4\mathbf{j})$ respectively, which have the common factor of $(3\mathbf{i} + 4\mathbf{j})$.

Alternatively, when a vector is written in terms of one or more single letters such as $\mathbf{a}$, $3\mathbf{b}$ or $\mathbf{a} - 2\mathbf{b}$ with a bold font or by drawing a wavy line underneath such as $\underset{\sim}{a}$, you will use the common factor to prove that two vectors such as $(2\mathbf{a} - 5\mathbf{b})$ and $(4\mathbf{a} - 10\mathbf{b})$ are parallel.

> **KEY INFORMATION**
>
> If two vectors are parallel, then they must contain a common factor.
>
> For example, if you are told that a vector is parallel to $(3\mathbf{i} + 2\mathbf{j})$, it can be written as $k(3\mathbf{i} + 2\mathbf{j})$.

Example 3

The vectors $\mathbf{m}$ and $\mathbf{n}$ are given by $(5\mathbf{i} - 2\mathbf{j})$ and $(-4\mathbf{i} + \mathbf{j})$ respectively.

If $w\mathbf{m} + 7\mathbf{n}$ is parallel to $(6\mathbf{i} - \mathbf{j})$, find the value of w.

Solution

Substitute $\mathbf{m}$ and $\mathbf{n}$.

$$w(5\mathbf{i} - 2\mathbf{j}) + 7(-4\mathbf{i} + \mathbf{j})$$

Write in the form $(a\mathbf{i} + b\mathbf{j})$.

$$5w\mathbf{i} - 2w\mathbf{j} - 28\mathbf{i} + 7\mathbf{j} = (5w - 28)\mathbf{i} + (-2w + 7)\mathbf{j}$$

For this to be parallel to $(6\mathbf{i} - \mathbf{j})$, it must be equal to $k(6\mathbf{i} - \mathbf{j})$ for some value of k.

$$(5w - 28)\mathbf{i} + (-2w + 7)\mathbf{j} = k(6\mathbf{i} - \mathbf{j})$$

Equate coefficients for $\mathbf{i}$ and $\mathbf{j}$.

$\mathbf{i}$: $5w - 28 = 6k$

$\mathbf{j}$: $-2w + 7 = -k$

Hence $k = 2w - 7$.

Substitute for k in the equation $5w - 28 = 6k$.

$$5w - 28 = 6(2w - 7)$$

Solve the equation.

$$5w - 28 = 12w - 42$$

$$14 = 7w$$

$$w = 2$$

Check your answer by substituting $w = 2$ into the expression $(5w - 28)\mathbf{i} + (-2w + 7)\mathbf{j}$.

$$(5 \times 2 - 28)\mathbf{i} + (-2 \times 2 + 7)\mathbf{j} = -18\mathbf{i} + 3\mathbf{j} = -3(6\mathbf{i} - \mathbf{j})$$

which is parallel to $(6\mathbf{i} - \mathbf{j})$.

Exercise 10.2A

Answers page 536

1 a Find:

     i $(4\mathbf{i} + 7\mathbf{j}) + (-2\mathbf{i} + 5\mathbf{j})$ **ii** $(-\mathbf{i} + \mathbf{j}) + (8\mathbf{i} - 5\mathbf{j})$ **iii** $\begin{bmatrix} 0 \\ -6 \end{bmatrix} + \begin{bmatrix} 13 \\ 2 \end{bmatrix}$

   b Find:

     i $(-3\mathbf{i} + 4\mathbf{j}) - (\mathbf{i} + 6\mathbf{j})$ **ii** $\begin{bmatrix} 4 \\ 17 \end{bmatrix} - \begin{bmatrix} 4 \\ 7 \end{bmatrix}$ **iii** $\begin{bmatrix} -5 \\ -1 \end{bmatrix} - \begin{bmatrix} -3 \\ 10 \end{bmatrix}$

   c Find:

     i $5(2\mathbf{i} + 4\mathbf{j})$ **ii** $\frac{1}{2}\begin{bmatrix} 22 \\ 9 \end{bmatrix}$ **iii** $-4\begin{bmatrix} 5 \\ -3 \end{bmatrix}$

   d Find:

     i $(-3\mathbf{i} + 4\mathbf{j}) - 3(-2\mathbf{i} + 5\mathbf{j})$ **ii** $3\begin{bmatrix} 7 \\ -4 \end{bmatrix} + 5\begin{bmatrix} -4 \\ 6 \end{bmatrix}$ **iii** $\frac{3}{4}\begin{bmatrix} -12 \\ 11 \end{bmatrix} - 2\begin{bmatrix} -19 \\ 2 \end{bmatrix}$

 2 The vectors **a** and **b** are given by $\mathbf{a} = \begin{bmatrix} 4 \\ 1 \end{bmatrix}$ and $\mathbf{b} = \begin{bmatrix} -1 \\ 3 \end{bmatrix}$.

This diagram shows the geometric representation of **a** + **b**.

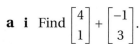

   a i Find $\begin{bmatrix} 4 \\ 1 \end{bmatrix} + \begin{bmatrix} -1 \\ 3 \end{bmatrix}$.

     ii Explain how your answer to **part i** is represented in the diagram.

   b i Find **a** – **b**.

     ii Represent **a** – **b** in a diagram.

     iii Explain the geometric significance of **a** – **b**.

   c i Find 2**a**.

     ii Represent 2**a** in a diagram.

     iii Explain the geometric significance of 2**a**.

3 Find the values of p and q given that:

a $5\begin{bmatrix} p \\ 2 \end{bmatrix} + q\begin{bmatrix} 8 \\ -10 \end{bmatrix} = \begin{bmatrix} 26 \\ 15 \end{bmatrix}$

b $p\begin{bmatrix} 2 \\ 7 \end{bmatrix} + q\begin{bmatrix} 4 \\ 5 \end{bmatrix} = \begin{bmatrix} 10 \\ 53 \end{bmatrix}$

c $p(4\mathbf{i} - \mathbf{j}) + q(3\mathbf{i} + 5\mathbf{j}) = 18\mathbf{i} + 7\mathbf{j}$

4 Given that $\mathbf{a} = 4\mathbf{i} - \mathbf{j}$ and $\mathbf{b} = \mathbf{i} + 2\mathbf{j}$, find:

a p if $p\mathbf{a} + 2\mathbf{b}$ is parallel to $(5\mathbf{i} + \mathbf{j})$

b q if $3\mathbf{a} + q\mathbf{b}$ is parallel to $(\mathbf{i} - \mathbf{j})$.

(PS) 5 Find values for a and b such that $a\begin{bmatrix} 2 \\ -1 \end{bmatrix} + b\begin{bmatrix} 8 \\ -3 \end{bmatrix}$:

a is equal to $\begin{bmatrix} 6 \\ -4 \end{bmatrix}$

b is parallel to the x-axis and has a magnitude of 12

c is parallel to $y = -x$.

> **Stop and think**
>
> Why does **question 5 part c** have an *infinite* number of answers but **parts a** and **b** only have one each?
>
> Can you find values for a and b such that the vector is parallel to other straight lines that pass through the origin, such as $y = x$, $y = 2x$ and the y-axis?

10.3 Vector geometry

A vector can be labelled in various ways. As discussed in **Section 10.1**, if a vector connects two points such as A and B, then it is often written as $\overrightarrow{AB}$ and the magnitude as $|\overrightarrow{AB}|$. The arrow shows that the vector is the journey from A to B, whereas $\overrightarrow{BA}$ would be the journey from B to A.

Alternatively, as discussed in **Section 10.2**, a vector can be written in terms of one or more single letters such as $\mathbf{a}$, $3\mathbf{b}$ or $\mathbf{a} - 2\mathbf{b}$. Here the magnitude would be written as $|\mathbf{a}|$ or in italics as a. In this case, the vector is not fixed between two specific points and in a parallelogram OABC where $\overrightarrow{CB} = \mathbf{a}$ then the vector $\overrightarrow{OA}$ will also be equal to $\mathbf{a}$, as is any other vector which is parallel and the same length.

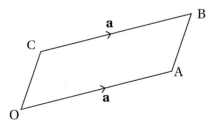

Triangle and parallelogram laws

Vectors can be added using the **triangle law**. If the vector $\overrightarrow{AB}$ (**a**) is followed by the vector $\overrightarrow{BC}$ (**b**), then the result is the vector $\overrightarrow{AC}$ (**a** + **b**).

$$\overrightarrow{AB} + \overrightarrow{BC} = \overrightarrow{AC}$$

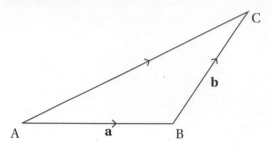

By drawing the two vectors in the opposite order on the same diagram (**b** followed by **a**), you complete a parallelogram. This is called the **parallelogram law**. $\overrightarrow{AC}$ is the diagonal of the parallelogram. The parallelogram is useful because it highlights the fact that the order of adding vectors is *not* important, as **a** + **b** gives the same result as **b** + **a**.

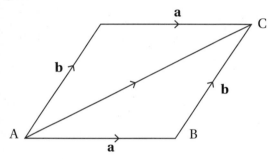

Example 4

OMN is a triangle. $\overrightarrow{OM} = 14\mathbf{m}$. $\overrightarrow{ON} = 21\mathbf{n}$.

P is a point such that OP and PM are in the ratio $2:5$.

Q is a point such that $\overrightarrow{ON} = \frac{7}{2}\overrightarrow{OQ}$.

Prove that PQ is parallel to MN.

Solution

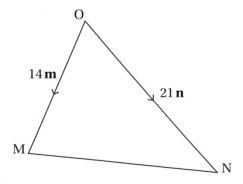

TECHNOLOGY

Plotting diagrams using graphics software enables you to visualise the situation and see how the vectors will still satisfy the rules even if the orientation of the vectors is not identical.

In order to prove that the vectors $\overrightarrow{PQ}$ and $\overrightarrow{MN}$ are parallel, you need to show that they have a common factor.

From the diagram, the vector $\overrightarrow{MN}$ is the same as $\overrightarrow{MO}$ followed by $\overrightarrow{ON}$.

$\overrightarrow{MO}$ is the opposite of the vector $\overrightarrow{OM}$, so $\overrightarrow{MO} = -14\mathbf{m}$.

$$\overrightarrow{MN} = \overrightarrow{MO} + \overrightarrow{ON} = -14\mathbf{m} + 21\mathbf{n}$$

Factorising, $\overrightarrow{MN} = 7(-2\mathbf{m} + 3\mathbf{n})$.

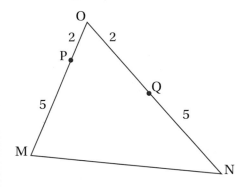

The ratio 2 : 5 splits the line OM into seven parts, where OP is two parts and PM is five parts.

Hence $\overrightarrow{OP} = \frac{2}{7}\,\overrightarrow{OM} = \frac{2}{7} \times 14\mathbf{m} = 4\mathbf{m}$.

Also, since $\overrightarrow{ON} = \frac{7}{2}\,\overrightarrow{OQ}$, $\overrightarrow{OQ} = \frac{2}{7}\,\overrightarrow{ON} = \frac{2}{7} \times 21\mathbf{n} = 6\mathbf{n}$.

$$\overrightarrow{PQ} = \overrightarrow{PO} + \overrightarrow{OQ} = -4\mathbf{m} + 6\mathbf{n}$$

Factorising, $\overrightarrow{PQ} = 2(-2\mathbf{m} + 3\mathbf{n})$.

Since $\overrightarrow{MN}$ and $\overrightarrow{PQ}$ have a common factor, they are parallel.

> ### PROOF
>
> To prove that two vectors are parallel, you need to show that they have a common factor.
>
> In **Example 4**, this means finding both vectors in terms of **m** and **n** and then factorising each one to show what the common factor is.

Annotate the diagram.

The ratio theorem

Consider the triangle OAB with $\overrightarrow{OA} = \mathbf{a}$ and $\overrightarrow{OB} = \mathbf{b}$. C divides AB in the ratio 1 : 2.

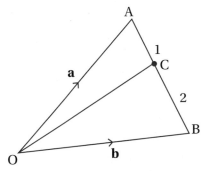

How can you find the vector $\overrightarrow{OC}$ in terms of **a** and **b**? Start with $\overrightarrow{AB}$.

$$\overrightarrow{AB} = \overrightarrow{AO} + \overrightarrow{OB} = -\mathbf{a} + \mathbf{b}$$

$\overrightarrow{AC} = \frac{1}{3}\,\overrightarrow{AB}$, so $\overrightarrow{AC} = \frac{1}{3}(-\mathbf{a} + \mathbf{b})$.

$\overrightarrow{OC} = \overrightarrow{OA} + \overrightarrow{AC} = \mathbf{a} + \frac{1}{3}(-\mathbf{a} + \mathbf{b})$

$$= \mathbf{a} - \tfrac{1}{3}\mathbf{a} + \tfrac{1}{3}\mathbf{b} = \tfrac{2}{3}\mathbf{a} + \tfrac{1}{3}\mathbf{b} = \tfrac{1}{3}(2\mathbf{a} + \mathbf{b})$$

This can be generalised using the **ratio theorem**.

Let C split AB into the ratio $m : n$.

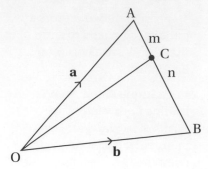

$\overrightarrow{AB}$ is still equal to $(-\mathbf{a} + \mathbf{b})$, but $\overrightarrow{AC} = \dfrac{m}{m+n}\overrightarrow{AB} = \dfrac{m}{m+n}(-\mathbf{a} + \mathbf{b})$.

$$\overrightarrow{OC} = \overrightarrow{OA} + \overrightarrow{AC} = \mathbf{a} + \frac{m}{m+n}(-\mathbf{a} + \mathbf{b})$$

$$= \mathbf{a} - \frac{m}{m+n}\mathbf{a} + \frac{m}{m+n}\mathbf{b}$$

$$= \frac{(m+n) - m}{m+n}\mathbf{a} + \frac{m}{m+n}\mathbf{b}$$

$$= \frac{n}{m+n}\mathbf{a} + \frac{m}{m+n}\mathbf{b}$$

$$= \frac{1}{m+n}(n\mathbf{a} + m\mathbf{b})$$

When C is the midpoint of AB and hence divides AB in a 1 : 1 ratio, you obtain the **midpoint rule**, a special case of the ratio theorem:

$$\overrightarrow{OC} = \tfrac{1}{2}(\mathbf{a} + \mathbf{b})$$

Example 5

Triangle OAB has $\overrightarrow{OA} = \mathbf{a}$ and $\overrightarrow{OB} = \mathbf{b}$.

C divides AB in the ratio 3 : 4.

M is the midpoint of OB.

OC and AM intersect at X.

Find the vector $\overrightarrow{OX}$ in terms of $\mathbf{a}$ and $\mathbf{b}$.

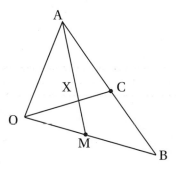

Solution

Use the ratio theorem to find the vector $\overrightarrow{OC}$.

$\overrightarrow{OC} = \frac{1}{7}(4\mathbf{a} + 3\mathbf{b})$.

Since X lies on the vector $\overrightarrow{OC}$, $\overrightarrow{OX}$ is parallel to $\overrightarrow{OC}$.

$\overrightarrow{OX}$ and $\overrightarrow{OC}$ share a common factor so $\overrightarrow{OX}$ can be written as $\lambda(4\mathbf{a} + 3\mathbf{b})$ for some value of λ.

$\overrightarrow{OX}$ is also equal to $\overrightarrow{OA} + \mu\,\overrightarrow{AM}$ for some value of μ, since $\overrightarrow{AX}$ and $\overrightarrow{AM}$ share a common factor.

Find $\overrightarrow{AM}$ in terms of $\mathbf{a}$ and $\mathbf{b}$.

$\overrightarrow{AM} = \overrightarrow{AO} + \overrightarrow{OM} = -\mathbf{a} + \frac{1}{2}$

Use $\overrightarrow{OX} = \overrightarrow{OA} + \mu\,\overrightarrow{AM}$.

$\overrightarrow{OX} = \mathbf{a} + \mu(-\mathbf{a} + \frac{1}{2}\mathbf{b})$

Put the two expressions for $\overrightarrow{OX}$ equal to each other.

$$\lambda(4\mathbf{a} + 3\mathbf{b}) = \mathbf{a} + \mu(-\mathbf{a} + \frac{1}{2}\mathbf{b})$$

Equate the coefficients for $\mathbf{a}$ and $\mathbf{b}$.

$\mathbf{a}$: $4\lambda = 1 - \mu$

$\mathbf{b}$: $3\lambda = \frac{1}{2}\mu$

Hence $\mu = 6\lambda$.

Substitute $\mu = 6\lambda$ into the first equation.

$$4\lambda = 1 - 6\lambda$$

$$10\lambda = 1$$

$$\lambda = \frac{1}{10}$$

Hence $\overrightarrow{OX} = \frac{1}{10}(4\mathbf{a} + 3\mathbf{b})$.

Exercise 10.3A

Answers page 536

 1 The triangle RST is such that $\overrightarrow{RS} = 4\mathbf{a}$ and $\overrightarrow{RT} = 6\mathbf{b}$.

U is the midpoint of RS and V is the midpoint of RT.
Prove that $\overrightarrow{ST}$ is parallel to $\overrightarrow{UV}$.

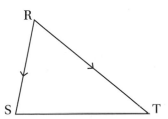

2 OABC is a parallelogram. $\overrightarrow{OA} = 6\mathbf{a}$ and $\overrightarrow{OC} = 6\mathbf{c}$.

S is the centre of OABC.

T is the midpoint of the side BC.

The point U divides AB in the ratio 2 : 1.

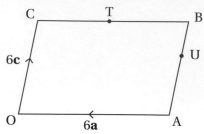

a Find each vector in terms of **a** and **c**.

   i $\overrightarrow{OB}$ **ii** $\overrightarrow{AC}$ **iii** $\overrightarrow{OU}$

   iv $\overrightarrow{TA}$ **v** $\overrightarrow{OS}$ **vi** $\overrightarrow{US}$

   vii $\overrightarrow{UT}$ **viii** $\overrightarrow{ST}$

b What is the relationship between the vectors $\overrightarrow{ST}$ and $\overrightarrow{AB}$?

3 Triangle OAB has $\overrightarrow{OA} = 5\mathbf{a}$ and $\overrightarrow{OB} = 6\mathbf{b}$.

The point X is such that it divides AB in the ratio 4 : 9.

Find the vector $\overrightarrow{OX}$ in terms of **a** and **b**.

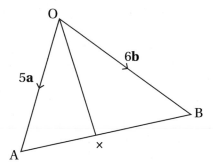

PF **4** DEF is a triangle with $\overrightarrow{DE} = 12\mathbf{a}$ and $\overrightarrow{DF} = 8\mathbf{b}$.

M divides DE in the ratio 1 : 2 and N divides DF in the ratio 1 : 2.

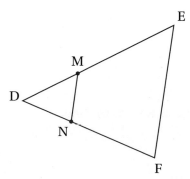

a Prove that FEMN is a trapezium.

b If DEF has an area of 72 units^2, find the area of FEMN.

(PS) **5** A has position vector $(7\mathbf{i} - 2\mathbf{j})$.

The vectors AB and AC are given by $(\mathbf{i} - \mathbf{j})$ and $(6\mathbf{i} + 4\mathbf{j})$ respectively.

The point D divides the line segment BC in the ratio $3 : 2$.

Find the position vector of D.

6 Triangle OAB has $\overrightarrow{OA} = 6\mathbf{a}$ and $\overrightarrow{OB} = 6\mathbf{b}$.

M divides OA such that OM is half of MA and N divides OB such that ON is half of NB.

AN and BM meet at X.

Find the vector $\overrightarrow{OX}$ in terms of $\mathbf{a}$ and $\mathbf{b}$.

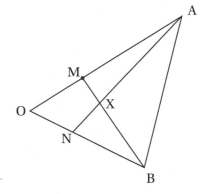

(PS) **7** Triangle OAB has $OM : MA = ON : NB = 2 : 1$.

(PF) $\overrightarrow{OA} = \mathbf{a}$ and $\overrightarrow{OB} = \mathbf{b}$.

X is the intersection of MB and NA.

a Show that $\overrightarrow{OX} = \frac{2}{5}(\mathbf{a} + \mathbf{b})$.

It is now given that $OM : MA = ON : NB = p : q$.

b Prove that $\overrightarrow{OX} = \dfrac{p}{2p + q}(\mathbf{a} + \mathbf{b})$.

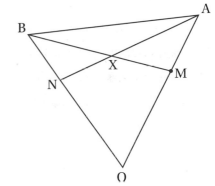

(PS) **8** In the diagram, $\overrightarrow{JK} = 4\mathbf{a}$, $\overrightarrow{LM} - 6\mathbf{a}$ and $\overrightarrow{JM} - 15\mathbf{b}$.

(CM)

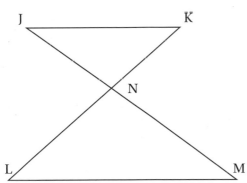

a Explain why JKN and MLN are similar triangles.

Given that P is on the line MN, and the line segment MP is twice as long as the line segment PN:

b find the two possible vectors $\overrightarrow{LP}$.

Three points on the same straight line are described as **collinear**.

PROOF

To prove that three points are collinear, it is sufficient to show that two vectors joining any of the three points are parallel, since any two of these vectors must contain a common point.

Example 6

UVWX is a parallelogram with $\overrightarrow{UV} = 3\mathbf{a}$ and $\overrightarrow{UX} = 8\mathbf{b}$.

Y is the point between V and W such that VY is three times as long as YW.

Z is the point such that $\overrightarrow{XZ} = \frac{4}{3}\overrightarrow{XW}$.

Prove that U, Y and Z are collinear.

Solution

Start by drawing a diagram.

Since UVWX is a parallelogram, $\overrightarrow{VW} = \overrightarrow{UX} = 8\mathbf{b}$.

Y is on the line segment VW and splits the line in the ratio 3 : 1.

Hence $\overrightarrow{VY} = \frac{3}{4}\overrightarrow{VW} = \frac{3}{4} \times 8\mathbf{b} = 6\mathbf{b}$.

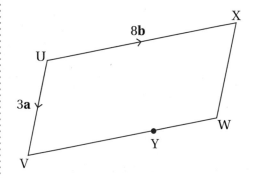

You need to find any two vectors from $\overrightarrow{UY}$, $\overrightarrow{UZ}$ and $\overrightarrow{YZ}$.

$$\overrightarrow{UY} = \overrightarrow{UV} + \overrightarrow{VY} = 3\mathbf{a} + 6\mathbf{b} = 3(\mathbf{a} + 2\mathbf{b})$$

Add point Z to the diagram. Z lies on the line segment XW extended such that $\overrightarrow{XZ} = \frac{4}{3}\overrightarrow{XW}$. Since U, Y and Z are collinear it will also lie on the line UY when that is extended.

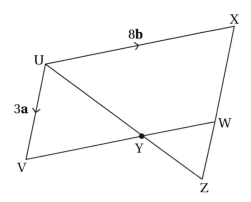

Since UVWX is a parallelogram, $\overrightarrow{XW} = \overrightarrow{UV} = 3\mathbf{a}$.

$$\overrightarrow{UZ} = \overrightarrow{UX} + \overrightarrow{XZ} = 8\mathbf{b} + \tfrac{4}{3} \times 3\mathbf{a} = 8\mathbf{b} + 4\mathbf{a} = 4(\mathbf{a} + 2\mathbf{b})$$

Since $\overrightarrow{UY}$ and $\overrightarrow{UZ}$ have a common factor, they are parallel.

Since $\overrightarrow{UY}$ and $\overrightarrow{UZ}$ also have a common point, U, Y and Z are collinear.

Exercise 10.3B

Answers page 537

(PF) **1** The parallelogram OABC has $\overrightarrow{OA} = \mathbf{a}$ and $\overrightarrow{OC} = \mathbf{c}$.

D is the midpoint of the side OA. E is the midpoint of the side AB.

a Prove that $\overrightarrow{DE}$ is parallel to $\overrightarrow{OB}$.

The side OA is extended to F such that OA = AF.

b Prove that C, E and F are collinear.

(PF) **2** The diagram in **Exercise 10.3A question 2** is reproduced below.

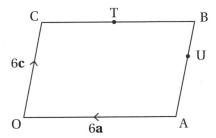

Given that X is the point such that $\overrightarrow{OA} - \overrightarrow{AX}$, prove that the points T, U and X are collinear.

(PF) **3** In the diagram, ABCD is a quadrilateral such that $\overrightarrow{AB} = 15\mathbf{a}$, $\overrightarrow{DC} = 5\mathbf{a}$ and $\overrightarrow{DA} = 14\mathbf{h}$.

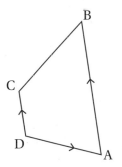

AD is extended to the point E such that the line segments AD and DE are in the ratio 2 : 1.

Prove that the points B, C and E are collinear.

(PS) **4** DEFG is a parallelogram such that $\overrightarrow{DE} = 12\mathbf{a}$ and $\overrightarrow{DG} = 8\mathbf{b}$.

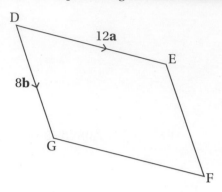

J is the midpoint of DG.

$$\overrightarrow{FG} = 3\overrightarrow{KG}$$

Given that JKQ and EFQ are both straight lines, find the vector $\overrightarrow{DQ}$ in terms of **a** and **b**.

(PS) **5** OPQR is a parallelogram. T divides the side RQ in the ratio 5 : 1.

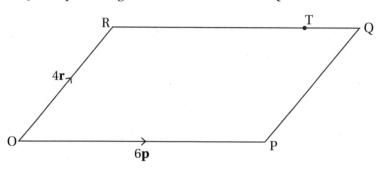

S is the centre of the parallelogram.

$\overrightarrow{OP} = 6\mathbf{p}$. $\overrightarrow{OR} = 4\mathbf{r}$.

Given that PQU and STU are straight lines, find the vector $\overrightarrow{OU}$ in terms of **p** and **r**.

10.4 Position vectors

In **Section 10.1** you saw that the relative displacement of B from A describes how to get from point A to point B. This section will show how to use position vectors to find the same result, then extend it to solve problems in pure mathematics.

The **position vector** of a point is its position relative to the **origin**. For A(5, 6), the position vector is (5**i** + 6**j**) and for B(−19, 16), the position vector is (−19**i** + 16**j**).

The relative displacement of B from A ($\overrightarrow{AB}$) can be found by subtracting the position vector of A from the position vector of B in the same way as you subtracted vectors in **Section 10.2**.

$$(-19\mathbf{i} + 16\mathbf{j}) - (5\mathbf{i} + 6\mathbf{j}) = -19\mathbf{i} - 5\mathbf{i} + 16\mathbf{j} - 6\mathbf{j}$$

$$= -24\mathbf{i} + 10\mathbf{j}$$

KEY INFORMATION

To find the relative displacement of B from A, $\overrightarrow{AB}$, subtract the position vector of B from the position vector A.

Hence you can find the distance between A and B by finding the magnitude of the relative displacement using Pythagoras' theorem, which as you have seen is 26.

Example 7

Point A has coordinates $(-3, 14)$ and point B has position vector $\begin{bmatrix} u \\ -1 \end{bmatrix}$.

a Write A as a position vector.

b Given that the magnitude of the vector $\overrightarrow{AB}$ is 17, find both possible values of u.

Solution

a The position vector of A is $\begin{bmatrix} -3 \\ 14 \end{bmatrix}$.

b Find the vector $\overrightarrow{AB}$.

$$\overrightarrow{AB} = \mathbf{b} - \mathbf{a} = \begin{bmatrix} u \\ -1 \end{bmatrix} - \begin{bmatrix} -3 \\ 14 \end{bmatrix} = \begin{bmatrix} u+3 \\ -15 \end{bmatrix}$$

The magnitude of $\overrightarrow{AB}$ is given by

$$\sqrt{(u+3)^2 + (-15)^2} = \sqrt{u^2 + 6u + 9 + 225}$$
$$= \sqrt{u^2 + 6u + 234}$$

Since you are told that the magnitude of $\overrightarrow{AB}$ is 17, put the expression $\sqrt{u^2 + 6u + 234}$ equal to 17.

$$\sqrt{u^2 + 6u + 234} = 17$$

Square both sides.

$$u^2 + 6u + 234 = 17^2$$
$$= 289$$

Put the equation equal to zero by subtracting 289.

$$u^2 + 6u - 55 = 0$$

Solve the equation by factorisation.

$$(u + 11)(u - 5) = 0$$
$$u = -11 \text{ or } 5$$

> Use Pythagoras' theorem to find the magnitude of the vector.

Stop and think

Why are there two possible position vectors for B?

If both answers for u were plotted as points B and C, what could you say about the triangle ABC?

You can use position vectors and the distances between them to identify quadrilaterals. In order to do this, you need to know the properties of different quadrilaterals:

> a square has four equal sides and four right angles

> a rectangle has two pairs of equal sides and four right angles

> a rhombus has four equal sides and two pairs of parallel sides

> a parallelogram has two pairs of parallel equal sides

> a trapezium has one pair of parallel sides

> a kite has two pairs of equal adjacent sides.

Example 8

A quadrilateral RSTU has vertices R(5, 1), S(–1, 8), T(–3, –1) and U(3, –8).

Use vectors to show that RSTU is a rhombus.

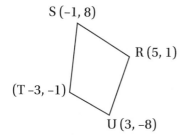

Solution

Begin by writing each point as a position vector.

$$\mathbf{r} = 5\mathbf{i} + \mathbf{j}, \ \mathbf{s} = -\mathbf{i} + 8\mathbf{j}, \ \mathbf{t} = -3\mathbf{i} - \mathbf{j} \text{ and } \mathbf{u} = 3\mathbf{i} - 8\mathbf{j}$$

Find the vector $\overrightarrow{RS}$ by subtracting $\mathbf{r}$ from $\mathbf{s}$.

$$\overrightarrow{RS} = \mathbf{s} - \mathbf{r} = (-\mathbf{i} + 8\mathbf{j}) - (5\mathbf{i} + \mathbf{j}) = -\mathbf{i} + 8\mathbf{j} - 5\mathbf{i} - \mathbf{j}$$
$$= -\mathbf{i} - 5\mathbf{i} + 8\mathbf{j} - \mathbf{j} = -6\mathbf{i} + 7\mathbf{j}$$

Find the vectors of the other sides in the same way.

$$\overrightarrow{ST} = \mathbf{t} - \mathbf{s} = (-3\mathbf{i} - \mathbf{j}) - (-\mathbf{i} + 8\mathbf{j}) = -3\mathbf{i} - \mathbf{j} + \mathbf{i} - 8\mathbf{j}$$
$$= -3\mathbf{i} + \mathbf{i} - \mathbf{j} - 8\mathbf{j} = -2\mathbf{i} - 9\mathbf{j}$$

$$\overrightarrow{UT} = \mathbf{t} - \mathbf{u} = (-3\mathbf{i} - \mathbf{j}) - (3\mathbf{i} - 8\mathbf{j}) = -3\mathbf{i} - \mathbf{j} - 3\mathbf{i} + 8\mathbf{j}$$
$$= -3\mathbf{i} - 3\mathbf{i} - \mathbf{j} + 8\mathbf{j} = -6\mathbf{i} + 7\mathbf{j}$$

$$\overrightarrow{RU} = \mathbf{u} - \mathbf{r} = (3\mathbf{i} - 8\mathbf{j}) - (5\mathbf{i} + \mathbf{j}) = 3\mathbf{i} - 8\mathbf{j} - 5\mathbf{i} - \mathbf{j}$$
$$= 3\mathbf{i} - 5\mathbf{i} - 8\mathbf{j} - \mathbf{j} = -2\mathbf{i} - 9\mathbf{j}$$

So the vectors $\overrightarrow{RS}$ and $\overrightarrow{UT}$ are the same, as are $\overrightarrow{ST}$ and $\overrightarrow{RU}$. Because opposite sides are equal, they are not only the same length but they are parallel, as required.

TECHNOLOGY

You may find it helpful to plot the points using a graphical software package. You will be able to see the relative positions of the vertices and count squares to help find the horizontal and vertical distances required to apply Pythagoras' theorem.

PROOF

For RSTU to be a rhombus, you need to prove that:

> opposite sides RS and UT are the same (parallel and equal magnitude)

> opposite sides ST and RU are the same (parallel and equal magnitude)

> adjacent sides, such as RS and RU, have the same magnitude

> at least one angle, such as RST, is not a right angle.

All that is left is to ensure that adjacent sides are also the same length.

For RS:

$$d^2 = (5 - (-1))^2 + (1 - 8)^2$$

$$= (6)^2 + (-7)^2$$

$$= 36 + 49$$

$$= 85$$

The formula for the distance between two points is
$$d^2 = (x_1 - x_2)^2 + (y_1 - y_2)^2$$

For RU:

$$d^2 = (5 - 3)^2 + (1 - (-8))^2$$

$$= (2)^2 + (9)^2$$

$$= 4 + 81$$

$$= 85$$

Since d^2 is the same for RS and RU, the sides are the same length.

All the conditions have been satisfied, so RSTU is a rhombus.

Stop and think What other pairs of adjacent sides could be used?

Exercise 10.4A

Answers page 538

 1 Four buildings have position vectors as follows:

Hospital:	$(-\mathbf{i} + 7\mathbf{j})$ km
Supermarket:	$(5\mathbf{i} + 2\mathbf{j})$ km
Church:	$(3\mathbf{i} - 9\mathbf{j})$ km
Museum:	$(-2\mathbf{i} - 4\mathbf{j})$ km

a For which two buildings is the relative displacement

i $(2\mathbf{i} + 11\mathbf{j})$ km　　**ii** $(7\mathbf{i} + 6\mathbf{j})$ km　　**iii** $(4\mathbf{i} - 16\mathbf{j})$ km?

b Find the relative displacement of the hospital from the museum.

c Which two buildings are $5\sqrt{2}$ km apart?

 **2** A triangle has vertices A(4, −2), B(−1, 3) and C(−3, −3).

a Write the position vectors of A, B and C, each in the form $(a\mathbf{i} + b\mathbf{j})$.

b Find the vectors $\overrightarrow{AB}$, $\overrightarrow{AC}$ and $\overrightarrow{BC}$, each in the form $(a\mathbf{i} + b\mathbf{j})$.

c Find the lengths of AB, AC and BC, each in simplified surd form.

d What type of triangle is ABC? Give a reason for your answer.

(PS) 3 The position vector of S is $(q\mathbf{i} - 6\mathbf{j})$ m.

The position vector of T is $(-2\mathbf{i} + 3\mathbf{j})$ m.

   a If $q = 4$, calculate the distance between S and T, giving the answer in the form $a\sqrt{13}$ m.

   b Given that ST = 15 m, find both possible values of q.

(PF) (PS) 4 Given the points A(−1, −4), B(8, 4), C(3, 6), D(−6, −2), E(−5, −14) and F(4, −6), use vectors to show that:

   a ABCD is a parallelogram

   b CDEF is a rhombus

   c DFBC is a trapezium.

(PS) 5 Points P, Q and R have position vectors $\mathbf{p} = 12\mathbf{i} - \mathbf{j}$, $\mathbf{q} = 6\mathbf{i} + 7\mathbf{j}$ and $\mathbf{r} = -2\mathbf{i} + \mathbf{j}$. PQRS is a square.

   a Find the position vector of the point S.

   b Find the position vector of the centre of the square.

   c Find the area of the square.

(PF) (PS) 6 A triangle has vertices J(−3, 1), K(9, −4) and L(3, 5).

Use vectors to prove that the triangle is right-angled.

Stop and think	How else can you demonstrate that a triangle is right-angled?
	How could you find the gradient of each side?

SUMMARY OF KEY POINTS

› Vectors can be written in **i j** notation or as column vectors.

› Use Pythagoras' theorem to calculate the *magnitude* of a vector.

› Use trigonometry to find the *direction* of a vector. This may be measured from the positive **i** vector or as a bearing.

› A unit vector is a vector with a magnitude of one.

› The unit vector can be used to find vectors of other magnitudes in the same direction.

› A vector connecting A and B can be written as $\vec{AB}$ (with a magnitude of $|\vec{AB}|$).

› A vector can also be written using lower-case letters, for example **a** or *a* (with a magnitude of *a*).

› Vectors can be added using the triangle law or the parallelogram law.

› Two vectors are equal if they have the same magnitude and the same direction.

› Two vectors are opposite if they have the same magnitude but point in opposite directions.

› Two vectors are parallel if they point in the same direction.

› Three points are collinear if they all lie on the same straight line.

› The position vector of a point is its position relative to the origin.

› To find the relative displacement of B from A, $\vec{AB}$, subtract the position vector of B from the position vector of A.

EXAM-STYLE QUESTIONS 10

Answers page 539

 1 A car is driven at $(-11\mathbf{i} + 21\mathbf{j})\,\mathrm{m\,s^{-1}}$.

    a Calculate the speed of the car. [2 marks]

    b Calculate the bearing the car is travelling along. [2 marks]

    Give both answers correct to 3 s.f.

2 A rambler sets off from a tavern at 14:20 p.m. on a course of $(2\mathbf{i} + \mathbf{j})\,\mathrm{km\,h^{-1}}$ for a hotel. He arrives at the hotel at 17:05 p.m.

    Find the distance between the tavern and the hotel, correct to 1 d.p. [3 marks]

3 The vectors $[(p+3)\mathbf{i} - 4\mathbf{j}]$ and $[8\mathbf{i} + (p-5)\mathbf{j}]$ have the same magnitude.

    a Find the value of *p*. [3 marks]

    b Find the acute angle between the vectors. [3 marks]

4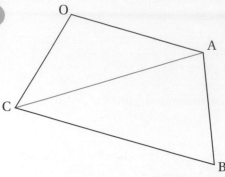

Quadrilateral OABC has $\overrightarrow{OA} = 8\mathbf{a}$, $\overrightarrow{OC} = 7\mathbf{c}$ and $\overrightarrow{CB} = 12\mathbf{a}$.

P divides CA in the ratio 3 : 2.

a Find the vector $\overrightarrow{OP}$. [2 marks]

b Prove that O, P and B are collinear. [2 marks]

c State the ratio OP : PB. [1 mark]

5 The vector $\mathbf{a}$ is given by $\begin{bmatrix} 5 \\ -6 \end{bmatrix}$. The vector $\mathbf{b}$ is given by $\begin{bmatrix} k \\ 1 \end{bmatrix}$.

The vector $\mathbf{c}$ is given by $\mathbf{c} = \mathbf{a} - 2\mathbf{b}$.

a Find, in terms of k:

   i the vector $\mathbf{c}$ **ii** the magnitude of the vector $\mathbf{c}$. [4 marks]

b Given that the magnitude of $\mathbf{c}$ is greater than $2\sqrt{17}$, find the set of possible values for k. [4 marks]

6 A rhombus ABCD has vertices with position vectors $\mathbf{a} = p\mathbf{i} + 5\mathbf{j}$, $\mathbf{b} = 2\mathbf{i} + 12\mathbf{j}$, $\mathbf{c} = (q + 4) + q\mathbf{j}$ and $\mathbf{d} = 22\mathbf{i} + 2\mathbf{j}$.

Find the values of p and q. [6 marks]

7 Three of the vertices of a rectangle have position vectors $(-8\mathbf{i} - 4\mathbf{j})$, $(-\mathbf{i} + 10\mathbf{j})$ and $(3\mathbf{i} + 8\mathbf{j})$.

a Find the area of the rectangle. [4 marks]

b Find the position vector of the fourth vertex of the rectangle. [3 marks]

8 The parallelogram ABCD is such that $\overrightarrow{AB} = 10\mathbf{b}$ and $\overrightarrow{AD} = 10\mathbf{d}$.

BX : BC = 2 : 3 and DY : YC = 3 : 2.

AD and XY are extended, meeting at the point P.

Find the vector $\overrightarrow{BP}$ in terms of $\mathbf{b}$ and $\mathbf{d}$.

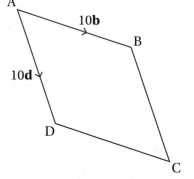

[6 marks]

9 A parallelogram CDEF has position vectors $\mathbf{c} = -4\mathbf{i} + \mathbf{j}$, $\mathbf{d} = -5\mathbf{i} - 6\mathbf{j}$, $\mathbf{e} = 4\mathbf{i} - 3\mathbf{j}$ and $\mathbf{f} = 5\mathbf{i} + 4\mathbf{j}$.

a Find the length of CF, giving the answer in the form $a\sqrt{10}$. [2 marks]

b Show that $\overrightarrow{CD}$ is equal to $\overrightarrow{FE}$. [2 marks]

X is a point on DE such that DX : XE = 1 : 2.

c Find the position vector of X. [2 marks]

d Show that XC is perpendicular to XD. [4 marks]

e Find the area of CDEF. [3 marks]

f Given that G has position vector $\mathbf{g} = 2\mathbf{i} + 3\mathbf{j}$, show that DXGC is a trapezium. [2 marks]

11 PROOF

If you are asked to 'prove' (or disprove) a mathematical statement, what does this actually mean? The *Oxford English Dictionary* defines proof as 'The action or an act of testing something; a test, a trial, an experiment … An operation to check the correctness of an arithmetical calculation.'

Can you show that if x and y are even integers then the sum of x and y will also be an even integer? For a group of mathematical statements to constitute a proof you need to ensure that the proof will work for all possibilities and not just one or two. For this proof, how could you ensure that x and y are even integers? Well, an even integer is a whole number that is divisible by 2. So if we say that m, n, p … are integers, then $2m$ will be an even integer and so will $2n$, $2p$, etc. Now you can let $x = 2m$ and $y = 2n$. If you now sum x and y you get $x + y = 2m + 2n$. This simplifies to $2(m + n)$, which is an even number. You have now proved that if x and y are even integers then the sum of x and y will also be an even integer. Some written proofs use the following connectives: $\equiv$ means 'is identical to or is congruent to', $p \Rightarrow q$ means 'p implies q' (if p then q), $p \Leftarrow q$ means 'p is implied by q' (if q then p) and $p \Leftrightarrow q$ means 'p implies and is implied by q' (p is equivalent to q).

If you are asked to 'show', 'prove' or 'demonstrate', you are being asked to prove that something is true. The word 'prove' may not actually appear in what you are being asked to do but you need to understand that this is what is required of you. You may also be asked to 'disprove' something, or to 'show by using a counter example' that something is false. Throughout this book, questions involving proof are flagged as such and proof is discussed in context in the margin.

LEARNING OBJECTIVES

You will learn how to:

> understand and use the structure of mathematical proof, proceeding from given assumptions through a series of logical steps to a conclusion

> use proof by deduction

> use proof by exhaustion

> use disproof by counter example.

TOPIC LINKS

Most chapters of this book demonstrate proofs related to specific topics within those chapters. For example, **Chapter 2 Algebra and functions 2: Equations and inequalities** includes the proof for the quadratic formula by completing the square. In this chapter you will practise using different types of proof.

PRIOR KNOWLEDGE

You should already know how to:

> argue mathematically to show that algebraic expressions are equivalent, and use algebra to support and construct arguments and proofs

> apply angle facts, triangle congruence, similarity and properties of quadrilaterals to conjecture and derive results about angles and sides, including Pythagoras' theorem and the fact that the base angles of an isosceles triangle are equal, and use known results to obtain simple proofs

> use vectors to construct geometric arguments and proofs.

You should be able to complete the following questions correctly:

1 Show that the sum of any three consecutive odd numbers is always a multiple of 3.
2 Show that the sum of two consecutive positive integers is always an odd number.
3 Show that the difference between the squares of any two consecutive positive integers is equal to the sum of the two integers.
4 Prove that the sum of the squares of any two even integers is always a multiple of 4.
5 Prove that $(2n-1)^2 - (2n+1)^2$ is a multiple of 8.

11.1 Proof by deduction

Proof by deduction is the most commonly used form of proof throughout this book – for example, the proofs of the sine and cosine rules in **Chapter 6 Trigonometry**. Proof by deduction is the drawing of a conclusion by using the general rules of mathematics and usually involves the use of algebra. Often, $2n$ is used to represent an even number and $2n + 1$ to represent an odd number.

KEY INFORMATION

Proof by deduction is the drawing of a conclusion by using the general rules of mathematics.

Example 1

Show that $n^2 - 8n + 17$ is positive for any integer n.

Solution

Complete the square.

$$n^2 - 8n + 17 = (n-4)^2 + 1$$

which is always positive for any integer n.

PROOF

The phrase 'show that' indicates that a proof is required.

Example 2

A bag has x green balls and 6 red balls. Two balls are taken from the bag at random. The probability that the balls are different colours is $\frac{1}{2}$. Show that $x^2 - 13x + 30 = 0$.

Solution

There are two possibilities for the balls to be different. The first one could be green and the second red or the first one could be red and the second green. The probabilities of each will be the same.

Since there are x green balls and $(x+6)$ balls in total, the probability that the first ball is green is $\frac{x}{x+6}$.

If a green ball is taken (and not replaced), the probability that the second ball is red is $\frac{6}{x+5}$, since there is 1 fewer ball in the bag.

The probability that the balls are different is therefore given by $\frac{x}{x+6} \times \frac{6}{x+5} \times 2$ since there are two ways this can happen. The question states that this probability is equal to $\frac{1}{2}$.

Hence $\frac{x}{x+6} \times \frac{6}{x+5} \times 2 = \frac{1}{2}$

Multiply both sides by 2.

$$\frac{x}{x+6} \times \frac{6}{x+5} \times 4 = 1$$

Multiply both sides by $(x+6)(x+5)$.

$$x \times 6 \times 4 = (x+6)(x+5)$$

Simplify.

$$24x = x^2 + 11x + 30$$

Subtract $24x$ from both sides.

$$0 = x^2 - 13x + 30, \text{ as requested.}$$

Since there are $(x+6)$ balls to start with, when one ball is removed there will be $(x+5)$ balls in the bag.

PROOF

When you are asked to show a result, the result should be the conclusion of your working and be the last line of working of a series of logical steps which follow on from one another.

Example 3

Demonstrate that the sum of the squares of two **consecutive** odd positive integers is always even.

PROOF

The words 'demonstrate that' have been used to indicate that a proof is required.

Solution

Let $2n$ be an even number. Then $2n - 1$ is the odd number which precedes $2n$ and $2n + 1$ is the odd number which succeeds $2n$.

$$(2n-1)^2 + (2n+1)^2 = 4n^2 - 4n + 1 + 4n^2 + 4n + 1$$
$$= 8n^2 + 2$$
$$= 2(4n^2 + 1)$$

which has a factor of 2 and so is even.

Alternatively:

Let $2n$ be an even number. Then $2n + 1$ is the odd number which immediately succeeds $2n$ and $2n + 3$ is the second odd number which succeeds $2n$.

$$(2n+1)^2 + (2n+3)^2 = 4n^2 + 4n + 1 + 4n^2 + 12n + 9$$
$$= 8n^2 + 16n + 10$$
$$= 2(4n^2 + 8n + 5)$$

which has a factor of 2 and so is even.

Example 4

Prove that the sum of any four consecutive integers a, b, c and d, where $a < b < c < d$, is equal to $cd - ab$.

Solution

Since the integers are consecutive, they can all be written in terms of the same variable.

Let the smallest integer, a, be equal to n.

Hence $b = n + 1$, $c = n + 2$ and $d = n + 3$. •────────

The sum of the four integers, $a + b + c + d$, is given by $n + (n + 1) + (n + 2) + (n + 3)$.

$$n + (n + 1) + (n + 2) + (n + 3) = n + n + 1 + n + 2 + n + 3$$
$$= 4n + 6$$

The expression $cd - ab$ can be rewritten as $(n + 2)(n + 3) - n(n + 1)$.

$$(n + 2)(n + 3) - n(n + 1) = n^2 + 5n + 6 - (n^2 + n)$$
$$= n^2 + 5n + 6 - n^2 - n$$
$$= 4n + 6$$

Since both resulting expressions are equal to $4n + 6$, the sum of any four consecutive integers a, b, c and d, where $a < b < c < d$, is equal to $cd - ab$.

> Consecutive integers increase by 1 from term to term.

Exercise 11.1A

Answers page 540

  1 Prove that the expression $n^3 - n$ is always the product of three consecutive integers for $n \geqslant 2$.

2 The diagram shows a shape with an area of $201\ \text{cm}^2$.

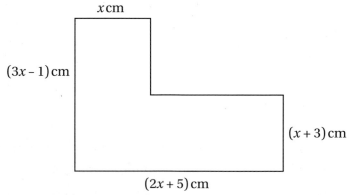

Prove that $4x^2 + 7x - 186 = 0$ and hence deduce that the perimeter of the shape is $68\ \text{cm}$.

3 Using completion of the square, prove that $n^2 - 6n + 10$ is positive for all
values of n.

 4 Prove that any square number is either a multiple of 4 or one more than a
multiple of 4.

PF **5** A cyclic quadrilateral has all four vertices on the circumference of a circle. Prove that the
PS opposite angles of a cyclic quadrilateral have a sum of 180°.

PF **6** Six people, Alf, Bukunmi, Carlos, Dupika, Elsa and Frederique, are to be seated around a
PS circular table for a dinner party. Alf is to sit next to Bukunmi. Carlos must *not* sit next to
Dupika but must sit next to Elsa. Ignoring rotations and reflections, show that there are eight
possible seating arrangements.

PF **7** The following diagram shows a large square with side length a containing four **congruent**
PS right-angled triangles and a smaller square. The base of each triangle is of length b.

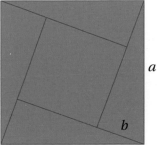

Show that the area of the four right-angled triangles plus the smaller square is equal to the
area of the large square.

 8 Look at this 'proof' which seems to show that $1 = 2$. Identify the flaw in the workings.

Let $a = b$ where $a, b > 0$.

Then	$ab = b^2$
So	$ab - a^2 = b^2 - a^2$
Factorising:	$a(b - a) = (b + a)(b - a)$
Dividing leaves	$a = b + a$
As $b = a$, substituting gives	$a = a + a$
So	$a = 2a$
If $a = 1$ then	$1 = 2$.

11.2 Proof by exhaustion

Proof by exhaustion is sometimes called the 'brute force method'.
It is called this because the steps involved are as follows:

› divide the statement into a finite number of cases

› prove that the list of cases identified is exhaustive – that is, that
there are no other cases

› prove that the statement is true for all of the cases.

KEY INFORMATION

Proof by exhaustion requires
splitting the statement into
a finite, exhaustive set of
cases, all of which are tested.

Example 5

Suppose n is an integer between 24 and 28 inclusive.
Prove that n is not a prime number.

Solution

A prime number is an integer which has two and only two distinct factors. 1 is not a prime number.

$n = \{24, 25, 26, 27, 28\}$

Case 1, $n = 24$: the factors of 24 are 1, 2, 3, 4, 6, 8, 12, 24 so 24 is not prime.

Case 2, $n = 25$: the factors of 25 are 1, 5, 25 so 25 is not prime.

Case 3, $n = 26$: the factors of 26 are 1, 2, 13, 26 so 26 is not prime.

Case 4, $n = 27$: the factors of 27 are 1, 3, 9, 27 so 27 is not prime.

Case 5, $n = 28$: the factors of 28 are 1, 2, 4, 7, 14, 28 so 28 is not prime.

If n is an integer between 24 and 28 then n is not a prime number.

PROOF

To use proof by exhaustion, you need to list all of the possible integers and test each case, one by one.

Example 6

Elif has been asked to list all the possible arrangements of the word PEEP. She has written EEPP and PEPE.

Prove by exhaustion that there are 6 distinct ways of arranging the letters in the word PEEP.

Solution

If the first letter is P and the second letter is P, then the final two letters must both be E.

Elif writes PPEE.

If the first letter is P and the second letter is E, then the final two letters must be P and E.

Elif writes PEPE and PEEP.

If the first letter is E and the second letter is P, then the final two letters must be P and E.

Elif writes EPEP and EPPE.

If the first letter is E and the second letter is E, then the final two letters must both be P.

Elif writes EEPP.

Elif's final solution is PPEE, PEPE, PEEP, EPEP, EPPE, EEPP.

Elif has considered all the possibilities and shown that there are 6 distinct ways of arranging the letters.

Example 7

Suppose x and y are even positive integers less than 8. Prove that their sum is divisible by 2.

Solution

$x = \{2, 4, 6\}$ and $y = \{2, 4, 6\}$

Case 1: $x + y = 2 + 2 = 4 = 2 \times 2$

Case 2: $x + y = 2 + 4 = 6 = 2 \times 3$

Case 3: $x + y = 2 + 6 = 8 = 2 \times 4$

Case 4: $x + y = 4 + 2 = 6 = 2 \times 3$

Case 5: $x + y = 4 + 4 = 8 = 2 \times 4$

Case 6: $x + y = 4 + 6 = 10 = 2 \times 5$

Case 7: $x + y = 6 + 2 = 8 = 2 \times 4$

Case 8: $x + y = 6 + 4 = 10 = 2 \times 5$

Case 9: $x + y = 6 + 6 = 12 = 2 \times 6$

As 2 is a factor in each case then if x and y are even integers less than 8, their sum is divisible by 2.

PROOF

Proof by exhaustion is really only a feasible method to use when the number of cases to be tested is small.

Example 8

Prove by exhaustion that there are only two distinct triangles with sides of integer length that have a perimeter of 10.

Solution

The question requires you to show that of all the possible sets of three integers that have a sum of 10, only two sets can be used to draw a triangle. It is simpler to start by listing all the possible ways that three integers can have a sum of 10 before considering whether or not each set can represent the sides of a triangle. The order of the integers does not matter.

To do this systematically, begin by considering possibilities including a 1. If one integer is 1, then the other two add up to 9. You can have 1, 1, 8 (two 1s and an 8), 1, 2, 7 (a 1, a 2 and a 7), 1, 3, 6 and 1, 4, 5. There are no other possibilities with a 1.

Case 1: 1, 1, 8

Case 2: 1, 2, 7

Case 3: 1, 3, 6

Case 4: 1, 4, 5

Now consider possibilities including a 2 but no 1. If one integer is 2, then the other two add up to 8. You can have 2, 2, 6 or 2, 3, 5 or 2, 4, 4.

Case 5: 2, 2, 6

Case 6: 2, 3, 5

Case 7: 2, 4, 4

Now consider possibilities including a 3 but no 1 or 2. If one integer is 3, then the other two add up to 7. You can have 3, 3, 4 only.

Case 8: 3, 3, 4

Hence there are eight sets of three integers with a sum of 10: {1, 1, 8}, {1, 2, 7}, {1, 3, 6}, {1, 4, 5}, {2, 2, 6}, {2, 3, 5}, {2, 4, 4} and {3, 3, 4}.

However, in order to be the sides of a triangle, any pair of sides must sum to more than the third side. Therefore, if any pair does not sum to more than the third side, then that set of numbers cannot represent the sides of a triangle. If the sides add up to less than the third side then they cannot reach to meet. If the sides add up to the same as the third side then they all lie on a straight line.

Consider each set of integers individually.

Case 1, {1, 1, 8}: $1 + 1 < 8$, so two sides sum to less than the third side.

Case 2, {1, 2, 7}: $1 + 2 < 7$, so two sides sum to less than the third side.

Case 3, {1, 3, 6}: $1 + 3 < 6$, so two sides sum to less than the third side.

Case 4, {1, 4, 5}: $1 + 4 = 5$, so two sides sum to the third side.

Case 5, {2, 2, 6}: $2 + 2 < 6$, so two sides sum to less than the third side.

Case 6, {2, 3, 5}: $2 + 3 = 5$, so two sides sum to the third side.

Case 7, {2, 4, 4}: $2 + 4 > 4$ and $4 + 4 > 2$, so any pair of sides sums to more than the third side.

Case 8, {3, 3, 4}: $3 + 3 > 4$ and $3 + 4 > 3$, so any pair of sides sums to more than the third side.

Therefore {2, 4, 4} and {3, 3, 4} are the only sets of numbers that give two distinct triangles with sides of integer length that have a perimeter of 10.

There cannot be any possibilities which only have integers of 4 or more because the sum would be at least 12.

Alternatively, you could observe that a triangle which satisfies this property cannot have a side length of 5 or more since this is half the perimeter or more. {2, 4, 4} and {3, 3, 4} are the only two sets for which all three integers are below 5.

Example 9

Suppose n is an integer between 2 and 10 inclusive. Prove that $n^2 - 2$ is not divisible by 5.

Solution

Case 1, $n = 2$: $n^2 - 2 = 2^2 - 2 = 2$, which is not a multiple of 5.

Case 2, $n = 3$: $n^2 - 2 = 3^2 - 2 = 7$, which is not a multiple of 5.

Case 3, $n = 4$: $n^2 - 2 = 4^2 - 2 = 14$, which is not a multiple of 5.

Case 4, $n = 5$: $n^2 - 2 = 5^2 - 2 = 23$, which is not a multiple of 5.

Case 5, $n = 6$: $n^2 - 2 = 6^2 - 2 = 34$, which is not a multiple of 5.

Case 6, $n = 7$: $n^2 - 2 = 7^2 - 2 = 47$, which is not a multiple of 5.

Case 7, $n = 8$: $n^2 - 2 = 8^2 - 2 = 62$, which is not a multiple of 5.

Case 8, $n = 9$: $n^2 - 2 = 9^2 - 2 = 79$, which is not a multiple of 5.

Case 9, $n = 10$: $n^2 - 2 = 10^2 - 2 = 98$, which is not a multiple of 5.

If n is an integer between 2 and 10 then $n^2 - 2$ is not divisible by 5.

TECHNOLOGY

Using a spreadsheet software package, you could generate the values for more cases and explore what the value of n is when $n^2 - 2$ is first a multiple of 5.

Exercise 11.2A   Answers page 541

(PF) (PS) **1** Three people, Alf, Bukunmi and Carlos, are to be seated along one edge of a table for a dinner party. Alf is to sit next to Bukunmi. Show using proof by exhaustion that there are four possible seating arrangements.

(PF) (PS) **2** Prove that there are two prime numbers between 20 and 30 inclusive.

(PF) (PS) **3** Prove that no square number below 100 ends in 7.

(PF) (PS) **4** Show that the expression $n^2 + 3n + 19$ is prime for all positive integer values of n below 10.

(PF) (PS) **5** Suppose x and y are odd positive integers less than 7. Prove that their sum is divisible by 2.

(PF) (PS) **6** Prove by exhaustion that there are only two distinct triangles with sides of integer length that have a perimeter of 11.

(PF) (PS) **7** Prove by exhaustion that there are five distinct quadrilaterals with sides of integer length that have a perimeter of 10.

8 A tetronimo is a 2D shape made from four identical squares joined by their edges. Show that, ignoring rotations and reflections, there are five possible tetronimos.

 9 Prove that there are only six possible half-time scores for a match which finishes 2–1.

10 Prove that if n is a positive integer less than 10 then $n^6 - n$ is a multiple of 2.

11.3 Disproof by counter example

To prove something is false it is only necessary to find *one* exception (even though there may be more). This is called a counter example. For example, when learning to spell you might have been told 'i before e except after c'. Turning this into a mathematical proof you could say '**disprove**, by counter example, that i does not always precede e after letters other than c in the English language'. A counter example would be the word 'weird'.

> **KEY INFORMATION**
>
> A counter example is an exception to the rule.

> **Example 10**
>
> Sawda notes that $1^2 - 1 + 5 = 5$ and $2^2 - 2 + 5 = 7$, which are both prime.
>
> Show that it is untrue that $n^2 - n + 5$ is prime for all positive integer values of n.
>
> **Solution**
>
> If Sawda continues with $n = 3$ and $n = 4$, the results are:
>
> $$3^2 - 3 + 5 = 11$$
>
> $$4^2 - 4 + 5 = 17$$
>
> Both of these are prime so are not counter examples.
>
> However, $5^2 - 5 + 5 = 25$, and 25 is *not* prime so $n = 5$ is a counter example.
>
> Note that if $n = 5$, then all three terms are multiples of 5, so this was likely to give an answer which was not prime (unless it was a negative answer, 0 or 5 itself).
>
> Also note that $n^2 + n + 41$ is prime for all integer values of n below 40.

> **TECHNOLOGY**
>
> Using a spreadsheet software package you could use the formula $n^2 - n + 5$ to generate the terms in the sequence.

Stop and think   What is the first integer value of n for which $n^2 + n + 41$ is not prime? Can you show this mathematically? Construct similar equations but with different prime numbers with similar qualities.

Example 11

Disprove that if x and y are consecutive odd numbers then one of x or y must be prime.

Solution

Let $x = 25$ and $y = 27$; then x and y are consecutive odd numbers.

A prime number is an integer which has two and only two distinct factors. 1 is not a prime number.

Let $x = 25$ – the factors of 25 are 1, 5, 25 so 25 is not prime.

Let $y = 27$ – the factors of 27 are 1, 3, 9, 27 so 27 is not prime.

If x and y are consecutive odd numbers then one of x or y is not necessarily a prime number.

In this example you could have methodically worked through each pair of consecutive odd numbers until you found a pair where one of the numbers was not prime.

Example 12

Georgina draws two regular polygons. The pentagon has an interior angle of 108° and the decagon has an interior angle of 144°. Georgina conjectures that the interior angle of any regular polygon will be an integer number of degrees. Find a counter example to show that Georgina's conjecture is false.

Solution

The interior angle (in degrees) of a regular polygon is given by $180 - \dfrac{360}{n}$, where n is the number of sides.

For the pentagon and decagon that Georgina drew, $n = 5$ and $n = 10$ respectively.

For a regular triangle (i.e. an equilateral triangle), $n = 3$, and the interior angle is 60°.

For a regular quadrilateral (i.e. a square), $n = 4$, and the interior angle is 90°.

For a regular hexagon, $n = 6$, and the interior angle is 120°.

None of these is a counter example to Georgina's conjecture that the interior angle is an integer.

However, for a regular heptagon with $n = 7$, $180 - \dfrac{360}{7} = 128\frac{4}{7}$, which is not an integer, so the heptagon is a counter example.

Note that, since the formula is $180 - \dfrac{360}{n}$, any value of n which is not a factor of 360 will be a counter example.

Stop and think How many regular polygons have an interior angle which is an integer number of degrees? How can you check that you have found them all?

Example 13

If x is irrational and y is irrational and $x \neq y$, disprove the statement that $\frac{x}{y}$ is irrational.

Solution

An irrational number is a real number that cannot be expressed as a fraction.

Let $x = \sqrt{2}$

Let $y = \sqrt{8}$

y can also be expressed as $2\sqrt{2}$.

$$\frac{x}{y} = \frac{\sqrt{2}}{2\sqrt{2}}$$
$$= \frac{1}{2}$$

$\frac{1}{2}$ is rational so if x is irrational and y is irrational and $x \neq y$

then $\frac{x}{y}$ is not necessarily irrational.

Stop and think   If x and y are irrational, can you generalise the qualities of x and y so that xy and $\frac{x}{y}$ are rational? Can you prove your generalisation?

Exercise 11.3A

Answers page 543

1 Simon says that:

    a all rectangles are squares

    b every integer less than 10 is positive

    c when a number is doubled the answer is larger

    d for every integer n, n^3 is positive

    e all prime numbers are odd

    f $\log(A + B) = \log A + \log B$

    g $(a + b)^2 = a^2 + b^2$

    Find a counter example for each of Simon's statements.

2 Patrick has noticed that $2^0 > 0^2$, $2^5 > 5^2$ and $2^{10} > 10^2$. Initially he makes a generalisation that $2^n > n^2$ for positive integer values of n but then notices that $2^2 = 2^2$. He then amends his generalisation to $2^n \geqslant n^2$. Prove that Patrick's generalisation is still incorrect.

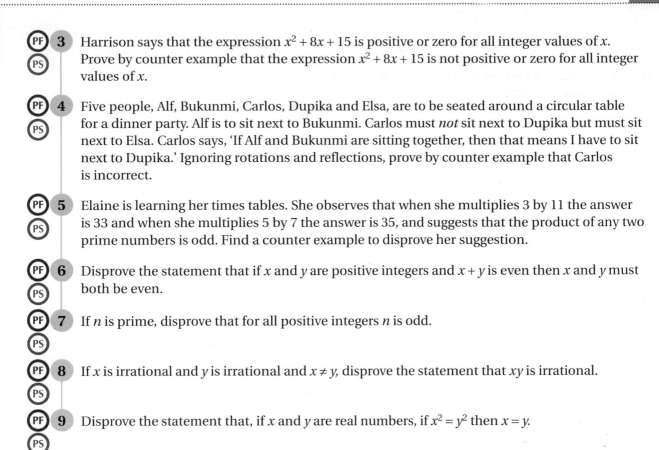

(PF) (PS) 3 Harrison says that the expression $x^2 + 8x + 15$ is positive or zero for all integer values of x. Prove by counter example that the expression $x^2 + 8x + 15$ is not positive or zero for all integer values of x.

(PF) (PS) 4 Five people, Alf, Bukunmi, Carlos, Dupika and Elsa, are to be seated around a circular table for a dinner party. Alf is to sit next to Bukunmi. Carlos must *not* sit next to Dupika but must sit next to Elsa. Carlos says, 'If Alf and Bukunmi are sitting together, then that means I have to sit next to Dupika.' Ignoring rotations and reflections, prove by counter example that Carlos is incorrect.

(PF) (PS) 5 Elaine is learning her times tables. She observes that when she multiplies 3 by 11 the answer is 33 and when she multiplies 5 by 7 the answer is 35, and suggests that the product of any two prime numbers is odd. Find a counter example to disprove her suggestion.

(PF) (PS) 6 Disprove the statement that if x and y are positive integers and $x + y$ is even then x and y must both be even.

(PF) (PS) 7 If n is prime, disprove that for all positive integers n is odd.

(PF) (PS) 8 If x is irrational and y is irrational and $x \neq y$, disprove the statement that xy is irrational.

(PF) (PS) 9 Disprove the statement that, if x and y are real numbers, if $x^2 = y^2$ then $x = y$.

SUMMARY OF KEY POINTS

> ❯ Questions involving proof are flagged throughout this book.

> ❯ Proof by deduction is the drawing of a conclusion by using the general rules of mathematics.

> ❯ Proof by exhaustion requires splitting the statement into a finite, exhaustive set of cases, all of which are tested:

>> ❯ In any proof by exhaustion, it is necessary to be systematic and to list all the possibilities in as structured an order as you can.

> ❯ Disproof by counter example:

>> ❯ A counter example is an exception to the rule.

EXAM-STYLE QUESTIONS 11 Answers page 543

(PF)(PS) 1 Prove that $(3n + 5)^2 - (3n - 5)^2$ is a multiple of 12 for all positive integer values of n. **[4 marks]**

(PF)(PS) 2 Elif has been asked to arrange the letters in the word SUMS. She has written MUSS and SMUS. Prove by exhaustion that there are 12 distinct ways of arranging the letters in the word SUMS. **[3 marks]**

(PF)(PS) 3 Suppose x and y are even positive integers less than 8. Prove by exhaustion that their difference is divisible by 2. **[5 marks]**

(PF)(PS) 4 Prove that the sum of n consecutive integers is a multiple of n when n is odd but not when n is even. **[5 marks]**

(PF)(PS) 5 A number is palindromic if it reads the same backwards and forwards, for example 252 and 1001. Show by exhaustion that 343 is the only palindromic three-digit cube number. **[3 marks]**

(PF)(PS) 6 Jayne writes that $\sin 2\theta = 2 \sin \theta$ for any angle θ. Prove by counter example that $\sin 2\theta$ does not always equal $2 \sin \theta$. **[2 marks]**

(PF)(PS) 7 Show that the statement '$n^2 - n + 1$ is a prime number for all values of n' is untrue for $n > 1$. **[4 marks]**

(PF)(PS) 8 Use the diagram to prove Pythagoras' theorem algebraically.

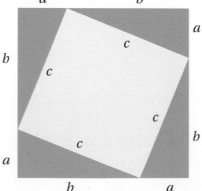

[6 marks]

(PF)(PS) 9 Disprove the statement that if x and y are real numbers, if $x^2 > y^2$ then $x > y$. **[3 marks]**

12 DATA PRESENTATION AND INTERPRETATION

Gender inequality in pay remains high on the agenda in most countries. A map showing the pay gaps is produced by Eurostat, the equivalent of the UK's Office for National Statistics. The figures show that pay levels throughout the whole of Europe differ hugely, with some areas seeing women earning more than men on average, while other areas see men earning more than their female counterparts. Here, 'average' is used to describe the mean earnings.

Data presentation allows you to analyse the data related to the gender pay gap. Using various measures of displacement and location, you can calculate the range and mean pay, and select the best measure to represent your findings about wages. It is also helpful to be able to deal with grouped data, both in calculations and graphically. A graphical representation is often used as it is visual, easier to interpret and allows for comparisons to be made between data sets, making it easier to discover any anomalies. It is also important to be able to find out if a result fits a trend or if it is an outlier, because outliers might skew results and make them unrepresentative.

Another important aspect of data interpretation is making inferences from your findings and using these to come to a conclusion. You can use data to report that differences in the mean wages favour males in certain areas within the UK. This is a powerful conclusion as it leads the reader to understand that men earn more.

LEARNING OBJECTIVES

You will learn how to:

> use discrete, continuous, grouped or ungrouped data

> use measures of central tendency: mean, median, mode

> use measures of variation: variance, standard deviation, range and interpercentile ranges

> use linear interpolation to calculate percentiles from grouped data

> use the statistic $S_{xx} = \sum (x - \bar{x})^2 = \sum x^2 - \dfrac{(\sum x)^2}{n}$

> use standard deviation $\sigma = \sqrt{\dfrac{S_{xx}}{n}}$

> understand and use linear coding

> interpret and draw histograms, frequency polygons, box and whisker plots (including outliers) and cumulative frequency diagrams

> use and interpret the equation of a regression line and understand how to make predictions within the range of values of the explanatory variable and understand the dangers of extrapolation

> describe correlation using terms such as positive, negative, zero, strong and weak.

TOPIC LINKS

In this chapter you will use the skills and techniques that you developed in **Chapter 1 Algebra and functions 1: Manipulating algebraic expressions**. Learning about displaying and interpreting data will help you to analyse data critically, including that in any large data set. You will work with the large data set provided by your chosen exam board in this chapter and in **Chapter 14 Statistical sampling and hypothesis testing**. Knowing how to display data will help in **Book 2, Chapter 13 Statistical distributions**.

PRIOR KNOWLEDGE

You should already know how to:

> interpret and construct tables, charts and diagrams, including frequency tables, bar charts, pie charts and pictograms for categorical data, vertical line charts for ungrouped discrete numerical data, tables and line graphs, and know their appropriate use

> construct and interpret diagrams for grouped discrete data and continuous data (i.e. histograms with equal and unequal class intervals and cumulative frequency graphs), and know their appropriate use

> interpret, analyse and compare the distributions of data sets

> use appropriate graphical representation involving discrete, continuous and grouped data, including box plots

> use appropriate measures of central tendency and spread

> apply statistics to describe a population

> use and interpret scatter graphs of bivariate data, recognise correlation and know that it does not indicate causation

> draw estimated lines of best fit, make predictions, interpolate and extrapolate apparent trends while knowing the dangers of doing this.

You should be able to complete the following questions correctly:

1 A test was given to 50 students and the following marks were awarded:

13	22	41	36	32	26	31	41	31	41
41	14	26	41	41	26	39	39	45	45
34	23	36	23	47	23	47	40	41	29
15	39	36	41	27	46	16	40	19	31
12	27	39	27	28	41	47	28	41	30

    a Is this data qualitative or quantitative? If it is quantitative, is it discrete or continuous?

    b Calculate the mean, median, mode and range of the data.

    c Display the data and make a statement about your findings.

12.1 Measures of central tendency and spread

Types of data
Information obtained from various sources is called **data**. There are two distinct types of data – qualitative and quantitative.

Qualitative data are usually descriptive data, given as categories – such as hair colour, car type or favourite chocolate bar. No numerical value means no numerical meaning can be calculated. Another name for qualitative data is **categorical data**.

Quantitative data, or **numerical data**, are given in numerical form and can be further split down in two categories – discrete and continuous. If all possible values can be listed, the data is **discrete**. Examples of discrete data are shoe size, clothes size and number of marks in a test. **Continuous** data are called continuous because they can be represented at any point on a scale. For instance, height has meaning at all points between any values, e.g. a student's height could be measured as 1.7 m, or 1.67 m, or even 1.668 m. Continuous data can be shown on a number line, and all points on the line have meaning and are different, whereas discrete data can only have a particular selection of values.

> **KEY INFORMATION**
>
> Qualitative data are not numerical, but categorical.
>
> Quantitative data, or numerical data, can be subdivided into discrete and continuous.
>
> Discrete data may only take separate values, for example whole numbers.
>
> Continuous data can be shown on a number line, and all points on the line have meaning and are different from the others.

Example 1

Are the following sets of data qualitative or quantitative, and if quantitative, is the data discrete or continuous?

A Eye colours of students in a class: {2 hazel, 7 blue, 6 green, etc}

B Temperatures of water in an experiment: {34.56, 45.61, 47.87, 56.19, etc}

C Totals of numbers shown on two dice in a board game: {2, 6, 7, 7, 8, 9, 12, etc}

D Numbers of spectators at football matches: {12 134, 2586, 6782, 35 765, etc}

E Favourite types of cereal: {cornflakes, oatflakes, rice crisps, etc}

F Times in seconds between 'blips' of a Geiger counter in a physics experiment: {0.23, 1.23, 3.03, 0.21, 4.51, etc}

G Scores out of 50 in a maths test: {20, 24, 43, 45, 49, etc}

H Sizes of epithelial cells: {1.2×10^{-5} m, 1.21×10^{-5} m, etc}

I Shoe sizes in a class: {6, 7, 7, 7, 8, 9, 10, 10, 11, etc}

Solution

A and E are qualitative data; all the others are quantitative.

C, D, G, I – discrete

B, F, H – continuous

> For example, different eye colours are separate categories without numerical values, so they are qualitative data.

> Discrete data can only take certain values. Continuous data can take any value within a range.

Measures of central tendency

Measures of central tendency – otherwise referred to as averages or measures of location – are often the first tools for comparing data sets and interpreting data. In your GCSE course you will have encountered three measures of central tendency – the median, the mode and the mean. At AS-level you need to be confident in deciding which measure is the most appropriate to use to answer a specific question.

Median

Data is arranged in numerical order and the **median** is the item of data in the middle. When there is an odd number of data items, the median is simply the middle number. However, when there is an even number of items, the median lies between two middle values, and you use the mean of these two values for the median. Use the formula $\frac{n+1}{2}$ to find the position of the median, where n is the number of data items.

You should use the median for quantitative data, particularly when there are extreme values (values that are far above or below most of the other data) that may skew the outcome.

Mode

The **mode** is the most commonly occurring item of data. It is the item with the highest **frequency**. So for the data set {1, 3, 5, 5, 7}, the mode is 5, as this item appears twice. There may be more than one mode, if more than one item has the highest frequency – for instance {1, 2, 2, 5, 5, 7} has modal values of 2 and 5.

You should use the mode with qualitative data (car models, etc) or with quantitative data (numbers) with a clearly defined mode (or which are bi-modal). The mode is not much use if the distribution is evenly spread, as any conclusions based on the mode will not be meaningful.

Mean

When people are discussing the average, they are usually referring to the **mean**. This is the sum of all of the items of data divided by the number of items of data.

The formula is normally written as $\bar{x} = \frac{\sum x}{n}$.

› $\bar{x}$ stands for the mean and is pronounced 'x bar'.

› The Greek letter sigma (Σ) means 'the sum of'. It gives the total of all the data.

› The number of data items is n.

You use the mean for quantitative data (numbers). As the mean uses all the data, it gives a true measure, but it can be affected by extreme values (outliers).

TECHNOLOGY

You can use calculators to calculate many useful statistics. Put your calculator in statistics mode and it may show a table for you to enter your raw data. Use the following set of data to calculate the mean:

2, 4, 6, 8, 10

When a salary increase is being negotiated, the management may well have a different opinion to the majority of workers. The following figures were collected to compare the salaries in a small fast-food company:

£3500, £3500, £3500, £3500, £4500, £4500, £4500, £8000, £10 000, £10 000, £10 000, £12 000, £12 000, £18 000, £30 000

Who do you think earns what? The median salary is £8000, the modal salary is £3500 and the mean salary is £9166.67. These figures can be used in a variety of ways, but which is the most appropriate measure? If you were the manager, you might quote the mean of £9166.67, but fewer than half of the employees earn this amount.

Workers leading the pay negotiations who want to criticise the current wage structure may choose to quote the mode (£3500), as this is the lowest average. This would highlight issues in the wage structure.

The mean takes account of the numerical value of every item of data. It is higher because of the effect of the £30 000 salary, which is an extremely large value in comparison to the others. The median and mode are not affected by extreme values.

The mean is a good measure to use as you use all your data to work it out. It can, however, be affected by extreme values and by distributions of data which are not symmetrical. The median is not affected by extremes so is a good measure to use if you have extremes in your data, or if you have data which isn't symmetrical. The mode can be used with all types of data, but some data sets can have more than one mode, which isn't helpful at all.

> **KEY INFORMATION**
>
> The median is the middle value when the data items are placed in numerical order.
>
> The mode is the most common or frequent item of data.
>
> The mean ($\bar{x}$) is found by adding the data values together and dividing by the number of values: $\bar{x} = \frac{\sum x}{n}$.

Using the large data set 12.1

a   Calculate the median and the mean of one category in one location. Comment on your findings.

b   Repeat for the same category but in another location. Does the location affect the mean and the median? Comment critically on your findings.

c   State any assumptions you have made whilst using your chosen category.

Measures of spread

Measures of spread show how spread out, or scattered, data items are. In your GCSE course you will have encountered two measures of spread – the range and the interquartile range.

Range

The simplest measure of spread is the difference between the highest and smallest data items, known as the **range**. This is straightforward to calculate, but can be highly sensitive to extreme values.

Range = highest value − lowest value

Interquartile range

A more trustworthy measure of spread is the range in the middle half of the data. The upper **quartile** is the median of the upper half of the data, and the lower quartile is the median of the lower half of the data.

The **interquartile range** measures the range of the middle 50% of the data.

In **Section 12.2** you will encounter another measure of spread – the standard deviation.

Quartiles split the data into quarters. If the data items are arranged in numerical order, the lower quartile is one quarter of the way through the data and the upper quartile is three quarters of the way through.

See **Section 12.3** for examples of the interquartile range in the context of box and whisker plots.

Using frequency tables to calculate measures of central tendency and spread

When data contains items that are repeated it is easier to record them using a frequency table. You used frequency tables for categorical data in your GCSE course – now you will use them to deal with numerical data. When you use frequency tables you can calculate measures of spread more quickly.

Example 2

Here are the scores when a four-sided spinner was spun repeatedly:

$\{3, 4, 3, 3, 1, 2, 1, 2, 2, 2, 2, 3, 4, 2, 3, 4, 2, 2, 1, 3\}$

The scores are recorded in the frequency table. Calculate the mean of the data set.

Score	Frequency
1	3
2	8
3	6
4	3

Solution

The mean of the data shown in the frequency table above can be written as $\bar{x} = \dfrac{\sum fx}{\sum f}$.

The frequencies are added to find the total number of data items. Representing the score by x and its frequency by f, the calculation of the mean starts by multiplying each score by its corresponding frequency to find the total value of all the scores.

Score, x	Frequency, f	fx
1	3	$1 \times 3 = 3$
2	8	$2 \times 8 = 16$
3	6	$3 \times 6 = 18$
4	3	$4 \times 3 = 12$
Total	$\sum f = 20$	$\sum fx = 49$

The total score is 49 and there are 20 data items. Dividing the total by the number of data items gives the mean.

Add up the two columns, f and fx, and use their totals to calculate the mean.

$$\bar{x} = \frac{\sum fx}{\sum f} = \frac{49}{20} = 2.45$$

TECHNOLOGY

Repeat the above calculation using a frequency table on your calculator. You may have to go into Set-up on your calculator and select 'Frequency on' before you start.

Finding the median using a frequency table is straightforward as the data is already ordered. There are 20 numbers in **Example 2**, so if the data were written out in a line the median would lie between the 10th and 11th values ($20 + 1 = 21$, then divide by 2 to give 10.5). The next step is to find the class that contains the 10.5th value.

Score, x	Frequency, f	Cumulative frequency
1	3	3 (not enough – cumulative frequency needs to be between 10 and 11)
2	8	$3 + 8 = 11$ (must be this score as the combined frequencies are more than 10.5)

The median is within the data associated with a score of 2. Therefore the median score is 2.

You can identify the mode as it is the score with the highest frequency. From the data in **Example 2** the highest frequency is 8, so the mode has to be 2.

Discrete grouped data

Grouped frequency tables are used when data is widely spread. The downside to this is that raw data is lost, since now you will only know the frequency of each grouping. Consider the following data on gross salaries for 427 professions, drawn from the Office for National Statistics' 2010 salary tables (*Annual Survey of Hours and Earnings*).

Salary, £	Frequency
0–9999	21
10000–19999	129
20000–29999	172
30000–39999	74
40000–49999	24
50000–59999	4
60000–69999	1
70000–79999	1
80000–89999	0
90000–99999	1

Immediately, you can see that the modal wage is between £20000 and £29999, as this grouping has the highest frequency, and that salaries above £60000 are unusual.

KEY INFORMATION

For discrete data, the groupings should not overlap.

Continuous grouped data

The PTA have given some money for 20 hockey players from a Year 7 school team to get their new team uniforms. Their heights in metres are:

| 1.45 | 1.48 | 1.46 | 1.52 | 1.46 | 1.61 | 1.60 | 1.51 | 1.55 | 1.56 |
| 1.61 | 1.64 | 1.53 | 1.51 | 1.48 | 1.70 | 1.70 | 1.62 | 1.45 | 1.50 |

Height, h (metres)	Frequency, f
$1.45 \leqslant h < 1.50$	6
$1.50 \leqslant h < 1.55$	5
$1.55 \leqslant h < 1.60$	2
$1.60 \leqslant h < 1.65$	5
$1.65 \leqslant h < 1.70$	0
$1.70 \leqslant h < 1.75$	2
Total	20

KEY INFORMATION

Measurements are frequently recorded to the nearest unit. So a height of 1.65 m could actually be in the range $1.645 \leqslant h < 1.655$.

Estimating the mean from grouped data

The reason this is now an estimate is because you are estimating the middle value of each class by assuming that the data is evenly distributed throughout each interval.

Salary (£)	Frequency, f	Estimate of x	fx
0–9999	21	5000	105 000
10 000–19 999	129	15 000	1 935 000
20 000–29 999	172	25 000	4 300 000
30 000–39 999	74	35 000	2 590 000
40 000–49 999	24	45 000	1 080 000
50 000–59 999	4	55 000	220 000
60 000–69 999	1	65 000	65 000
70 000–79 999	1	75 000	75 000
80 000–89 999	0	85 000	0
90 000–99 999	1	95 000	95 000
Total	$\Sigma f = 427$		$\Sigma fx = 10\,465\,000$

$$\bar{x} = \frac{\Sigma fx}{\Sigma f} = \frac{10\,465\,000}{427} = £24\,508.20 \ (2 \ \text{d.p.})$$

Exercise 12.1A

Answers page 544

1 State whether the following data are discrete or continuous:

  a daily rainfall in Lincoln

  b monthly texts you send on your mobile

  c the number of burgers sold in a fast-food restaurant

  d the duration of a marathon

  e the ages of the teachers in your school.

2 Classify the following as qualitative or quantitative, discrete or continuous:

  a gender

  b height

  c GCSE grades in maths

  d examination score in maths

  e waist size

  f whether people are car owners or not

  g weekly self-study time.

(CM) **3** Write down two quantitative variables about your class. Identify each variable as discrete data or continuous data.

(CM) **4** The numbers of visits to a library made by 20 children in one year are recorded below.

  0, 2, 6, 7, 5, 9, 12, 43, 1, 0, 45, 2, 7, 12, 9, 9, 32, 11, 36, 13

  a What is the modal number of visits?

  b What is the median number of visits?

  c What is the mean number of visits?

  d Comment on the best measure of central tendency to use and why.

5 Fiona records the amount of rainfall, in mm, at her home, each day for a week. The results are:

  2.8, 5.6, 2.3, 9.4, 0.0, 0.5, 1.8

Fiona then records the amount of rainfall, x mm, at her house for the following 21 days. Her results are:

  $\Sigma x = 84.6 \, \text{mm}$

  a Calculate the mean rainfall over the 28 days.

  b Fiona realises she has transposed two of her figures. The number 9.4 should be 4.9 and the 0.5 should be 5.0. She corrects these figures. What effect will this have on the mean?

(CM) **6** An employee has to pass through 8 sets of traffic lights on her way to work. For 100 days she records how many sets of lights she gets stopped at. Here are her results:

Number of times stopped	Number of journeys
≤1	3
2	5
3	11
4	21
5	22
6	17
7	14
8	7

Find the median, mode and mean number of times stopped at traffic lights and comment on the best measure of central tendency to represent the data.

(M) **Modelling**   (PS) **Problem solving**   (PF) **Proof**   (CM) **Communicating mathematically**   **307**

7 During a biological experiment, fish of various breeds were measured, to the nearest cm, one year after being released into a pond. The lengths were recorded in the following table:

Length, x (cm)	Frequency
$7.5 \leqslant x < 10$	30
$10 \leqslant x < 15$	70
$15 \leqslant x < 20$	100
$20 \leqslant x < 30$	80
$30 \leqslant x < 35$	40

Calculate an estimate of the mean length of one of these fish.

8 The following data was collected but some information was missed out. Complete the missing parts and confirm the estimate of the mean.

Height (cm)	Frequency	
$100 < h \leqslant 120$	5	
$120 < h \leqslant 140$	4	
	12	1800
$160 < h \leqslant 180$	13	
$180 < h \leqslant 200$	8	
		6600

Estimate of the mean $= 157.14$

12.2 Variance and standard deviation

One limitation with quartiles and the interquartile range is that they do not take all the data items into account. Variance and standard deviation are measures that involve determining the spread of *all* the data items.

Consider a small set of data: {1, 2, 3, 4, 5}

The mean of this data is 3.

The **deviation** is the difference between each data item and the mean, usually notated as $x - \bar{x}$.

The set of deviations for this set of data is:

$$1 - 3 = -2$$
$$2 - 3 = -1$$
$$3 - 3 = 0$$
$$4 - 3 = 1$$
$$5 - 3 = 2$$

Adding up the deviations:

$$-2 + -1 + 0 + 1 + 2 = 0$$

Why do the deviations add up to zero?

The deviations, $x - \bar{x}$, give a measure of spread. Combining the deviations, by squaring each deviation and then adding them together, gives the sum of their squares, notated as S_{xx}.

The sum of the squares is usually written

$$S_{xx} = \Sigma(x - \bar{x})^2 = \Sigma x^2 - n\bar{x}^2$$

The second formula may be easier to work with if you are given raw data.

To derive this formula:

$$S_{xx} = \Sigma(x - \bar{x})^2$$
$$= \Sigma\left(x^2 - 2x\bar{x} + \bar{x}^2\right)$$
$$= \Sigma x^2 - \Sigma 2x\bar{x} + \Sigma\bar{x}^2$$
$$= \Sigma x^2 - 2\bar{x}\Sigma x + \Sigma\bar{x}^2$$
$$= \Sigma x^2 - 2n\bar{x}^2 + n\bar{x}^2$$
$$= \Sigma x^2 - n\bar{x}^2$$

Taking another small set of data, {1, 4, 5, 6, 14}:

$$S_{xx} = (1-6)^2 + (4-6)^2 + (5-6)^2 + (6-6)^2 + (14-6)^2$$
$$= -5^2 + -2^2 + -1^2 + 0^2 + 8^2 = 94$$

The mean of the sum of the squares is a measure of spread called the **variance**, $\dfrac{\Sigma(x - \bar{x})^2}{n}$.

The square root of the variance is called the **standard deviation**,

$\sqrt{\dfrac{\Sigma(x - \bar{x})^2}{n}}$, which is usually denoted by the symbol σ.

Therefore the variance for this set of data is $\dfrac{94}{5} - 18.8$

and the standard deviation is $\sqrt{18.8} = 4.34$.

An easier formula, which involves doing fewer subtractions, is:

$$\sigma = \sqrt{\dfrac{\Sigma x^2}{n} - \bar{x}^2}$$

Standard deviation is especially useful when comparing different sets of data and when analysing the position of an item of data in a population. An advantage of standard deviation is that it uses all of the data. A disadvantage is that it takes longer to calculate than other measures of spread.

You can divide by n or $(n-1)$ when calculating either the variance of a population or an estimate for the population from sample data. Either divisor will be accepted unless a question specifically requests an unbiased estimate of a population variance, in which case you would use $(n-1)$.

Example 3

Calculate the variance and standard deviation for this small data set: {2, 3, 5, 8, 13}.

Solution

Calculate the mean.

$$\bar{x} = \frac{2 + 3 + 5 + 8 + 13}{5} = 6.2$$

Determine the deviations from the mean.

$$= 6.2 - 2, 6.2 - 3, 6.2 - 5, 6.2 - 8, 6.2 - 13$$

$$= 4.2, 3.2, 1.2, -1.8, -6.8$$

Square these deviations.

$$(x - \bar{x})^2 = 17.64, 10.24, 1.44, 3.24, 46.24$$

Find the mean of these squared deviations, or the variance.

$$\Sigma(x - \bar{x})^2 = 17.64 + 10.24 + 1.44 + 3.24 + 46.24 = 78.8$$

$$= \frac{78.8}{5} = 15.76$$

Find the standard deviation by taking the square root of the variance.

$$\sqrt{\frac{\Sigma(x - \bar{x})^2}{n}} = \sqrt{15.76} = 3.97$$

Stop and think   How could you use the alternative formula in the above example?

Calculating variance and standard deviation using frequency tables

Example 4

The table below shows the numbers of pieces of fruit eaten by sixth form students in one day. Find the standard deviation of the number of pieces of fruit eaten.

Number, x	Frequency, f
1	2
2	12
3	45
4	114
5	41
6	16
Total	230

Solution

First multiply each number of pieces of fruit by its corresponding frequency to give you fx, then sum to find out the total number of pieces of fruit, $\sum fx$.

Next, square each value (x^2) and then sum to find out the total $\sum x^2$.

Finally, multiply x^2 by the frequency to give fx^2, then sum to find out the total, $\sum fx^2$.

Number, x	Frequency, f	fx	x^2	fx^2
1	2	2	1	2
2	12	24	4	48
3	45	135	9	405
4	114	456	16	1824
5	41	205	25	1025
6	16	96	36	576
	230	918	91	3880

For discrete frequency distributions, the formulae are:

$$\sigma = \sqrt{\frac{\sum f(x - \bar{x})^2}{\sum f}} \quad \text{or} \quad \sigma = \sqrt{\frac{\sum fx^2}{\sum f} - \bar{x}^2}.$$

$\sum f$, the sum of the frequencies, is used instead of n.

For grouped frequency distributions, you use the midpoint of the group for x.

$$\text{Mean} = \frac{918}{230}$$

$$\text{Variance} = \frac{3880}{230} - \left(\frac{918}{230}\right)^2 = 0.939$$

$$\text{Standard deviation} = \sqrt{0.939} = 0.969$$

$$\sigma = 0.969$$

TECHNOLOGY

Check that you can work out the mean and standard deviation for this example using the frequency table option on your calculator.

Exercise 12.2A

Answers page 544

 1 Calculate the mean and standard deviation of the following.

  a 2, 4, 6, 8, 10, 12, 14

  b 50, 60, 70, 80, 90

  c 12, 15, 18, 16, 7, 9, 14

  Check your results using a calculator.

TECHNOLOGY

In all these questions you can use the statistical buttons on a calculator.

2 For each of the following sets of data, find the mean and standard deviation.

  a 2, 2, 4, 4, 4, 5, 6, 6, 8, 9

  b 13.1, 20.4, 17.4, 16.5, 21.0, 14.8, 12.6

(CM) **3** Here are the shoe sizes of a class of Year 7 students:

Size	3	4	5	6	7	8	9
Frequency	1	0	8	14	6	2	1

  a Calculate the mean and standard deviation shoe size.

  b What can you say about the mean shoe size compared with the median and modal sizes?

4 $\Sigma x = 27$, $\Sigma x^2 = 245$, $n = 3$

  Find the mean and standard deviation of the data.

5 The lengths of time, t minutes, taken for a bus journey is recorded on 15 days and summarised by

$$\Sigma x = 102, \Sigma x^2 = 1181$$

  Find the mean and variance of the times taken for this bus journey.

6 For a set of 20 data items, $\Sigma x = 12$ and $\Sigma x^2 = 144$.
  Find the mean and standard deviation of the data.

(CM) **7** Mrs Moat has a choice of two routes to work. She times her journeys along each route on several occasions and the times in minutes are given below.

Town route	15	16	20	28	21
Country route	19	21	20	22	18

  a Calculate the mean and standard deviation for each route.

  b Which route would you recommend? Give a reason.

(CM) **8** A machine is supposed to produce ball bearings with a mean diameter of 2.0 mm.
  A sample of eight ball bearings was taken from the production line and the diameters measured.
  The results in millimetres were:

$$2.0, 2.1, 2.0, 1.8, 2.4, 2.3, 1.9, 2.1$$

  a Calculate the mean and standard deviation of the diameters.

  b Do you think the machine is set correctly?

Data cleaning

Any large data set is likely to be missing some data. If values are left blank, it is likely that the software package you use to work out statistics or draw graphs may not work, or may interpret missing values as zero. This will cause errors in your calculations, leading to misleading outcomes. Often, unusual data entries are classed as **outliers**. The process of dealing with missing and unusual values in data is called **data cleaning**.

Data cleaning involves identifying and then removing invalid data points from a data set. You can then calculate your statistics and draw your graphs using the remaining data. If data has been inputted as 'n/a', most packages will not understand this, so errors will appear in your calculations. You need to use your judgment to decide which points are valid and which are not.

The points to be cleaned are generally either missing data points or outliers. One way of detecting points which do not fit the trend is to plot the data and then inspect for points that lie far away from the trend.

The importance of reliable data in any statistical analysis cannot be over-emphasised. This is one reason why larger data sets are generally more reliable.

Using the large data set 12.2	The large data set has a lot of raw data so should be cleaned before you make any analysis. You need to be able to use spreadsheets to perform the calculations and produce the diagrams.
	a Check through and clean the data for the category you used in **Using the large data set 12.1**, giving valid reasons for your omissions as you progress through the data.
	b Does having clean data change the median and mean you calculated in **12.1**? Explain your reasons.

Standard deviation and outliers

Data sets may contain extreme values and you need to be able to deal with these. Many data sets are samples drawn from larger populations which are normally distributed. In cases like this, approximately:

ᐳ 68% of data items lie within 1 standard deviation of the mean

ᐳ 95% of data items lie within 2 standard deviations of the mean

ᐳ 99.75% of data items lie within 3 standard deviations of the mean.

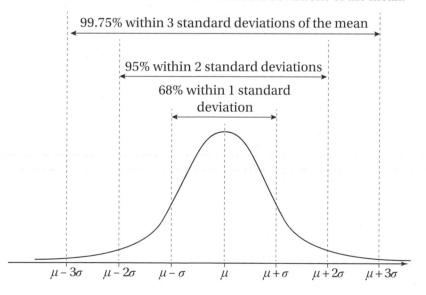

You will cover normal distribution in **Book 2, Chapter 13 Statistical distributions**. However, you should know that if a value is more than a certain number of standard deviations away from the mean it should be treated as an outlier. In your course, any value more than 2 standard deviations away from the mean will be treated as an outlier.

Example 5

The times of journeys to work, in minutes, were recorded by one person for two weeks. The results are:

$$51, 65, 48, 41, 56, 65, 45, 42, 41, 87$$

Determine whether any of the data items are outliers.

Solution

The mean is $\dfrac{\sum x}{n} = 54.1$ minutes.

The standard deviation is $\sqrt{\dfrac{\sum (x - \bar{x})^2}{n}} = 13.94$ minutes.

$$54.1 + 2(13.94) = 81.98 \text{ minutes}$$

$$54.1 - 2(13.94) = 26.22 \text{ minutes}$$

The longest time to get to work was 87 minutes. You should class this as an outlier because $81.98 < 87$ minutes.

Using the large data set 12.3

a Calculate the mean and standard deviation of one of the categories in your large data set.

b Describe how you would clean the data and how this may affect the mean and standard deviation.

c Clean the data for your category and evaluate any changes it has made to the mean and standard deviation.

Linear coding

Linear coding can be used to make certain calculations easier. It is used to simplify messy arithmetic in calculations and also to convert between different units.

The data set {12, 3, 8, 7, 11} has mean $\bar{x} = 8.2$. Using a calculator you can find the standard deviation, $\sigma = 3.19$ (2 d.p.).

Add 10 to all the data points: the data set is now {22, 13, 18, 17, 21}. You can work out the new mean and the new standard deviation using a calculator: $\bar{x} = 18.2$, $\sigma = 3.19$ (2 d.p.). The data points have all increased by 10 but the spread in the data has remained the same.

If the original data set is multiplied by 7, you would get {84, 21, 56, 49, 77}. Using a calculator you can see that the mean is $\bar{x} = 57.4$ and the standard deviation is $\sigma = 22.31$ (2 d.p.). This time the size and the spread in the data have increased.

The following table summarises what happens to the mean and standard deviation of x when various transformations are applied to the data set {2, −2, −8, 0, 2.3}.

	x	$x + 3$	$x - 2$	$10x$	$\frac{x}{2} + 2$
	2	5	0	20	3
	−2	1	−4	−20	1
	−8	−5	−10	−80	−2
	0.8	3.8	−1.2	8	2.4
	2.3	5.3	0.3	23	3.15
Mean	−1.0	2.0	−3.0	−9.8	1.5
Standard deviation	3.8	3.8	3.8	38.24	1.91

Stop and think   What do you notice about the mean and standard deviation in each case?

If an number is added to or subtracted from a data set, the mean increases or decreases by that amount, but the standard deviation is not affected as the spread in the data remains the same.

If the data set is multiplied or divided by a number, the mean increases or decreases by that factor, as does the standard deviation.

Five people were asked how far they worked from home. Here are their results:

x	24	28	30	33	35

$\bar{x} = 30$ and $\sigma = 3.8$

If you add 40 miles onto each person's journey, the mean will increase but the standard deviation remains unchanged because the spread is still the same.

Y	64	68	70	73	75

$\bar{Y} = 70$ and $\sigma = 3.8$

Similarly, if you subtracted 20 miles from everyone's journey the mean would decrease but the standard deviation would remain the same.

z	4	8	10	13	15

$\bar{z} = 10$ and $\sigma = 3.8$

If all the journey times doubled, the mean would double but so would the standard deviation, as the data would now be more spread out.

a	48	56	60	66	70

$\bar{a} = 60$ and $\sigma = 7.7$

Notice that the mean and standard deviation change by the magnitude of the x coefficient. The mean also changes by the number being added or subtracted from it, but the standard deviation remains unchanged when a number is simply added to or subtracted from x.

There is no multiplier to affect the standard deviation.

There is no multiplier to affect the standard deviation.

The multiplier affects the standard deviation.

If the journey time doubled and then 20 more miles were added on, the mean would increase by a multiple of 2 plus 20 for everyones time:

b	68	76	80	86	90

$\bar{b} = 80$ and $\sigma = 7.7$

The multiplier affects the standard deviation, but the addition does not.

In general if x is coded to $Y = ax + b$:

Mean = $\bar{x}$	Coded mean = $a\bar{x} + b$
Standard deviation = σ	Coded standard deviation = $a\sigma$

Example 6

A data set has been coded using $Y = \frac{x}{12}$. The coded standard deviation is 1.41.

Find the standard deviation of the original data.

Solution

This example shows x being divided by 12. This will affect the standard deviation. The original data was coded, so to find the original you need to multiply by 12.

$$1.41 \times 12 = 16.92$$

Example 7

A data set has been coded using $Y = x - 2.7$. The coded standard deviation is 3.641.

Find the standard deviation of the original data.

Solution

This time a number is subtracted from the x values. This does not affect the standard deviation as the spread remains the same.

The original standard deviation was 3.641, as the standard deviation does not change.

Example 8

A data set has been coded using $Y = \frac{x}{7} + 1.79$. The coded standard deviation is 12.342.

Find the standard deviation of the original data.

Solution

1.79 has been added to the data which is then divided by 7. Adding 1.79 has no effect but the division by 7 does affect it. So you need to multiply this by 7.

The original standard deviation was 86.39.

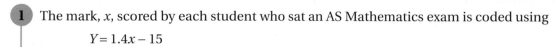

| Using the large data set 12.4 | Using a category within your large data set, write a linear code and use this to find the mean and standard deviation of the data. |

Exercise 12.2B

Answers page 545

1 The mark, x, scored by each student who sat an AS Mathematics exam is coded using

$$Y = 1.4x - 15$$

The coded marks have a mean of 60.2 and a standard deviation of 4.5.

Find the mean and the standard deviation of x.

2 A system is used in schools to predict students, A-level grades using their GCSE results. The GCSE score is g and the predicted A-level score is a. For students taking their maths GCSE in 2016, the coding equation was given by:

$$a = 3.1g - 9.53$$

In 2017 there are 97 students in their second year. Their GCSE scores are summarised as:

$$\Sigma g = 418.3 \text{ and } \Sigma g^2 = 2312.19$$

a Find the mean and standard deviation of the GCSE scores.

b Find the mean and standard deviation of the predicted A-level scores.

(M) 3 On her summer holiday, Farida recorded the temperature at noon each day for use in a statistics project. The values recorded, in degrees Fahrenheit, were as follows (correct to the nearest degree):

47, 59, 68, 62, 49, 67, 66, 73, 70, 68, 74, 84, 80, 72

a Find the mean and standard deviation of Farida's data.

b The formula for converting temperatures from f degrees Fahrenheit to c degrees Celsius is

$$c = \frac{5}{9}(f - 32)$$

Use this formula to estimate the mean and standard deviation of the temperatures in degrees Celsius.

4 The mean weekly cheese consumption per household is 139 grams. The standard deviation is 5.7 grams. Assuming 1 ounce is approximately equal to 28.35 grams, calculate the mean and standard deviation in ounces.

12.3 Displaying and interpreting data

Different visual representations can be used to present and interpret data – including **box and whisker plots**, cumulative frequency diagrams, **histograms** and scatter diagrams. You will remember these types of representations from your GCSE course – at A-level you need to be able to make judgments about which method of presentation is most appropriate for particular data sets and requirements.

Other visual representations that you may come across are dot plots, which are similar to bar charts but with stacks of dots in lines to represent frequency, and frequency charts, which resemble histograms with equal width bars but the vertical axis is frequency.

Box and whisker plots

The median and quartiles can be displayed graphically by means of a box and whisker plot, sometimes just referred to as a box plot. This can be used to compare sets of data.

The diagram shows a generic box and whisker plot. A box is drawn between the lower and upper quartiles and a line is drawn in the box showing the position of the median. Whiskers extend to the lowest value and to the highest value. If an outlier is recorded but not used, this is displayed as a cross.

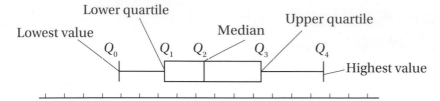

The lowest item of data is referred to as Q_0, the lower quartile as Q_1, the median as Q_2, the upper quartile as Q_3 and the highest data item as Q_4.

The range is the difference between the highest and smallest values in the data set.

$$\text{Range} = Q_4 - Q_0$$

The mid-range is the value halfway between the upper and lower extreme values. It is easy to calculate but is only useful if the data is reasonably symmetrical and free from outliers.

$$\text{Mid-range} = \frac{(Q_4 + Q_0)}{2}$$

As you saw in **Section 12.1**, the upper quartile is the median of the upper half of the data, the lower quartile is the median of the lower half of the data and the interquartile range measures the range of the middle 50% of the data.

$$\text{Interquartile range} = Q_3 - Q_1$$

It is useful to be able to compare sets of data using their box and whisker plots. The following two box and whisker plots represent the annual salaries of 40-year-olds in 2010.

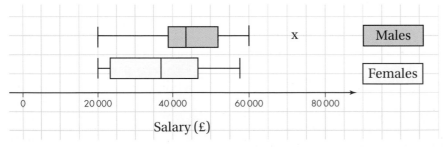

The ranges of salary are similar, shown by the distance between the whiskers. The males have a smaller interquartile range than the females, shown by the size of the boxes, which suggests that the majority of the pay is less spread out for males. The median and quartiles for males are higher than those for females, so on average males earn more than females at age 40. The cross on the plot for the males indicates an anomaly or rogue data item which doesn't seem to fit the trend – an outlier.

Identifying outliers using quartiles

The interquartile range is simply the difference between the quartiles, or $Q_3 - Q_1$. An outlier can be identified as follows (IQR stands for interquartile range):

- any data which are $1.5 \times$ IQR below the lower quartile

- any data which are $1.5 \times$ IQR above the upper quartile.

> ### Example 9
>
> The following data set lists the heights of employees.
>
> 1.45 1.48 1.46 1.52 1.46 1.61 1.60 1.51 1.55 1.56
>
> 1.61 1.64 1.53 1.51 1.48 1.70 1.70 1.62 1.45 1.50
>
> Are there any outliers in this data set?
>
> ### Solution
>
> Q_1 = lower quartile = 1.48
>
> Q_2 = median = 1.525
>
> Q_3 = upper quartile = 1.61
>
> The interquartile range (IQR) is $1.61 - 1.48 = 0.13$ m.
>
> $$1.5 \times \text{IQR} = 1.5 \times 0.13 = 0.195$$
>
> $1.5 \times$ IQR below the lower quartile = $1.48 - 0.195 = 1.285$ m, so there are no low outliers as everyone is taller than this.
>
> $1.5 \times$ IQR above the upper quartile = $1.61 + 0.195 = 1.805$ m, so there are no high outliers as everyone is shorter than this.

Cumulative frequency diagrams

These are used when the data is grouped. It is easy to find the median and quartiles using **cumulative frequency** diagrams.

Here is a grouped frequency table showing the time spent queuing for rides at a theme park:

Time, t (minutes)	$0 < t \leqslant 5$	$5 < t \leqslant 10$	$10 < t \leqslant 20$	$20 < t \leqslant 30$	$30 < t \leqslant 60$
Frequency	3	24	41	17	15

It shows, for example, that 41 people queued for between 10 and 20 minutes.

You can now work out the cumulative frequencies by adding up the frequencies as you go along.

Time, t (minutes)	$0 < t \leqslant 5$	$0 < t \leqslant 10$	$0 < t \leqslant 20$	$0 < t \leqslant 30$	$0 < t \leqslant 60$
Frequency	3	24	41	17	15
Cumulative frequency	3	27	68	85	100

Notice that the time intervals have also changed as the frequencies have accumulated from the previous time.

You can now plot these cumulative frequencies on graph paper, making sure that you plot the cumulative frequencies at the upper bound of the interval, that is, at (5, 3), (10, 27), (20, 68), (30, 85) and (60, 100).

You should also plot the point (0, 0).

Join the points with a smooth curve (not straight line segments) to create the graph shown here:

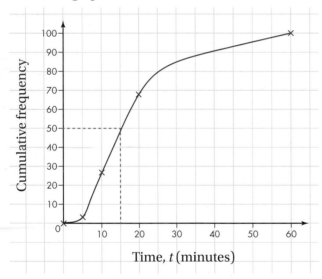

Find the median by drawing a line from 50 (half of 100) on the cumulative frequency axis to the graph, then down to the time axis: here you can see that the median is 15 minutes.

TECHNOLOGY
You could do this quickly and accurately using a graph-drawing software package.

Estimating percentiles from grouped data
Percentiles give the value below which a given percentage of observations fall. They are often used in the reporting of scores. For example, if you achieved a score in the 40th percentile, this would mean your score is higher than 40% of the other scores.

You have already encountered some specific percentiles:

> the 25th percentile, known as the first quartile (Q_1)

> the 50th percentile, known as the median or second quartile (Q_2)

> the 75th percentile, known as the third quartile (Q_3).

If data is only available as a grouped frequency distribution, then it is not possible to find the exact values of the median, quartiles or other percentiles. However, it is possible to estimate values using linear **interpolation**.

> Interpolation is the process of finding a value between two points.

To find the 50th percentile (the median) you need to know the total. There are 427 data items in the data set on gross salaries, so you need to find the 214th data item.

> $\frac{427 + 1}{2} = 214$th data item

Salary (£)	Frequency	Cumulative frequency
0–9999	21	21
10000–19999	129	150
20000–29999	172	322
30000–39999	74	396
40000–49999	24	420
50000–59999	4	424
60000–69999	1	425
70000–79999	1	426
80000–89999	0	426
90000–99999	1	427

To do this, create cumulative frequencies from the data in the table. This shows clearly that the 214th data item lies in the class interval 20000–29999. However, you want to estimate what the value of the 50th percentile is, not which class it is in.

To do this you need to find out where in the class the 214th value is. $214 - 150 = 64$, so it is the 64th item in this class interval (which contains 172 items). As a fraction, it is $\frac{64}{172}$ within the class, and the class is 10000 wide.

So in order to estimate the 50th percentile:

$$\frac{64}{172} \times 10\,000 = 3720.93$$

> This value is the median within the class.

You must then add this value onto the lower bound of the class that the 50th percentile lies in.

$$£20\,000 + £3720.93 = £23\,270.93$$

If data is only available as a grouped frequency distribution, then it is not possible to find the median using exact values of the quartiles or other percentiles. However, it is possible to estimate values for these.

KEY INFORMATION

Use this formula to find a percentile:

$$\left(\frac{\text{position in class}}{\text{class frequency}}\right) \times \text{class width} + \text{lower boundary of class}$$

Suppose you were asked to estimate the 45th percentile of the data set shown below.

Age of employees	Frequency
$16 \leqslant a < 25$	2
$25 \leqslant a < 30$	6
$30 \leqslant a < 40$	10
$40 \leqslant a < 50$	8
$50 \leqslant a < 65$	5
over 65	0

There are 31 data items, and $\frac{45}{100} \times 31 = 13.95$, so you need to find the 14th data item. It is helpful to add the cumulative frequencies to this table. This shows clearly that the 14th data item lies in the class interval $30 \leqslant w < 40$; it is the 6th item in this class interval (which contains 10 items).

Age of employees	Frequency	Age of employees	Cumulative frequency
$16 \leqslant a < 25$	2	$a \leqslant 25$	2
$25 \leqslant a < 30$	6	$a \leqslant 30$	8
$30 \leqslant a < 40$	10	$a \leqslant 40$	18
$40 \leqslant a < 50$	8	$a \leqslant 50$	26
$50 \leqslant a < 65$	5	$a \leqslant 65$	31
over 65	0	$a > 65$	31

$\frac{6}{10} \times 10 = 6$, so the 6th item is $30 + 6 = 36$.

When you are dealing with discrete data, you must ensure that cumulative frequencies less than or equal to a value, as it is cumulating the results together.

Example 10

The percentage marks scored in a driving theory test sat by 200 members of the public in one day were as follows:

Mark (%)	Frequency
1–10	3
11–20	11
21–30	13
31–40	18
41–50	26
51–60	33
61–70	45
71–80	34
81–90	11
91–100	6

Only 45% of the people who sat the test that day passed. Estimate the pass mark.

Solution

Mark (%)	Frequency	Mark, m	Cumulative frequency
1–10	3	$m < 10.5$	3
11–20	11	$m < 20.5$	14
21–30	13	$m < 30.5$	27
31–40	18	$m < 40.5$	45
41–50	26	$m < 50.5$	71
51–60	33	$m < 60.5$	104
61–70	45	$m < 70.5$	149
71–80	34	$m < 80.5$	183
81–90	11	$m < 90.5$	194
91–100	6	$m \leqslant 100$	200

The 45th percentile is needed. This is the 90th data item, which lies in the 51–60 class interval.

> Although the data are discrete, they are taken to be continuous because a mark of 40.6% would be rounded to 41%, and so the class interval is taken to be $40.5 \leqslant m < 50.5$.

The 45th percentile is the 19th data item of the 33 data items in this class interval.

45th percentile $= 50.5 + \dfrac{19}{33} \times 10 = 56.26$

So, an estimate of the pass mark is 56%.

Exercise 12.3A

Answers page 545

1 This is a frequency table for the number of people in a household. Calculate the median number of people in a household.

Number in household	Number of households
1	15
2	20
3	22
4	23
5	11
6	4

2 A football coach measured the distance a random sample of 120 eleven-year-old children could kick a football. The lengths are summarised in the table.

  a Display this data as a cumulative frequency graph.

  b Use interpolation to estimate the distance of the kick from the 40th child.

  c Calculate an estimate for the mean.

Kick distance, l (m)	Number of children
$5 \leqslant l < 10$	5
$10 \leqslant l < 20$	53
$20 \leqslant l < 30$	29
$30 \leqslant l < 50$	15
$50 \leqslant l < 70$	11
$70 \leqslant l < 100$	7

3 The number of aphids on a farmer's strawberry field were counted. The results are presented in the table.

  a Construct a cumulative frequency curve and find the median.

  b Using linear interpolation, estimate the median.

  c Find the estimate of the mean.

Number of strawberry plants	Number of aphids
0–19	38
20–29	97
30–39	173
40–49	225
50–69	293
70–99	174

4 Here are the numbers of runs which Jenson and Molly scored in 11 cricket games.

Jenson: 3, 21, 45, 66, 12, 4, 12, 65, 46, 55, 31

Molly: 34, 57, 12, 98, 17, 22, 17, 43, 23, 76, 44

  a Draw box and whisker plots showing their scores.

  b Compare their scores, using your diagrams and calculations to help you.

5 Consider this grouped frequency distribution:

  a Identify the upper class values, if the measurements were recorded to the nearest mm.

  b Construct a cumulative frequency diagram of the data.

  c Using your graph, estimate the median.

  d Use linear interpolation to estimate the median correct to 2 decimal places.

Length (cm)	Frequency
17.5–17.9	15
18.0–18.4	27
18.5–18.9	18
19.0–19.9	12
20.0–24.9	15
25.0–29.9	4

  e Compare your arithmetic findings to your graphical estimate.

  f Compare the median and the mean.

CM **6** The marks of 45 students randomly selected from those who sat a statistics test are displayed below:

36, 39, 39, 40, 41, 42, 42, 43, 44, 45, 46, 46, 46, 48, 50, 52, 53, 53, 54, 54, 55, 55, 56, 57, 57, 59, 60, 60, 60, 60, 61, 63, 64, 64, 64, 65, 65, 66, 67, 68, 69, 71, 72, 73, 73

  a What is the modal mark?

  b What are the lower quartile, the median and the upper quartile?

  c Represent the data as a box and whisker plot.

d Are there any outliers in the data set?

e Represent the data as a cumulative frequency diagram.

f What is the range in marks between the 90th percentile and 10th percentile? Why might this be a useful measure?

7 This diagram shows the raw test marks of two student groups for the same test.

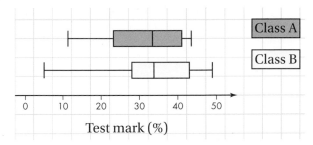

a Use it to determine which of the following statements are true.

    A The median is the same for both classes.

    B Q_1 for Class A is 5 marks less than Q_1 for Class B.

    C The interquartile range for Class A is $\frac{2}{3}$ the interquartile range of Class B.

    D $P_{75} - P_{50}$ is the same for both classes.

b Write a comparison of the student groups.

8 This diagram shows the times that 200 17-year-olds spent using mobile phone apps.

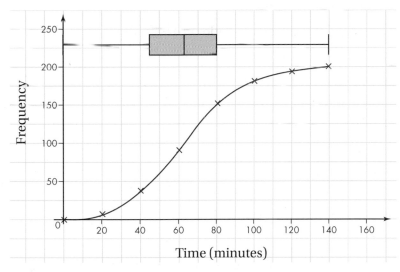

a Use the diagram to estimate the following statistics:

    i the median time spent using apps

    ii the range of times spent using apps

    iii $Q_3 - Q_1$

    iv The amount of time 20% of 17-year-olds spend using mobile phone apps.

b How many students spent more than 1.6 hours using mobile phone apps?

Histograms

Histograms are best used for large sets of data when the data has been grouped into classes. They are most commonly used for continuous data and often have bars of varying width, representing unequal class intervals. The frequency of the data is shown by the area of the bars and not the height.

The vertical axis of a histogram is labelled Frequency density, which is calculated by the following formula:

$$\text{Frequency density} = \frac{\text{frequency}}{\text{class width}}$$

The table below shows data on gross salaries for 427 professions, drawn from the Office for National Statistics' 2010 salary tables (*Annual Survey of Hours and Earnings*).

Salary (£)	Frequency	Class width	Frequency density
0–9999	21	10	2.1
10000–19999	129	10	12.9
20000–29999	172	10	17.2
30000–39999	74	10	7.4
40000–49999	24	10	2.4
50000–59999	4	10	0.4
60000–69999	1	10	0.1
70000–79999	1	10	0.1
80000–89999	0	10	0
90000–99999	1	10	0.1

TECHNOLOGY

You could use graphing software to create a histogram of this data.

Plotting the frequency against the classes gives the following diagram:

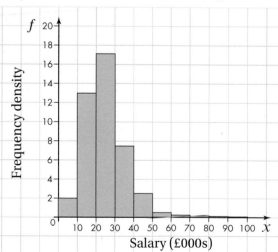

An alternative is to use unequal classes to display the same data.

Salary (£)	Frequency	Class width	Frequency density
0–9999	21	10	2.1
10 000–19 999	129	10	12.9
20 000–29 999	172	10	17.2
30 000–49 999	98	20	4.9
50 000–99 999	7	50	0.14

This diagram gives an impression of the overall distribution of the data which tallies with that given by the first diagram. The data is now fairly represented, even though it is grouped into intervals with different widths.

Look at the bar shown in blue. The width is 10 and the frequency density is 12.9. So the area of the bar is $10 \times 12.9 = 129$, which equals the frequency.

On all histograms the vertical axis should either be labelled as Frequency density, or with the units of the frequency density.

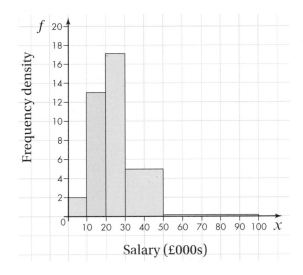

Salary (£000s)

Using the large data set 12.5

a Choose a category from your large data set, group the data using unequal intervals and display using a histogram.

b Using your table from **part a**, use linear interpolation to find the median of your category.

c Could the data gathered from your category be used to make predictions about other categories? Explain your reasons.

Exercise 12.3B

Answers page 546

1 Mr Hardy selects a random sample of 40 students and records, to the nearest hour, the time they spent gaming in a particular week.

Time (hours)	1–5	6–10	11–15	16–20	21–30	31–49
Frequency	3	7	11	14	4	1
Midpoint		8	13	18		40

a Find the midpoints of the 1–5 hour and 21–30 hour groups.

On a histogram representing this data set, the 6–10 hour group is represented by a bar of width 2 cm and height 5 cm.

b Find the width and height of the 31–49 hour group.

c Estimate the mean.

d Using linear interpolation, estimate the median time spent gaming by these students.

(CM) **2** Danielle and Hannah decide to weigh cookies served in the canteen.

Here is a grouped frequency table showing the masses, in grams, of the cookies.

Mass, m (grams)	$30 \leqslant m < 40$	$40 \leqslant m < 50$	$50 \leqslant m < 60$	$60 \leqslant m < 80$	$80 \leqslant m < 120$
Frequency	13	37	56	8	6

a Draw a cumulative frequency graph and use it to estimate:

  i the median mass of a cookie

  ii the upper and lower quartile masses of the cookies.

b Construct a frequency density table and use it to draw a histogram displaying the masses.

c Danielle wants to know the maximum weight a cookie can be before it is deemed an outlier. Show how she would do this.

(CM) **3** The lengths of runner beans were measured to the nearest whole cm. Eighty observations are given in the table below.

Length (cm)	3–8	9–13	14–25
Frequency	47	22	11

On a histogram representing this data set, the bar representing the 3–8 class has a width of 2 cm and a height of 4 cm. For the 9–13 class find:

a the width

b the height of the bar representing this class.

(CM) **4** The following table summarises the distances, to the nearest km, that 7359 people travelled to attend a small festival.

Distance (km)	Number of people
40–49	67
50–59	124
60–64	4023
64–69	2981
70–84	89
85–149	75

a Give a reason to justify the use of a histogram to represent these data.

b Calculate the frequency densities needed to draw a histogram for these data.

5 An agriculturalist is studying the mass, m kg, of courgette plants. The data from a random sample of 70 courgette plants are summarised in the table.

(You may use $\Sigma fx = 750$ and $\Sigma fx^2 = 11\,312.5$)

A histogram has been drawn to represent these data.

The bar representing the mass $5 \leqslant m < 10$ has a width of 1.5 cm and a height of 6 cm.

Yield, m (kg)	Frequency, f
$0 \leqslant m < 5$	11
$5 \leqslant m < 10$	29
$10 \leqslant m < 15$	18
$15 \leqslant m < 25$	8
$25 \leqslant m < 35$	4

a Calculate the width and the height of the bar representing the mass $25 \leqslant m < 35$.

b Use linear interpolation to estimate the median mass of the courgette plants.

c Estimate the mean and the standard deviation of the mass of the courgette plants.

6 Here is a frequency table for the variable t, which represents the time taken, in minutes, by a group of people to run 3 km.

a Copy and complete the frequency table for t.

t	5–10	10–14	14–18	18–25	25–40
Frequency		15	22		18
Frequency density	2			3	

b Estimate the number of people who took longer than 20 minutes to run 3 km.

c Find an estimate for the mean time taken.

d Find an estimate for the standard deviation of t.

e Estimate the median and quartiles for t.

 7 A survey of 100 households gave the following results for their monthly shopping bills, $£y$.

Shopping, y (£)	Midpoint	Frequency, f
$0 \leqslant y < 200$	100	6
$200 \leqslant y < 240$	220	20
$240 \leqslant y < 280$	260	30
$280 \leqslant y < 350$	315	24
$350 \leqslant y < 500$	425	12
$500 \leqslant y < 800$	650	8

A histogram was drawn and the class $200 \leqslant £y < 240$ was represented by a rectangle of width 2 cm and height 7 cm.

a Calculate the width and the height of the rectangle representing the class $280 \leqslant £y < 350$.

b Use linear interpolation to estimate the median shopping bill to the nearest pound.

c Estimate the mean and the standard deviation of the shopping bill for these data.

Scatter diagrams

You may remember meeting scatter diagrams at GCSE level in both maths and other subjects. You use scatter diagrams when you are comparing two variables (**bivariate data**), such as scores in maths and scores in physics.

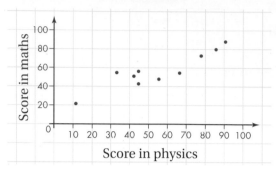

You used them to suggest if there was a **correlation** in data. You will also have drawn lines of best fit by eye to predict values. You now need to be able to state the relationship between the variables and calculate the equation of the line of best fit for a set of bivariate data. Although the calculations you will learn give you more reliable information than the 'by eye' methods, scatter diagrams are still an important part of the process.

Correlation in scatter diagrams

Correlation describes the relationship or link between two variables. A scatter diagram visually shows the relationship between the two variables. If the values of both variables are increasing, then they are *positively* correlated.

If one variable is increasing but the other is decreasing, there is a *negative* correlation.

If there is no pattern, and the points are scattered about the axes, there is no correlation.

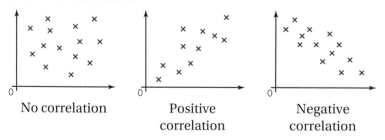

No correlation   Positive correlation   Negative correlation

The positions of the data items within a scatter diagram may show linear correlation. A line of best fit can be drawn by eye, but only when the points lie almost on a straight line. The closer to a straight line the points lie, the stronger the correlation. The line of best fit is a representation of all the data. There is, however, one point on the scatter diagram that the line of best fit *must* go through, which is the point $(\bar{x}, \bar{y})$, where $\bar{x}$ is the mean of x and

$\bar{y}$ is the mean of y. This helps the line of best fit to be central to all the data points.

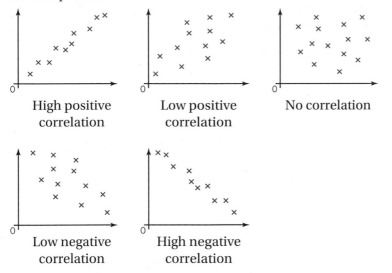

High positive correlation | Low positive correlation | No correlation

Low negative correlation | High negative correlation

The following diagram illustrates why it is important to interpret scatter diagrams with caution.

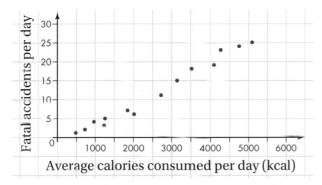

The scatter diagram shows the number of fatal accidents plotted against the average calories consumed on a single day, for different regions. Does the diagram indicate that the more calories you consume the more likely you are to have a fatal accident?

In fact, the average calories consumed and the number of fatal accidents in a day for a particular region both also correlate with a third variable – the number of units of alcohol consumed. The more alcohol is consumed, the more calories a person absorbs. The more alcohol consumed, the more fatal accidents happen on roads. Combining just the average calories consumed against the number of fatal accidents gives a false impression about the relationship between the two variables – this is spurious correlation.

In general, correlation does *not* imply causation.

<table>
<tr><td rowspan="4">

Using the large data set 12.6

</td><td>

a Choose two variables to compare from your large data set. Calculate the standard deviation of each of your categories.

</td></tr>
<tr><td>

b For each of your categories, what would be deemed an outlier?

</td></tr>
<tr><td>

c Display the data as a scatter diagram, clearly indicating the mean point.

</td></tr>
<tr><td>

d Using the diagram, comment on any trends.

</td></tr>
</table>

Exercise 12.3C

Answers page 547

 1 The table below shows the height and weight of each of 10 students.

Name	Height (cm)	Weight (kg)
Chloe	172	68
Arron	157	66
Melbin	190	75
Dufia	141	50
Emily	155	74
Cody	169	74
Tahlia	151	60
Thomas	177	70
Zineb	183	68
Joseph	139	52

a Plot the data on a scatter diagram.

b Describe what you notice from the diagram. Is there any correlation?

c Draw a line of best fit. Make sure it goes through the mean point.

d How heavy would you expect Uche to be if he is 196 cm tall?

2 Write down a real-life example scenario and sketch a scatter diagram for each of the following cases:

a a weak positive correlation

b a strong negative correlation

c no correlation

d a nonsense correlation.

3 The following table shows information about six students' test results in two subjects. Draw a scatter diagram and draw on a line of best fit for the data.

Subject	A	B	C	D	E	F
biology	43	64	35	57	69	92
geography	46	81	31	55	58	87

4 **a** Plot the following additional data on the scatter diagram you drew in **question 3** and draw on a new line of best fit.

Subject	G
biology	72
geography	78

b Is there an outlier within the data?

5 The table shows information about the number of cups of coffee sold in a café on 14 consecutive days in January.

Temperature (°F)	Cups of coffee sold
62	37
67	62
66	27
69	12
58	80
62	62
68	19
64	39
65	32
56	96
68	11
64	38
69	5
60	72

a Plot the data on a scatter diagram.

b Describe what you notice from the diagram. Is there any correlation?

c Draw a line of best fit. Make sure it goes through the mean point.

d On another day the temperature was 58 °F. How many cups of coffee would you expect to be sold on this day?

Regression lines

A correlation coefficient provides you with a measure of the level of association between two variables in a bivariate distribution. If a relationship is indicated, you will want to know what that means. Drawing a line of best fit by eye is one way to spot a linear correlation but it is not a very accurate method. A more accurate way to find the line of best fit is to find and use the linear equation of the line.

The general form of an equation of a straight line is:

$$y = mx + c$$

> m is the gradient (the steepness of the line)

> c is the y-intercept (where the line crosses the y-axis).

See **Chapter 4 Coordinate geometry 1: Equations of straight lines** for more about gradients and intercepts.

An accurate way of plotting a line of best fit is to draw the **regression line**. This ensures that the distance between each point and the line of best fit is reduced to a minimum – these amounts that are left over are known as residuals. The regression line goes through the middle of the points plotted.

You use the same equation but with different letters:

$$y = a + bx$$

where

> b is the gradient

> a is the y-intercept.

If b is positive then there is a positive correlation.

If b is negative then there is a negative correlation.

The line of best fit with the equation $y = a + bx$ is called the regression line.

KEY INFORMATION

Regression line: a line of best fit for a given set of values, using the equation of a straight line, $y = a + bx$.

Variables and the line of best fit

In bivariate data, the **independent (explanatory) variable** is plotted on the x-axis and is independent of the other variable. The **dependent (response) variable** is plotted on the y-axis and is determined by the independent variable. The regression line is the line of best fit.

For example, your two variables could be sales of a particular book and the number of bookshops selling it. The book's sales will be dependent (y) on the number of bookshops selling it (x) – but the number of bookshops is independent.

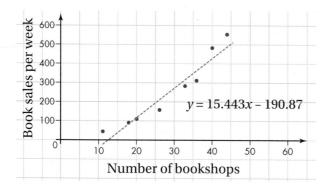

The equation of a regression line

The equation of the regression line of y on x is:

$$y = a + bx$$

The values of a and b will be given to you, you need to know how to apply these in the equation.

Subject	A	B	C	D	E	F
biology	76	84	75	54	62	86
geography	65	43	73	54	58	77

In this data set the values of a and b are $a = 41.931$ and $b = 0.271$, so the equation of the regression line is:

$$y = 41.931 + 0.271x$$

where x is the biology result and y is the geography result.

You can now use the equation of the regression line and the mean point to draw the regression line on the scatter diagram.

➤ From the equation, you know the y-intercept is at $(0, 41.931)$.

➤ You can calculate the coordinates of the mean point to be $(72.83, 61.67)$.

➤ Drawing a line between these two points give you the regression line.

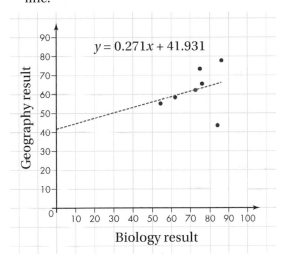

TECHNOLOGY

You can use a spreadsheet software package to do the calculation. Enter the data into a spreadsheet with one column for the x values and another column for the corresponding y values. Then use the function INTERCEPT to calculate a and SLOPE to calculate b and follow the onscreen instructions.

Interpolation and extrapolation

Once you have drawn a line of best fit, either by eye or more accurately by plotting the equation of the regression line, you can start using the line of best fit to make predictions.

A regression equation can be used to predict the value of the dependent variable, based on a chosen value of the independent variable.

➤ **Interpolation** is estimating a value that is within the range of the data you have.

➤ **Extrapolation** is estimating a value outside the range of the data that you have. As the value is outside the range of the data you have, extrapolated values can be unreliable.

Generally, avoid extrapolating values unless asked and, even then, treat answers with caution.

Example 11

An experiment in which different masses were hung on a metal spring and the resulting length of the spring was measured produced the following results. The equation of the regression line is $y = 32.7 + 0.402x$.

Mass, x (kg)	10	20	30	50	100
Length, y (cm)	24	44	53	59	68

a Estimate the value for y when $x = 35$ kg. Is this interpolation or extrapolation?

b Estimate the value for y when $x = 120$ kg. Is this interpolation or extrapolation?

Solution

a Using the equation of the regression line:

$$y = 32.7 + 0.402x$$

Substitute in the x value of 35 kg.

$$y = 32.7 + (0.402 \times 35)$$

$$= 46.77 \text{ cm}$$

This is interpolation as $x = 35$ is within the range of the data you have.

b Setting x to be 120 kg, you get:

$$y = 32.7 + 0.402x$$

$$= 32.7 + (0.402 \times 120)$$

$$= 80.94 \text{ cm}$$

This is extrapolation as $x = 120$ is outside the range of the data you have.

Using the large data set 12.7

In **Using the large data set 12.6** you drew a scatter diagram.

a Describe the correlation of your scatter diagram.

b Using a spreadsheet, find the equation of the regression line. Explain what it means.

Exercise 12.3D

Answers page 548

1 The weight, w grams, and the length, l cm, of each of 10 randomly selected new-born babies are given in the table below.

l	49.0	52.0	53.0	54.5	44.1	53.4	50.0	41.6	49.5	51.2
w	3424	3234	3479	3910	2855	3596	3001	2108	2906	1954

a The equation of the regression line of w on l is written in the form $w = a + bl$. Explain what a and b represent in this instance.

b The equation of the regression line is $w = 97.25l - 1799.4$. Use the equation of the regression line to estimate the weight of a new-born baby of length 60 cm.

c Comment on the reliability of your estimate, giving a reason for your answer.

2 The table below gives the number of hours spent studying for a science exam (x) by seven students, and their final exam percentage (y). The equation of the regression line is $y = 48.9 + 7.93x$. Predict the score of a student who studies for 6.5 hours.

x	3	5	1	0	4	2	6
y	72	91	60	43	78	71	94

3 The table below shows the lengths in cm (nose to tail) and corresponding weights (in kg) of cats. If a cat was 78 cm in length, predict its weight. State if this prediction is reliable or not.

Length (cm)	60	62	64	66	68	70	72
Weight (kg)	1.81	1.76	1.24	2.32	1.98	2.68	3.24

4 A football coach is doing a study on inside leg length and the distance a football can be kicked. This model allows him to determine the length of a kick when only the leg length is given. The inside leg length and kick distance for 10 males are given in the table.

Inside leg length (cm)	Kick distance (m)
72.5	41.9
73.0	44.7
74.8	43.8
77.8	47.1
79.4	52.4
79.9	50.2
80.4	57.3
81.6	58.9
83.4	54.1
87.5	50.8

a Draw a scatter diagram to represent the data.

b The equation of the regression line that models the data is written in the form $y = a + bx$. If the intercept of the line is -18.8 and the gradient of the line is 0.87, write out the equation of the regression line.

c The coach measures a kick at 56 m. How long was the person's inside leg?

d If a person has an inside leg length of 70 cm, what does the model predict for their kick distance? Is this an accurate model?

CM 5 The amount, in grams, of different types of bread products consumed weekly per household was monitored in 2001 and 2014. The results are presented in the scatter diagram below.

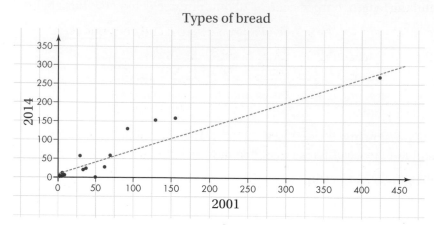

Types of bread

a What correlation does this show?

b Using the scatter graph, what amount in grams would you expect to be consumed in 2014, if the amount consumed in 2001 was 90 grams?

c Was this interpolation or extrapolation?

The equation of the regression line is $y = 0.6427x + 5.7408$.

d What does this mean?

In 2014 the amount per week of one type of product rose to 300 grams.

e What amount would you expect to have been consumed in 2001?

f Was this interpolation or extrapolation? What are the dangers of this?

6 Jen studied the relationship between rainfall, r mm, and humidity h%, for a random sample of 11 days from a city in the UK. She obtained the following results.

h	99	96	98	93	99	99	96	97	83	83	96
r	1.7	8.9	6.3	8.2	0	0.6	5.8	7.3	8.8	7.5	12.1

Jen examined the figures for rainfall and found the following statistics:

$Q_1 = 1.8$

$Q_2 = 5.8$

$Q_3 = 8.25$

a Determine if there are any outliers in her sample.

b Present her data as a graph.

c Describe the correlation between rainfall and humidity.

d Jen found the the gradient of her line was −0.7255 and the intercept of her line was 98.88. Write the equation of the regression line in the form $r = a + bh$.

e Use your equation to estimate the humidity with a rainfall of 9.3 mm.

SUMMARY OF KEY POINTS

> Qualitative data are not numerical, but categorical.

> Quantitative data, or numerical data, can be subdivided into discrete and continuous.

> Discrete data may only take separate values, for example whole numbers.

> Continuous data may take any value, usually measured; these are usually within a range.

> The median is the middle value when the data items are placed in numerical order.

> The mode is the most common or most frequent item of data.

> The mean ($\bar{x}$) is found by adding the data values together and dividing by the number of values
$$\bar{x} = \frac{\sum x}{n}$$

> Variance $= \dfrac{\sum(x - \bar{x})^2}{n}$

> Standard deviation $= \sqrt{\dfrac{\sum(x - \bar{x})^2}{n}}$

> An outlier can be identified as follows (IQR stands for interquartile range):

> > any data which are $1.5 \times$ IQR below the lower quartile

> > any data which are $1.5 \times$ IQR above the upper quartile

> > any data which are more than 2 standard deviations away from the mean.

> > Sometimes statisticians use mean $\pm 3 \times$ standard deviation to define outliers.

> For scatter diagrams, positive correlation means that as one variable increases, so does the second variable. If one variable is increasing, but the other is *decreasing*, there is a negative correlation.

> A regression line is a line of best fit for a given set of values, using the equation of a straight line, $y = a + bx$.

> Box and whisker plots display the quartiles and minimum and maximum points visually.

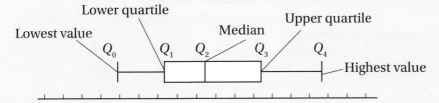

EXAM-STYLE QUESTIONS 12

Answers page 548

1 The amount of money invested, I, measured in £1000, and the amount of profit returned, P, measured in pounds, is modelled below for six different saving accounts using a risky stocks and shares account.

Savings account	A	B	C	D	E	F
I (£1000)	4.3	2.8	6.0	0.3	6.8	9.5
P (£)	1564	2098	2592	988	3200	4805

a Draw a scatter diagram to display this data and comment upon any correlation. [2 marks]

b The equation of the regression line is $y = 597 + 393x$. Explain what the equation shows. [4 marks]

c If you had £3500 to invest, what profit would you expect to make? [1 mark]

CM **2** A city in the UK had a mean annual salary for graduates of £28 500 with standard deviation of £9000.

What salaries would be outliers for the graduates? [3 marks]

3 Powerlifters have a certain amount of combined weight to lift in their category. The total of the combined weights, w, lifted by 17 powerlifters is 4930 kg and the standard deviation is 10.5.

a What is the maximum weight a powerlifter could lift before a judge would become concerned that it was too much? [1 mark]

Another powerlifter joins the competition and lifts a combined weight of 292 kg.

b Find the new mean. [2 marks]

4 It is reported in the news that teenagers use social media for a long time each day. A random sample of 11 students were interviewed and asked how long they spent using social media in an average week.

The total duration, y minutes, for the 11 students were:

7, 98, 121, 132, 151, 187, 204, 255, 260, 277, 357

a Find the median and quartiles for these data. [2 marks]

b Show that there are no outliers. [2 marks]

CM **5** Sixth-form students carry out an investigation on the lengths of children's feet before they have their teenage growth spurt. The sixth-form students measured the foot lengths of a random sample of 100 Year 7 students. The lengths are shown in the table.

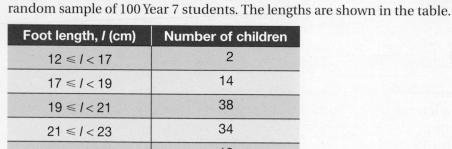

Foot length, l (cm)	Number of children
$12 \leqslant l < 17$	2
$17 \leqslant l < 19$	14
$19 \leqslant l < 21$	38
$21 \leqslant l < 23$	34
$23 \leqslant l < 25$	12

a Using a written method, find an estimate of the median foot length. [4 marks]

b Using a cumulative frequency curve, find an estimate of the median foot length. [2 marks]

c What is the percentage error between your two estimates of the medians? [2 marks]

d Using a calculator, estimate the mean and standard deviation. [3 marks]

e What type of average would you use to best describe this data? [1 mark]

6 A group of students are asked to do a blindfolded reaction test. When they hear a beep, they have to press a button which records the speed of their answer. A random sample of 104 17-year-olds are asked to take part and the results are recorded to the nearest millisecond. The results are summarised in this table.

Time (milliseconds)	Midpoint	Frequency
0–9	4.5	6
10–19	14.5	14
20–29	24.5	34
30–39	34.5	27
40–49	44.5	19
50–99	74.5	4

In a histogram, the group '10–19 milliseconds' is represented by a rectangle 2.5 cm wide and 4 cm high.

a Calculate the width of the rectangle representing the group '50–99 milliseconds'. **[2 marks]**

b Calculate the height of the rectangle representing the group '20–29 milliseconds'. **[2 marks]**

c Using a written method, estimate the median and interquartile range. **[3 marks]**

d The estimate of the mean for the data is 30.2. Would you use the mean or the median when reporting your findings? **[2 marks]**

 7 A car manufacturer monitored the number of times, x, 20 cars went back in to the garage for warranty work on engine parts. The distribution was as follows.

Number of times back, x	0	1	2	3	4	5	6	7	8 or more
Number of cars, f	5	8	2	0	2	1	1	1	0

a Use a suitable diagram to display the data. **[2 marks]**

b Find the median of the data set. **[2 marks]**

c Calculate the mean and the standard deviation of the data set. **[3 marks]**

d An outlier is a value that can be defined as outside 'the mean $\pm$ 2 $\times$ the standard deviation'. What is the maximum number of times a car could return for warranty work before the manufacturer should be concerned? **[4 marks]**

 8 Peter invests in a free holding and wants to make homemade ice cream. He decides to buy a Jersey cow and he monitors the amount of milk, to the nearest litre, that she produces each day during June. His results are as follows:

24	11	22	21	15	8	19	16	15	26
20	17	31	17	19	25	21	11	8	22
18	27	16	12	14	19	19	21	22	23

a Draw a box and whisker plot to display the data. **[4 marks]**

b An outlier is a value that is more than 1.5 times the interquartile range below the lower quartile or more than 1.5 times the interquartile range above the upper quartile. Show that there are no outliers in the data set. **[3 marks]**

c Compare the mean and median amounts of milk produced and comment on your answer. **[3 marks]**

13 PROBABILITY AND STATISTICAL DISTRIBUTIONS

Uncertainty surrounds many real-life situations when you have to make a choice from the options that are available. These include questions such as 'Will it be sunny today? Do I need to take a hat?' or 'Should I dive left or right to save this penalty?' These situations demand a decision and you can use probability to help you make it. You use the concept of probability every day, such as the probability of all the traffic lights being green on your journey home, the likelihood of passing your driving theory test first time, or even your chances of winning the lottery.

Knowing about chance can help make you make predictions. If the probability of traffic lights at rush hour being green was 0.3, you would know it was unlikely that you could go straight through it as it would be more likely to be red. From this, you would be able to calculate that the probability of passing through two sets of green lights would be even more unlikely. Traffic controllers use these sorts of calculations to programme traffic light timings, to ensure that rush hour traffic runs smoothly and does not get too backed up.

LEARNING OBJECTIVES

You will learn how to:

> use both Venn diagrams and tree diagrams to calculate probabilities

> calculate binomial probabilities using the notation $x \sim \mathrm{B}(n, p)$

> identify a discrete uniform distribution

> use a calculator to find individual or cumulative binomial probabilities.

TOPIC LINKS

You will need the skills and techniques that you learnt in **Chapter 1 Algebra and functions 1: Manipulating algebraic expressions** to manipulate algebraic expressions. Knowing how to calculate binomial probabilities will help in **Chapter 14 Statistical sampling and hypothesis testing**. Learning about Venn diagrams and tree diagrams will help you in **Book 2, Chapter 12 Probability**.

PRIOR KNOWLEDGE

You should already know how to:

> complete tables and grids to show outcomes and probabilities

> complete a trcc diagram to show outcomes and probabilities

> interpret P(A) as the probability of event A occurring

> interpret P(A') as the probability of event A *not* occurring

> interpret a Venn diagram.

You should be able to complete the following questions correctly:

1   A gym club has 120 members. 89 of the members train for boxing. 54 of the members train for kickboxing. 30 of the members train for both.

   a   Use this information to complete the Venn diagram.

   B represents those members who box.

   K represents those members who kickbox.

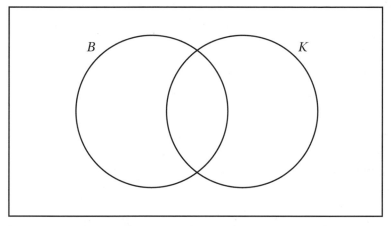

   b   How many members train for neither boxing nor kickboxing?

2   A bag contains 15 counters – 5 are pink, 6 are red and 4 are black. If one counter is selected from the bag at random, find the probability that the counter is:

   a   pink

   b   black

   c   black or pink.

13.1 Calculating and representing probability

In maths, **probability** means thinking about the likelihood of an **event** happening (or not happening). An event could be anything from 'obtaining a head when flipping a coin' to 'it will rain next Friday'. In your GCSE course you expressed probabilities as values from 0 to 1 and calculated them for single and combined trials (or experiments). You also presented the outcomes of trials in various ways, including Venn diagrams and tree diagrams. Now you will need to express probabilities in formal notation.

A trial has various outcomes. For example, an unbiased die has six faces. The outcomes are 1–6. Events are one or more of the outcomes. The probability of any given face being on the top when the die is rolled is $\frac{1}{6}$. When all the possible outcomes are equally likely, you can use a formula to calculate the probability of an event happening.

The probability that an event, *A*, will happen is written as P(*A*). The probability that event *A* will *not* happen is called the **complement** of *A* and is written as P(*A′*).

A **sample space** diagram can be a useful way of determining the probability of an event as it lists all the possible outcomes. In a situation where all the equally likely outcomes are identified, the sum of the relevant outcomes' probabilities gives the probability of that event. If a trial consists of two separate activities, a sample space diagram is a good way to show your results.

If two dice are rolled simultaneously and their scores added, the sample space diagram would look like this:

6	7	8	9	10	11	12
5	6	7	8	9	10	11
4	5	6	7	8	9	10
3	4	5	6	7	8	9
2	3	4	5	6	7	8
1	2	3	4	5	6	7
+	1	2	3	4	5	6

It is easy to see the most common occurring number would be 7. The number 7 appears 6 times out of a total of 36 outcomes, so the probability of rolling a 7 is $P(7) = \frac{6}{36} = \frac{1}{6}$.

Example 1

A game involves flipping a coin and picking a card at random, from four cards numbered 1 to 4.

a Draw a sample space diagram to show all the possible outcomes.

b Find the probability of flipping a head and picking a card with a square number on it.

Solution

a First you need to draw your sample space diagram, thinking about the labels for your headings. Then list all the possible outcomes of the combined events.

Tails	(T, 1)	(T, 2)	(T, 3)	(T, 4)
Heads	(H, 1)	(H, 2)	(H, 3)	(H, 4)
	1	2	3	4

b Using the sample space diagram, identify the successful outcomes.

Tails	(T, 1)	(T, 2)	(T, 3)	(T ,4)
Heads	(H, 1)	(H, 2)	(H, 3)	(H, 4)
	1	2	3	4

$$P(\text{heads and square number}) = \frac{2}{8} = \frac{1}{4}$$

Mutually exclusive and independent events

Two events are **mutually exclusive** if one event occurring excludes the other event from occurring – they cannot happen simultaneously.

For instance, when a fair die is rolled, the probability of rolling a 2 or a 1 is calculated as:

$$P(2 \text{ or } 1) = P(2) + P(1) = \frac{1}{6} + \frac{1}{6} = \frac{2}{6} = \frac{1}{3}$$

For mutually exclusive events A and B, then:

$$P(A \text{ or } B) = P(A) + P(B)$$

If two events A and B are **independent** of each other, event A can happen whether or not event B has happened, and vice versa.

When two events A and B are independent, the probability of both occuring is $P(A \text{ and } B) = P(A) \times P(B)$.

Let A be the probability of catching a cold and B be the probability of getting a cough.

If $P(A) = 0.6$, $P(B) = 0.5$ and the $P(A \text{ and } B) = 0.35$. Show that the two events are not independent.

$$P(A \text{ and } B) = P(A) \times P(B)$$

$$0.35 = 0.5 \times 0.6$$

0.35 does not equal 0.3

This shows they are not independent.

Example 2

There are 30 students in a class. 14 of the students are boys and 7 of the students have blue eyes. Of the students with blue eyes, 3 are girls.

a Show this information on a Venn diagram.

b A student is selected from the class at random. Find the probability that the student is a boy with blue eyes.

Solution

a First, identify the groups. Let B be boys and BE blue eyes. Draw a Venn diagram and label each group.

In the group BE, 3 are girls. This must be in the region which doesn't overlap with the boys.

The overlap (intersection) of the BE and B groups must be 4, as 7 students have blue eyes, and you already know that 3 of them are girls.

There must be 10 boys without blue eyes, as 4 of them have blue eyes and there are 14 boys altogether.

The union of B and BE includes 17 students. So that means 13 students are not within either the B or BE groups (that is, they are girls without blue eyes), as there are 30 students altogether.

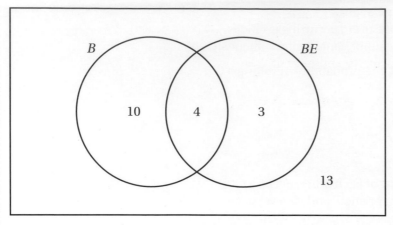

b There are 14 boys altogether. Using the Venn diagram, you can see that 4 of them have blue eyes. Altogether there are 30 students in the class.

So P(boy with blue eyes) $= \frac{4}{30} = \frac{2}{15}$

Example 3

The Venn diagram below represents two events, *P* and *Q*. The numbers show how many outcomes correspond to each of the two events.

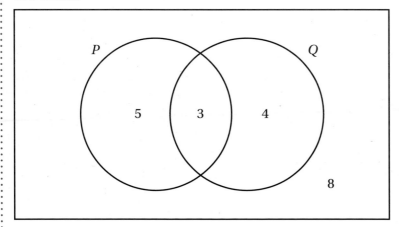

a Use the diagram to find the following probabilities:

   i P(*Q*)   **ii** P(*Q*′)   **iii** P(*P* and *Q*)   **iv** P(*P* or *Q*)

b Are events *P* and *Q* independent of each other?

Solution

a i Use the Venn diagram to work out the total number of outcomes:

      $5 + 3 + 4 + 8 = 20$ possible outcomes

Now use the Venn diagram to work out the number of outcomes in *Q*.

$3 + 4 = 7$ outcomes, out of 20 possible outcomes

So $P(Q) = \dfrac{7}{20}$

ii Use the Venn diagram to work out the number of outcomes not in Q:

$5 + 8 = 13$ outcomes out of the 20 possible outcomes

So $P(Q') = \dfrac{13}{20}$

iii This is the probability of P and Q both happening. This is the intersection of the two circles:

$P(P \text{ and } Q) = \dfrac{3}{20}$

iv This is the probability of either event P or event Q (or both) happening, or the union of both events. Add together the numbers in all three sections:

$5 + 3 + 4 = 12$ so $P(P \text{ or } Q) = \dfrac{12}{20}$

b $P(Q) = \dfrac{7}{20}$ and $P(P) = \dfrac{8}{20}$

Independent events would mean $P(P \text{ and } Q) = P(P) \times P(Q)$.

$P(P \text{ and } Q) = \dfrac{3}{20}$, but $P(P) \times P(Q) = \dfrac{56}{400}$.

$\dfrac{3}{20} \neq \dfrac{56}{400}$, so P and Q are not independent.

Example 4

If a fair die is rolled, what is the probability that it shows:

a Event A: a prime number

b Event B: a number less than 3

c Event A or B: a prime or a number less than 3?

Solution

a Three out of the six numbers on a die are prime: 2, 3 and 5.

So $P(A) = \dfrac{3}{6} = \dfrac{1}{2}$

b Two out of the six numbers on a die are less than 3.

So $P(B) = \dfrac{2}{6} = \dfrac{1}{3}$

c Four numbers on a die are prime or less than 3, or both: 1, 2, 3 and 5.

So $P(A \text{ or } B) = \dfrac{4}{6} = \dfrac{2}{3}$

There are several different approaches you could use for questions like this. Sample space diagrams are useful if you are asked to list the outcomes of two events. Venn diagrams are useful if you are working out combined probabilities of two or more events.

Example 5

In a small college with 100 students, maths and economics are the two most popular subjects.

55 students are studying maths. 27 students are studying economics.

17 students are studying maths and economics.

Find the probability that a student selected at random is studying maths or economics.

Subject	Number of students	Probability
maths	55	0.55
economics	27	0.27
both	17	0.17

Solution

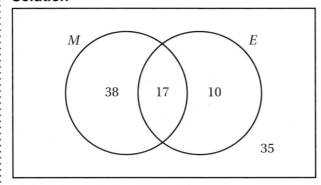

$P(M \text{ or } E) = 0.38 + 0.17 + 0.10 = 0.65$

Another way to solve problems involving a combination of exclusive and independent events is to draw a tree diagram. These show probabilities for sequences of two or more events.

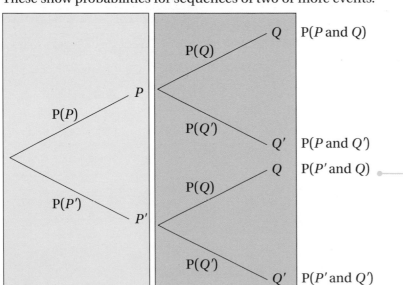

Probabilities on each branch add to 1.

To find the probabilities for a sequence of events you multiply along the branches.

The total of the probabilities of the end products is always 1.

The first box indicates one set of exclusive events. The branches on a tree diagram are all mutually exclusive – they cannot happen at the same time. The second box indicates an independent event, event Q, which can happen whether or not event P happens.

Example 6

The probability of scoring a strike first time in ten pin bowling is 0.21. If a strike is achieved, the probability of another strike is 0.53. However, if a strike wasn't scored first time, the probability of getting a strike the second time is 0.39.

a Show this information on a tree diagram.

b Find the probability of getting at least one strike.

Solution

a

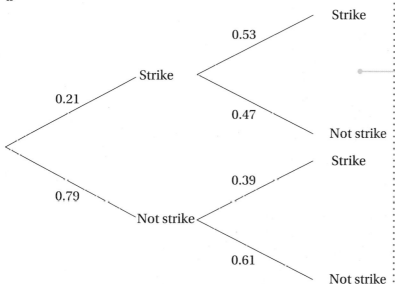

Always multiply across a tree diagram to find probabilities of each event.

Each pair of branches sum to 1 (they are exclusive). Events in the first and second sets of branches are independent of each other.

KEY INFORMATION

The sum of all the probabilities of the end products in a tree diagram is 1.

b A simpler and faster way of answering this question is to use the complement.

$$P(\text{at least 1 strike}) = 1 - P(\text{no strikes})$$
$$= 1 - (0.79 \times 0.61)$$
$$= 1 - 0.4819$$
$$= 0.5181$$

Exercise 13.1A

Answers page 549

1 A school has four houses: Rio, York, Gretna and Brighton. Of the 200 students in Year 9, 40 are in 'Rio' house, 62 in 'York' house and 52 in 'Gretna' house.

    a What is the probability that a student selected at random is in 'Brighton' house?

    b What is the probability that a student selected at random is not in either 'York' house or 'Gretna' house?

2 Tickets numbered 1 to 100 are placed in a hat and a single ticket is picked at random.

    a What is the probability of selecting a ticket with an even number?

    b What is the probability of selecting a ticket with a number which is not a square number?

    c What is the probability of selecting a ticket with a number containing the digit 5?

3 A bag contains 20 balls of which x are black, $2x$ are white and 8 are red. A ball is picked at random from the bag. What is the probability that it is not black?

4 For events A and B, $P(A) = 0.5$, $P(B) = 0.3$ and $P(A \text{ and } B) = 0.2$.

    a What is $P(A \text{ or } B)$?

    b What is $P(A' \text{ and } B)$?

5 For events S and T, $P(S) = 0.3$, $P(T) = 0.5$ and $P(S \text{ and } T) = 0.18$.

    a What is $P(S \text{ or } T)$?

    b What is $P(S' \text{ and } T')$?

6 A and B are 2 events such that: $P(A) = 0.3$, $P(B) = 0.5$ and $P(A \text{ and } B) = 0.15$. Find:

    a $P(A')$ **b** $P(B')$ **c** $P(A \text{ or } B')$

7 A spinner is split into 7 equal sections: 4 red, 2 blue and 1 green. When it lands on a red section, a biased coin with probability $\frac{2}{3}$ of landing on heads is flipped. When the spinner lands on a blue or a green section, a fair coin is flipped.

    a Draw and label a tree diagram to show the possible outcomes and associated probabilities.

    b Danny spins the spinner and flips the appropriate coin. Find the probability he obtained a head.

13.2 Discrete and continuous distributions

From your GCSE course and **Chapter 12 Data presentation and interpretation**, you will remember that **discrete** data can only take certain values, whereas **continuous** data can take values within a range.

A **discrete random variable** is a random variable which can only take certain values. It is usually represented by an uppercase letter such as X. The particular values that X can take are represented by the lowercase letter x. A **probability distribution** is a table showing all the possible values a discrete random variable can take, plus the probability of each value occurring.

The probabilities of all the possible values that a discrete random variable can take add up to 1.

For a discrete random variable X:

$$\sum_{\text{all } x} P(X = x) = 1$$

Flipping a fair coin twice and counting the number of heads would mean event X 'the number of heads' and the values it can take, x, could be 0, 1 or 2. You can check that this works using the formula:

$$\sum_{x=0,1,2} P(X = x) = P(X = 0) + P(X = 1) + P(X = 2)$$

$$= \tfrac{1}{4} + \tfrac{1}{2} + \tfrac{1}{4} = 1$$

KEY INFORMATION

The probabilities of all the possible values that a discrete random variable can take add up to 1.

Example 7

A biased coin is flipped. The probability of getting heads is $\frac{1}{3}$ and tails is $\frac{2}{3}$. If the coin is flipped once, write the probability function of X.

Solution

The outcome can be either heads or tails, so X can be either 0 or 1.

The probability of heads, $P(X = 0)$, is $\frac{1}{3}$ and the probability of tails, $P(X = 1)$, is $\frac{2}{3}$.

As there are two different probabilities, use two formulae, one for each x.

$$P(X = x) = \begin{cases} \frac{1}{3} & X = 0 \\ \frac{2}{3} & X = 1 \end{cases}$$

Example 8

Draw the probability distribution table for X, where X is the score on a fair six-sided die.

Solution

X is rolling a die, x are the random values it can land on. List all the values X can take in the table.

x	1	2	3	4	5	6
$P(X = x)$	$\frac{1}{6}$	$\frac{1}{6}$	$\frac{1}{6}$	$\frac{1}{6}$	$\frac{1}{6}$	$\frac{1}{6}$

$P(X = x)$ is the notation for the probability that X has the value x.

The corresponding probabilities are entered underneath each variable; it is always useful to check they add up to 1.

In **Example 8**, every value of X is equally likely. This is called a **uniform distribution**. The probability distribution of a random variable with a discrete uniform distribution always has the same shape.

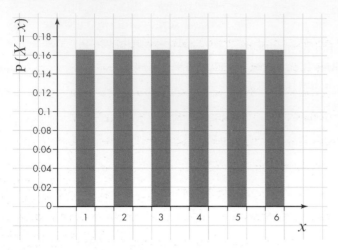

Example 9

A random variable A has a discrete probability distribution given by the probability function:

$$P(A = a) = k(a + 1) \text{ where } a = 0, 1, 2, 3, 4$$

Draw a table to show the distribution of A.

Solution

First you need to draw a table and substitute the values of a into the given expression to find k.

a	0	1	2	3	4
$k(a + 1)$	$k(0 + 1)$ $= 1k$	$k(1 + 1)$ $= 2k$	$k(2 + 1)$ $= 3k$	$k(3 + 1)$ $= 4k$	$k(4 + 1)$ $= 5k$

The probabilities sum to 1, so you can calculate the value of k:

$$1k + 2k + 3k + 4k + 5k = 1$$

$$15k = 1$$

$$k = \frac{1}{15}$$

Now you can go back to the table and put the corresponding probability in for each value:

a	0	1	2	3	4
$k(a + 1)$	$1k$	$2k$	$3k$	$4k$	$5k$
$P(A = a)$	$\frac{1}{15}$	$\frac{2}{15}$	$\frac{3}{15}$	$\frac{4}{15}$	$\frac{5}{15}$

Check your probabilities sum to 1.

In **Example 9**, every value of A is *not* equally likely. The probability distribution of a random variable would look like this:

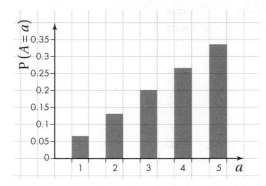

Example 10

The discrete random variable X has the following probability distribution:

x	0	1	2	3	4
$P(X = x)$	0.1	a	0.3	a	0.1

Find:

a the value of a

b $P(1 \leqslant X < 3)$

c the mode.

Solution

a Use
$$\sum_{\text{all } x} P(X = x) = 1.$$
$$0.1 + a + 0.3 + a + 0.1 = 1$$
$$2a + 0.5 = 1$$
$$2a = 0.5$$
$$a = 0.25$$

b This is asking for the probability between 1 and 3, including 1 but not including 3. In other words $X = 1$ and $X = 2$.

Add up the corresponding probabilities.
$$P(1 \leqslant X < 3) = P(X = 1) + P(X = 2)$$
$$= 0.25 + 0.3$$
$$= 0.55$$

> As the events are mutually exclusive, you add the probabilities.

c The mode is the most likely (most frequent) value. The highest probability is 0.3, so the mode is 2.

Continuous random variables are similar to discrete random variables but now they can take any value within a range. They represent such measurements as height, rainfall or speed, all of which would be measured on a continuous scale.

You can still draw a graph to represent a continuous variable, but as the data are continuous you use a continuous line rather than bars, to show that it can represent the full range of values.

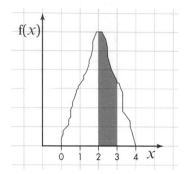

The graph is called a **probability density function** (pdf), represented by f(x). The area under a pdf shows the probability of X, and sums to 1. In the diagram above, the shaded region shows the probability that the continuous variable will take a value between 2 and 3.

The **cumulative distribution function**, f(x), gives the probability that X will be less than or equal to a particular value of x. It is a running total of probabilities.

> **KEY INFORMATION**
>
> Probabilities can never take a negative value. The total probability of a random variable is 1.

Exercise 13.2A

Answers page 550

CM 1 The random variable X is given by the sum of the scores when two fair dice are rolled.

    a Find the probability distribution of X.

    b Draw a diagram to show the distribution and describe the shape.

CM 2 The random variable A is given by the product of the scores when two ordinary dice are rolled.

    a Find the probability distribution of A.

    b Draw a diagram to show the distribution and describe the shape.

CM 3 Two tetrahedral dice labelled 1, 2, 3 and 4 are rolled. The random variable X represents the sum of the numbers shown on the dice.

    a Find the probability distribution of X.

    b Draw a diagram to show the distribution and describe the shape.

    c Calculate the probability that any roll of the dice results in the value of X being a prime number.

4 The probability distribution of a random variable B is given by

$\qquad$ $P(B = r) = kr$ for $r = 1, 2, 3, 4$

$\qquad$ $P(B = r) = 0$ otherwise

a Find the value of the constant k.

b Find the probability distribution of B.

c Draw a diagram to show the distribution and describe the shape.

5 The random variable X has the probability distribution:

x	2	3	5	7	11	13
$P(X = x)$	k	k	k	k	k	k

a Find k.

b Find $P(X < 7)$.

c Find $P(X \geqslant 3)$.

d Find $P(2 \leqslant X < 11)$.

6 For each of the following probability functions, $x = 1, 2, 3, 4, 5$.

Find k and then write down the probability distribution of X.

a $P(X = x) = kx^2$

b $P(X = x) = \dfrac{k}{x}$

c $P(X = x) = \dfrac{x}{k}$

As with cumulative frequency, seen in **Chapter 12 Data presentation and interpretation**, cumulative distribution involves adding up previous probabilities to make a running total. To find the cumulative probability of $F(x_0)$, for any given value of x_0, you add up all of the probabilities of the values of X which are less than or equal to x_0.

This can be summarised by the formula:

$$F(x_0) = P(X \leqslant x_0) = \sum_{x \leqslant x_0} P(X = x)$$

Example 11

The probability distribution for a biased die, D, is shown below. Draw a table to show the cumulative distribution function D.

d	1	2	3	4	5	6
$P(D = d)$	0.1	0.15	0.1	0.2	0.25	0.2

Solution

There are six values of d, so you have to find the probability that D is less than or equal to each of them in turn.

Start with the smallest value of d:

$F(1) = P(D \leq 1)$ This is the same as $P(D = 1)$ since there are no smaller values. So $F(1) = 0.1$

$F(2) = P(D \leq 2)$ This is the same as $P(D = 1) + P(D = 2)$.
So $F(2) = 0.1 + 0.15 = 0.25$

$F(6) = P(D \leq 6)$ This is the same as

$P(D = 1) + P(D = 2) + P(D = 3) + P(D = 4) + P(D = 5) + P(D = 6)$.

So $F(6) = 0.1 + 0.15 + 0.1 + 0.2 + 0.25 + 0.2 = 1$

> You know already that probabilities add up to 1.

Finally, you need to put the results into a table:

d	1	2	3	4	5	6
$F(d) = P(D \leq d)$	0.1	0.25	0.35	0.55	0.8	1

Example 12

The formula below gives the cumulative distribution function $F(x)$ for a discrete random variable X:

$$F(x) = kx^2, \text{ for } x = 1, 2, 3, 4$$

Find k and the probability distribution for X.

Solution

You must use the fact that probabilities add up to 1. To find k, X has to be equal to or less than 4.

$$P(X \leq 4) = 1$$

Therefore $F(4) = P(X \leq 4) = 1$

So $\qquad F(4) = 1$

This implies $16k = 1$, so $k = \dfrac{1}{16}$

> Substitute 4 for x in kx^2.

You can now use this to work out the probabilities of X being less than or equal to 1, 2, 3 and 4 by substituting $k = \dfrac{1}{16}$ into $F(x) = kx^2$ for each x value.

$$F(1) = P(X \leq 1) = 1 \times \frac{1}{16} = \frac{1}{16}$$

$$F(2) = P(X \leq 2) = 4 \times \frac{1}{16} = \frac{4}{16}$$

$$F(3) = P(X \leq 3) = 9 \times \frac{1}{16} = \frac{9}{16}$$

$$F(4) = P(X \leq 4) = 16 \times \frac{1}{16} = 1$$

The above probabilities are all cumulative. In the probability distribution table you want to know exact values. By subtracting the previous cumulative value you are left with the exact value.

$$P(X=4) = P(X \leqslant 4) - P(X \leqslant 3) = 1 - \frac{9}{16} = \frac{7}{16}$$

$$P(X=3) = P(X \leqslant 3) - P(X \leqslant 2) = \frac{9}{16} - \frac{4}{16} = \frac{5}{16}$$

$$P(X=2) = P(X \leqslant 2) - P(X \leqslant 1) = \frac{4}{16} - \frac{1}{16} = \frac{3}{16}$$

$$P(X=1) = P(X \leqslant 1) = \frac{1}{16}$$

The probability distribution of x is:

x	1	2	3	4
$P(X = x)$	$\frac{1}{16}$	$\frac{3}{16}$	$\frac{5}{16}$	$\frac{7}{16}$

KEY INFORMATION

$P(X = x_i) = F(x_i) - F(x_{i-1})$

Exercise 13.2B

Answers page 551

1 For each of the following probability distributions for a discrete random variable, X, draw up a table to show the cumulative distribution $F(x)$.

a

X	1	2	3	4	5
$P(X = x)$	0.15	0.2	0.5	0.1	0.05

b

X	1	2	4	8
$P(X = x)$	0.25	0.25	0.25	0.25

c

X	3	−1	1	3
$P(X - x)$	0.8	0.1	0.02	0.08

d

X	1	2	3	4	5
$P(X = x)$	0.2	$2a$	0.4	a	0.1

2 The discrete random variable X has the probability function:

$$P(X = x) = \frac{1}{5} \text{ for } x = 1, 2, 3, 4, 5$$

a Draw a table showing the cumulative distribution function $F(x)$.

b Find $P(X < 3)$.

c Find $P(1 \leqslant X < 4)$.

3 Each table shows a cumulative distribution function of a discrete random variable X. Write down the probability distribution for each table.

a

x	2	4	6	8
$F(x)$	0.3	0.6	0.8	1

b

x	−5	0	5	10
F(x)	0.05	0.4	0.5	1

c

x	2	5	6	22
F(x)	0.1	0.5	0.75	1

4 The discrete random variable X has the cumulative distribution function:

x	1	2	3	4
F(x)	0.32	a	0.78	1

Given that $P(X=2) = P(X=3)$, draw a table to show the probability distribution of X.

5 The cumulative distribution function for a random variable r is given in the table. Work out the probability distribution of R.

r	0	1	2
F(r)	$\frac{1}{8}$	$\frac{1}{2}$	1

(M) 6 A fair octahedral die has faces numbered 1, 1, 2, 2, 4, 4, 6 and 6. The random variable S represents the score when the die is rolled.

a Write down the probability distribution of S.

b State the name of this probability distribution.

A second die is yellow and the random variable Y represents the score when the yellow die is rolled.

The probability distribution of Y is:

y	1	3	5	7	9
P(Y = y)	$\frac{1}{12}$	$\frac{1}{6}$	$\frac{1}{2}$	$\frac{1}{6}$	$\frac{1}{12}$

c What type of distribution is this?

d What is the probability of rolling a prime number on both dice?

7 The random variable X has the probability function:

$$P(X=x) = x - \frac{6}{36} \text{ for } x = 1, 2, 3, 4, 5, 6$$

a Construct a table giving the probability distribution of X.

b Find $P(2 < X \leqslant 5)$.

8 A discrete random variable X has the probability function:

$$P(x) = kx^3 \text{ for } x = 1, 3, 5, 7, 9$$

where k is a positive constant.

a Show that $k = \frac{1}{1225}$.

b Find $P(X \geqslant 7)$.

9 The discrete random variable Y has the probability distribution:

Y	2	4	6	8
$P(Y=y)$	a	b	0.3	c

where a, b and c are constants.

The cumulative distribution function $F(y)$ of Y is given in the following table

Y	2	4	6	8
$F(y)$	0.1	0.5	d	1.0

where d is a constant.

Find the value of a, the value of b, the value of c and the value of d.

PF 10 The discrete random variable X can take only the values 1, −1, −3 and −5. For these values, the probability distribution function is given by

x	1	−1	−3	−5
$P(X=x)$	$\dfrac{2}{18}$	$\dfrac{2k}{18}$	$\dfrac{7}{18}$	$\dfrac{k}{18}$

where k is a positive integer.

a Show that $k=3$.

b Show the cumulative probability distribution.

13.3 The binomial distribution

The **binomial distribution** is a discrete probability distribution that describes discrete random variables.

In **Chapter 1 Algebra and Functions 1: Manipulating algebraic expressions** you learnt about arrangements of various objects using $n!$ ('n factorial'). If there are n different trials, k are identical to each other, so $n-k$ would also be identical to each other. The **binomial coefficient** tells you how many ways there are of choosing unordered outcomes from all of the possibilities.

> If there are k events of one type and $n-k$ events of another type, they can be arranged in $\dfrac{n!}{k!(n-k)!}$ different orders.

The binomial distribution is about 'success' and 'failure', as there are only two possible outcomes to each trial. For instance, if you flip a coin once it can only land on a heads or a tails. The 'success' would be heads, whilst the 'failure' would be tails (or vice versa, as long as you refer to each outcome consistently).

If you were to flip a coin twice and wanted to know the probability of getting exactly one head, success would be heads and failure would be tails. The possible outcomes are HH, HT, TH or TT. However, you need to look at the number of arrangements for the heads. You could get a head and then a tail or a tail and then a head. You have 2 possible arrangements for getting one head out of 4 possible outcomes. The probability of flipping one head would be $\frac{1}{2}$.

If you were to flip a coin three times and wanted to know the probability of getting one head, success would be heads and failure would be tails. You could flip HHH, HHT, HTH, THH, HTT, THT, TTH or TTT. However, you need to look at the number of arrangements for the heads. You could get HTT, TTH or THT, so there are 3 possible arrangements for getting one head, out of 8 possible outcomes. The probability for one head would be $\frac{3}{8}$.

It is important that you can identify situations that can be modelled using the binomial distribution. You can use the binomial distribution:

› if there are a fixed number of n independent trials

› if there are just two possible outcomes to each trial, success and failure, with fixed probabilities of p and q respectively, where $q = 1 - p$

› if all trials are independent of each other

› if the probability of success, p, is the same in every trial.

The discrete random variable X is the number of successes in the n trials. X is modelled by the binomial distribution B(n, p). You can write $X \sim$ B(n, p).

The probability of success in a single trial is usually taken as p and the probability of failure as q. The probability of achieving exactly k successes in n trials is:

$$P(k \text{ successes in } n \text{ trials}) = {}^{n}C_{k}\,p^{k}q^{n-k}$$

where n = number of trials

k = number of successes

$n - k$ = number of failures

p = probability of success in one trial

$q = 1 - p$ = probability of failure in one trial.

If a coin was flipped 5 times and you wanted to know the probability of getting 3 heads you would need to use the binomial distribution, as a written method would be time-consuming and very cumbersome.

A success is heads and a failure is tails.

The number of arrangements of 3 heads in 5 coin flips can be represented using ${}^{5}C_{3} = 10$.

The probability of success is $\frac{1}{2}$.

The probability of failure is $\frac{1}{2}$.

$$P(3 \text{ successes}) = \frac{1}{2} \times \frac{1}{2} \times \frac{1}{2} = \left(\frac{1}{2}\right)^{3}$$

$$P(2 \text{ failures}) = \frac{1}{2} \times \frac{1}{2} = \left(\frac{1}{2}\right)^{2}$$

Combining all of this together gives $10 \times \frac{1}{8} \times \frac{1}{4} = 0.3125$

TECHNOLOGY

Your calculator will be able to perform exact binomial distribution calculations. Look for a menu labelled 'distribution' and select 'binomial PD'. Input the figures required in the calculations.

Example 13

A coin is flipped 15 times. Find the probability of getting 5 heads.

Solution

The probability of getting a head is $\frac{1}{2}$ and the probability of getting a tail is $\frac{1}{2}$. Therefore, the probability of throwing 5 tails is $\left(\frac{1}{2}\right)^5$.

The total number of ways of throwing 5 heads in 15 trials is $^{15}C_5$, which gives 3003 arrangements.

$$P(5 \text{ heads}) = \left(\frac{1}{2}\right)^5$$

$$P(10 \text{ tails}) = \left(\frac{1}{2}\right)^{10}$$

Hence the probability of throwing 5 heads and 15 trials is $^{15}C_5 \times \left(\frac{1}{2}\right)^5 \times \left(\frac{1}{2}\right)^{10} = 0.092$ (3 d.p.).

This can also be written as $X \sim B(15, 0.5)$.

Example 14

You are taking a multiple choice test containing ten questions. If each question has four choices and you guess the answer to each question, what is the probability of getting exactly seven questions correct?

Solution

This is a binomial distribution because outcomes are either a success or a failure, independent of each other, therefore $X \sim B(10, 0.25)$. Using the formula:

$$n = 10$$

$$k = 7$$

$$n - k = 3$$

$p = 0.25$
= probability of guessing the correct answer to a question

$q = 0.75$
= probability of guessing the wrong answer to a question

P(exactly 7 correct guesses) $= {}^{10}C_7 (0.25)^7 (0.75)^3 \approx 0.0031$

Example 15

A particular set of traffic lights is on green 60% of the time. X is the number of drivers who pass through the traffic lights out of 30. What is the probability that exactly 11 drivers pass through?

Solution

Each driver represents a trial, so $n = 30$. The probability of success is 60% so $p = 0.6$.

$$X \sim B(30, 0.6)$$

$P(X = 11) = {}^{30}C_{11} \times 0.6^{11} \times 0.4^{19} = 0.00545$

Example 16

$X \sim B(11, 0.41)$

Find the following probabilities:

a $P(X = 1)$ **b** $P(X = 0)$ **c** $P(X \geqslant 2)$

Solution

$X \sim B(11, 0.41)$. So $n = 11$, $p = 0.41$ and $q = 0.59$.

a $P(X = 1) = {}^{11}C_1 \times 0.41^1 \times 0.59^{10} = 0.0231$ (3 s.f.)

b $P(X = 0) = {}^{11}C_0 \times 0.41^0 \times 0.59^{11} = 0.00302$ (3 s.f.)

c $P(X \geqslant 2) = 1 - P(X = 0) + P(X = 1)$

$\qquad = 1 - (0.00302 + 0.0231)$

$\qquad = 0.974$ (3 s.f.)

TECHNOLOGY

Your calculator will be able to perform cumulative binomial distribution calculations. Look for a menu labelled 'distribution' and select 'binomial CD'. Input the figures required in the calculations.

'Greater than or equal to 2' is the same as saying '1 minus the probability of 1 or less'.

Example 17

In a game of chance the probability of winning is 0.73 and of losing is 0.27. Find the probability of winning more than 7 out of 10 games.

Solution

The number of successes is a random variable, $X \sim B(10, 0.73)$, assuming independence of trials.

$P(X > 7) = P(X = 8) + P(X = 9) + P(X = 10)$

$\qquad = {}^{10}C_8 \times 0.73^8 \times 0.27^2 + {}^{10}C_9 \times 0.73^9 \times 0.27^1 + {}^{10}C_{10} \times 0.73^{10}$

$\qquad = 0.46648877$

$\qquad = 0.466$ to 3 s.f.

So P(more than 7 wins in 10 games) $= 0.466$.

'More than 7 games' means winning 8, 9 or 10 games.

TECHNOLOGY

You can do this on your calculator using the 'binomial CD' menu.

Could you use the complement of an event to answer **Example 17**?

Example 18

For $X \sim B(24, 0.84)$, find the probability that $19 < X \leqslant 21$.

Solution

Earlier in the chapter we looked at the cumulative distribution function. Using the same logic,

$$P(X \leqslant 21) = P(X = 0) + P(X = 1) \ldots + P(X = 21)$$

$19 < X \leqslant 21$ shows that you need

$$P(X = 20 \text{ or } 21) = P(X \leqslant 21) - P(X \leqslant 19) = 0.763 - 0.337$$

$$= 0.426 \text{ to } 3 \text{ d.p.}$$

Use your calculator to find the values

Example 19

A bag contains a large number of red and white discs, of which 59% are red. 132 discs are taken from the bag. Find the probability that the number of red discs is between 80 and 110 inclusive.

Solution

You can make an assumption in this question: as there is a large number of discs in the bag, you can assume that the probability of a red disc remains the same for each trial, $p = 0.59$.

Let X be the number of red discs $\Rightarrow X \sim B(132, 0.59)$.

You now want $P(80 \leqslant X \leqslant 110)$.

$$\Rightarrow P(80 \leqslant X \leqslant 110) = P(X \leqslant 110) - P(X \leqslant 80)$$

$$= 0.999\,999 - 0.676\,97$$

$$= 0.323 \text{ to } 3 \text{ d.p.}$$

As you want between 80 and 110, you need to subtract $P(X \leqslant 80)$ (as the lowest number of red discs is 80) from $P(X \leqslant 110)$ (as the number of red discs can be 110).

Exercise 13.3A

Answers page 552

 1 Find the probabilities below using your calculator. Give your answer to 3 d.p. where appropriate.

    a For $X \sim B(10, 0.2)$

        i $P(X = 2)$ **ii** $P(X = 5)$ **iii** $P(X = 10)$

    b For $X \sim B(7, 0.8)$

        i $P(X = 2)$ **ii** $P(X = 5)$ **iii** $P(X < 3)$

    c For $X \sim B(20, \frac{1}{8})$

        i $P(X = 18)$ **ii** $P(X \leqslant 5)$ **iii** $P(3 \leqslant X < 15)$

    d For $X \sim B(5, 0.4)$

        i $P(X < 4)$ **ii** $P(X = 5)$ **iii** $P(0 \leqslant X < 3)$

    e For $X \sim B(17, 0.16)$

        i $P(X \leqslant 12)$ **ii** $P(5 \leqslant X < 8)$ **iii** $P(2 \leqslant X < 13)$

2 A fair six-sided die is rolled 8 times. What is the probability of rolling exactly 3 sixes?

 3 In basketball, the probability of scoring from a shot is s. Each basket scored is independent of another.

    Write the probability formula to represent scoring 6 times from 10 shots.

4 A random variable X has the distribution $B(12, p)$. Given that $p = 0.25$, find:

    a $P(X < 5)$

    b $P(X \geqslant 7)$

 5 In a maths class, an average of 3 out of every 5 students ask for paper to write on. A random sample of 30 students is selected.

    Find the probability that:

    a exactly 10 students ask for paper

    b fewer than 20 students ask for paper.

6 An airport has poor visibility 25% of the time in winter. A pilot flies to the airport 10 times during winter.

    a What is the probability that she encounters poor visibility exactly 3 times?

    b What is the probability that she encounters poor visibility at least 3 times?

7 There are 20 students in a maths class.

    a Find the probability that exactly one has a birthday in January.

    b Find the probability that at most four have a birthday in January.

SUMMARY OF KEY POINTS

> Probability, P, is a number: $0 \leqslant P \leqslant 1$

> The complement of an event A is shown by $P(A') = 1 - P(A)$

> If A and B are mutually exclusive (do not overlap) then $P(A \cup B) = P(A) + P(B)$

> If A and B are independent then $P(A \cap B) = P(A) \times P(B)$

> Each individual outcome in the sample space is equally likely.

$$\sum_{\text{all } x} P(X = x) = 1$$

> The binomial distribution is a discrete probability distribution that describes discrete random variables.

> You can use the binomial distribution:

> if there are a fixed number of n independent trials

> if there are just two possible outcomes to each trial, success and failure, with fixed probabilities of p and q respectively, where $q = 1 - p$

> if all trials are independent of each other

> if the probability of success, p, is the same in every trial.

> The discrete random variable X is the number of successes in n trials. X is modelled by the binomial distribution $B(n, p)$. You can write $X \sim B(n, p)$.

EXAM-STYLE QUESTIONS 13

Answers page 552

1   A karting venue has 2 red, 1 blue and 1 green karts. Two karts are to be raced at random.

   a   Draw a tree diagram to illustrate all the possible outcomes and associated probabilities.   **[3 marks]**

   b   Find the probability that a blue kart and a green kart are raced.   **[2 marks]**

2   300 students at a local university were interviewed to find out how many takeaways they have each week. The results are shown below.

	Number of takeaways				
	0	1	2	More than 2	Total
Male	25	40	38	37	140
Female	40	70	41	9	160
Total	65	110	79	46	300

   a   Find P(Male).   **[1 mark]**

   b   Find P(Male and 1 takeaway).   **[2 marks]**

   c   Find P(Female and more than 2 takeaways).   **[2 marks]**

3 There are 185 students at a youth club. Students can choose three activities.

        112 do climbing.

        70 do table tennis.

        81 do bowling.

        35 do climbing and bowling.

        16 do table tennis and bowling.

        40 do climbing and table tennis.

        11 do all three activities.

  a Draw a Venn diagram to represent this information.   **[5 marks]**

A student is selected at random. Find the probability that this student

  b does none of the activities   **[2 marks]**

  c does bowling only.   **[1 mark]**

4 Jessica, Demelza and Kamila attend a dance class. As exams are approaching, the teacher is concerned about their punctuality. For each class the probabilities of them arriving late are 0.1, 0.5 and 0.25 respectively, independent of each other.

  a Calculate the probability that all three girls are late for a particular class.   **[1 mark]**

  b Calculate the probability that at least one girl is late.   **[3 marks]**

Sophie now joins the class. The probability that she is late is 0.75 when Demelza is late. If Demelza is on time, the probability that Sophie is on time is 0.15.

  c Calculate the probability that for a particular session:

    i both of them are late   **[2 marks]**

    ii at most one of them is late.   **[3 marks]**

5 Uche, Eli and Ellis play darts. The independent probabilities that Uche, Eli and Ellis hit a bull's eye are 0.62, 0.17 and 0.68 respectively.

Find the probability that:

  a none of them hits a bull's eye   **[2 marks]**

  b at least one of them hits a bull's eye   **[2 marks]**

  c exactly one of them hits a bull's eye   **[3 marks]**

  d either one or two of them hits a bull's eye.   **[3 marks]**

6 Badges are made in various colours. The number of red badges may be modelled by a binomial distribution with a probability of 0.2.

The contents of packets of 50 badges of various colours may be considered to be random samples.

Determine the probability that a packet contains:

  a less than or equal to 15 red badges   **[2 marks]**

  b exactly 10 red badges   **[2 marks]**

  c more than 5 but fewer than 15 red badges.   **[3 marks]**

7 Kartik and Khari decide to have a competition in the common room playing each other at pool. They decide to play 10 frames of pool to see who is better.

The probability that Kartik will win any frame is 0.21. The outcome of each frame is independent of the outcome of every other frame.

Find the probability that, during 10 frames, Kartik wins:

a exactly 2 frames [3 marks]

b at least 4 frames [3 marks]

c more than 3 but fewer than 9 frames. [3 marks]

M **PS** **8** A conservatory saleswoman is in a busy garden centre. She knows that the probability of a passing person taking one of her leaflets is 0.25. During a random 20-minute period, 40 people pass her.

a Suggest a suitable model to describe the number of people who pass her. [1 mark]

b What is the probability that more than 10 people take a leaflet? [3 marks]

M **9** Pawel plants 15 randomly selected seeds in a planter. He knows that 10% of the seeds will yield red flowers while the others will yield white flowers. All the seeds grow successfully and Pawel counts the number of red flowers and the number of white flowers.

a Find the probability of 3 flowers being red. [2 marks]

b Find the probability of all the flowers being red. [3 marks]

PS **10** At a local gym, 150 members were surveyed to see what facilities they use.

13 people used the pool, gym and went to classes.

35 people went to the gym and pool but did not attend classes.

8 people went to classes and the gym but did not use the pool.

24 people used the pool and went to classes but did not use the gym.

18 people only attended classes.

78 people used the gym.

2 people did not use any of the facilities.

a Copy and complete the Venn diagram to show these results.
One member is picked at random. Find the probability that:

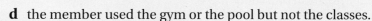

b the member only used the gym [1 mark]

c the member did not use the gym [2 marks]

d the member used the gym or the pool but not the classes. [2 marks]

e Given that someone used the gym, find the probability that they also used the pool and classes. [3 marks]

11 The number, T, of empty trays in a school lunch queue at 12.30 pm on a weekday has the following probability distribution:

$$P(T = t) = k(4 - t) \text{ when } t = 1 \text{ to } 5$$

where k is constant.

Show this information in a probability distribution table and show that $k = \frac{1}{5}$. [3 marks]

14 STATISTICAL SAMPLING AND HYPOTHESIS TESTING

A new medicine company believes they have developed a better medicine compared to the one currently used. They wish to compare their medicine to the current treatment as they believe they will see a better outcome for the patients. The medicines and healthcare regulatory agency inform the new company that they must demonstrate that their medicine is better than the current treatment to receive approval. In order to do this the new company must run a clinical trial. Patients are selected at random, then some would receive the new medicine whilst other patients receive the previous medicine. A prediction (called a hypothesis) would be drawn up prior to treatment being given out. The new company claims that the new medicine is more effective, so in the results a better response to the treatment should be visible for the new medicine and a standard response visible for the current medicine.

LEARNING OBJECTIVES

You will learn how to:

- use sampling techniques, including simple random sampling and opportunity sampling

- use simple random sampling, stratified sampling, systematic sampling, cluster sampling, quota sampling and opportunity (or convenience) sampling

- select or critique sampling techniques in the context of solving a statistical problem, including understanding that different samples can lead to different conclusions about the population

- apply the language of statistical hypothesis testing, developed through a binomial model: null hypothesis, alternative hypothesis, significance level, test statistic, 1-tail test, 2-tail test, critical value, critical region, acceptance region, p-value

- have an informal appreciation that the expected value of a binomial distribution as given by np may be required for a 2-tail test

- conduct a statistical hypothesis test for the proportion in the binomial distribution and interpret the results in context

- understand that a sample is being used to make an inference about the population and hypotheses should be expressed in terms of the population parameter p

- appreciate that the significance level is the probability of incorrectly rejecting the null hypothesis.

TOPIC LINKS

You will need the skills and techniques that you learnt in **Chapter 13 Probability and statistical distributions** to help you calculate probabilities in hypothesis testing. From that same chapter, the ability to use binomial expansions will help you calculate exact and cumulative probabilities, which are needed in hypothesis testing.

PRIOR KNOWLEDGE

You should already know how to:

> infer properties of populations or distributions from a sample, while knowing the limitations of sampling

> apply statistics to describe a population.

You should be able to complete the following questions correctly:

1 What is meant by a random sample?

2 Describe how you would find a random sample of 30 students in a year group of 100.

3 What are the advantages and disadvantages of using a sample?

14.1 Populations and samples

For any statistical investigation, there will be a group that you want to find out about. A **population** is every item or person within a group that you want to find information about. It might be a group of people or it could be simply a group of numbers. A population can be **finite** or **infinite**. With a finite group, it is possible to count every member of the group. With an infinite group it is impossible to know the exact number of members.

When you want to collect information about your population you would carry out a survey by means of questioning or examining every item. A **census** is when information about every member of the population is collected. This means the information gathered is a true and accurate reflection of the whole population and is therefore unbiased. The disadvantage of this method is that if the population is large, it can be difficult to collect and process so much information. It is also expensive to do and hard to ensure that every member of the population has been surveyed.

Stop and think How is data collected and used about people in the UK? Look at the website for the Office for National Statistics.

Collecting information about every member of a population may be impractical or even impossible. If so, a selection or sample of items or people within the population is taken. From your GCSE course, you should know that a **sample survey** involves information being collected from a small representative part of the population. For example, if you are trying to discover the favourite holiday destination of 17-year-olds in the UK, it would be impractical to set up a census to ask every 17-year-old in the country. Instead, a group of 500 17-year-olds might be chosen and asked.

A **sampling unit** is an individual member of the population that can be sampled. Sampling units ensure that any one member of a population is not sampled more than once. If you wanted to find out how many pets the average family has, you might use a sampling unit of a household rather than a single person, because you would not want to ask two people from the same household.

A **sampling frame** is the collection of all of the sampling units. This should be the whole population, but this is often impractical, especially with infinite populations. This list must ensure that every sampling unit has a unique name or reference number so that the collection of samples is fair and unbiased.

The **target population** is the group of individuals from which the sample might be taken. The target population might be as broad as all teenagers but it might be a smaller group, such as 17-year-olds, people with allergies to pollen or people who wear glasses. You need to make sure that the people in your sample are similar to the other members of the target population because you want to generalise from the sample to the target population. Your sample should be as representative as possible of all of the members of the target population. The more representative the sample, the more confident you can be that the results can be generalised to the target population.

When selecting your sample you will need to be careful to avoid **bias**. Sampling bias occurs where the sample does not reflect the target population. To avoid bias, ensure that none of the population is excluded. An accurate sampling frame must be used to ensure that all the sample members are used.

There are several different types of sampling you could use.

Random sampling
With this type of sampling, everyone in the entire target population has an equal chance of being selected, and each selection is independent of the others. For example, the population in a school is all of the students. Each student has an equal chance of being selected for a 'student voice' opinion.

KEY INFORMATION

In a random sample every member of the population has an equal chance of selection.

Example 1
From a population of 86 students studying maths, form a simple random sample of size 9 to conduct a survey about exams.

Solution
Begin by assigning a number to each of the students. Since there are a total of 86 students, and 86 is a two-digit number, every individual in the population is assigned a two-digit number: 01, 02, 03, … , 84, 85, 86.

Use a table of random numbers to determine which of the 86 students should be chosen in your sample.

Start at any place in the random number table and write the random digits in groups of two. For example:

22	43	91	82	15	41	86	82	41	26	23	19	83

Select the first 9 numbers that are in the range from 01 to 86. You should omit 91 as this number is greater than the total number of students in your population. The numbers 82 and 41 are repeated so use them only once.

The students who should be contacted for the survey are those with the following numbers:

22	43	82	15	41	86	26	23	19

> Random numbers can be generated from a table of random numbers, using a computer or by using a calculator.

The advantages of random sampling are that your target population is represented well and sampling bias should be eliminated. The disadvantage is that it is very difficult to achieve a true random sample as it will take a lot of planning, and it can take a lot of time, effort and money.

Stratified sampling

The strata are the different groups within the target population and the sample is in the proportions needed for the sample to be representative of every stratum.

For example, the amount of money that college students spend on sports may be of interest to you. The main sport each student plays may be an important variable. Students who go skiing may spend more money on their sport than students who play football or netball. If you use a very large percentage of students who ski, your results will not be accurate: they will be inflated. You need to work out the proportion of each group in the target population – for example, the percentages of the student groups may look like this:

skiing 10%	football 55%	netball 30%	others 5%

The sample must then contain all these groups in the same proportion as in the target population of college students.

The advantage of a **stratified sample** is that it should be representative of the target population and therefore you can generalise from the results obtained. However, gathering such a sample is extremely time-consuming and difficult.

Cluster sampling

The population is divided into subgroups (called clusters), but unlike stratified sampling, each cluster represents the population. A random sample is then taken from each cluster.

> **KEY INFORMATION**
>
> In a stratified sample, the size of each stratum is in proportion to its size in the population.

For example, if you wanted a sample of people that use a gym, you would take a sample from a selection of different gyms. When choosing the gyms, you must ensure that the information you want to gather is represented in all gyms.

The advantage of cluster sampling is that it is usually a cost-effective method. However, it may produce less accurate results than random sampling or stratified sampling.

Opportunity sampling

This method of sampling uses people from the target population who are available there and then, and are willing to take part. It is a convenient sample.

An **opportunity sample** is obtained by asking members of your target population if they will take part in your research – for example, selecting a sample of diners coming out of a restaurant.

This type of sample is quick and easy to gather, but the downside is that it may *not* provide a representative sample, so it could be biased.

Quota sampling

Quota sampling involves splitting the population into groups and sampling a given number of people from each group.

Quota sampling is easy to implement. For example, if someone is interviewing people at a gym, they may have been told to interview 50 men and 50 women. It doesn't matter how they choose each set of 50 participants.

If there is no sampling frame (list of sampling units), the previous sampling methods in this section couldn't be implemented. Quota sampling might be the only real possibility.

There must always be enough participants to make the sample representative of the target population. The sample must not be so large that the study takes too long or is too expensive, or so small that it doesn't provide good representation.

Systematic sampling

This method involves gathering your sample in a systematic way from the target population – for example, every 5th person from a list of names.

To take a **systematic sample** you need to number all the members of the population and then decide upon a sample size. Divide the number of people in the population by the number of people you want in your sample to get a number, n. If you take every nth name, you will get a systematic sample of the correct size. If, for example, you wanted to sample 81 students from a college of 1377 students, you would choose every 17th name.

The advantage of systematic sampling is that you get a representative sample, but the downside is that it is difficult to achieve as it requires time, effort and money.

> **KEY INFORMATION**
> Opportunity samples are convenient and can be taken from any group from the population at any time.

> **KEY INFORMATION**
> Quota samples take a certain number of the population in any order.

> See **Chapter 12 Data presentation and interpretation** for more on data cleaning.

> **KEY INFORMATION**
> Systematic samples use a repetitive system to obtain the sample from the population.

> Bias cannot be removed by increasing the size of the sample.

> **Using the large data set 14.1**
>
> The large data set contains a lot of raw data. Remember that your data needs to be cleaned before performing any viable calculations.
>
> **a** Select a random sample of 30 items from a chosen category and find the mean and standard deviation.
>
> **b** Using the same category, take a systematic sample of 30 items, find the mean and standard deviation.
>
> **c** Compare the two samples using the mean and standard deviation. Explain ways in which the samples are similar or different.
>
> **d** Using your knowledge of the large data set, explain why a sample of 30 may not always be possible.

Exercise 14.1A

Answers page 554

1 Explain what you understand by the terms:

   a sampling unit

   b sampling frame

   c sampling distribution.

(CM) 2 A cricket club decides to find out how supporters might react to a new kit.

   a Explain why the club might decide to take a random sample of the supporters rather than ask everyone.

   b Suggest a suitable sampling frame.

   c Identify the sampling units.

(CM) 3 Each vacuum cleaner produced at factory is stamped with a unique serial number. The factory produces vacuum cleaners in batches of 500, out of which they do a quality control test on a random sample of 5 vacuum cleaners to see how long they can be used before breaking down.

   a Explain what you understand by the term 'census'.

   b Give one reason, other than to save time and cost, why a sample is taken rather than a census.

   c Suggest a suitable sampling frame from which to obtain this sample.

   d Identify the sampling units.

(CM) 4 A group of students conduct a blindfolded taste test to see if teenagers can tell the difference between shops' own brand and commercially produced fizzy drinks. They take a sample of 242 teenagers from a local town and ask them to take part in their experiment. The proportion of teenagers who say they can tell the difference is 35%.

   a State the population in this context.

   b Explain what you understand by the sampling distribution of this statistic.

 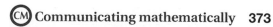

5 A random sample of 5 teachers is to be selected from 50 to discuss the application process for university. The following extract from a table of random numbers, starting at the beginning, is to be used to obtain the sample. Write down the process you would use to select your sample.

| 13 | 15 | 41 | 87 | 95 | 46 | 13 | 10 | 11 | 98 |

| 45 | 65 | 75 | 98 | 03 | 65 | 01 | 11 | 19 | 25 |

 6 The editor of a school newspaper is attempting to find out the proportion of students who are interested in getting a part-time job. One issue of the newspaper contains a questionnaire which readers are invited to complete and return.

a Give two reasons why the results obtained may be biased.

b Describe briefly an unbiased method of obtaining the information.

7 A sample of 10 was produced from a population of 100. Explain what the differences could be between using the following pairs of sampling techniques:

a random or stratified

b random or quota

c systematic or opportunity.

14.2 Hypothesis testing

Hypothesis testing involves using data from your sample to see if a statement about the population is believable or not. The word 'hypothesis' means a theory or prediction. The hypothesis is put forward because it is suspected to be true, unless evidence is found to indicate otherwise.

Null and alternative hypotheses

In any experiment, you usually have your own hypothesis as to how the results will turn out.

A **null hypothesis** states what has always happened beforehand – the expected or theoretical outcome. An **alternative hypothesis** is what you now believe is true or what you are attempting to prove. The notation H_0 is used for the null hypothesis and H_1 for the alternative hypothesis.

If you wanted to find out if a die was fair or not, you would carry out a hypothesis test. The theoretical probability of a die landing on any side is $\frac{1}{6}$. If the die is fair, the probability of it landing on any side, for instance a 6, will be $\frac{1}{6}$. If the die is biased, the probability of it landing on a 6 would be anything other than $\frac{1}{6}$.

The null hypothesis for rolling a die and it landing on a 6 would be set out as:

$$H_0: p = \frac{1}{6}$$

There are two variations on the alternative hypothesis for the die:

> A **1-tail** alternative hypothesis specifies whether the parameter you are investigating is either greater than or less than the value you are using in H_0.

$H_1 : p > \frac{1}{6}$, 1-tail test to show whether the die lands on more 6s than other sides.

$H_1 : p < \frac{1}{6}$, 1-tail test to show whether the die lands on fewer 6s than other sides.

> A **2-tail** alternative hypothesis does not specify the parameter, it just states it is not equal to H_0.

$H_1 : \quad p \neq \frac{1}{6}$, 2-tail test to show whether the die is biased.

Once you have carried out your experiment, you need to decide if there was enough evidence to support or reject the null hypothesis. If you find that the probability of the die landing on a 6 is $\frac{1}{6}$, you would accept the null hypothesis as the evidence suggests that this is the case. However, if you find that the probability is not $\frac{1}{6}$, you would reject the null hypothesis as the evidence suggests it is unlikely to be true. Notice that you say you 'reject H_0', rather than 'accept H_1'. The formal way to present your findings is to relate them back to whether or not to accept or reject H_0.

> The 1-tail test gets its name from testing the region under one of the tails (sides) of the distribution.

> The 2-tail test gets its name from testing the region under both of the tails (sides) of the distribution.

KEY INFORMATION

A 1-tail test looks at only one part of the distribution, either the upper or the lower region. A 2-tail test looks at both regions.

Example 2

Simon rolls a tetrahedral die, which he thinks is biased towards 1. He wants to do a hypothesis test to test his theory.

a Write down a suitable null hypothesis.

b Write down a suitable alternative hypothesis.

c Is this a 1-tail or 2-tail test?

Solution

a The null hypothesis must state a specific value of the theoretical probability. The null hypothesis shows what the probability would be if the die were unbiased.

$$H_0 : p = 0.25$$

b Simon thinks his die is biased towards 1, so there would be a greater probability of it landing on a 1 each time.

$$H_1 : p > 0.25$$

c The alternative hypothesis specifies that p is greater than 0.25, so the test is 1-tail.

KEY INFORMATION

You can only accept or reject H_0 and state the evidence that supports this. As this is only an experiment from a sample of the population, you *cannot* make a categorical conclusion from your findings.

> The null hypothesis, H_0, is what you expect. In this case, the probability of rolling a 1 would be 0.25.

> The alternative hypothesis, H_1, is what you believe is going to happen. It also determines whether you use a 1-tail or 2-tail test.

Answers page 554

> **Using the large data set 14.2**
>
> Select a category from your large data set.
>
> **a** If you were asked to test if there was any evidence to suggest a change in your data either over time or between locations, what null and alternative hypotheses would you use and why? Would the test be 1-tail or 2-tail?
>
> **b** Take a stratified sample of 30 items of data from the population of the category you are investigating. Repeat for the same category either in a different location or different year.
>
> **c** Using the mean from your samples, would you suggest there is enough evidence to indicate change? Explain your reasons.

Exercise 14.2A

CM 1 A child can catch a ball with success of 0.55. His parents practise with him for 10 minutes every day. The parents want to see if he is now better at catching the ball.

    a Write down a suitable null hypothesis.

    b Write down a suitable alternative hypothesis.

    c State whether this is a 1-tail or 2-tail test.

CM 2 A cheese stall at a Christmas market sells 15 packs of cheese per hour. The probability of selling the cheese to a passing customer is 0.68. The owner decides to change the labels and packing to see if this has a difference on the amount she sells.

    a Write down a suitable null hypothesis.

    b Write down a suitable alternative hypothesis.

    c State whether this is a 1-tail or 2-tail test.

CM 3 The probability of winning a computer game is 0.65. Stan practises and believes he is now much better. He wants to test his theory.

    a Write down a suitable null hypothesis.

    b Write down a suitable alternative hypothesis.

    c State whether this is a 1-tail or 2-tail test.

Significance levels

To reject H_0, your data must suggest the probability is unlikely. The significance level, α, is a way of showing how far you are willing to trust that the results are down to chance instead of just indicating that H_0 is incorrect. In other words, the significance level is the probability of mistakenly rejecting the null hypothesis.

> The significance level (α) of a test determines how unlikely your data needs to be according to the null hypothesis before you reject H_0.

The level of statistical significance is often expressed as the p-value. This is the probability for your population, calculated from your sample, assuming that the null hypothesis is true.

The lower the significance level, the stronger the evidence needs to be to reject H_0. The most common significance levels are 1%,

5% and 10%. If your significance level is $\alpha = 0.01$ (1%), you would only reject H_0 if your data fell into the most extreme 1% of possible outcomes.

A lower significance level suggests that you would perform a much more rigorous test – but it can be a disadvantage. You would be much less likely to be able to reject H_0, which would result in your experiment failing to conclude anything.

In order to check whether your results are significant, calculate their probability using H_0, then choose whether to reject or accept the null hypothesis. This result is known as the **test statistic**, X. If your result lies in the acceptable region then you accept your null hypothesis, H_0.

The region outside this acceptable region is called the **critical region**. If your test statistic lies in the critical region you reject the null hypothesis, H_0, and your conclusion will be based on the alternative hypothesis, H_1.

A fair tetrahedral die should have the same chance of landing on each side. If you had a tetrahedral die which you suspected to be biased towards 1, your suspicion would be proved if it had a greater probability of landing on a 1 each time. $H_1: p > 0.25$. This test is 1-tail.

To test whether the die was biased you would carry out a hypothesis test at a 5% **significance level**. If you rolled the die 60 times, the test statistic X would be the number of 1s rolled. X would follow a binomial distribution $B(n, p)$ as the number, n, is stated and the probability of success, p, is also stated and independent for each roll.

> **KEY INFORMATION**
>
> The significance level is the probability of rejecting H_0 when it is in fact true. The p-value, or calculated probability, is the probability of finding the observed results when the null hypothesis (H_0) is true.

> **KEY INFORMATION**
>
> The test statistic, X, is a statistic calculated from sample data which is used to reject or accept H_0.

> **KEY INFORMATION**
>
> The significance level, α, is a pre-chosen probability and the p-value is a probability that you calculate after a given study.

Stop and think	What is the probability of success? Why is this a binomial distribution?

As the test statistic X has a binomial distribution with the probability of success being 0.25 and a sample size of 60, you can represent this using formal notation, $X \sim B(60, 0.25)$.

If you rolled 21 1s, then $X = 21$.

As you have rolled 21 1s, you would want to know the probability of a result at least as extreme this. This means you want to find out the probability of rolling 21 or more.

This is $1 - P(X \leqslant 20) = 1 - 0.946 = 0.0541$

This value is bigger than the significance level as $0.0541 > 0.05$, so you would accept H_0 – there is not enough evidence to support the die being biased in favour of 1 at this significance level.

Suppose that you rolled 22 1s instead. The probability of a result at least as extreme as $X = 22$ is $P(X \geqslant 22)$.

Under H_0 this is $1 - P(X \leqslant 21) = 1 - 0.9702 = 0.0298$

> **TECHNOLOGY**
>
> Some mathematics courses have binomial cumulative probability tables in a formulae booklet, but the sample size may not be what you require. You can use your calculator to work out the probability for any sample size. Open the binomial cumulative probability distribution mode and input $n = 60$, $x \leqslant 20$ and $p = 0.25$.

This value is less than the significance level as $0.0298 < 0.05$, so you would reject H_0 – there is enough evidence to support the die being biased in favour of 1 at this significance level.

The graph opposite shows the probability function for the test statistic. The blue bars form the distribution of the 1-tail test statistic, where the values of the test statistic would lead you to reject H_0 in favour of H_1. Notice that the blue bars are all at the upper end, due to H_1: $p > 0.25$.

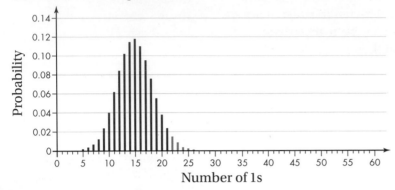

Suppose you suspected a different die of being biased against 1. Your alternative hypothesis would now be H_1: $p < 0.25$. The blue bars would have all been at the lower end, as in this graph:

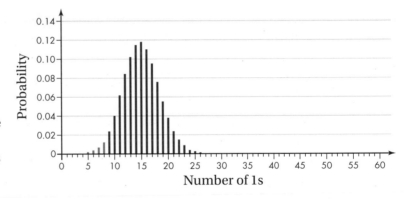

If you suspected your tetrahedral die of being biased, without knowing which number it was biased towards or against, you would need to carry out a 2-tail test. In this case, you are not saying whether the die lands more or less on one side than the other, only that it is biased. The alternative hypothesis would be H_1: $p \neq 0.25$.

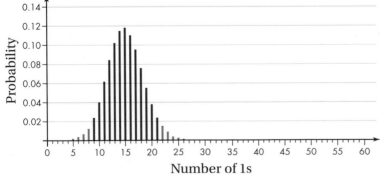

The outcomes would now be at either the high or the low end of the distribution. The significance level is the total probability of results. So, for a 2-tail test you need to divide by 2 and use half the significance level at each end of the distribution.

KEY INFORMATION

For a 2-tail test the probability is split between the two ends. If the significance was 5%, you would split this up, with 2.5% in the lower end and 2.5% in the upper end. This can be written as $\frac{\alpha}{2}$.

Finding the critical region

Finding the **critical region** is a different method of carrying out a hypothesis test. You work out in advance all the values of the test statistic that would lead you to reject H_0. This allows you to quickly say whether any observed value of the test statistic, X, is significant.

A 1-tail test has a single critical region. The critical region is the area where you would reject the null hypothesis and accept the alternative hypothesis.

Returning to the tetrahedral die, you had the following hypotheses when the die was biased towards 1 with a significance level, α, of 5%:

$H_0 : p = 0.25$

$H_1 : p > 0.25$

$\alpha = 0.05$

$P(X \geqslant 21) = 1 - 0.9459 = 0.0541 > \alpha$

$P(X \geqslant 22) = 1 - 0.9702 = 0.0298 < \alpha$

So the critical region is $P(X \geqslant 22)$.

When $X = 21$ you can't reject H_0 as it is not in the critical region. When $X = 22$, you can reject H_0 as it is in the critical region, so it is significant.

$X = 22$ is called the **critical value** as it is on the starting point of the critical region.

For the tetrahedral die, you had the following hypotheses when the die was biased against 1 with a significance level, α, of 5%:

$H_0 : p - 0.25$

$H_1 : p < 0.25$

$\alpha = 0.05$

$P(X \leqslant 9) = 0.04517 < \alpha$

$P(X \leqslant 10) = 0.08589 > \alpha$

So the critical region is $P(X \leqslant 9)$.

When $X = 10$, you can't reject H_0 as it is not in the critical region. When $X = 9$, you can reject H_0 as it is in the critical region, so it is significant.

In a 2-tail test, you are looking for either an increase or a decrease, and the critical region has two parts. Remember that the significance level is shared between the upper and lower ends of the distribution.

> **KEY INFORMATION**
> The critical region is the region where H_0 can be rejected.

> Everything after $P(X \geqslant 22)$ was below 5% and therefore significant.

> **KEY INFORMATION**
> The critical region is the set of all values of the test statistic that would cause you to reject H_0.

For the tetrahedral die, you had the following hypotheses when the die was biased with a significance level, α, of 5%:

$P(X \geqslant 22) = 1 - 0.9702 = 0.0298 > \alpha$

$P(X \geqslant 23) = 1 - 0.9846 = 0.0154 < \alpha$

So the critical region is $P(X \geqslant 23)$.

$P(X \leqslant 8) = 0.0212 < \alpha$

$P(X \leqslant 9) = 0.0452 > \alpha$

So the second critical region is $P(X \leqslant 8)$.

The two critical regions are $P(X \geqslant 23)$ and $P(X \leqslant 8)$.

You need to know how to find the cumulative probabilities on your calculator. Some courses have formulae booklets with tables of binomial cumulative probabilities in them. Using the tables is quick and easy – however, the tables do not cover all values of n, which is why you also need to know how to find the cumulative probabilities on your calculator.

TECHNOLOGY

Use your calculator to work out values.

Example 3

A biologist is investigating the gender of new-born kittens and predicts that there will be fewer females than males. She investigates the next 20 kittens born. Find the critical region, at the 5% significance level, where the null hypothesis should be rejected.

Solution

$H_0 : p = 0.5$

$H_1 : p < 0.5$

$n = 20$

5% significance level

$\alpha = 0.05$

$P(X \leqslant 6) = 0.0577 > \alpha$

$P(X \leqslant 5) = 0.0207 < \alpha$

So the critical region is $P(X \leqslant 5)$.

Using the tables, identify $p = 0.5$ and $n = 20$.

You are looking for 'less than' this time, so everything that has a probability of less than 5% will be significant.

$p =$	0.05	0.10	0.15	0.20	0.25	0.30	0.35	0.40	0.45	0.50
$n = 20, x = 0$	0.3585	0.1216	0.0388	0.0115	0.0032	0.0008	0.0002	0.0000	0.0000	0.0000
1	0.7358	0.3917	0.1756	0.0692	0.0243	0.0076	0.0021	0.0005	0.0001	0.0000
2	0.9245	0.6769	0.4049	0.2061	0.0913	0.0355	0.0121	0.0036	0.0009	0.0002
3	0.9841	0.8670	0.6477	0.4114	0.2252	0.1071	0.0444	0.0160	0.0049	0.0013
4	0.9974	0.9568	0.8298	0.6296	0.4148	0.2375	0.1182	0.0510	0.0189	0.0059
5	0.9997	0.9887	0.9327	0.8042	0.6172	0.4164	0.2454	0.1256	0.0553	0.0207
6	1.0000	0.9976	0.9781	0.9133	0.7858	0.6080	0.4166	0.2500	0.1299	0.0577
7	1.0000	0.9996	0.9941	0.9679	0.8982	0.7723	0.6010	0.4159	0.2520	0.1316
8	1.0000	0.9999	0.9987	0.9900	0.9591	0.8867	0.7624	0.5956	0.4143	0.2517
9	1.0000	1.0000	0.9998	0.9974	0.9861	0.9520	0.8782	0.7553	0.5914	0.4119
10	1.0000	1.0000	1.0000	0.9994	0.9961	0.9829	0.9468	0.8725	0.7507	0.5881
11	1.0000	1.0000	1.0000	0.9999	0.9991	0.9949	0.9804	0.9435	0.8692	0.7483
12	1.0000	1.0000	1.0000	1.0000	0.9998	0.9987	0.9940	0.9790	0.9420	0.8684
13	1.0000	1.0000	1.0000	1.0000	1.0000	0.9997	0.9985	0.9935	0.9786	0.9423
14	1.0000	1.0000	1.0000	1.0000	1.0000	1.0000	0.9997	0.9984	0.9936	0.9793
15	1.0000	1.0000	1.0000	1.0000	1.0000	1.0000	1.0000	0.9997	0.9985	0.9941
16	1.0000	1.0000	1.0000	1.0000	1.0000	1.0000	1.0000	1.0000	0.9997	0.9987
17	1.0000	1.0000	1.0000	1.0000	1.0000	1.0000	1.0000	1.0000	1.0000	0.9998
18	1.0000	1.0000	1.0000	1.0000	1.0000	1.0000	1.0000	1.0000	1.0000	1.0000

Identify the area where the probability is below 0.05, and where it is above. Using a calculator, go to the binomial cumulative probability function and select the sample size as 20, the probability of success as 0.5 and try different values of x systematically until you see the change at 5%. Ensure that you write down each stage so you can see the change.

The area below $p = 0.05$ is the critical region and $x = 5$ is the critical value as this is the first value which swaps to being significant.

At the 5% significance level the biologist should accept H_1 if the number of female kittens born is 5 or less.

Example 4

A ten-sided spinner is spun. Aleena thinks the spinner is not fair towards the square numbers. She spins the spinner a further 50 times. Find the critical region for this hypothesis test. The square numbers on the spinner are 1, 4 and 9. If the spinner is fair, the probability of it landing on one of these faces will be 0.3. This is the null hypothesis.

Solution

$H_0 : p = 0.3$

$H_1 : p \neq 0.3$

$n = 50$

5% significance level

$\alpha = 0.05$

Remember that you need to divide the significance by 2 as you are looking at the two ends of the distribution, so you need to look for a change at 2.5%. Ensure that your calculator is in binomial cumulative probability mode.

Set $n = 50$

Set $p = 0.3$

Now try different x values systematically until you see a change at the significance level.

$$P(X \geqslant 22) = 1 - 0.9749 = 0.0251 > \frac{\alpha}{2}$$

$$P(X \geqslant 23) = 1 - 0.9877 = 0.0123 < \frac{\alpha}{2}$$

So the critical region is $P(X \geqslant 23)$.

$$P(X \leqslant 9) = 0.0402 > \frac{\alpha}{2}$$

$$P(X \leqslant 8) = 0.0183 < \frac{\alpha}{2}$$

So the critical region is $P(X \leqslant 8)$.

Combined, the critical regions are $P(X \geqslant 23)$ and $P(X \leqslant 8)$.

Aleena should be concerned and accept H_1 when the spinner lands on fewer than 8, or more than 22, square numbers in 50 spins.

Example 5

Heather has 40 pens in her pencil case. She thinks that the probability of picking a pen that works has improved since she started using a new brand of stationery. In the past the probability of one working has been 40%. Find the critical region for this hypothesis test at the 2% significance level.

A variety of different significance levels may be specified and used.

Solution

$H_0 : p = 0.4$

$H_1 : p > 0.4$

$n = 40$

2% significance level

$\alpha = 0.02$

Ensure that your calculator is in binomial cumulative probability mode.

Set $n = 40$

Set $p = 0.4$

Now try different x values systematically until you see a change at the significance level.

This time, you are looking for 'greater than', so everything that has a probability of over 98% will be significant.

$$P(X \geqslant 22) = 1 - 0.9608 = 0.0392 > \alpha$$
$$P(X \geqslant 23) = 1 - 0.9811 = 0.0189 < \alpha$$

So the critical region is $P(X \geqslant 23)$.

If Heather picks 22 or more working pens out of the 40 pens in her pencil case, this is significant and within the critical region so she would accept H_1.

Example 6

Past records show that 15% of customers who buy chocolate bars from express supermarkets buy them as single bars rather than in multiple packs. During a particular day a random sample was taken of 20 customers who bought chocolate. Two of them bought single bars.

Use the data to test, at the 5% level of significance, whether or not the percentage of customers who bought a single chocolate bar that day was lower than usual. State your hypotheses clearly.

Solution

The type of distribution '15% of customers who buy chocolate bars from express supermarkets buy them as single bars rather than multiple packs' is binomial so $X \sim B(n, p)$ where $p = 0.15$

The value of n is indicated by the phrase 'random sample of 20 customers' so $n = 20$.

The test statistic is 'less than or equal to 2'.

'15% of customers' so $H_0 : p = 0.15$.

The alternative hypothesis is 'was lower than usual' so $H_1 : p \leqslant 0.15$, this is 1-tail

The significance level is '5% level of significance' so $\alpha = 0.05$

The predicted probability (check critical region) is 'a random sample of 20 customers who had bought chocolates was taken and two of them had bought single bars' so $P(X \leqslant 2)$.

If $P(X \leqslant 2) < \alpha = 0.05$ you would reject H_0, accept H_1 and comment for example, 'At the 5% significance level, there is significant evidence that this is lower than usual'.

If $P(X \leqslant 2) \geqslant \alpha = 0.05$ you would reject H_1, accept H_0 and comment for example, 'At the 5% significance level, there is not significant evidence that this is lower than usual'.

The final calculation would look like this:

X = number buying single bars of chocolate

$X \sim B(20, 0.15)$

$H_0 : p = 0.15$

$H_1 : p \leqslant 0.15$, this is 1-tail

$P(X \leqslant 2) = 0.4049 > 5\%$, so not significant.

At the 5% significance level, there is not significant evidence that this is lower than usual, so accept H_0.

TECHNOLOGY

Select 'distribution' on your calculator and select the mode for binomial cumulative probability.

On your calculator input the number in your sample (often referred to as N) as 20. Input the probability (often referred to as p) as 0.15. Input the sample size (often referred to as x) as less than or equal to 2.

Example 7

A teacher thinks that 35% of the students in a college complete their homework regularly. He selects 20 students from Year 13 at random.

Calculate, at the 5% level of significance, what the critical region would be for students completing homework. State your hypotheses clearly.

Solution

The type of distribution 'thinks that 35% of the students in a school complete homework regularly' is binomial so $X \sim B(n, p)$ where $p = 0.35$.

To find the sample size use '20 students from Year 13 at random' so $n = 20$.

'35% of the students in a school complete homework regularly' so $H_0 : p = 0.35$.

'what the critical region would be for students completing homework' so $H_1 : p \neq 0.35$; this is 2-tail.

The significance level is '5% level of significance' so $\alpha = 0.05$.

It is a 2-tail test, so look at 2.5% at each end of the data.

The final calculation would look like this:

$X =$ number completing homework

$X \sim B(20, 0.35)$

$H_0 : p = 0.35$

$H_1 : p \neq 0.35$, this is 2-tail

$P(X \leq 2) = 0.0121 < 2.5\%$, so significant

$P(X \leq 3) = 0.0444 > 2.5\%$, so not significant

$P(X \geq 11) = 1 - 0.9468 = 0.0532 > 2.5\%$, so not significant

$P(X \geq 12) = 1 - 0.9804 = 0.0196 < 2.5\%$, so significant

At 5% significance level, the critical region would be $P(X \leq 2)$ and $P(X \geq 12)$ for completing homework.

Exercise 14.2B

Answers page 554

Find the critical region for each of the following hypothesis tests:

1 $H_0 : p = 0.8$

$H_1 : p \neq 0.8$

$n = 40$

5% significance level

2 $H_0 : p = 0.2$

$H_1 : p > 0.2$

$n = 10$

5% significance level

3 $H_0 : p = 0.15$

$H_1 : p < 0.15$

$n = 20$

10% significance level

4 $H_0 : p = 0.45$

$H_1 : p < 0.45$

$n = 20$

5% significance level

5 What is incorrect in this solution? Redo it using the correct method.

$H_0 : p = 0.10$

$H_1 : p \neq 0.10$

$n = 50$

10% significance level

Using a calculator in binomial probability mode, select $n = 50$ and $p = 0.1$.

Select $x = 7$, then select $x = 8$.

$P(X \geqslant 7) = 1 - 0.8779 = 0.1221 > \alpha$

$P(X \geqslant 8) = 1 - 0.9421 = 0.0579 < \alpha$

So the critical region is $P(X \geqslant 8)$.

Using a calculator in binomial probability mode, select $n = 50$ and $p = 0.1$.

Select $x = 2$, then select $x = 1$.

$P(X \leqslant 2) = 0.1117 > \alpha$

$P(X \leqslant 1) = 0.0338 < \alpha$

So the critical region is $P(X \leqslant 2)$.

Combined, the critical regions are $P(X \geqslant 8)$ and $P(X \leqslant 2)$.

 6 What is incorrect in this solution? Redo it using the correct method.

$H_0 : p = 0.35$

$H_1 : p > 0.35$

$n = 30$

1% significance level

$P(X \leqslant 5) = 0.0233 > \alpha$

$P(X \leqslant 4) = 0.0075 < \alpha$

So the critical region is $P(X \leqslant 4)$.

 7 What is incorrect in this solution? Redo it using the correct method.

$H_0 : p = 0.25$

$H_1 : p \neq 0.25$

$n = 20$

10% significance level

Using a calculator in binomial probability mode, select $n = 20$ and $p = 0.25$.

$P(X \geqslant 7) = 1 - 0.8982 = 0.1018 > \alpha$

$P(X \geqslant 8) = 1 - 0.9591 = 0.0409 < \alpha$

So the critical region is $P(X \geqslant 8)$.

$$P(X \leq 3) = 0.2252 > \alpha$$

$$P(X \leq 2) = 0.0913 < \alpha$$

So the critical region is $P(X \leq 2)$.

Combined, the critical regions are $P(X \geq 8)$ and $P(X \leq 2)$.

8 If $X \sim B(10, 0.35)$ find

   a $P(X = 2)$

   b $P(X = 4)$

   c $P(X < 5)$

   d $P(X > 4)$

9 The probability of a student being late for a lesson is 0.1. For a period covering 10 lessons, find the probability that he is late

   a for exactly 3 lessons

   b for fewer than 3 lessons.

10 In a laboratory producing genetically modified wheat, the probability of a seed not germinating is 0.125. A batch of 200 seeds are tested. Find the probability that:

   a 20 do not germinate

   b more than 30 do not germinate

   c between 20 and 30 do not germinate.

11 A calculator is used to generate random digits, 0 to 9 inclusive. Each digit is equally likely.

   a What is the probability that the first digit generated is greater than 5?

   b If ten digits are generated, find the probability there will be fewer than four digits greater than 5.

CM 12 The probability that patients have to wait more than 15 minutes in a health practice is 0.3. One of the doctors claims that there has been a fall in the number of patients who have to wait more than 15 minutes. He records the waiting times for the next 20 patients and three of them wait more than 15 minutes.

Is there evidence at the 5% level to support the doctor's claim?

CM 13 The probability that James wins at chess against his computer is 0.4. He says he has improved his game. In the next 8 games he wins 6.

Set up hypotheses to test James's claim at the 5% significance level.

CM 14 A company has mixed sweets delivered with 70% red and the rest blue. The company wishes to test whether there has been a change in the ratio of blue to red sweets. To do this, they take a sample of 20 sweets of which 10 are red. Does this indicate a change at the 10% significance level?

Do a hypothesis test to find out whether or not there has been a change.

SUMMARY OF KEY POINTS

> In a random sample every member of the population has an equal chance of selection.

> In a stratified sample the size of each stratum is in proportion to its size in the population.

> Opportunity samples are convenient and can be taken from any groups from the population at any time.

> Quota samples take a certain number of the population in any order.

> Systematic samples use a repetitive system to obtain the sample from the population.

> A hypothesis test is a mathematical procedure to examine a value of a population parameter proposed by the null hypothesis, H_0, compared to the alternative hypothesis, H_1.

> The null hypothesis, H_0, is the 'working hypothesis' (i.e. what you assume to be truc for the purpose of the test). The alternative hypothesis, H_1, is what you conclude if you reject the null hypothesis: it also determines whether you use a 1-tail or a 2-tail test.

> In order to check whether your results are significant, calculate their probability using H_0, then choose whether to reject or accept the null hypothesis. If your result lies in the 'acceptable' region then you accept the null hypothesis, H_0.

> The region outside this 'acceptable' region is called the critical region. If your statistic lies in the critical region you reject the null hypothesis, H_0, and your conclusion will be based on the alternative hypothesis, H_1.

> The significance level is the probability of rejecting H_0 when it is in fact true. If the significance level is 5% then the 'acceptable' region is 95% of all possible outcomes and the critical region is the remaining 5% of all possible outcomes.

EXAM-STYLE QUESTIONS 14

Answers page 555

 1 For each of the following descriptions of methods of sampling, state whether it is a definition of method A, B, C or D.

a A sample which has no order to it, but is quick and easy to gather.

    A Quota   **B** Systematic   **C** Stratified   **D** Random   **[1 mark]**

b Every member of the population has the same chance of being selected.

    A Quota   **B** Systematic   **C** Stratified   **D** Random   **[1 mark]**

c A proportional representation of each group is selected.

    A Quota   **B** Systematic   **C** Stratified   **D** Random   **[1 mark]**

2 A school contains 1200 students. The number of students in each year group is shown below:

Year	7	8	9	10	11
Number of students	228	264	240	252	216

A stratified sample of 200 students is to be taken. How many students from each year group should be selected? **[2 marks]**

CM **3** In a local driving test centre the probability of passing your driving test first time is 54%. Stan claims that his driving school is better than this. He records the first-time pass rate for 50 randomly selected learners he teaches. 31 of these pass.

Test Stan's claim at the 5% level of significance. State your hypotheses clearly. **[7 marks]**

CM **4** A manager of a gym knows that 40% of clients will buy a protein shake from a vending machine in reception. After a refurbishment, a random sample of 30 clients were surveyed. Only 11 had bought a protein shake from the machine.

a Using a 5% level of significance, test whether or not there has been a change in the proportion of clients buying protein shakes from the vending machine. State your hypotheses clearly. **[6 marks]**

During the refurbishment, a selection of fruit is added to the vending machine. Before the refurbishment only 35% of clients bought fruit from the café. The manager claims that the proportion of clients buying fruit has now increased. The manager selects a sample of 100 clients and finds that 18 of them have bought fruit.

b Using a suitable approximation, test the manager's claim. Use a 10% significance level and state your hypotheses clearly. **[10 marks]**

CM **5** Explain briefly what is meant by:

a the critical region of a test statistic **[1 mark]**

b the level of significance of a hypothesis test. **[1 mark]**

CM **6** At Stan's driving school the learners take a multiple-choice theory test. The test has 20 questions each with 5 possible answers, of which only one is correct. One of Stan's learners gets 6 questions correct. Stan claims she was guessing the answers.

Using a 2-tail test, at the 5% level of significance, test whether or not there is evidence to reject Stan's claim. State your hypotheses clearly. **[7 marks]**

7 A food packaging company has discovered a fault that may affect the quality of one of their products. The company wishes to know how widespread the problem might be.

a What would the population be? **[1 mark]**

b Suggest a possible sampling frame. **[2 marks]**

c Describe how you would take a sample. **[2 marks]**

PS **8**

CM A class of 20 Year 12 students studying mathematics wish to test if a die is biased towards 6. They each roll the die once and record whether or not it lands on a 6. Six students record a 6. Use their results to carry out a hypothesis test at the 10% significance level, stating the null and alternate hypotheses clearly. **[7 marks]**

9 A company working to improve young children's spellings advertises that on average 90% of children will improve. Suppose that 25 children are chosen at random.

a Find the probability that 24 children improve if the company's claim is correct. **[3 marks]**

b Find the probability that 24 children improve if the company's claim is incorrect, and only 75% improve. **[3 marks]**

A school does not think the company is making any difference, and thinks the improvement rate is below 90%. The head teacher decided to carry out a hypothesis test at a 5% significance level.

c Write down the null and alternate hypothesis, explaining why the alternate hypothesis takes the form it does. **[3 marks]**

d Find the critical region for the test, explaining your reasons. **[3 marks]**

15 KINEMATICS

On 20th July 1969, Neil Armstrong uttered the immortal words, 'That's one small step for man, one giant leap for mankind' as he stepped onto the Moon. It was the culmination of years of space exploration – fruit flies had been sent into space in 1947, then the first satellite, Sputnik 1, in 1957, and the first man, Yuri Gagarin, in 1961. Space exploration continued with further missions to the outer reaches of the Solar System and the setting up of manned space stations such as Mir and the International Space Station. Now the focus is turning towards establishing permanent human settlements to populate other planets, starting with Mars. None of this would have been possible without an understanding of kinematics.

Kinematics is concerned with motion. The equations of constant acceleration (often referred to as the SUVAT equations) describe the relationship between displacement, velocity, constant acceleration and time. When an object undergoes a succession of constant accelerations, it is helpful to illustrate the motion in a graph. When the acceleration is not constant but is dependent on time, calculus can be used to model the motion.

The gravitational acceleration at the Earth's surface is approximately $9.8\,\mathrm{m\,s^{-2}}$, but on the surface of the Moon it is only a sixth of that. As a result, astronauts can jump higher on the Moon than on the Earth. An understanding of kinematics was essential in developing the technology and techniques required for space exploration.

LEARNING OBJECTIVES

You will learn how to:

› use the SI units of displacement, velocity and acceleration

› derive and use the equations of constant acceleration, including for situations with gravity

› draw and interpret a displacement–time graph and a velocity–time graph

› use calculus to find displacement, velocity and acceleration when they are functions of time.

TOPIC LINKS

This chapter is concerned with particles in motion. The theory of vector quantities from **Chapter 10 Vectors** is applied to the equations of constant acceleration, which are then used in **Chapter 16 Forces** where forces produce accelerations. The topics of differentiation and integration from **Chapter 8 Differentiation** and **Chapter 9 Integration** are then applied to situations where displacement, velocity and acceleration are functions of time.

PRIOR KNOWLEDGE

You should already know how to:

- substitute values and rearrange formulae
- use equations of constant acceleration
- plot straight line graphs and curved graphs
- find the gradient of a straight line
- find the area of shapes (e.g. triangles and trapezia)
- solve linear and quadratic equations
- solve cubic equations using the factor theorem
- find the gradient by differentiation
- find the area under a curve by integration.

You should be able to complete the following questions correctly:

1 One of the equations of constant acceleration is $v^2 = u^2 + 2as$.
   a Find v when $u = 40$, $a = 18$ and $s = 49$.
   b Find s when $v = 148$, $u = 48$ and $a = 9.8$.

2 Find the area of a trapezium with parallel sides of 19 cm and 34 cm and a perpendicular height of 22 cm.

3 a Solve the equation $2(T + 2) = \frac{1}{2}(3T + 14)$.
   b Solve the equation $2T^2 = 5T + 12$.

4 A function is given by $f(x) = 6x^2 - 7x + 8$.
   a Differentiate $f(x)$ with respect to x.
   b Integrate $f(x)$ with respect to x.

15.1 The language of kinematics

Consider a car waiting at traffic lights. When the lights turn green, the initial speed of the car is zero. As the driver pushes down on the accelerator pedal, the car starts to move. The faster the **acceleration**, the faster the speed increases and the further the car travels in the same amount of time. Here, **distance, speed**, acceleration and time are all interrelated.

These quantities are connected by the equations of constant acceleration (often referred to as SUVAT). The letters of SUVAT represent displacement (s), initial velocity (u), final velocity (v), acceleration (a) and time (t). These equations are described in greater depth in **Section 15.2**. **Displacement, velocity** and acceleration are **vector** quantities.

Note that distance and speed are scalar quantities whereas displacement and velocity are vector quantities. Acceleration is a vector quantity but the same word is also used to describe the scalar quantity. In **Book 2**, motion will occur in two dimensions, so it is important to understand how to find the magnitude and direction of a vector. However, in **Book 1** motion is only ever in one dimension (along a straight line), so all that matters is that the sign of the displacement, velocity or acceleration (positive or negative) is correct.

SI units

SI units are an internationally recognised system of units from which all other units can be derived. The SI unit for distance is the metre and for time it is the second. The units for all the quantities in the table below can be described using just these two units: metres (m) and seconds (s).

Term	Definition	SI unit
Position	The position of an object is where it is compared with the origin, O.	The SI unit for distance travelled, displacement and position is the metre (m).
Displacement	The displacement of an object is how far it is from its original position. For example, if an object starts at 2 metres from O and finishes at −5 metres from O, then its displacement is −7 metres. If the motion is along a horizontal line, you usually define left to right as the positive direction.	
Distance travelled	The distance travelled is how far the object has moved. For example, if an object travels 9 metres from O in the negative direction, then 5 metres back, then the displacement is −4 metres but the distance travelled is 14 metres.	
Velocity	The velocity is how quickly an object is travelling in a certain direction. For example, if an object travels a displacement of −10 metres in 5 seconds then its velocity is −2 metres per second (ms^{-1}).	The SI unit for both speed and velocity is the metre per second (ms^{-1}) since they are the rate of change of distance and displacement respectively, as you met in **Chapter 8 Differentiation**.
Speed	Speed is how quickly an object is travelling but the direction does not matter. Hence the speed of the object above is 2 metres per second (ms^{-1}) even though the object is travelling in a negative direction. Note that two objects with velocities of $5ms^{-1}$ and $-5ms^{-1}$ have the same speed but are travelling in opposite directions.	The unit ms^{-1} means the same as m/s (metres per second). The index -1 shows that you are dividing by time.
Acceleration	Acceleration describes how quickly the speed (or velocity) is changing. For example, if an object increases in speed from $4ms^{-1}$ to $10ms^{-1}$ in 2 seconds, then its acceleration is 3 metres per second per second (ms^{-2}). Note that an object with a positive velocity but a negative acceleration will come to instantaneous rest before travelling in the opposite direction. When the velocity and acceleration have different signs, the object will be **decelerating**.	The SI unit for acceleration is the metre per second per second (ms^{-2}) since it is the rate of change of speed or velocity, as you met in **Chapter 8 Differentiation**. The unit ms^{-2} means the same as m/s² (metres per second squared). This represents the rate of change of velocity in ms^{-1} each second.

Example 1

A girl walks along a straight line. She walks 35 m in the positive direction, then turns round and walks 56 m in the negative direction.

a Find the total distance travelled by the girl.

b Find the girl's displacement at the end of her walk from her starting point.

Solution

Draw a diagram.

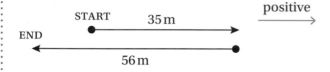

a The girl has walked 35 m in one direction and then 56 m in the other.

The total distance is given by $35 + 56 = 91$ m.

> Why does the direction not matter when working out the total distance?

b The girl has walked 35 m in the positive direction and then 56 m in the negative direction.

The overall displacement is given by $35 - 56 = -21$ m.

> If you travel further in the negative direction than the positive direction, why will your overall displacement be negative?

Exercise 15.1A

Answers page 557

(M) 1 A particle travels 30 m in the negative direction, then 20 m in the positive direction.

Find the overall distance travelled and the displacement from its original position.

(M) 2 A ball is launched upwards from the top of a 15 m tall building to a maximum height of 40 m above the ground, before falling to the ground. Taking upwards as the positive direction, find the overall distance travelled by the ball and the displacement of the ball from its original position.

(M) 3 A boy walks 32 m west, then 24 m south.

a Find the total distance travelled by the boy.

b Find the displacement of the boy from his starting point.

c If it took the boy 28 seconds, what was his speed?

(M) 4 A snail crawls around the perimeter of a rectangle with sides of 7 m and 4 m.

a Find the total distance travelled by the snail.

b Find the overall displacement of the snail at the end of his journey from his starting point.

15.2 Equations of constant acceleration

Consider a car travelling at a constant acceleration. It starts with an initial velocity of $u\,\mathrm{m\,s^{-1}}$ and after t s it has a final velocity of $v\,\mathrm{m\,s^{-1}}$. In the velocity–time graph below you can see that this is represented as a straight line with a constant gradient. u is the initial velocity, v is the final velocity and t is the time taken. Its constant acceleration is given by $a\,\mathrm{m\,s^{-2}}$ and it has travelled s m.

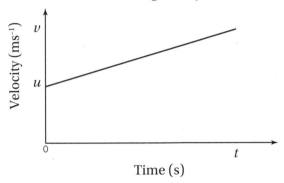

Time (s)

Note that of the five quantities (s, u, v, a and t), time is the only variable that is *scalar*, not a vector. Whereas the other quantities can take negative values, and motion can take place forwards or backwards, time can only take positive values.

These five quantities (displacement (s), the initial velocity (u), the final velocity (v), the constant acceleration (a) and the time (t)) are related by the five equations of constant acceleration. Each equation can be derived from the definition of acceleration (the rate of change of velocity). •

Because the variables are s, u, v, a and t, the equations of constant acceleration are often referred to as SUVAT.

From the graph, the acceleration is given by

$$a = \frac{v - u}{t}$$

This can be rearranged to make v the subject.

$$at = v - u$$

$$v = u + at$$

Acceleration is the rate of change of velocity, the gradient of the graph.

Another equation can be derived by considering the average speed.

The average speed can be written in terms of the displacement (s) and the time (t).

$$\text{Average speed} = \frac{s}{t}$$

The average speed can also be written in terms of the initial velocity (u) and the final velocity (v).

$$\text{Average speed} = \frac{1}{2}(u + v)$$

Put these expressions equal to each other.

$$\frac{s}{t} = \frac{1}{2}(u + v)$$

Multiply by t.

$$s = \frac{1}{2}(u + v)t$$

By substituting for $v = u + at$ into $s = \frac{1}{2}(u + v)t$, you can derive a third equation.

$$s = \frac{1}{2}(u + v)t$$

Substitute for v.

$$s = \frac{1}{2}(u + u + at)t$$

Simplify.

$$s = \frac{1}{2}(2u + at)t$$

Expand.

$$s = \frac{1}{2} \times 2ut + \frac{1}{2} \times at^2$$

Simplify.

$$s = ut + \frac{1}{2}at^2$$

By making t the subject of $s = \frac{1}{2}(u + v)t$ and substituting for t in the equation $v = u + at$, you can derive the fourth equation.

$$s = \frac{1}{2}(u + v)t$$

Multiply by 2.

$$2s = (u + v)t$$

Divide by $(u + v)$.

$$t = \frac{2s}{u + v}$$

Substitute for t in the equation $v = u + at$.

$$v = u + \frac{2as}{u + v}$$

Subtract u.

$$v - u = \frac{2as}{u + v}$$

Multiply by $(u + v)$.

$$(v - u)(u + v) = 2as$$

Expand.

$$v^2 - u^2 = 2as$$

Add u^2.

$$v^2 = u^2 + 2as$$

A final equation can be derived by making u the subject of $v = u + at$ and substituting for u in the equation $v^2 = u^2 + 2as$.

$$v = u + at$$

Subtract at to make u the subject.

$$u = v - at$$

Substitute $u = v - at$ in the equation $v^2 = u^2 + 2as$.

$$v^2 = (v - at)^2 + 2as$$

Expand the brackets.

$$v^2 = v^2 - 2vat + a^2t^2 + 2as$$

Subtract v^2 from both sides.

$$0 = -2vat + a^2t^2 + 2as$$

Divide through by a.

$$0 = -2vt + at^2 + 2s$$

Divide through by 2.

$$0 = -vt + \frac{1}{2}at^2 + s$$

Make s the subject.

$$s = vt - \frac{1}{2}at^2$$

In summary, the equations are:

$$v = u + at \qquad \text{(no } s)$$

$$v^2 = u^2 + 2as \qquad \text{(no } t)$$

$$s = \frac{1}{2}(u + v)t \qquad \text{(no } a)$$

$$s = ut + \frac{1}{2}at^2 \qquad \text{(no } v)$$

$$s = vt - \frac{1}{2}at^2 \qquad \text{(no } u)$$

KEY INFORMATION

You should learn these five equations and make sure you can rearrange them.

Stop and think   How can you use SI units to show that the terms v and at have the same dimensions and that the terms v^2 and as have the same dimensions?

Example 2

A ball travels 143 m in 13 s at a constant acceleration. Given that its final velocity is $14 \, \text{m s}^{-1}$, find its initial velocity.

Solution

You are told that:

$$s = 143 \, \text{m}$$
$$t = 13 \, \text{s}$$
$$v = 14 \, \text{m s}^{-1}$$

u is unknown.

Write down the equation you are going to use.

$$s = \frac{1}{2}(u + v)t$$

Substitute for s, t and v.

$$143 = \frac{1}{2}(u + 14) \times 13$$

What information have you been given? Write it out using s, u, v, a and t.

What is the only variable you do not need for this question? How does that help you choose which equation to use?

Rearrange the equation to make u the subject.

Divide both sides by 13.

$$11 = \frac{1}{2}(u + 14)$$

Multiply both sides by 2.

$$22 = u + 14$$

Subtract 14.

$$u = 8 \, \text{m s}^{-1}$$

Remember to include units with your answer.

Example 3

A train travelling at an initial velocity of $130 \, \text{m s}^{-1}$ decelerates for half a minute with a constant deceleration. It travels $2.1 \, \text{km}$ during this time.

a Find the magnitude of the deceleration.

b Find the final velocity of the train.

Solution

a You are told that:

$$u = 130 \, \text{m s}^{-1}$$

$$t = 30 \, \text{s}$$

$$s = 2100 \, \text{m}$$

a is unknown.

Ensure you convert measurements into SI units before you start.

Write down the equation.

$$s = ut + \frac{1}{2}at^2$$

Substitute.

$$2100 = 130 \times 30 + \frac{1}{2} \times a \times 30^2$$

Choose the equation which has the letters u, t, s and a.

Simplify the terms.

$$2100 = 3900 + 450a$$

Make a the subject.

$$2100 - 3900 = 450a$$

$$-1800 = 450a$$

$$a = -4 \, \text{m s}^{-2}$$

b Since you know the values of u, t, s and a, you can now use any of the equations which contains a v to find v. The simplest is $v = u + at$.

$$v = u + at$$

$$= 130 - 4 \times 30$$

$$= 130 - 120$$

$$= 10 \, \text{m s}^{-1}$$

How can you check that your answer makes sense? Would you expect a negative answer for the acceleration? How do you know that your answer should be negative?

What type of equation would you have to solve if you were given the values of s, u and a and needed to find the value of t? Is there a way to avoid having to solve this type of equation? What might you need to remember if you were going to use a different method?

Exercise 15.2A

Answers page 557

1 **a** Find s given that $u = 70\,\mathrm{m\,s^{-1}}$, $a = -3\,\mathrm{m\,s^{-2}}$ and $t = 10\,\mathrm{s}$.

b Find s given that $u = 15\,\mathrm{m\,s^{-1}}$, $v = 29\,\mathrm{m\,s^{-1}}$ and $t = 9\,\mathrm{s}$.

c Find a given that $u = 3\,\mathrm{m\,s^{-1}}$, $v = 38\,\mathrm{m\,s^{-1}}$ and $t = 7\,\mathrm{s}$.

d Find s given that $u = 22\,\mathrm{m\,s^{-1}}$, $a = 6\,\mathrm{m\,s^{-2}}$ and $v = 28\,\mathrm{m\,s^{-1}}$.

e Find t given that $v = -28\,\mathrm{m\,s^{-1}}$, $a = -7\,\mathrm{m\,s^{-2}}$ and $s = 0\,\mathrm{m}$.

(M) 2 A particle has an initial velocity of $24\,\mathrm{m\,s^{-1}}$ and a constant acceleration of $5\,\mathrm{m\,s^{-2}}$.

a Find its velocity after it has travelled $10\,\mathrm{m}$.

b Find the time taken to travel $10\,\mathrm{m}$.

(M) 3 A particle has an initial velocity of $60\,\mathrm{m\,s^{-1}}$ and constant deceleration of $8\,\mathrm{m\,s^{-2}}$.

a Find the velocity of the particle after 10 seconds.

b Find the time at which the particle is at instantaneous rest.

(M) 4 From a standing start, a cyclist wins a sprint race in 40 seconds.

If she has a constant acceleration of $0.5\,\mathrm{m\,s^{-2}}$:

a how fast was she travelling when she crossed the finish line?

b what distance was the race?

(M) 5 A bullet is fired from a gun at $50\,\mathrm{m\,s^{-1}}$ and travels $1500\,\mathrm{m}$ before it stops.

(CM) a Find the deceleration of the bullet and the length of time it is in flight.

b What assumptions must you make to answer this question?

(PF) 6 **a** A particle at rest suddenly accelerates at $4\,\mathrm{m\,s^{-2}}$. How long does it take the particle to
(M) travel $128\,\mathrm{m}$?

b A particle travelling at $180\,\mathrm{m\,s^{-1}}$ starts to decelerate at a constant $3\,\mathrm{m\,s^{-2}}$. How many minutes does it take to return to its original position?

c A particle with an initial velocity of $5\,\mathrm{m\,s^{-1}}$ and a constant acceleration of $3\,\mathrm{m\,s^{-2}}$ takes T seconds to travel $84\,\mathrm{m}$. Show that $3T^2 + 10T - 168 = 0$ and hence find the value of T.

(M) 7 A coin is rolling down a hill with a constant acceleration of $0.25\,\mathrm{m\,s^{-2}}$. It travels $930\,\mathrm{m}$ in a minute. What was the initial speed of the coin and what was its speed after one minute?

(M) 8 A car accelerates uniformly from rest to $60\,\mathrm{km\,h^{-1}}$ in 8 seconds, then maintains a steady speed.

(PS) a How long does it take the car to travel $200\,\mathrm{m}$?

b How long does it take the car to travel one kilometre?

Example 4

A car travels along a straight line ABC with a constant acceleration. AB = 27 m and BC = 85 m. The car takes 3 seconds to get from A to B and another 5 seconds to get from B to C.

a Find the acceleration of the car.

b Find the speed of the car at A.

Solution

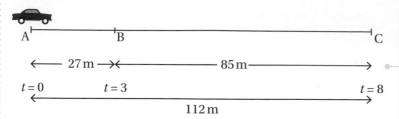

Drawing a diagram is often the best way to understand a situation.

a Write out what you know for AB and AC.

For AB: $s = 27$ m and $t = 3$ s

For AC: $s = 27 + 85 = 112$ m and $t = 3 + 5 = 8$ s

Use the equation which has s, t, u and a.

Why is it best to consider the journeys from A to B and from A to C rather than from B to C?

For AB: $s = ut + \dfrac{1}{2}at^2$

Substitute $s = 27$ and $t = 3$ into the equation.

$$27 = u \times 3 + \frac{1}{2} \times a \times 3^2$$

Simplify the equation.

$$27 = 3u + \frac{9}{2}a$$

Multiply by 2 so that a does not have a fractional coefficient.

$$54 = 6u + 9a$$

Divide through by common factor 3. Label this equation ①.

$$18 = 2u + 3a \qquad\qquad ①$$

For AC: $s = ut + \dfrac{1}{2}at^2$

Substitute $s = 112$ and $t = 8$ into the equation.

$$112 = u \times 8 + \frac{1}{2} \times a \times 8^2$$

Simplify.

$$112 = 8u + 32a$$

Divide though by common factor 8. Label this equation ②.

$$14 = u + 4a \qquad\qquad ②$$

Double equation ② so that the equations ① and ② have the same coefficient for u. Label this equation ③.

$$28 = 2u + 8a \qquad\qquad ③$$
$$18 = 2u + 3a \qquad\qquad ①$$

Since you only have two pieces of information (displacement and time), you need to use simultaneous equations.

Subtract equation ① from equation ③.

$$10 = 5a$$

$$a = 2\,\text{m\,s}^{-2}$$

b Substitute $a = 2$ into equation ① (but any equation will do).

$$18 = 2u + 3 \times 2$$

Solve the equation.

$$18 = 2u + 6$$

$$12 = 2u$$

$$u = 6\,\text{m\,s}^{-1}$$

Exercise 15.2B

Answers page 557

(M) 1 A particle moves with a constant acceleration along a straight line PQRS, passing Q five seconds after P and passing R fifteen seconds after Q. PQ is 100 m and QR is 720 m.

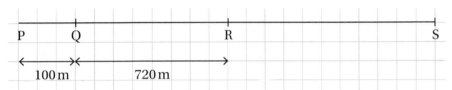

a Find the acceleration of the particle.

b Find the velocity of the particle at Q.

c Given that RS is 830 m, how long does it take the particle to travel from Q to S?

(M) 2 After a windy night, a road has a pile of debris blocking the way. A motorist driving towards the debris notices it when he is 80 m away, travelling at $20\,\text{m\,s}^{-1}$ and with a constant acceleration of $2\,\text{m\,s}^{-2}$. As soon as he engages the brakes, he will decelerate at a constant $4\,\text{m\,s}^{-2}$.

a If the motorist reacts immediately, how far will he stop from the debris?

b If he takes two seconds to react, at what speed will he be driving when he crashes into the debris?

(PS) (M) 3 Particles A and B are at rest at either end of a 100 m track. They set off instantaneously towards each other with constant accelerations of $6\,\text{m\,s}^{-2}$ and $2\,\text{m\,s}^{-2}$ respectively.

a Where will they meet?   **b** When will they meet?

c At what speed will each be travelling when they meet?

(M) 4 A ball is launched along a straight line with an initial velocity of $13\,\text{m\,s}^{-1}$ and a constant acceleration of $-2\,\text{m\,s}^{-2}$. At what times is the ball 30 m from its starting point?

(M) 5 J passes O at $2\,\text{m\,s}^{-1}$ and with a constant acceleration of $1\,\text{m\,s}^{-2}$. When J is 30 m from O, K sets off from rest from O with a constant acceleration of $0.8\,\text{m\,s}^{-2}$. When K reaches a velocity of $4\,\text{m\,s}^{-1}$, L passes O at $3\,\text{m\,s}^{-1}$ and with a constant acceleration of $0.5\,\text{m\,s}^{-2}$. If J, K and L are all travelling in the same direction along the same straight line, find the distance between J and K when L has a velocity of $7\,\text{m\,s}^{-1}$.

6 An object moves along the straight line EFGH with a constant acceleration. It takes the object seven seconds to travel between E and F, three seconds to travel the 66 m between F and G and four seconds to travel between G and H, where it comes to rest.

   a Find the deceleration of the object.

   b Find the velocity of the object at E.

7 A cyclist passes a garage, a restaurant and a library, in that order, along a straight road, maintaining a constant acceleration of $0.15\,\text{m s}^{-2}$ throughout and passing the library at $9\,\text{m s}^{-1}$. The library is 240 m from the restaurant. The cyclist takes twice as long to cycle from the restaurant to the library as she does to cycle from the garage to the restaurant. How far is the library from the garage?

15.3 Vertical motion

One specific example of a constant acceleration is the acceleration due to **gravity**, which pulls objects towards the centre of the Earth. At the surface of the Earth this is approximately $9.8\,\text{m s}^{-2}$. All objects are pulled towards the Earth at the same acceleration, regardless of their mass.

Because the force of gravitational attraction (gravity) always acts towards the centre of the Earth, you will always represent it in a diagram as acting vertically downwards. Because displacement, velocity and acceleration are vector quantities, you will need to choose which direction to take as the positive direction. You usually define the direction in which the object starts moving as positive, so if an object is launched vertically upwards then that will be the positive direction (and $a = -9.8\,\text{m s}^{-2}$), whereas if the object is dropped or thrown vertically downwards then downwards will be the positive direction (and $a = 9.8\,\text{m s}^{-2}$).

> Although the actual value of the acceleration due to gravity varies with the distance from the centre of the Earth, you can assume it is constant over small distances.

> **KEY INFORMATION**
>
> You are expected to use $9.8\,\text{m s}^{-2}$ for the value of g unless you are told otherwise.

Example 5

A ball is thrown vertically upwards at $30\,\text{m s}^{-1}$ from 6 m above the ground.

   a Find the maximum height reached by the ball.

   b Find the time taken for the ball to reach its maximum height.

   c Find the time taken for the ball to hit the ground.

> **KEY INFORMATION**
>
> If an object is thrown *upwards*, it will slow down because gravity is acting downwards. It will slow down and eventually come to a stop (instantaneous rest), and then it will fall to the ground, getting faster and faster.

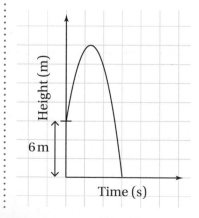

Solution

a You wish to find the maximum height, which is the distance from the ground, so you need to find the value of s. You are given u, and you know that $v = 0\,\text{m}\,\text{s}^{-1}$ and $a = -9.8\,\text{m}\,\text{s}^{-2}$, so you choose the equation which has s, u, v and a. That equation is $v^2 = u^2 + 2as$.

$$v^2 = u^2 + 2as$$

Substitute the values into the equation.

$$0 = 30^2 + 2 \times -9.8 \times s$$

Simplify and solve for s.

$$0 = 900 - 19.6s$$
$$19.6s = 900$$
$$s = 45.9\,\text{m}$$

Since the ball was thrown from 6 m above the ground, you need to add that to your answer, so the maximum height is 51.9 m.

b For the time taken, the simplest equation to use is $v = u + at$.

$$v = u + at$$

Substitute and solve for t.

$$0 = 30 - 9.8t$$
$$9.8t = 30$$
$$t = 3.06\,\text{s}$$

c For the time taken for the ball to hit the ground, the displacement is −6 m. You know the values of s, u and a, so use $s = ut + \frac{1}{2}at^2$.

Write down the equation.

$$s = ut + \frac{1}{2}at^2$$

Substitute.

$$-6 = 30t + \frac{1}{2} \times -9.8 \times t^2$$

Simplify.

$$-6 = 30t - 4.9t^2$$

Because this is a quadratic equation, rearrange the terms so that they are in the form $at^2 + bt + c = 0$.

$$4.9t^2 - 30t - 6 = 0$$

Solve the equation using the quadratic formula.

$$t = \frac{30 \pm \sqrt{(-30)^2 - 4 \times 4.9 \times (-6)}}{2 \times 4.9}$$

$$= 6.32\,\text{s or} -0.194\,\text{s}$$

Because the time must be positive, it takes the ball 6.32 s to hit the ground.

Start by choosing which direction is positive. Since the motion starts upwards, let that be the positive direction. Hence acceleration (gravity) is acting in the negative direction, and for all parts of this question, $u = 30\,\text{m}\,\text{s}^{-1}$ and $a = -9.8\,\text{m}\,\text{s}^{-2}$.

KEY INFORMATION

At the maximum height, $v = 0\,\text{m}\,\text{s}^{-1}$.

It is important to make sure you are using the correct sign (negative or positive) for each value. Check the direction of each vector carefully.

Alternative method: Use $v^2 = u^2 + 2as$ to find v, then use $v = u + at$ to find t. Remember that the ball will be travelling in the opposite direction to the direction it was thrown.

Example 6

A ball is dropped from the top of a building.

It takes the ball 2.1 s to reach the ground.

a How tall is the building?

Janet's office is two-thirds of the way up the building.

b With what velocity did the ball pass Janet's office?

> Since the ball is dropped, what do you know about the initial velocity?

Solution

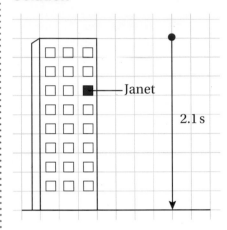

Janet

2.1 s

a You are told that $t = 2.1$ s. To find the height of the building you need to find the value of s. The equation which has u, a, t and s is $s = ut + \frac{1}{2}at^2$.

$$s = ut + \frac{1}{2}at^2$$

Substitute.

$$s = 0 \times 2.1 + \frac{1}{2} \times 9.8 \times 2.1^2$$
$$= 21.6\,\text{m}$$

b Janet's office is two-thirds of the way up the building, so the ball will have travelled one-third of its total distance.

$$s = \frac{1}{3} \times 21.6 = 7.2\,\text{m}$$

The equation you need is the one with s, u, a and v, which is $v^2 = u^2 + 2as$.

$$v^2 = u^2 + 2as$$

Substitute into the equation.

$$v^2 = 0^2 + 2 \times 9.8 \times 7.2$$
$$= 141.12$$
$$v = \sqrt{141.12} = 11.9\,\text{m s}^{-1}$$

> Start by choosing which direction is positive. Since the motion starts downwards, let that be the positive direction. Hence acceleration (gravity) is acting in the positive direction, and for all parts of this question, $u = 0\,\text{m s}^{-1}$ and $a = 9.8\,\text{m s}^{-2}$.

Stop and think If a ball is dropped, how will this simplify the equation $s = ut + \frac{1}{2}at^2$? What will be the relationship between displacement and the time when the ball is dropped? If it takes twice as long to hit the ground, how much taller will the building be?

Exercise 15.3A **Answers page 557**

(M) **1** A pebble is dropped down a well. It takes 1.8 seconds until the pebble hits the water.
Find the distance travelled by the pebble before it hits the water.

(M) **2** A plate falls from a table. The table is 1.2 m high.
Find the time taken for the plate to hit the ground.

(M) **3** A ball is thrown vertically downwards at $7\,\mathrm{m\,s^{-1}}$ and strikes the ground at $15\,\mathrm{m\,s^{-1}}$.
From how far above the ground was the ball thrown?

(M) **4** A stone is dropped from the summit of a 48 m cliff.
Find, correct to 3 s.f.:

  a the speed of the stone when it hits the ground.

  b the time taken until the stone hits the ground.

(M) **5** A rocket is launched vertically upwards from a cliff at a speed of $39.2\,\mathrm{m\,s^{-1}}$.
It hits the ground at the foot of the cliff 10 seconds later.

  a Find the height of the cliff.

  b Find the maximum height reached by the rocket.

  c Find the time taken to reach the maximum height.

  d Find the speed at which the rocket hits the ground.

  e Find the time taken for the rocket to return to the same height as it was launched.

(M) **6** A ball of mass 0.8 kg is launched vertically upwards into the air from the ground at $25\,\mathrm{m\,s^{-1}}$.

(CM) **a** Find the time taken for the ball to return to the ground.

  b Find the maximum height the ball reaches above the ground.

  c Find the time taken for the ball to first reach three-quarters of its maximum height.

  d Is the maximum height an underestimate or an overestimate? Explain your answer.

  e How would the answers change if the ball had a mass of 1.6 kg instead?

(PS) **7** A ball is projected vertically upwards from an 11.4 m high balcony. It takes the ball T
(M) seconds to reach its maximum height and 4.3 seconds to land on the ground. Find the
value of T.

(M) **8** A golf ball is thrown vertically upwards from the ground with a velocity of $42\,\mathrm{m\,s^{-1}}$.
(PS) How long is the golf ball more than 87.5 m above the ground?

Example 7

A ball is dropped from the top of a tower of height Hm. If the
ball had been launched vertically upwards at $16\,\mathrm{m\,s^{-1}}$, it would
have taken 2 seconds longer to reach the ground.

Find the value of H.

KEY INFORMATION

If two objects are launched
in different ways, look for
what their journeys have in
common.

Solution

Let the positive direction be upwards, and let the time taken for the dropped ball to reach the ground be T.

For the dropped ball, $u = 0 \, \text{m s}^{-1}$, $s = -H \, \text{m}$, $a = -9.8 \, \text{m s}^{-2}$, $t = T \, \text{s}$.

Use $s = ut + \frac{1}{2}at^2$.

Substitute.

$$-H = 0 \times T + \frac{1}{2} \times -9.8 \times T^2$$

Simplify.

$$-H = -4.9T^2 \qquad \text{①}$$

For the launched ball, $u = 16 \, \text{m s}^{-1}$, $s = -H \, \text{m}$, $a = -9.8 \, \text{m s}^{-2}$, $t = (T + 2) \, \text{s}$.

Use $s = ut + \frac{1}{2}at^2$.

Substitute.

$$-H = 16 \times (T + 2) + \frac{1}{2} \times -9.8 \times (T + 2)^2$$

Simplify.

$$-H = 16(T + 2) - 4.9(T^2 + 4T + 4)$$

$$= 16T + 32 - 4.9T^2 - 19.6T - 19.6$$

$$= -4.9T^2 - 3.6T + 12.4 \qquad \text{②}$$

Put ① = ②.

$$-4.9T^2 = -4.9T^2 - 3.6T + 12.4$$

Subtract $-4.9T^2$ from both sides.

$$0 = -3.6T + 12.4$$

Solve the equation.

$$3.6T = 12.4$$

$$T = \frac{12.4}{3.6} = 3.44 \, \text{s}$$

Since $-H = -4.9T^2$, you can find H by substituting $T = 3.44$.

$$-H = -4.9T^2$$

Divide both sides by -1.

$$H = 4.9T^2$$

$$= 4.9 \times 3.44^2$$

$$= 58.1$$

Hence, the tower is 58.1 m tall.

> The ball will move in different directions depending upon the situation, so there is no clear choice as to which direction to take as positive.

Exercise 15.3B

Answers page 558

  1 **a** An object is dropped from above the ground. If h is the initial height and t is the time until the object hits the ground, prove that $t = \sqrt{\dfrac{2h}{g}}$.

b If the same object had been dropped on another planet from a point 12 times as high and the gravitational acceleration had been 3 times as great, how many times as long would it take the object to hit the ground?

 2 A stone is dropped from the top of a castle. One second later another stone is thrown downwards from the same place at $14\,\mathrm{m\,s^{-1}}$.

Given that both stones hit the moat at the same time, find the height of the castle.

 3 Two students, Sawda and Katie, were given this question:

'A ball is launched vertically upwards at $15\,\mathrm{m\,s^{-1}}$ from $20\,\mathrm{m}$ above the ground. Find the velocity with which the ball hits the ground.'

Sawda's solution:

$$s = ut + \frac{1}{2}at^2$$

$$-20 = 15t - 4.9t^2$$

$$4.9t^2 - 15t - 20 = 0$$

$$t = \frac{15 \pm \sqrt{(-15)^2 - 4 \times 4.9 \times (-20)}}{2 \times 4.9} = \frac{15 \pm \sqrt{617}}{9.8} = 4.065 \text{ or } -1.004$$

Since t is positive, $t = 4.065$ seconds

$$v = u + at = 15 - 9.8 \times 4.065 = -24.8\,\mathrm{m\,s^{-1}}$$

Katie's solution:

$$v^2 = u^2 + 2as$$

$$= 15^2 + 2 \times (-9.8) \times (-20)$$

$$= 617$$

$$v = 24.8\,\mathrm{m\,s^{-1}}$$

Whose solution is correct? Evaluate the two students' approaches and solutions.

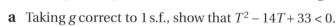 **4** An object is launched vertically upwards from the ground at $70\,\mathrm{m\,s^{-1}}$. T seconds after launching, the object is more than $165\,\mathrm{m}$ above the ground.

a Taking g correct to 1 s.f., show that $T^2 - 14T + 33 < 0$.

b Hence find the length of time for which the object is more than $165\,\mathrm{m}$ above the ground.

 5 A weight is pulled vertically upwards from rest by a rope from the top of a table at $0.8\,\mathrm{m\,s^{-2}}$. After 4 seconds, the rope snaps, at which point the weight travels only under the influence of gravity.

How long after the rope snaps does the weight return to the table?

6 A stone is dropped from the top of a 37 m tall building. When it is halfway down the building, another stone is thrown vertically downwards.

At what speed must the second stone be thrown for both stones to hit the ground at the same time?

 7 Holly can only catch a ball travelling at $5\,\text{m}\,\text{s}^{-1}$ or slower. She is standing at the top of a building, her hands 40 m above the ground. Ruby throws a ball vertically upwards from the ground at $U\,\text{m}\,\text{s}^{-1}$.

a What are the minimum and maximum values of U if Holly is to catch the ball on her first attempt?

b If Holly misses the ball on her first attempt, what is the maximum time before she gets a second chance?

  8 A ball is dropped from the top of a tower of height H m. If the ball had been launched vertically upwards at $U\,\text{m}\,\text{s}^{-1}$, it would have taken X seconds longer to reach the ground.

Show that H is given by $H = 4.9X^2\left(\dfrac{U - 4.9X}{9.8X - U}\right)^2$.

15.4 Displacement–time and velocity–time graphs

You saw in **Section 15.1** that velocity is the rate of change of displacement and acceleration is the rate of change of velocity. In **Chapter 8 Differentiation** and **Chapter 9 Integration** you saw how differentiation and integration could be used to find the gradient and area under a curve.

Differentiating any function gives a rate of change (whether it is with respect to distance x or time t, or any other variable). Hence there is a connection between displacement–time and velocity–time graphs and the results found by calculus.

For a displacement–time graph, the gradient is the velocity (because velocity is the rate of change of displacement with respect to time).

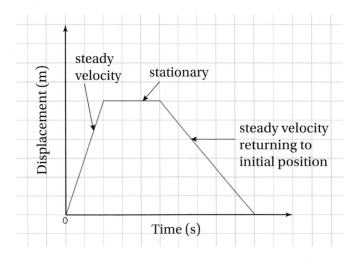

KEY INFORMATION

For a displacement–time graph, the gradient is the velocity.

For a velocity–time graph, the gradient is the acceleration (because acceleration is the rate of change of velocity with respect to time) and the area under the graph is the displacement (because by the fundamental theorem of calculus, differentiation and integration are inverse processes).

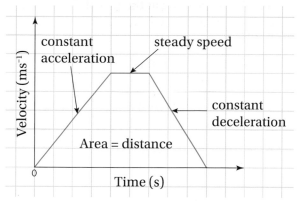

KEY INFORMATION

For a velocity–time graph, the gradient is the acceleration and the area under the graph is the displacement.

In this section, you will apply these concepts to graphs constructed from straight lines, where the constant acceleration SUVAT equations apply. Then, in **Section 15.5**, you will use calculus to solve problems where displacement, velocity and acceleration are functions of time.

Example 8

A boy is on his way to school. He leaves home at 7:40 a.m. and walks 0.5 km at 5 km h^{-1} to the bus stop. The bus arrives at the stop 9 minutes after he does, and travels the 12 km journey to school at an average speed of 20 km h^{-1}. School starts at 8:30 a.m.

a Draw a displacement–time graph for this journey.

b Does the boy arrive on time?

Solution

a The boy walks at 5 km h^{-1}. 5 kilometres per hour is the same as 1 kilometre every 12 minutes, so it will take the boy 6 minutes to walk 0.5 km.

The boy leaves home at 7:40 a.m. and arrives at the bus stop 6 minutes later at 7:46 a.m.

The bus arrives 9 minutes later at 7:55 a.m.

The bus travels at 20 km h^{-1}. 20 kilometres per hour is the same as 1 kilometre every 3 minutes, so it will take the bus 36 minutes to travel 12 kilometres.

The bus arrives 36 minutes after 7:55 a.m. at 8:31 a.m.

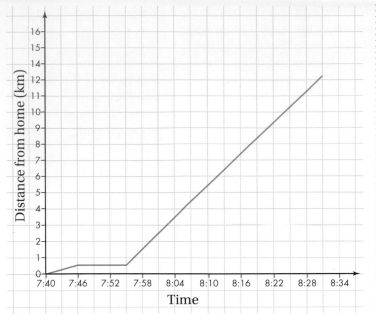

b Since school starts at 8:30 a.m., the boy arrives 1 minute late.

Exercise 15.4A

Answers page 558

(M) 1 A particle is moving in a straight line. It starts by moving at a velocity of $4\,\text{m}\,\text{s}^{-1}$ for 10 s. It then remains stationary for 15 s. The particle then moves with a velocity of $-3\,\text{m}\,\text{s}^{-1}$ until its displacement is 25 m from its starting point. After it remains stationary for another 20 s, the particle returns to its starting point at a velocity of $-2.5\,\text{m}\,\text{s}^{-1}$.

Complete the sketch of the displacement–time graph to show the displacement and time at each part of the particle's journey.

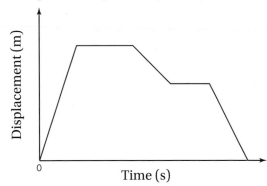

(M) 2 A coach sets off from Brighton at 11:20 a.m. and travels at $45\,\text{km}\,\text{h}^{-1}$ for 40 minutes. It stays at its destination for an hour, then returns to Brighton at $50\,\text{km}\,\text{h}^{-1}$.

a Sketch a displacement–time graph to show the motion of the coach.

b At what time does the coach return to Brighton?

3 A particle moves at a steady velocity of $3\,\text{m}\,\text{s}^{-1}$ for 12 seconds followed immediately by a steady velocity of $-2\,\text{m}\,\text{s}^{-1}$ for 15 seconds.

    a Sketch a displacement–time graph to show the displacement of the particle from its initial position.

A second particle starts moving from the same initial position three seconds after the first particle starts to move. This particle moves at a steady velocity of $1.6\,\text{m}\,\text{s}^{-1}$.

    b Sketch the motion of the second particle on the same displacement–time graph.

    c At what time and displacement do the particles meet?

 4 Two cars, A and B, are travelling in opposite directions along a straight road between two towns 42 km apart. Both cars set off at 4:45 p.m. Car A travels at $75\,\text{km}\,\text{h}^{-1}$ for 20 minutes, stops for 10 minutes and then completes its journey at $51\,\text{km}\,\text{h}^{-1}$. Car B travels at $72\,\text{km}\,\text{h}^{-1}$ until 4:55 p.m., stops for 20 minutes and then completes its journey at $72\,\text{km}\,\text{h}^{-1}$.

    a Sketch a displacement–time graph to show the motion of the cars.

    b At what time are the cars in the same place?

5 Two particles, P and Q, start moving along the same straight line at the same time. P starts at O and moves at $1.25\,\text{m}\,\text{s}^{-2}$ for T seconds until it is D m from O. Q starts at a displacement of D m from O and moves at $-0.5\,\text{m}\,\text{s}^{-2}$ for T seconds until it is a displacement of 30 m from O.

    a Sketch a displacement–time graph to show the motion of the particles.

    b Find the value of T.

    c Find the value of D.

This velocity–time graph shows the motion of a car.

There are three stages to the journey:

› during the first stage, the car gets faster

› during the second stage, the car travels at a steady speed

› during the third and final stage, the car decelerates to rest.

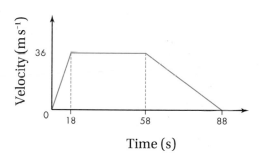

You are now going to see how finding the gradient and area under the graph give the same results for the acceleration and distance travelled as using the SUVAT equations.

For the first stage of the journey, the car travels at a constant acceleration from rest to a velocity of $36\,\text{m}\,\text{s}^{-1}$ in 18 seconds.

Using $v = u + at$, you can find the acceleration.

$$a = \frac{v-u}{t} = \frac{36-0}{18} = 2\,\text{m}\,\text{s}^{-2}$$

Note that this is the same as the gradient of the line segment joining (0, 0) and (18, 36).

Using $s = \frac{1}{2}(u + v)t$, you can find the displacement.

$$s = \frac{1}{2}(u + v)t = \left(\frac{36+0}{2}\right) \times 18$$

$$= \frac{1}{2} \times 18 \times 36$$

$$= 324\,\text{m}$$

Note that this is the same as the area of the triangle with a base of 18 and a height of 36.

For the second stage of the journey, the car travels at a steady speed of $36\,\mathrm{m\,s^{-1}}$ for 40 seconds. Note that both the acceleration and the gradient are zero.

Given that $a = 0$, you can rewrite $s = ut + \frac{1}{2}at^2$ as $s = ut$.

$$s = 36 \times 40$$
$$= 1440\,\mathrm{m}$$

$s = ut$ is equivalent to saying that for a constant speed

distance = speed × time

Note that this is the same as the area of the rectangle with a base of 40 and a height of 36.

For the final stage of the journey, the car travels at a constant deceleration from $36\,\mathrm{m\,s^{-1}}$ to rest in 30 seconds.

Using $v = u + at$, you can find the deceleration.

$$a = \frac{v - u}{t} = \frac{0 - 36}{30} = -1.2\,\mathrm{m\,s^{-2}}$$

Note that this is the same as the gradient of the line segment joining $(58, 36)$ and $(88, 0)$.

Using $s = \frac{1}{2}(u + v)t$, you can find the displacement.

$$s = \frac{1}{2}(u + v)t = \left(\frac{0 + 36}{2}\right) \times 30$$
$$= \frac{1}{2} \times 30 \times 36$$
$$= 540\,\mathrm{m}$$

Note that this is the same as the area of the triangle with a base of 30 and a height of 36.

Stop and think   How can you find the total distance from the whole graph without splitting it into two triangles and a rectangle?

Example 9

A car accelerates from rest at $1.6\,\mathrm{m\,s^{-2}}$ for 25 seconds. It then maintains a steady speed for 42 seconds before decelerating to rest in 20 seconds.

a Draw a velocity–time graph for this journey.

b Find the deceleration of the car as it comes to rest.

c Find the total distance travelled.

Solution

a There are three stages to this journey.

For the first stage, you are told that the car starts at rest (so $u = 0\,\mathrm{m\,s^{-1}}$) and accelerates at $1.6\,\mathrm{m\,s^{-2}}$ (so $a = 1.6\,\mathrm{m\,s^{-2}}$) for 25 seconds (so $t = 25\,\mathrm{s}$).

Substitute into $v = u + at$.

$$v = u + at = 0 + 1.6 \times 25 = 40\,\mathrm{m\,s^{-1}}$$

For the second stage, the car travels at a steady $40\,\mathrm{m\,s^{-1}}$ for 42 seconds, from 25 seconds to 67 seconds.

For the third stage, the car decelerates to rest in 20 s, from 67 s to 87 s.

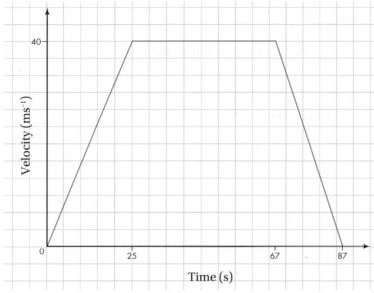

b For the third stage, the car has $u = 40\,\mathrm{m\,s^{-1}}$, $v = 0\,\mathrm{m\,s^{-1}}$ and $t = 20\,\mathrm{s}$, from which:

$$a = \frac{v - u}{t} = \frac{0 - 40}{20} = -2\,\mathrm{m\,s^{-2}}$$

c The total distance travelled is given by the area under the graph. In this example, the graph is in the shape of a trapezium.

The total journey time is given by $25 + 42 + 20 = 87\,\mathrm{s}$.

The area is given by Area $= \frac{1}{2}(a + b)h$

$$= \frac{1}{2} \times (87 + 42) \times 40 = 2580$$

Total distance travelled $= 2580\,\mathrm{m}$.

Example 10

A bus departs from a bus stop with a constant acceleration for 15 seconds until it reaches a speed of $20\,\mathrm{m\,s^{-1}}$. It then travels at $20\,\mathrm{m\,s^{-1}}$ until it is X seconds from its destination, at which point it decelerates at $0.8\,\mathrm{m\,s^{-2}}$, coming to rest at the next bus stop. The bus stops are $1100\,\mathrm{m}$ apart and the total journey between the bus stops takes T seconds.

a Draw a velocity–time graph.

b Calculate the value of X.

c Calculate the value of T.

Solution

a There are three stages to this journey.

For the first part, the bus accelerates from $0\,\text{m s}^{-1}$ to $20\,\text{m s}^{-1}$ in 15 seconds.

For the second part, the bus travels at $20\,\text{m s}^{-1}$ until it is $(T - X)$ seconds from its destination.

For the third part, the bus decelerates to rest in X seconds.

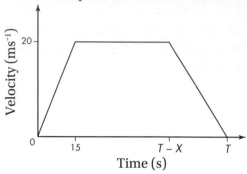

b When the bus is decelerating you have $a = -0.8\,\text{m s}^{-2}$, $u = 20\,\text{m s}^{-1}$, $v = 0\,\text{m s}^{-1}$ and $t = X\,\text{s}$.

$$v = u + at$$
$$t = \frac{v - u}{a}$$
$$X = \frac{0 - 20}{-0.8}$$
$$= 25\,\text{s}$$

c The total distance travelled is the area under the graph.

The second part of the journey takes $[(T - X) - 15]\,\text{s}$, which is $(T - 40)\,\text{s}$, since $X = 25$.

$$\text{Area} = \frac{1}{2}(a + b)h$$
$$1100 = \frac{1}{2}(T + T - 40) \times 20$$
$$= 10(2T - 40)$$
$$110 = 2T - 40$$
$$150 = 2T$$
$$T = 75\,\text{s}$$

> The distance $1100\,\text{m}$ is the area under the graph.
>
> The parallel sides of the trapezium are T and $(T - 40)$.

Exercise 15.4B

Answers page 559

(M) 1 The velocity–time graph shows the journey of a cyclist. Initially the cyclist was travelling at a steady speed of $12\,\text{m s}^{-1}$ for 10 seconds. He then decelerated for 20 seconds until he was travelling at $5\,\text{m s}^{-1}$, which he then maintained for the final 10 seconds.

a Find the cyclist's deceleration.

b Find the total distance travelled by the cyclist.

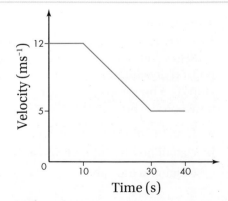

(M) 2 A car accelerates from rest to $10\,\mathrm{m\,s^{-1}}$ in $6\,\mathrm{s}$, then maintains this speed for $14\,\mathrm{s}$ before accelerating at $3.75\,\mathrm{m\,s^{-2}}$ to reach a speed of $25\,\mathrm{m\,s^{-1}}$. It maintains this speed for $500\,\mathrm{m}$, then decelerates to rest at $2.5\,\mathrm{m\,s^{-2}}$.

   a Find the total time taken.

   b Draw a velocity–time graph to represent this journey.

   c Find the total distance travelled.

   d State the initial acceleration.

(PS) (M) 3 A car accelerates from rest for $30\,\mathrm{s}$ at a constant acceleration, stays at a steady speed of $V\,\mathrm{m\,s^{-1}}$ for $X\,\mathrm{s}$, then decelerates at a constant acceleration, coming to a standstill $100\,\mathrm{s}$ after it set off. It travels a quarter of the total distance during the first $30\,\mathrm{s}$.

   a Sketch a velocity–time graph to show the motion of the car.

   b Find the value of X.

   c Given that the car decelerates at $0.9\,\mathrm{m\,s^{-2}}$, find the value of V.

(PS) (M) 4 A motorcyclist passes a parked police car at a steady $18\,\mathrm{m\,s^{-1}}$. Ten seconds later the police car gives chase, accelerating to $24\,\mathrm{m\,s^{-1}}$ in $10\,\mathrm{s}$, then pursuing the motorcyclist at this steady speed.

   a Represent this information in a velocity–time diagram, starting from the time the motorcycle passes the police car.

   b How long after the motorcyclist passed the police car did it take for the police car to catch up?

   c How far did the police car travel in pursuit?

(PS) (M) 5 Two athletes compete in a race. Both start from rest. The first athlete initially runs with a constant acceleration of $0.8\,\mathrm{m\,s^{-2}}$ for $10\,\mathrm{s}$, then stays at a constant speed until the end of the race. She takes a total of time of $105\,\mathrm{s}$ to complete the race. The second athlete initially runs with a constant acceleration of $0.225\,\mathrm{m\,s^{-2}}$ for $40\,\mathrm{s}$, then reduces to a different constant acceleration which she maintains until the end of the race. The second athlete wins the race, finishing $3\,\mathrm{s}$ sooner than her opponent.

   a Sketch a velocity–time graph to show the motion of the two athletes.

   b Find the length of the race (in metres).

   c Find the speed of the winning athlete as she finishes the race.

(PF) (M) 6 For a journey of total time T which involves a constant acceleration from rest for T_1 seconds to $V\,\mathrm{m\,s^{-1}}$, a steady speed of $V\,\mathrm{m\,s^{-1}}$ for T_2 seconds and a final constant deceleration to rest for T_3 seconds, show that the distance given by the equations of constant acceleration can be written as the area of a trapezium with parallel sides of T and T_2 and a perpendicular height of V.

(PS) (M) 7 Two vehicles start and finish at the same time and the same place. Both vehicles start from and finish at rest. Vehicle A accelerates at $1.5\,\mathrm{m\,s^{-2}}$ for $20\,\mathrm{s}$, then maintains a steady speed for $1\,\mathrm{min}\,10\,\mathrm{s}$ before decelerating. Vehicle B accelerates at $0.8\,\mathrm{m\,s^{-2}}$ for $50\,\mathrm{s}$, then maintains a steady speed for $20\,\mathrm{s}$ before decelerating.

   a Sketch a velocity–time graph to illustrate this situation.

   b Find the total time taken.

   c Find the total distance travelled by each vehicle.

   d Find the deceleration of each vehicle.

8 A skydiver leapt from a plane. She spent the first 45 seconds in free fall, then immediately deployed her parachute, descending to the ground at a steady speed of $8\,\mathrm{m\,s^{-1}}$ for 3 minutes.

    a Draw a velocity–time graph for her descent.

    b Draw a distance–time graph for her descent.

    c Draw an acceleration–time graph for her descent.

    d Calculate the total descent of the skydive in metres given by this model.

    e Will the actual descent be longer or shorter than your answer to **part d**? Explain your answer.

9 A van travels at a constant velocity, $V\,\mathrm{m\,s^{-1}}$, for three-quarters of a minute, then decelerates at $1.6\,\mathrm{m\,s^{-2}}$ to rest in T seconds. The van travels a total distance of 2.3 km.

    a Sketch a velocity–time graph to show the motion of the van.

    b Prove that $T^2 + 90T - 2875 = 0$.

    c Find the value of V.

15.5 Variable acceleration

As the name suggests, the equations of constant acceleration only apply when the acceleration is constant. When the acceleration is a function of time, then you will need to use calculus instead.

Differentiation

Velocity is the rate of change of displacement. If you have an expression for the displacement in terms of time (t), **differentiate** it to find an expression for the velocity. Similarly, acceleration is the rate of change of velocity. If you have an expression for the velocity in terms of time (t), differentiate it to find an expression for the acceleration.

Integration

Because differentiation and **integration** are inverse processes, you can also find the velocity and displacement if you know the acceleration. If you integrate an expression for the acceleration you will get an expression for the velocity and if you integrate an expression for the velocity you will get an expression for the displacement.

However, you need to remember to include a **constant of integration** (c), which you can then calculate by substituting values. For example, if $v = t^2 - 4t + c$ and you know that the velocity is $3\,\mathrm{m\,s^{-1}}$ after 2 seconds, then:

$$v = t^2 - 4t + c$$

$$3 = 2^2 - 4 \times 2 + c$$

$$3 = 4 - 8 + c$$

$$c = 7$$

In summary:

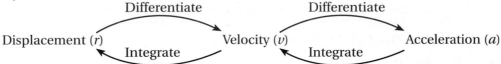

Differentiate   Differentiate

Displacement (r)   Velocity (v)   Acceleration (a)

Integrate   Integrate

Example 11

After t seconds, the displacement of a particle is given by $r = t^3 - 27t^2 + 246t + 16$, where r is in metres.

Find the acceleration of the particle at each time that the velocity is $15\,\mathrm{m\,s^{-1}}$.

> **KEY INFORMATION**
>
> r is used instead of s for displacement when it is a function of time.

Solution

Find an expression for v by differentiating the expression for r with respect to t.

$$v = \frac{\mathrm{d}r}{\mathrm{d}t} = 3t^2 - 54t + 246$$

Find an expression for a by differentiating the expression for v with respect to t.

$$a = \frac{\mathrm{d}v}{\mathrm{d}t} = 6t - 54$$

> **KEY INFORMATION**
>
> Differentiate r to get v and differentiate v to get a.

Find the times at which the velocity is $15\,\mathrm{m\,s^{-1}}$ by putting the expression for v equal to 15.

$$3t^2 - 54t + 246 = 15$$

Since this is a quadratic equation, subtract 15 from each side so you can write it in the form $at^2 + bt + c - 0$.

$$3t^2 - 54t + 231 = 0$$

Simplify the equation by dividing through by the common factor of 3.

$$t^2 - 18t + 77 = 0$$

Factorise and solve.

$$(t - 7)(t - 11) = 0$$

$t = 7$ seconds or 11 seconds.

Find the acceleration at each of the times by substituting for t.

When $t = 7$: $a = 6 \times 7 - 54 = -12\,\mathrm{m\,s^{-2}}$.

When $t = 11$: $a = 6 \times 11 - 54 = 12\,\mathrm{m\,s^{-2}}$.

Exercise 15.5A

Answers page 560

1   The displacement, in metres, of a particle is given by $r = 13t + 20 - 4t^2$, where t is the time in seconds.

     a   Find the velocity of the particle when $t = 2\,\mathrm{s}$.

     b   Show that the acceleration is constant.

(M) **2** After t seconds, a particle has a displacement of $r = (5t^3 - 7t^2)$ metres.

    a Find expressions for the velocity and acceleration in terms of t.

    b Find the velocity after 2 seconds.

    c Find the acceleration after 3 seconds.

(M) **3** At time ts, a particle is $(t^3 - 8t^2 + 5t + 10)$ m from the origin.

    a At what times is the particle instantaneously at rest?

    b At what times is the velocity positive?

    c At what times is the acceleration positive?

    d Find the velocity and displacement when the particle is accelerating at $2\,\mathrm{m\,s^{-2}}$.

(M) (PS) **4** The velocity, in $\mathrm{m\,s^{-1}}$, of a particle is given by $v = 18t - t^2$.

    Find the acceleration at both instants when the particle has a velocity of $65\,\mathrm{m\,s^{-1}}$.

(M) (PS) **5** A particle has a displacement of $r = (t^3 - 14t^2 + 48t)$ metres, where t is the time in seconds.

    a At what times does the particle return to its original position?

    b Find the displacement of the particle when it is travelling at its minimum velocity.

(M) (PF) **6** The displacement of a particle after t seconds is given by $r = 8\sqrt{t} - \dfrac{64}{t}$.

    a Show that the particle is at the origin when $t = 4$.

    b Find the velocity and acceleration of the particle when $t = 4$.

(M) **7** A particle has a displacement, in cm, of $r = (t^4 - 8t^3 + 22t^2 - 20t + q)$ from O after t seconds.

    After 4 seconds, the particle is 33 cm from O.

    a Find the value of q.

    b Find the displacement of the particle each time the particle is moving at $4\,\mathrm{cm\,s^{-1}}$.

    c Find the displacement of the particle from O when the particle is accelerating at $188\,\mathrm{cm\,s^{-2}}$.

Example 12

After t seconds, the acceleration of a particle is given by
$a = (36 - 12t)\,\mathrm{m\,s^{-2}}$.

After 5 seconds, the displacement of the particle is 11 m and the velocity is $2\,\mathrm{m\,s^{-1}}$.

Find the displacement of the particle when it is travelling at its maximum velocity.

Solution

Find an expression for v by integrating the expression for a with respect to t.

$$v = \int a\,\mathrm{d}t = \int (36 - 12t)\,\mathrm{d}t = 36t - 6t^2 + c$$

Substitute $t = 5$ and $v = 2$ to find the value of the constant of integration.

$$2 = 36 \times 5 - 6 \times 5^2 + c$$

$$2 = 180 - 150 + c$$

$$c = -28$$

Hence, $v = 36t - 6t^2 - 28$

Find an expression for r by integrating the expression for v with respect to t.

$$r = \int v \, dt = \int (36t - 6t^2 - 28) \, dt = 18t^2 - 2t^3 - 28t + c_2$$

Substitute $t = 5$ and $r = 11$ to find the value of the new constant of integration.

$$11 = 18 \times 5^2 - 2 \times 5^3 - 28 \times 5 + c_2$$

$$11 = 450 - 250 - 140 + c_2$$

$$c_2 = -49$$

Hence, $r = 18t^2 - 2t^3 - 28t - 49$

At maximum velocity, $a = 0 \, \text{m s}^{-2}$. Solve the expression for a equal to zero.

$$36 - 12t = 0$$

$$t = 3 \, \text{s}$$

Substitute $t = 3$ into the expression for r.

$$r = 18 \times 3^2 - 2 \times 3^3 - 28 \times 3 - 49 = -5 \, \text{m}$$

> **KEY INFORMATION**
>
> Integrate a to get v and integrate v to get r. Remember to include a constant of integration.

> **KEY INFORMATION**
>
> Use a different constant of integration when integrating for a second time.

Example 13

A particle P moves in a straight line such that after t seconds, its velocity, $v \, \text{m s}^{-1}$, is given by

$$v = \begin{cases} \dfrac{2}{3}(t^2 + 2) & 0 \leqslant t \leqslant 4 \\[2mm] 2t^{\frac{3}{2}} + \dfrac{1}{2}t & t > 4 \end{cases}$$

Initially the particle is displaced 3 m from O.

a Find the acceleration of the particle when:

  i $t = 2$ **ii** $t = 9$

b Find the total distance travelled during the first 9 seconds.

Solution

a **i** For $0 \leqslant t \leqslant 4$, differentiate $v = \dfrac{2}{3}(t^2 + 2)$.

$$a = \frac{dv}{dt} = \frac{2}{3}(2t) = \frac{4}{3}t$$

Substitute $t = 2$.

$$a = \frac{4}{3} \times 2 = \frac{8}{3} \,\mathrm{m\,s^{-2}}$$

ii For $t > 4$, differentiate $v = 2t^{\frac{3}{2}} + \frac{1}{2}t$.

$$a = \frac{dv}{dt} = 2 \times \frac{3}{2}t^{\frac{1}{2}} + \frac{1}{2} = 3t^{\frac{1}{2}} + \frac{1}{2}$$

Substitute $t = 9$ into this expression.

$$a = 3 \times 9^{\frac{1}{2}} + \frac{1}{2} = 3 \times 3 + \frac{1}{2} = 9\frac{1}{2} \,\mathrm{m\,s^{-2}}$$

b For $0 \leqslant t \leqslant 4$, integrate $v = \frac{2}{3}(t^2 + 2)$.

$$r = \int v \, dt = \frac{2}{3} \int (t^2 + 2) \, dt = \frac{2}{9}t^3 + \frac{4}{3}t + c$$

Substitute $t = 0$ and $r = 3$.

$$3 = \frac{2}{9} \times 0^3 + \frac{4}{3} \times 0 + c$$

$$3 = c$$

Hence, $r = \frac{2}{9}t^3 + \frac{4}{3}t + 3$

Substitute $t = 4$ into this expression.

> Find r when $t = 4$ so you can use it to find the constant of integration for the second part of the travel.

$$r = \frac{2}{9} \times 4^3 + \frac{4}{3} \times 4 + 3 = \frac{203}{9}$$

For $t > 4$, integrate $v = 2t^{\frac{3}{2}} + \frac{1}{2}t$.

> When the velocity is given as two parts, make sure you use the correct part.

$$r = \int v \, dt = \int \left(2t^{\frac{3}{2}} + \frac{1}{2}t\right) dt = \frac{4}{5}t^{\frac{5}{2}} + \frac{1}{4}t^2 + c_2$$

Substitute $t = 4$ and $r = \frac{203}{9}$

$$\frac{203}{9} = \frac{4}{5} \times 4^{\frac{5}{2}} + \frac{1}{4} \times 4^2 + c_2$$

$$\frac{203}{9} = \frac{128}{5} + 4 + c_2$$

$$c_2 = -\frac{317}{45}$$

$$r = \frac{4}{5}t^{\frac{5}{2}} + \frac{1}{4}t^2 - \frac{317}{45}$$

When $t = 9$, $r = \frac{4}{5} \times 9^{\frac{5}{2}} + \frac{1}{4} \times 9^2 - \frac{317}{45} = 208 \,\mathrm{m}$

Exercise 15.5B

Answers page 561

1 The velocity of a particle after t seconds is given by $v = (t - 2) \,\mathrm{m\,s^{-1}}$. After 6 seconds, the displacement is 11 m from O.

    a Find an expression for the displacement, r.

    b Find the displacement after 8 seconds.

    c Find the velocity when the particle has a displacement of 53 m from O.

2 The acceleration of a particle after t seconds is given by $a = (6t - 20) \,\mathrm{m\,s^{-2}}$. After 12 seconds, the velocity is 223 $\mathrm{m\,s^{-1}}$.

    a Find an expression for the velocity, v.

After 7 seconds, the displacement is 40 m.

b Find an expression for the displacement, r.

c Find the times at which the particle is at the origin.

d Find the times at which the particle is at travelling at $19\,\mathrm{m\,s^{-1}}$.

 3 A particle is accelerating at $(pt - 10)\,\mathrm{m\,s^{-2}}$. Two seconds after it sets off it has a velocity of $30\,\mathrm{m\,s^{-1}}$. Half a second after that, it has a velocity of $52\,\mathrm{m\,s^{-1}}$. If the particle is initially at the origin, find its position after three seconds.

 4 A particle has an acceleration of $(6t - 30)\,\mathrm{m\,s^{-2}}$. Initially the particle is at O with a velocity of $72\,\mathrm{m\,s^{-1}}$.

a Find the minimum velocity of the particle.

b When is the particle at instantaneous rest?

c Find the total distance travelled during the first ten seconds.

 5 A particle has an acceleration of $(9 - 2t)\,\mathrm{m\,s^{-2}}$. Initially the particle is at rest at the origin. Find the distance travelled by the particle:

a during the first two seconds

b during the fifth second

c during the eighth and ninth seconds.

 6 A particle P moves in a straight line such that after t seconds, its acceleration, $a\,\mathrm{m\,s^{-2}}$, is given by

$$a = \begin{cases} \dfrac{1}{2}(t+6) & 0 \leqslant t \leqslant 4 \\[2mm] \dfrac{320}{t^3} & 4 < t \leqslant 8 \end{cases}$$

Initially the particle is at the origin and travelling at a velocity of $2\,\mathrm{m\,s^{-1}}$.

a Find the speed of the particle when:

   i $t = 3$ **ii** $t = 5$

b Find the total distance travelled during the 8 seconds.

7 A particle P moves in a straight line such that after t seconds, its velocity, $v\,\mathrm{m\,s^{-1}}$, is given by

$$v = \begin{cases} 2t + 4 & 0 \leqslant t \leqslant 6 \\ 20 - (t-8)^2 & 6 < t \leqslant 10 \end{cases}$$

Initially the particle is at the origin.

a Draw a velocity–time graph to illustrate the journey.

b Find the maximum speed of P.

c Find the acceleration of the particle when:

   i $t = 5$ **ii** $t = 9$

d Find the displacement at $t = 5$ and $t = 9$.

 8 A particle is initially at the origin, O. It travels with a constant acceleration from an initial velocity of $6\,\mathrm{m\,s^{-1}}$ to $33\,\mathrm{m\,s^{-1}}$ in 18 seconds. Subsequently, its acceleration is given by $a = \frac{1}{4}(t - 12)\,\mathrm{m\,s^{-2}}$. Find the total distance travelled by the particle during the first 30 seconds.

SUMMARY OF KEY POINTS

❯ Distance and speed are scalar quantities. Displacement and velocity are vector quantities. Acceleration is a vector quantity but the same word is also used to describe the magnitude of the acceleration vector.

❯ Displacement, velocity, acceleration and time are related by the five SUVAT equations:

  ❯ $v = u + at$

  ❯ $v^2 = u^2 + 2as$

  ❯ $s = \frac{1}{2}(u + v)t$

  ❯ $s = ut + \frac{1}{2}at^2$

  ❯ $s = vt - \frac{1}{2}at^2$

❯ Make sure you write the quantities in SI units (m and s). You will need to decide which direction to take as positive.

❯ When an object is travelling vertically, it experiences an acceleration of $9.8\,\mathrm{m\,s^{-2}}$ downwards due to gravity.

❯ For a displacement–time graph, the gradient is the velocity.

❯ For a velocity–time graph, the gradient is the acceleration and the area under the graph is the distance travelled.

❯ When displacement, velocity or acceleration is a function of time, use calculus. Remember to use a constant of integration when integrating.

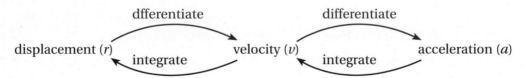

displacement (r) — dfferentiate → velocity (v) — differentiate → acceleration (a); velocity (v) — integrate → displacement (r); acceleration (a) — integrate → velocity (v)

EXAM-STYLE QUESTIONS 15 Answers page 561

 1 A car moves with constant acceleration along a straight horizontal road. The car passes point A with speed $1.5\,\mathrm{m\,s^{-1}}$ and 6 s later it passes point B, where AB = 90 m.

  a Find the acceleration of the car. **[3 marks]**

  When the car passes the point C, it has speed $37.5\,\mathrm{m\,s^{-1}}$.

  b Find the distance AC. **[3 marks]**

2 A small ball is projected vertically upwards from a point A. The greatest height reached by the ball is 90 m above A. Calculate:

  a the speed of projection **[3 marks]**

  b the time between the instant that the ball is projected and the instant it returns to A. **[3 marks]**

(PS) (M) 3 Three posts, P, Q and R, are fixed in that order at the side of a straight horizontal road. A car passes Q five seconds after passing P, and the car passes R ten seconds after passing Q. The distance from P to Q is 40 m and the distance from Q to R is 20 m. The car is moving along the road with constant acceleration a m s^{-2} and the speed of the car as it passes P is u m s^{-1}. Find:

   a the value of u **[5 marks]**

   b the value of a.   **[2 marks]**

(M) (CM) 4 A small ball is projected vertically upwards from ground level with speed u m s^{-1}. The ball takes 6 s to return to ground level.

   a Draw a velocity–time graph to represent the motion of the ball during the first 6 s.   **[2 marks]**

   b The maximum height of the ball above the ground during the first 6 s is 44.1 m. Find the value of u.   **[3 marks]**

   c State a physical factor that has been ignored in this model.   **[1 mark]**

(M) (PS) 5 A particle moving along the x-axis is initially at the origin O. At time t seconds, the velocity of the particle is given by $v = (3t^2 - 12t + 11)$ m s^{-1} in the positive x-direction. Find the distance of the particle from O when it is moving with minimum velocity.   **[8 marks]**

(M) 6 Two cars, A and B, are moving in the same direction along a straight horizontal road. At time $t = 0$, they are side by side, passing a point O on the road. Car A travels at a constant speed of 36 m s^{-1}. Car B passes O with a speed of 18 m s^{-1}, and has a constant acceleration of 3 m s^{-2}. Find:

   a the speed of B when it has travelled 42 m from O   **[2 marks]**

   b the distance from O of A when B is 42 m from O   **[4 marks]**

   c the time when B overtakes A.   **[5 marks]**

(PS) (M) 7 A car starts from rest at a point S on a straight race track. The car moves with constant acceleration for 30 s, reaching a speed of 24 m s^{-1}. The car then travels at a constant speed of 24 m s^{-1} for 150 s. Finally it moves with constant deceleration, coming to rest at a point F.

   a Sketch a velocity–time graph to illustrate the motion of the car.   **[2 marks]**

The distance between S and F is 4.5 km.

   b Calculate the total time the car takes to travel from S to F.   **[3 marks]**

A motorcycle starts at S, 20 s after the car has left S. The motorcycle moves with constant acceleration from rest and passes the car at a point P which is 1920 km from S. When the motorcycle passes the car, the motorcycle is still accelerating and the car is moving at a constant speed. Calculate:

   c the time the motorcycle takes to travel from S to P   **[5 marks]**

   d the speed of the motorcycle at P.   **[2 marks]**

PS **8** A runner is competing in a race. Starting from rest, he accelerates at $4\,\mathrm{m\,s^{-2}}$ for

M 2 seconds, then at $\frac{1}{3}\,\mathrm{m\,s^{-2}}$ for 3 seconds and $\frac{3}{13}\,\mathrm{m\,s^{-2}}$ for 13 seconds, before finally decelerating at $2.4\,\mathrm{m\,s^{-2}}$ until he comes to rest at the finish line. What is the runner's average speed for the race, correct to 3 s.f.? **[6 marks]**

M **9** A particle P moves along a straight line. After t seconds, the velocity v is given by:

PS

$$v = \begin{cases} 8t - t^2 & 0 \leqslant t \leqslant 6 \\ 15 - \frac{1}{2}t & t < 6 \end{cases}$$

The particle is initially at the origin O. Find:

a the acceleration of P when $t = 5$ **[3 marks]**

b the total distance travelled by P during the first 40 seconds. **[8 marks]**

PS **10** A car moves along a horizontal straight road, passing two points A and B.

M At A, the speed of the car is $18\,\mathrm{m\,s^{-1}}$. When the driver passes A, she sees a warning sign W ahead of her, 180 m away. She immediately applies the brakes and the car decelerates with uniform deceleration, reaching W with speed $6\,\mathrm{m\,s^{-1}}$.

At W, the driver sees that the road is clear. She then immediately accelerates the car with uniform acceleration for 18 s to reach a speed of $V\,\mathrm{m\,s^{-1}}$ ($V > 18$).

She then maintains the car at a constant speed of $V\,\mathrm{m\,s^{-1}}$. Moving at this constant speed, the car passes B after a further 33 s.

a Sketch a velocity–time graph to illustrate the motion of the car as it moves from A to B. **[3 marks]**

b Find the time taken for the car to move from A to B. **[3 marks]**

The distance from A to B is 1.2 km.

c Find the value of V. **[5 marks]**

M **11** A particle P moves on the x-axis. The acceleration of P at time t seconds is

PS $(t - 6)\,\mathrm{m\,s^{-2}}$ in the positive x-direction. The velocity of P at time t seconds is $v\,\mathrm{m\,s^{-1}}$. When $t = 0$, $v = 10$. Find:

a v in terms of t **[4 marks]**

b the values of t when P is instantaneously at rest **[3 marks]**

c the distance between the two points at which P is instantaneously at rest. **[4 marks]**

PS **12** A ball is projected vertically upwards at $35\,\mathrm{m\,s^{-1}}$ from the top of a building. 2 seconds later,

M a second ball is projected vertically upwards at $21.7\,\mathrm{m\,s^{-1}}$ from the same place. The balls land on the ground at the same time. Find the height of the building. **[9 marks]**

PS **13** A particle travels in a straight line with a constant acceleration through points

M A, B and C. BC is 3 times the distance of AB. The particle passes B 3 s after it passes A and passes C 7 s after it passes B. Given that the velocity of the particle at B is $4.5\,\mathrm{ms^{-1}}$ faster than at A, find the length AC. **[8 marks]**

PF **14** The displacement of a particle P from the origin after t s is given by $r = t^2(t + k)$, $k \neq 0$.

M Given that the particle comes to instantaneous rest after 6 s, show that:

a the acceleration of the particle after 13 s is $60\,\mathrm{m\,s^{-2}}$ **[6 marks]**

b the particle travels 316 m during the first 10 s. **[6 marks]**

16 FORCES

It may or may not have really happened, but the legend is that in the late 17th century, Isaac Newton was sitting beneath an apple tree. After an apple fell on his head, he devised his universal law of gravitation. Since the initial velocity of the apple was zero, but the final velocity was not, the apple had accelerated to the ground. According to Newton's own second law, this indicated that a force must be acting on the apple. Newton named this force gravity. He then surmised that the same force of gravity which made the apple accelerate towards the Earth must affect other objects, such as the Moon. In turn, the Moon exerts a gravitational force upon the Earth. In fact, any two objects exert a gravitational force on each other.

Newton's universal law of gravitation, along with the three laws of motion that bear his name, can be applied to all sorts of situations. The three main situations featured in this chapter are vertical motion (such as lifts), connected particles (such as cars towing caravans) and pulleys.

Although pulleys were invented long before Newton, and are often credited to Archimedes in the 3rd century BCE, without Newton's insights we would not fully understand the mathematics that underpins how they work. The earliest examples of pulleys could have been used to lift water from wells, but they can also be used to make lifting any heavy objects easier. You are likely to encounter pulleys in everyday life – for example in window blinds, escalators and lifts, garage doors, curtains, gym equipment, clothes lines and flagpoles.

LEARNING OBJECTIVES

You will learn how to:

- find a resultant force by adding vectors
- find the resultant force in a given direction
- apply Newton's laws and the equations of constant acceleration to problems involving forces
- distinguish between the forces of reaction, tension, thrust, resistance and weight
- model situations such as lifts, connected particles and pulleys
- represent problems involving forces in a diagram.

TOPIC LINKS

A force is a vector quantity and the resultant force on a particle is equal to its mass multiplied by its acceleration, as stated by Newton's second law. So, this chapter builds upon the material you covered on vectors in **Chapter 10 Vectors**, such as the magnitude and direction of a vector written in **i j** or column vector form, and upon the material in **Chapter 15 Kinematics**, such as the SUVAT equations.

PRIOR KNOWLEDGE

You should already know how to:

- use Pythagoras' theorem and trigonometry to find the magnitude and direction of a vector
- apply the equations of constant acceleration (SUVAT) from kinematics
- assign a direction to motion to decide whether a vector is positive or negative
- solve simultaneous linear equations
- solve quadratic equations.

You should be able to complete the following questions correctly:

1   A particle has an initial velocity of $4\,\text{m}\,\text{s}^{-1}$ and travels $55\,\text{m}$ in $5\,\text{s}$.

   a   Find the acceleration of the particle.

   b   Find the final velocity of the particle.

2   The force **R** is given by $(-3\mathbf{i} + 7\mathbf{j})\,\text{N}$. Find the magnitude of **R** and the angle that **R** makes with the vector **i**.

3   Solve the simultaneous equations.

$$30 - y = 3x \qquad \qquad ①$$
$$y - 10 = x \qquad \qquad ②$$

16.1 Forces

If an object experiences a push or a pull, it is said to be acted upon by a **force**. A force is a vector quantity, so it can be written in **i j** notation or as a column vector, and you can calculate its magnitude and direction, as you saw in **Chapter 10 Vectors**.

SI units

The SI unit of force is the newton (N). Newton's second law states that $F = ma$, from which you can find the base units of the newton. The SI units for mass and acceleration are the kilogram (kg) and $\text{m}\,\text{s}^{-2}$; hence the base units for the newton are $\text{kg}\,\text{m}\,\text{s}^{-2}$ (i.e. $1\,\text{N} = 1\,\text{kg}\,\text{m}\,\text{s}^{-2}$). One newton is defined as the force that will cause a mass of $1\,\text{kg}$ to accelerate at $1\,\text{m}\,\text{s}^{-2}$. Note that $1\,\text{kN} = 1000\,\text{N}$.

If two equal and opposite forces act upon a particle, such as $3\mathbf{i}\,\text{N}$ and $-3\mathbf{i}\,\text{N}$, the forces cancel each other out and the overall net force (known as the resultant force) is zero. When the resultant force is zero, this is known as equilibrium.

Similarly, if forces of $7\mathbf{i}\,\text{N}$ and $-2\mathbf{i}\,\text{N}$ act upon a particle, the resultant force is $5\mathbf{i}\,\text{N}$.

> **KEY INFORMATION**
>
> The sum of all the forces is called the resultant force.
>
> If the resultant force is equal to zero, the object is in equilibrium.

Example 1

Find the resultant force when the forces $(-25\mathbf{i} + 17\mathbf{j})$ N and $(8\mathbf{i} - 10\mathbf{j})$ N act upon a particle.

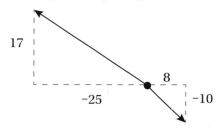

In order to find the resultant force, add the forces together.

Solution

$(-25\mathbf{i} + 17\mathbf{j}) + (8\mathbf{i} - 10\mathbf{j}) = (-25 + 8)\mathbf{i} + (17 - 10)\mathbf{j} = (-17\mathbf{i} + 7\mathbf{j})$ N

Add the $\mathbf{i}$ terms and the $\mathbf{j}$ terms separately.

Example 2

Forces of $\mathbf{F}_1$, $\mathbf{F}_2$ and $\mathbf{F}_3$ act upon a particle. The forces are given by $\mathbf{F}_1 = ((a + 1)\mathbf{i} - 9\mathbf{j}))$ N, $\mathbf{F}_2 = (-5\mathbf{i} + a\mathbf{j})$ N and $\mathbf{F}_3 = (b\mathbf{i} + 5\mathbf{j})$ N. The resultant force $\mathbf{R}$ is given by $(8\mathbf{i} - \mathbf{j})$ N.

Find:

a   the magnitude of $\mathbf{R}$, correct to 2 d.p.

b   the angle between $\mathbf{R}$ and the vector $\mathbf{j}$, correct to the nearest degree

c   the resultant force when $\mathbf{F}_3$ is replaced by the force $(10\mathbf{i} + 7\mathbf{j})$ N.

Solution

a   Magnitude $= \sqrt{8^2 + (-1)^2}$

$= \sqrt{64 + 1}$

$= \sqrt{65}$

$= 8.06$ N correct to 2 d.p.

To find the magnitude of a force written in $\mathbf{i}\,\mathbf{j}$ notation, apply Pythagoras' theorem.

b

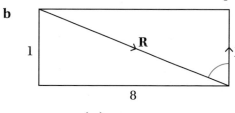

$\tan^{-1}\left(\dfrac{8}{1}\right) = 83°$

To find the angle of a force written in $\mathbf{i}\,\mathbf{j}$ notation, use trigonometry.

c   Before you can find the new resultant force, you need to find the values of a and b.

Add the $\mathbf{i}$ terms and the $\mathbf{j}$ terms separately to form equations in a and b.

For **i**: $(a+1)-5+b=8$

For **j**: $-9+a+5=-1$

Solve the second equation to find the value of a.

$$-9+a+5=-1$$
$$-4+a=-1$$
$$a=3$$

Substitute $a=3$ into the first equation and solve for b.

$$(3+1)-5+b=8$$
$$4-5+b=8$$
$$-1+b=8$$
$$b=9$$

Hence $\mathbf{F}_3 = (9\mathbf{i}+5\mathbf{j})$ N.

If $(9\mathbf{i}+5\mathbf{j})$ is replaced by the force $(10\mathbf{i}+7\mathbf{j})$ N, then the new resultant force will be given by:

$$(8\mathbf{i}-\mathbf{j})-(9\mathbf{i}+5\mathbf{j})+(10\mathbf{i}+7\mathbf{j}) = (8-9+10)\mathbf{i}+(-1-5+7)\mathbf{j}$$
$$= (9\mathbf{i}+\mathbf{j})\,\text{N}$$

Exercise 16.1A

Answers page 562

(M) **1** Find the resultant force when forces $\mathbf{F}_1$, $\mathbf{F}_2$ and $\mathbf{F}_3$ act upon a particle.

   a $\mathbf{F}_1 = (2\mathbf{i}+9\mathbf{j})$ N, $\mathbf{F}_2 = (-6\mathbf{i}+3\mathbf{j})$ N, $\mathbf{F}_3 = (-\mathbf{i}+\mathbf{j})$ N

   b $\mathbf{F}_1 = (6\mathbf{i}-7\mathbf{j})$ N, $\mathbf{F}_2 = (-8\mathbf{i}-5\mathbf{j})$ N, $\mathbf{F}_3 = (10\mathbf{i}-3\mathbf{j})$ N

   c $\mathbf{F}_1 = (-0.5\mathbf{i}+2.1\mathbf{j})$ N, $\mathbf{F}_2 = (2.4\mathbf{i}+5.6\mathbf{j})$ N, $\mathbf{F}_3 = (\mathbf{i}-9.5\mathbf{j})$ N

(M) **2** Forces $\mathbf{F}_1$, $\mathbf{F}_2$ and $\mathbf{F}_3$ act upon a particle. Given that the particle is in equilibrium, find the values of p and q when:

   a $\mathbf{F}_1 = (5\mathbf{i}+\mathbf{j})$ N, $\mathbf{F}_2 = (q\mathbf{i}+2\mathbf{j})$ N, $\mathbf{F}_3 = (-8\mathbf{i}+p\mathbf{j})$ N

   b $\mathbf{F}_1 = (p\mathbf{i}-2q\mathbf{j})$ N, $\mathbf{F}_2 = (p\mathbf{i}-12\mathbf{j})$ N, $\mathbf{F}_3 = (7\mathbf{i}-q\mathbf{j})$ N

   c $\mathbf{F}_1 = (-4\mathbf{i}+3p\mathbf{j})$ N, $\mathbf{F}_2 = (3q\mathbf{i}+q\mathbf{j})$ N, $\mathbf{F}_3 = (2p\mathbf{i}-13\mathbf{j})$ N

(M) **3** For each force, find:

   i the magnitude

   ii the angle that the force makes with the vector **i**.

   Give both values correct to 1 d.p. where appropriate.

   a $(-2\mathbf{i}+3\mathbf{j})$ N   **b** $(8\mathbf{i}-5\mathbf{j})$ N

   c $(-4\mathbf{i}-4\mathbf{j})$ N   **d** $(9\mathbf{i}+7\mathbf{j})$ N

(M) **4** A force is given by $(24a\mathbf{i}+7a\mathbf{j})$ N, where a is a negative constant.

   a Find the bearing along which the force acts.

   b Given that the magnitude of the force is 200 N, find the value of a.

5 Force $\mathbf{F}_1 = (16\mathbf{i} - 9\mathbf{j})$ N. Force $\mathbf{F}_2$ acts in the direction $(-2\mathbf{i} + 3\mathbf{j})$.

The resultant of the two forces is $\mathbf{R}$.

a Given that $\mathbf{R}$ is parallel to $\mathbf{i}$, find the magnitude of $\mathbf{R}$.

b Given instead that $\mathbf{R}$ is parallel to $\mathbf{j}$, find the magnitude of $\mathbf{R}$.

16.2 Newton's laws of motion

Newton's first law states that a particle will remain at rest or continue to move at a constant velocity in a straight line unless it is acted upon by an external force.

In the case of the apple falling from the tree, the apple is motionless until it becomes detached from the tree, at which point it accelerates towards the Earth. A rocket travelling in outer space after its initial launch, on the other hand, is not acted upon by any large external force, so will continue to travel at a steady speed.

Newton's second law states that the force, *F*, acting upon a particle is proportional to the particle's mass, *m*, and acceleration, *a*. This results in the formula $F = ma$.

This extends the first law by stating that when an external force is applied, the particle will no longer be stationary or travelling at a steady speed, but instead will accelerate. The rate of acceleration is dependent on the magnitude of the force and the mass of the particle.

Newton's third law states that every action has an equal and opposite reaction.

If Newton's third law was not true, then the apple would not stop once it hit the ground. From observation, you know that your weight is balanced by a force from the ground which prevents you from falling into it. This is true for any object on a stable, solid surface, such as a vase on a table or a car on a road.

> **KEY INFORMATION**
> You should know and be able to apply Newton's three laws of motion.

Example 3

Two forces, $\mathbf{F}_1$ and $\mathbf{F}_2$, act upon a particle of mass 4 kg.

$$\mathbf{F}_1 = (13\mathbf{i} - 17\mathbf{j})\,\text{N and } \mathbf{F}_2 = (7\mathbf{i} + 2\mathbf{j})\,\text{N}$$

a Find the magnitude of the acceleration produced by the resultant force.

b Given that the particle has an initial velocity of $(-2\mathbf{i} + 3\mathbf{j})\,\text{m s}^{-1}$, find its velocity 8 s later.

Solution

i j vector notation was introduced in **Chapter 10 Vectors**, alongside column vectors, as a way to describe the horizontal and vertical components of a vector. In this case, for example, the force $\mathbf{F}_1$ has a component of 13 N in the positive x-direction and a component of 17 N in the negative y-direction.

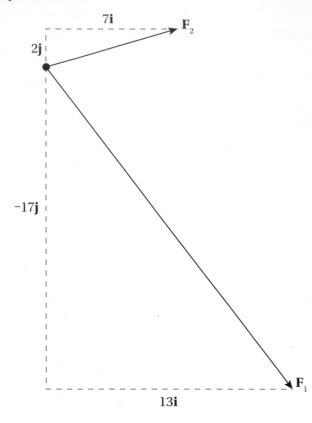

a Start by adding the **i** and **j** terms of $\mathbf{F}_1$ and $\mathbf{F}_2$ to find the resultant force.

$$(13\mathbf{i} - 17\mathbf{j}) + (7\mathbf{i} + 2\mathbf{j}) = (20\mathbf{i} - 15\mathbf{j})\,\text{N}$$

The resultant force is the sum of all the forces acting upon the particle. Add the **i** and **j** terms separately.

Apply the formula $\mathbf{F} = m\mathbf{a}$.

$$20\mathbf{i} - 15\mathbf{j} = 4\mathbf{a}$$

Hence

$$\mathbf{a} = (5\mathbf{i} - 3.75\mathbf{j})\,\text{m s}^{-2}$$

Use Pythagoras' theorem to find the magnitude of the acceleration.

$$\sqrt{5^2 + \left(-3.75\right)^2} = 6.25\,\text{m s}^{-2}$$

b As you are given **u** and t and have already found an expression for **a**, use the SUVAT equation $\mathbf{v} = \mathbf{u} + \mathbf{a}t$.

$$\mathbf{v} = \mathbf{u} + \mathbf{a}t$$
$$= (-2\mathbf{i} + 3\mathbf{j}) + (5\mathbf{i} - 3.75\mathbf{j}) \times 8$$
$$= -2\mathbf{i} + 3\mathbf{j} + 40\mathbf{i} - 30\mathbf{j}$$
$$= (38\mathbf{i} - 27\mathbf{j})\,\text{m s}^{-1}$$

Note that the equation $v = u + at$ has been written as $\mathbf{v} = \mathbf{u} + \mathbf{a}t$ to highlight the fact that you are substituting terms in **i j** notation.

Exercise 16.2A
Answers page 563

(M) **1** **a** Find the magnitude of the resultant force acting on a particle of mass 800 g that would produce an acceleration of $3.1\,\mathrm{m\,s^{-2}}$.

   b A particle of mass m kg has a deceleration of $4\,\mathrm{m\,s^{-2}}$ when acted upon by a resultant force of 5.2 kN. Find the value of m.

(M) **2** A particle of mass 2.5 kg experiences a resultant force of $\begin{bmatrix} 10 \\ -15 \end{bmatrix}$ N.

   a Find the acceleration in the form $\begin{bmatrix} a \\ b \end{bmatrix}\mathrm{m\,s^{-2}}$.

   A resultant force of $(14\mathbf{i} + 35\mathbf{j})$ N produces an acceleration of $(10\mathbf{i} + 25\mathbf{j})\,\mathrm{m\,s^{-2}}$ for a particle of mass M kg.

   b Find the value of M.

(M) **3** Which of these is the magnitude of the acceleration when a resultant force of $\begin{bmatrix} -45 \\ 24 \end{bmatrix}$ N acts upon a particle with a mass of 1.5 kg?

   A $-36\,\mathrm{m\,s^{-2}}$ **B** $\begin{bmatrix} -30 \\ 16 \end{bmatrix}\mathrm{m\,s^{-2}}$

   C $76.5\,\mathrm{m\,s^{-2}}$ **D** $34\,\mathrm{m\,s^{-2}}$

(M) **4** A particle of mass $\frac{2}{3}$ kg is acted upon by three forces: $\mathbf{F}_1 = (p\mathbf{i} - 8\mathbf{j})$ N, $\mathbf{F}_2 = (5\mathbf{i} - q\mathbf{j})$ N and $\mathbf{F}_3 = (6\mathbf{i} - 19\mathbf{j})$ N.

   The acceleration produced by the resultant force is $(36\mathbf{i} - 15\mathbf{j})\,\mathrm{m\,s^{-2}}$. The initial velocity of the particle is $(4\mathbf{i} + \mathbf{j})\,\mathrm{m\,s^{-1}}$.

   a Find the values of p and q.

   b Find the magnitude of the resultant force.

   c Find the angle the resultant force makes with the vector $\mathbf{i}$, correct to 1 d.p.

   d Find the velocity of the particle after $\frac{1}{3}$ s.

(M) **5** A toy car of mass 0.7 kg has an initial velocity of $\begin{bmatrix} 8 \\ -5 \end{bmatrix}\mathrm{m\,s^{-1}}$.

   A constant force is applied to the toy car.

   After six seconds, the velocity of the toy car is $\begin{bmatrix} -1 \\ 7 \end{bmatrix}\mathrm{m\,s^{-1}}$.

   a Find the acceleration of the toy car in the form $\begin{bmatrix} a \\ b \end{bmatrix}\mathrm{m\,s^{-2}}$.

   b Find the force applied to the toy car in the form $\begin{bmatrix} a \\ b \end{bmatrix}$ N.

   c Find the magnitude of the force applied to the toy car.

CM 6 Two students, Joseph and Liam, have been set this problem:

'A particle of mass 2 kg has an initial velocity of $(5\mathbf{i} + \mathbf{j})\,\mathrm{m\,s^{-1}}$. A force of $F\,\mathrm{N}$ is applied to the particle and after four seconds the velocity of the particle is $(17\mathbf{i} + 9\mathbf{j})\,\mathrm{m\,s^{-1}}$. Find the magnitude of the force.'

Joseph's solution:

Magnitude of initial velocity $= \sqrt{5^2 + 1^2} = 5.099\,\mathrm{m\,s^{-1}}$

Magnitude of final velocity $= \sqrt{17^2 + 9^2} = 19.235\,\mathrm{m\,s^{-1}}$

$a = \dfrac{v - u}{t} = \dfrac{19.235 - 5.099}{4} = 3.53\,\mathrm{m\,s^{-2}}$

$F = ma = 2 \times 3.53 = 7.07\,\mathrm{N}$

Liam's solution:

$\mathbf{a} = \dfrac{\mathbf{v} - \mathbf{u}}{t} = \dfrac{(17\mathbf{i} + 9\mathbf{j}) - (5\mathbf{i} + \mathbf{j})}{4} = (3\mathbf{i} + 2\mathbf{j})\,\mathrm{m\,s^{-2}}$

Magnitude of acceleration $= \sqrt{3^2 + 2^2} = 3.61\,\mathrm{m\,s^{-2}}$

$F = ma = 2 \times 3.61 = 7.21\,\mathrm{N}$

Whose solution was correct? Why was the other student's solution incorrect?

Types of force
The five types of force you need to know are weight, normal reaction, tension, thrust and resistance.

Weight
The **weight** of an object is a force caused by the gravitational acceleration experienced by the object and acts vertically downwards towards the centre of the Earth.

In this diagram, the weight is labelled as W. Forces are usually drawn coming out of the particle they are acting upon, with a line and a black-headed arrow showing the direction of action of the force. Because weight acts vertically downwards, the arrow in the diagram also points vertically downwards.

W

Based upon the formula $F = ma$, weight is given by the formula $W = mg$, where m is the mass in kg and g is the acceleration due to gravitational attraction. On Earth, this acceleration is commonly taken as $9.8\,\mathrm{m\,s^{-2}}$.

Weight is different from mass because it depends on the value of g. For example, the value of g on the surface of the Moon is $1.6\,\mathrm{m\,s^{-2}}$, which is why astronauts experience reduced weight. Make sure that you do not confuse g for grams and g for gravitational acceleration.

Normal reaction
If an object is in contact with a surface, the object will experience a **reaction force** perpendicular to the surface. This is often called the normal reaction, where 'normal' means 'perpendicular' (as it

does for the straight line perpendicular to a tangent – see **Chapter 5 Coordinate geometry 2: Circles**). The reaction force is the one often described as the force exerted on the particle by the ground, since it is equal and opposite to the weight of the particle when the ground is horizontal, in agreement with Newton's third law.

In a diagram, the normal reaction is usually labelled R. In this diagram with a horizontal surface, the weight acts vertically downwards (as always) and the reaction force acts vertically due to its always acting perpendicular to the surface. As a result, for an object on a horizontal table, the reaction force on a particle will always be equal to its weight. The forces are balanced, as expected from Newton's third law.

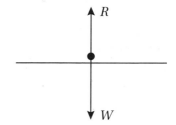

Tension

When an object is pulled by or suspended from a string or rope, the string will become **taut** and the object will experience the **tension** in the string.

In a diagram, tension is usually labelled T and acts away from the object.

Thrust (compression)

Consider a ball held in mid-air. If the ball is suspended from a string, so the string is above the ball and taut, the force in the string is tension. On the other hand, if the ball is held up on a wooden rod, then the force in the wooden rod is **thrust** (or compression). When an object is pushed rather than pulled, the object will experience a thrust force instead.

In this diagram, the particle is held up by a rod rather than being suspended from a string, so the T stands for thrust. In the case of thrust, since it is a pushing force, the force appears to be going *into* the particle rather than coming out of it.

In summary, tension always acts *away* from an object whereas thrust always acts *towards* the object.

Resistance

When a force opposes motion, it is often called a **resistive force** (or the resistance to motion). This will often be friction, although friction is not covered as a specific force until **Book 2**. It also includes forces such as air resistance. In a diagram, friction is usually labelled F but there is no particular letter used to represent resistance. XN has been used in this diagram to represent the resistive force when a particle is pulled by a string along a rough table.

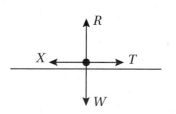

In this diagram, the reaction and weight forces are balanced and there is no vertical motion. If the particle is travelling at a steady speed along the horizontal line, then the tension and resistance are also balanced. This is confirmed by Newton's first law, where the particle remains at a steady speed unless an external force is applied (the resultant force is zero). If, however, the tension is greater than the resistance, then Newton's second law ($F = ma$) states that the resultant force ($T - X$) will equal ma and the particle will accelerate.

As well as the new content in this chapter, you can expect questions to test your knowledge of the **i j** notation from **Chapter 10 Vectors** and the SUVAT equations from **Chapter 15 Kinematics**, as you will see in **Example 6**.

Modelling assumptions

Throughout this chapter, you will use simplified models to represent real-life situations. Each section includes a brief explanation of the modelling assumptions you will be using. In this first section, for example:

› objects will be modelled as particles so that their mass is concentrated at a single point

› motion will take place horizontally in a straight line (such as along a straight horizontal road) so that angles need not be considered and so that weight has no effect in the direction of motion, since it will be perpendicular to the motion.

Example 4

A vase is at rest on a horizontal table. The weight of the vase is 25 N.

Find the force exerted on the vase by the table.

Solution

Start by drawing a diagram to represent the information. Model the vase as a particle. Draw a 25 N force acting vertically downwards for the weight of the vase. The force exerted on the vase is the reaction force, which acts perpendicular to the surface (i.e. vertically upwards).

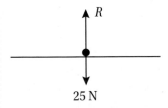

Since both forces act vertically and there is no acceleration, the two forces are balanced.

Hence $R = 25$ N.

How might the reaction and weight forces be related if the vase was on a table inclined at an angle of 1° to the horizontal? Would it matter whether the table was smooth or rough? What would happen if the angle was increased?

Example 5

A boy is pulling his sledge horizontally across the snow using a rope which is also horizontal. The resistance to motion is 40 N and the sledge is accelerating at $0.3\,\mathrm{m\,s^{-2}}$. The sledge has a mass of 18 kg. Find the tension in the rope.

Solution

Draw a diagram. Model the sledge as a particle. Again, there will be vertical weight and reaction forces but there will also be horizontal tension and resistance forces. Since the mass is given rather than the weight, use the formula $W = mg$ to write the weight as $18g\,\mathrm{N}$.

Because the sledge is accelerating, the tension and resistance are not balanced and instead there is a resultant force F equal to ma. Represent the acceleration $a\,\mathrm{m\,s^{-2}}$ using a double-headed arrow.

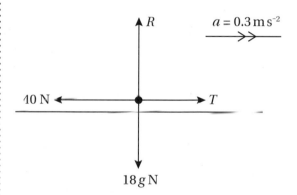

Consider the horizontal forces. Since the sledge is moving to the right, the tension is positive and the resistance is negative.

Substitute into $F = ma$.

$$T - 40 = 18 \times 0.3$$
$$= 5.4$$
$$T = 45.4\,\mathrm{N}$$

KEY INFORMATION

When you use the formula $F = ma$, where F is the resultant force, you should add up the forces in the direction of motion. In this example, the sledge is moving to the right.

Example 6

A car is towing a caravan along a straight horizontal road at $24\,\text{m s}^{-1}$. The caravan experiences a resistance to its motion of $340\,\text{N}$. The car decelerates at a constant rate for $112\,\text{m}$, which takes $8\,\text{s}$. Given that the caravan has a mass of $300\,\text{kg}$, find the thrust in the tow bar whilst the car is decelerating.

> When a car is pulling a caravan and accelerating, the towbar is described as experiencing tension. When a car is pulling a caravan and decelerating, the towbar is described as experiencing thrust.

Solution

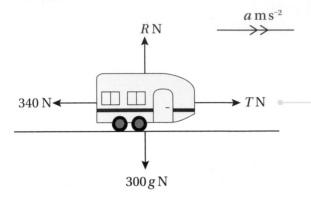

> It is a good idea to draw a diagram to represent the information.

> Remember that the thrust force acts *towards* the object.

You need to use the formula $F = ma$. The resultant force F will be a combination of the resistance to motion and the thrust in the tow bar. The mass m is $300\,\text{kg}$. The acceleration a can be found using SUVAT.

You are given $u = 24\,\text{m s}^{-1}$, $s = 112\,\text{m}$ and $t = 8\,\text{s}$.

The relevant SUVAT equation is $s = ut + \frac{1}{2}at^2$.

Substitute into the equation.

$$s = ut + \frac{1}{2}at^2$$

$$112 = 24 \times 8 + \frac{1}{2} \times a \times 8^2$$

$$= 192 + 32a$$

$$-80 = 32a$$

$$a = -2.5\,\text{m s}^{-2}$$

Substitute into $F = ma$. The forces should be considered in the direction of motion.

$$-T - 340 = 300 \times -2.5$$

$$= -750$$

$$-T = -410$$

The thrust is $410\,\text{N}$.

> **KEY INFORMATION**
>
> When you use the formula $F = ma$, where F is the resultant force, you should add up the forces in the direction of motion. In this example, the caravan is moving to the right, but both horizontal forces are pointing to the left, so both are negative.

Exercise 16.2B

Answers page 563

(M) 1 Find the weight in newtons of a wrecking ball that has a mass of 600 kg.

(M) 2 A toy train is pulled along a horizontal table at a steady speed by a taut string. The resistance to motion is given by X. The mass of the train is 0.6 kg. Label the missing forces in the diagram.

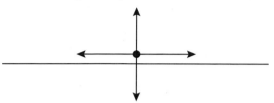

(M) 3 A box is pushed by a thrust force at a steady speed along a rough horizontal table. The resistance to motion is 12 N. The reaction force exerted on the box by the table is 58.8 N.

  a Draw a force diagram to represent the four forces acting upon the box.

  b Calculate the magnitude of the thrust force.

  c Calculate the mass of the box.

(M) 4 A car travelling along a straight horizontal road produces a driving force of 910 N whilst experiencing a resistance to motion of 190 N. The car takes 15 s to accelerate uniformly from $4 \, \mathrm{m \, s^{-1}}$ to $22 \, \mathrm{m \, s^{-1}}$. Find the mass of the car.

(M) 5 A 1.6 tonne coach is travelling at $33 \, \mathrm{m \, s^{-1}}$ along a straight horizontal lane when the driver spots a roadblock 260 m ahead. He immediately decelerates at a constant rate, coming to rest 20 m before the roadblock. The coach experiences a resistive force of 1.1 kN. Find the braking force required to bring the coach to a halt.

(M) 6 Starting from rest, a car is driven 460 m at a constant acceleration until it reaches $46 \, \mathrm{m \, s^{-1}}$ against a resistance to motion of 159 N. In order to achieve this, the engine of the car produces a driving force of $\begin{bmatrix} 756 \\ 192 \end{bmatrix}$ N. Find the mass of the car.

(M) 7 A particle at rest 2 m from the edge of a horizontal table is subjected to a thrust of X N. The resistance to motion is 250 N. The mass of the particle is 5 kg.

  Find the value of X given that:

  a the particle does not accelerate

  b the particle accelerates at $2 \, \mathrm{m \, s^{-2}}$

  c the particle falls off the edge of the table after 0.5 seconds.

(M) 8 A minibus of mass 800 kg produces a driving force of 600 N. Given that the initial velocity of the minibus is $4 \, \mathrm{m \, s^{-1}}$ and the minibus experiences a resistance of 280 N, find the time it takes the minibus to travel 385 m along a straight horizontal road.

(PS) (M) 9 A car of mass 320 kg accelerates from $15 \, \mathrm{m \, s^{-1}}$ to $30 \, \mathrm{m \, s^{-1}}$ along a straight 400 m stretch of a horizontal motorway at a constant rate. During this acceleration, the engine produces a driving force of 700 N. The motorist then spots a fault on the dashboard and decides to pull over on to the hard shoulder, removing the driving force. The car then decelerates at a constant rate with a braking force of 500 N. Assuming the resistance to motion is constant throughout, how much further does the car travel before it comes to rest?

16.3 Vertical motion

In **Section 16.2**, where cars and particles were travelling horizontally, you did not need to consider the weight of the object as a force because it acted perpendicularly (vertically) to the motion. However, when an object is travelling vertically then you do need to consider its weight. When an object travels upwards, the weight acts in the opposite direction and slows the object down, and when an object travels downwards, the weight is acting in the same direction and speeds up the object.

Recall that the formula for weight is $W = mg$, where m is the mass of the object and g is the gravitational acceleration. This will usually have the value $9.8\,\text{m}\,\text{s}^{-2}$, but this is an approximate value. You should expect to use $10\,\text{m}\,\text{s}^{-2}$, $9.8\,\text{m}\,\text{s}^{-2}$ or $9.81\,\text{m}\,\text{s}^{-2}$, depending on whether the values are given to 1, 2 or 3 s.f. Note that this value of g is only appropriate to objects on the surface of the Earth and that the value will vary according to the distance from the centre of the Earth. For example, $1000\,\text{km}$ above the Earth's surface, the value of g is $7.33\,\text{m}\,\text{s}^{-2}$. On the surface of Mars, g is $3.75\,\text{m}\,\text{s}^{-2}$ and on the Moon it is $1.6\,\text{m}\,\text{s}^{-2}$. On Jupiter, due to its huge mass, g is $26.0\,\text{m}\,\text{s}^{-2}$ – almost three times as great as on Earth!

KEY INFORMATION

$W = mg$

KEY INFORMATION

Usually g is taken as $9.8\,\text{m}\,\text{s}^{-2}$, but some questions may use values of $9.81\,\text{m}\,\text{s}^{-2}$ or $10\,\text{m}\,\text{s}^{-2}$.

Modelling assumptions

> Motion is vertical, so angles need not be considered and weight contributes fully to the resultant force.

> The value of g can be considered to be constant over small changes in altitude (height).

Example 7

A suitcase of mass $4\,\text{kg}$ is in free fall. It experiences air resistance of $0.2\,\text{N}$ as it falls. Find the acceleration of the suitcase.

Solution

Start by drawing a diagram. Both the air resistance and the weight of the suitcase act vertically.

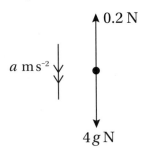

Consider the resultant force in the direction of motion (downwards).

$$F = ma$$
$$4g - 0.2 = 4a$$
$$4 \times 9.8 - 0.2 = 4a$$
$$39 = 4a$$
$$a = 9.75 \, \text{m s}^{-2}$$

Note that, in most questions, you assume that air resistance is negligible and hence acceleration in free fall is the same as g.

Example 8

A ball of mass 0.7 kg is raised vertically by a rope. Initially, the ball is stationary and at a height of 0.5 m above the ground. The tension in the rope is 10 N. After 1.5 s, the rope snaps. Find how long after the rope snaps it takes for the ball to return to the ground.

Solution

Draw a diagram.

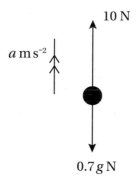

The motion of the ball goes through two stages: before and after the rope snaps. Before the rope snaps, the acceleration of the ball is determined by the tension in the rope and the weight of the ball. After the rope snaps, the acceleration is determined only by the weight of the ball, since the ball is in free fall.

During the first stage, the ball is acted upon by the upwards tension of 10 N and its weight, 0.7g N.

Hence the resultant force, F, is given by $(10 - 0.7g)$ N.

Substitute $F = (10 - 0.7g)$ N and $m = 0.7$ kg into the equation $F = ma$.

$$(10 - 0.7g) = 0.7a$$
$$a = \frac{10 - 0.7g}{0.7} = 4.49 \, \text{m s}^{-2}$$

In order to find the time during the second stage once the rope has snapped, it is necessary to know the initial velocity of the ball, which will be the same as the final velocity when the rope snaps, and the height of the ball.

$$u = 0\,\text{m s}^{-1}, t = 1.5\,\text{s}, a = 4.49\,\text{m s}^{-2}$$

$$v = u + at$$

$$= 0 + 4.49 \times 1.5 = 6.73\,\text{m s}^{-1}$$

$$s = ut + \frac{1}{2}at^2$$

$$= 0 \times 1.5 + \frac{1}{2} \times 4.49 \times 1.5^2 = 5.046\,\text{m}$$

The height above the ground is $5.046 + 0.5 = 5.546\,\text{m}$.

When the rope snaps, taking upwards as the positive direction, $u = 6.73\,\text{m s}^{-1}, s = -5.546\,\text{m}, a = -9.8\,\text{m s}^{-2}$.

$$v^2 = u^2 + 2as$$

$$= 6.73^2 + 2 \times -9.8 \times -5.546$$

$$= 154$$

$$v = -12.4\,\text{m s}^{-1}$$

Substitute into $v = u + at$ to find t.

$$-12.4 = 6.73 - 9.8t$$

$$t = 1.95\,\text{s}$$

> **KEY INFORMATION**
>
> When a string snaps, the final velocity as it snaps is the initial velocity after it snaps.
>
> The acceleration will also change when a force is removed.

> There are two answers for the equation $v^2 = 154$, which are $v = 12.4$. However, since the ball is in freefall, it is travelling downwards, so v is negative because upwards has been chosen as the upwards direction.

Exercise 16.3A Answers page 563

In this exercise, take g on the surface of the Earth as $9.8\,\text{m s}^{-2}$ unless it is stated otherwise.

(M) 1   **a**   Find the weight of a dog of mass 8.4 kg on the Earth's surface.

      b   Find the weight of an astronaut on the surface of Mars ($g = 3.75\,\text{m s}^{-2}$) who weighs 735 N on the surface of Earth.

(M) 2   A ball of mass 120 g is dropped from the top of a tower of height 25 m.

      The ball experiences an air resistance of 0.03 N as it falls.

      Find how much longer it takes for the ball to hit the ground than if air resistance were ignored.

(M) 3   A box is accelerated upwards at an acceleration of $a\,\text{m s}^{-2}$ by a rope. The mass of the box is M kg and the tension in the rope is T N.

      a   Find the value of a given that $M = 3$ and $T = 40$.

      b   Find the value of M given that $a = 0.6$ and $T = 8$.

      c   Find the value of T given that $a = 1.5$ and $M = 1.2$.

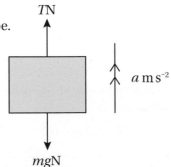

M **4** A load of mass 45 kg is moved vertically by means of an inextensible cord at a constant acceleration of $0.75\,\text{m s}^{-2}$.

Find the tension in the cord if:

a the acceleration is upwards

b the acceleration is downwards.

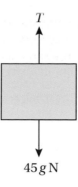

T

$45g\,\text{N}$

M **5** A toy of mass 350 g is dropped in a swimming pool and hits the surface of the water before falling vertically to the bottom of the pool. The depth of the pool is 1.9 m. The toy experiences a resistance to its motion of 2.8 N. The initial velocity of the toy at the surface is $3.7\,\text{m s}^{-1}$.

Find the time it takes to reach the bottom of the pool.

M **6** A box of mass M kg is held at the top of a vertical rod. When the box is lowered vertically with an acceleration of $0.2\,\text{m s}^{-2}$, the compression in the rod is 12 N.

Find the compression in the rod when the box is raised vertically with an acceleration of $0.2\,\text{m s}^{-2}$.

M **7** A 20 kg rock is pulled vertically upwards by a vertical cable from a speed of $2\,\text{m s}^{-1}$ to a speed of $5\,\text{m s}^{-1}$. The tension in the cable is 250 N.

Taking g as $10\,\text{m s}^{-2}$, calculate how far the rock is pulled.

M **8** A 3 N weight at rest on the floor is attached to a vertical string which pulls it vertically upwards. When the weight is 2.5 m above the floor, it is moving at $4\,\text{m s}^{-1}$.

a Find the tension in the string.

The string is then released.

b Find the maximum height achieved by the weight.

c Find the velocity of the weight when it returns to the floor.

Stop and think Why does the tension in a cord have a different magnitude depending upon whether the load is moving upwards or downwards?

16.4 Connected particles

A car is travelling along a straight horizontal road. If the car is accelerating then the engine exerts a driving force, D N. If the road is **rough** then there will be a frictional force in the opposite direction. If the road is **smooth** there will be *no* frictional force, but there may be other resistive forces to consider, such as air resistance. Assume that these resistive forces can be grouped as one overall resistance to motion, X N. The resultant force will be the difference between the driving force and this resistance to motion.

This is Newton's second law.

The frictional force will be considered in greater depth in **Book 2**. In this book, resistive forces will be considered together.

$$F = ma$$

$$D - X = ma$$

The car will also have a vertical weight force, $W = mg\,\mathrm{N}$, acting vertically downwards and a reaction force, $R\,\mathrm{N}$, exerted on the car by the road. These forces will be equal and opposite.

This is Newton's third law.

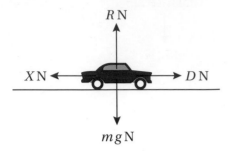

Now consider what would happen if the car was towing a caravan. As well as the driving force and any resistive forces opposing the motion, there will be a tension force in the tow bar.

$$F = ma$$

$$D - X_\mathrm{C} - T = m_\mathrm{C}a \qquad \text{①}$$

Taken individually, without knowing the magnitude of the tension force, it may not be possible to calculate the other forces or the acceleration.

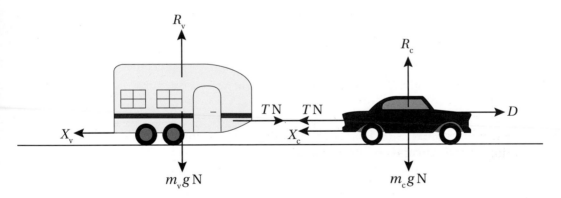

However, when you consider the horizontal forces acting upon the caravan, there will be an equal and opposite tension force as well as the caravan's own resistive forces.

$$F = ma$$

$$T - X_\mathrm{V} = m_\mathrm{V}a \qquad \text{②}$$

Hence you have two equations, ① for the car and ② for the caravan, which can be solved simultaneously. If you know the masses of the vehicles and the driving and resistive forces, then the two unknowns are T and a. Note that, because the tensions are equal and opposite, adding these equations together will eliminate T.

$$D - X_C - T = m_C a \qquad ①$$
$$T - X_V = m_V a \qquad ②$$

Add ① + ②.

$$D - X_C - T + T - X_V = m_C a + m_V a$$
$$D - X_C - X_V = m_C a + m_V a$$
$$D - (X_C + X_V) = (m_C + m_V)a$$

> ### KEY INFORMATION
> Simultaneous equations will usually be required to solve problems on connected particles.

Modelling assumptions

> Again, motion is strictly horizontal or vertical.

> Tensions are equal and opposite.

> Cables and tow bars are modelled as light inextensible strings or rods:

>> light, so that the tension is the same for both parts of the string

>> inextensible, so that the acceleration is the same for both particles.

Example 9

A car of mass 350 kg tows a trailer of mass 550 kg along a straight horizontal road. The engine of the car exerts a driving force of 1220 N. The car experiences a resistance to its motion of 200 N whilst the trailer experiences a resistance of 300 N. Find the acceleration of each vehicle and the tension in the tow bar.

Solution

As usual, it is best to represent the information in a diagram before you start.

The car is towing the trailer so in the diagram the vehicles are accelerating to the right.

Let the tension be T N, the acceleration be a m s^{-2} and the reaction forces be R_1 N and R_2 N. Note that, although they are not required in this question, they are equal and opposite to the weight of each vehicle.

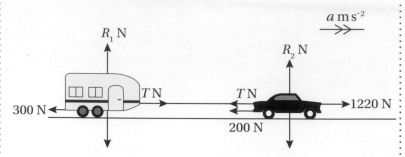

The resultant force on the car is given by
$(1220 - T - 200) = (1020 - T) \, \text{N}$.

Substitute into $F = ma$.

$$1020 - T = 350a \qquad \text{①}$$

The resultant force on the trailer is given by $(T - 300) \, \text{N}$.

Substitute into $F = ma$.

$$T - 300 = 550a \qquad \text{②}$$

Note that ① and ② are simultaneous equations and adding them together will eliminate T.

$$(1020 - T) + (T - 300) = 350a + 550a$$

$$1020 - T + T - 300 = 900a$$

$$720 = 900a$$

$$a = 0.8 \, \text{m s}^{-2}$$

You can now substitute $a = 0.8 \, \text{m s}^{-2}$ into equation ② to find the value of T.

$$T - 300 = 550 \times 0.8$$

$$T - 300 = 440$$

$$T = 740 \, \text{N}$$

Since both vehicles are accelerating at the same rate, the acceleration of each vehicle is $0.8 \, \text{m s}^{-2}$ and the tension in the tow bar is $740 \, \text{N}$.

Example 10

Two bricks, A and B, of masses 3 kg and 2 kg respectively are connected by a light cable. Initially at rest the bricks are held such that the cable is vertical as shown in the diagram (overleaf), with B 1 m above the ground. A vertical force of 66 N is applied to A, which causes the bricks to accelerate upwards. When B is 3 m above the ground the cable breaks. Show that B will travel a further $\frac{34}{49}$ m before coming to instantaneous rest.

Solution

In this example, the motion is vertical, so the weights will contribute to the resultant forces.

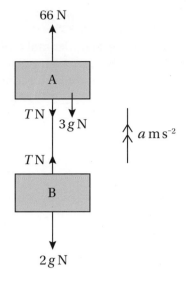

The resultant force on A is given by $(66 - T - 3g)\,\text{N}$.

Substitute into $F = ma$.

$$66 - T - 3g - 3a \quad \textcircled{1}$$

The resultant force on B is given by $(T - 2g)\,\text{N}$.

Substitute into $F = ma$.

$$T - 2g = 2a \quad \textcircled{2}$$

Again, as in the previous example, these are simultaneous equations and adding them eliminates T.

$$(66 - T - 3g) + (T - 2g) = 3a + 2a$$

$$66 - 3g - 2g = 5a$$

$$17 = 5a$$

$$a = 3.4\,\text{m s}^{-2}$$

The bricks move 2 m before the cable breaks.

$u = 0\,\text{m s}^{-1},\ s = 2\,\text{m},\ a - 3.4\,\text{m s}^{-2}$

$$v^2 = u^2 + 2as$$

$$= 0^2 + 2 \times 3.4 \times 2$$

$$= 13.6$$

$$v = \sqrt{13.6}\ \text{m s}^{-1}$$

> When the cable breaks, the tension force is removed.

Once the cable breaks, B will continue to move upwards but will slow down as the gravitational acceleration acts downwards.

$u = \sqrt{13.6}\,\text{m s}^{-1},\ v = 0\,\text{m s}^{-1},\ a = -9.8\,\text{m s}^{-2}$

$$v^2 = u^2 + 2as$$

$$s = \frac{v^2 - u^2}{2a}$$

$$= \frac{0^2 - \left(\sqrt{13.6}\right)^2}{2 \times -9.8}$$

$$= \frac{-13.6}{-19.6}$$

$$s = \frac{34}{49}\ \text{m as required.}$$

Exercise 16.4A   Answers page 563

(M) **1** Two particles, G and H, are at rest on a smooth horizontal table and connected by a light inextensible string. A force of 3 N is applied to H, which causes the particles to move. The mass of G is 150 g and the mass of H is 350 g. Find the tension in the string and the acceleration of G.

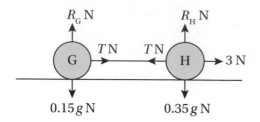

(M) **2** A particle P of mass $4m$ kg is suspended from a vertical rope. A particle Q of mass $3m$ kg is suspended from particle P by another vertical rope. A vertical force X N is applied to P such that the particles accelerate upwards at $\frac{4}{7}g$. Find the tension in the rope and the value of X, giving the answers in terms of m and g.

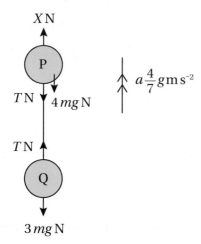

(M) **3** A van of mass 700 kg is pulling a trailer of mass 200 kg along a smooth straight horizontal lane. Given that the tension in the tow bar is 500 N, find the driving force exerted by the engine of the car.

(M) **4** A car of mass 300 kg is towing a caravan of mass 840 kg along a straight horizontal stretch of motorway. The engine of the car is exerting a driving force of 1.71 kN. Both vehicles experience resistive forces which are proportional to their masses. Find the tension in the tow bar.

(M) (PS) **5** A particle X, of mass 5 kg, is suspended from a vertical string. A second particle Y, of mass 3 kg, is suspended from X by a vertical cord. A force of 100 N is applied to the string and the two particles accelerate upwards. When the particles are travelling at 9 m s^{-1}, the cord snaps. How long does it take Y to return to its original position?

(PS) (PF) **6** A minibus of mass 950 kg is towing a small cart along a straight horizontal road. The resistances to motion are 475 N and 25 N respectively. Whilst the minibus is increasing in speed from 12 m s^{-1} to 32 m s^{-1} in 8 s, the driving force of the minibus is 3 kN. When the minibus reaches 32 m s^{-1}, the minibus and cart become uncoupled.

Show that the cart travels 1024 m before it comes to rest.

7 A car of mass M kg is towing a caravan of mass 750 kg along a straight horizontal road. The resistance on the car is X N and the resistance on the caravan is 400 N. When the engine of the car exerts a driving force of 1450 N, the car accelerates at $0.6\,\text{m s}^{-2}$; but when the engine of the car exerts a driving force 1 kN larger, the car accelerates at $1.4\,\text{m s}^{-2}$.

a Find the value of M.

b Find the value of X.

Another situation in which objects are connected is when a person is travelling in a lift. In this situation, the connection is the contact between the person and the floor of the lift, as discussed in **Example 11**.

Example 11

A woman is standing in a lift which is being pulled upwards by a cable. The woman has a mass of 80 kg and the lift has a mass of 400 kg. Given that the reaction force, R, exerted on the woman by the lift is 920 N, find the tension in the cable.

KEY INFORMATION

Problems involving lifts often require you to consider the reaction force acting on a person in the lift, as well as considering the lift as a whole.

Solution

First, represent the information in a diagram.

In this situation, there are two different resultant forces, one for the system as a whole and one for the interaction between the woman and the floor of the lift.

Considering the woman and the floor, the resultant force is given by $(R - 80g)$ N, where $R = 920$ N.

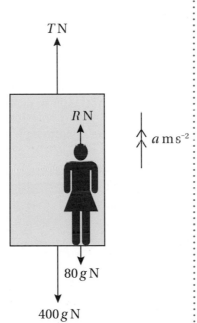

Substitute the resultant force into $F = ma$.

$$920 - 80g = 80a$$
$$920 - 80 \times 9.8 = 80a$$
$$136 = 80a$$
$$a = 1.7\,\text{m s}^{-2}$$

Considering the system as a whole, the force downwards is provided by the weights of both the woman and the lift.

$$80g + 400g = 480g$$

The resultant force is given by $(T - 480g)$ N.

Substitute into $F = ma$.

$$T - 480g = 480 \times 1.7$$
$$T = 480g + 480 \times 1.7$$
$$= 480(9.8 + 1.7)$$
$$= 5520\,\text{N}$$

Exercise 16.4B

Answers page 563

(M) **1** A teenager standing in a lift has a mass of 55 kg.

Find the reaction force exerted on the teenager by the lift when the lift is:

a accelerating upwards at $2.4\,\text{m}\,\text{s}^{-2}$

b accelerating downwards at $1.6\,\text{m}\,\text{s}^{-2}$

c decelerating upwards at $0.6\,\text{m}\,\text{s}^{-2}$.

R

$55\,g\,\text{N}$

(M) **2** A dog of mass 25 kg sits in a lift of mass 140 kg. The lift is descending at $1.8\,\text{m}\,\text{s}^{-2}$.

a Find the reaction force exerted by the lift on the dog.

b Find the tension in the lift cable.

(M) **3** Two adults, of masses 75 kg and 65 kg, enter a lift. The lift has a mass of 550 kg. The tension in the lift cable is 7480 N. Find the time it takes for the lift to ascend 24 m.

(PS) (M) **4** A lift of mass 400 kg has a notice stating that the maximum load is 160 kg. When the lift ascends with a load of this mass, the lift accelerates at $0.75\,\text{m}\,\text{s}^{-2}$. However, a caretaker carelessly puts a 200 kg crate in the lift whilst the lift is on the ground floor, without checking the notice. After the lift has been ascending for 30 s, the cable snaps and the lift returns to the ground floor. Assuming that the tension in the cable before it snaps is the same as when the load was 160 kg, how long after the cable snaps does it take for the lift to return to the ground floor?

(PS) (M) **5** A hotel has a lift of mass 280 kg.

a Margarita gets into the lift. As the lift ascends, the tension in the lift cable is 4320 N and Margarita experiences a reaction force of 960 N from the floor of the lift. Find:

i Margarita's mass

ii the acceleration of the lift.

b Margarita gets out of the lift and Guillaume and Josef get in. Now as the lift ascends, the tension in the lift cable is 5100 N. Guillaume and Josef experience reaction forces of 700 N and 900 N respectively from the floor of the lift. Find:

i Guillaume's mass

ii Josef's mass

iii the acceleration of the lift.

16.5 Pulleys

Another example of a system with two connected particles is the **pulley** system. Consider two particles, A and B, of masses 5 kg and 10 kg respectively, suspended vertically as shown in the diagram, connected by a string passing over a pulley.

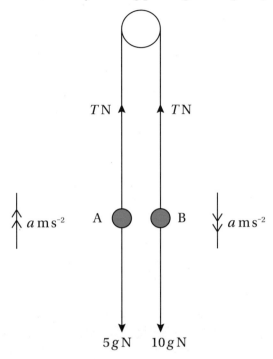

Since B is heavier than A, B will pull A. Both particles will experience a resultant force and B will accelerate downwards whilst A accelerates upwards.

The main difference between objects connected via a pulley and connected objects such as cars and trailers is that objects connected via a pulley will be travelling in different directions.

Two systems will be considered in this section:

- In the first system, both the objects travel vertically, but one is going upwards and the other is going downwards.

- In the second system, one object is suspended vertically from a pulley but the other is on a horizontal surface.

Modelling assumptions
- Motion is strictly horizontal or vertical.
- Tensions are equal and opposite.
- Strings and ropes are modelled as light and inextensible.
- Pulleys are assumed to be fixed and smooth (i.e. without friction).

KEY INFORMATION
Objects connected via a pulley will travel in different directions.

Example 12

Two particles, A and B, of masses 11 kg and 3 kg respectively, are connected to either end of a light inextensible string via a smooth pulley, as shown in the diagram. Initially, the particles are at rest. Find the tension in the string and the acceleration of the two particles when the particles are released.

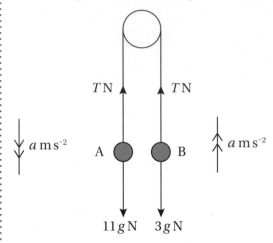

Solution

Because particle A is heavier than particle B, particle A will move downwards and particle B will move upwards.

For particle A moving downwards, the resultant force $= (11g - T)$ N.

Substitute the resultant force into the formula $F = ma$.

$$11g - T = 11a \qquad ①$$

For particle B moving upwards, the resultant force $= (T - 3g)$ N.

Substitute the resultant force into the formula $F = ma$.

$$T - 3g = 3a \qquad ②$$

Again, these are simultaneous equations. Adding them together will eliminate T.

$$11g - 3g = 11a + 3a$$

$$8g = 14a$$

$$a = \frac{8g}{14} = \frac{4g}{7}$$

$$= 5.6 \, \text{m s}^{-2}$$

Substitute $a = 5.6 \, \text{m s}^{-2}$ into equation ②.

$$T - 3g = 3a$$

$$T = 3g + 3a$$

$$= 3 \times 9.8 + 3 \times 5.6$$

$$= 46.2 \, \text{N}$$

> The resultant force for each particle will be in the direction that it is moving.

In **Example 12**, both objects travel vertically with the heavier particle, A, moving downwards and the lighter particle, B, moving upwards. In **Example 13**, although particle D travels vertically, particle C is on a horizontal table and is pulled along the table by particle D. Hence the weight of particle C will not feature in the resultant force on C.

Example 13

Particle C of mass 4 kg is connected to particle D of mass 5 kg by a light inextensible string which passes over a smooth pulley at the edge of a table. C is on the table 1.5 m from the pulley. D is suspended vertically below the pulley and 0.6 m above the floor. Both particles are held in position until they are released. In the subsequent motion, D hits the floor and C experiences a resistive force of 22 N as it is pulled across the table.

Find the distance of C from the pulley when it comes to rest.

Solution

Start by putting the information in a diagram. Let the tension $= T$ N, the acceleration $= a$ m s^{-2} and the reaction force exerted on C by the table $= R$ N.

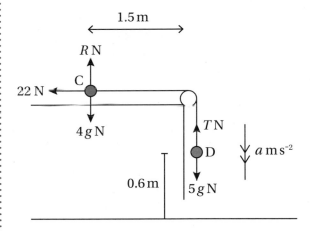

Since the question describes the motion of C and D, you can assume that the resistive force is not sufficient to prevent motion. Particle D will fall vertically, pulling C across the table.

For particle D, the resultant force is given by $(5g - T)$ N.

Substituting into $F = ma$ gives $5g - T = 5a$.   ①

For particle C moving across the table, the resultant force is given by $(T - 22)$ N.

Substituting into $F = ma$ gives $T - 22 = 4a$.   ②

Adding the simultaneous equations gives $5g - 22 = 9a$.

$$5 \times 9.8 - 22 = 9a$$

$$49 - 22 = 9a$$

$$9a = 27$$

$$a = 3\,\mathrm{m\,s^{-2}}$$

Particles C and D will both move with an acceleration of $3\,\mathrm{m\,s^{-2}}$ until D hits the floor, at which point the string will become slack and D will no longer be pulling C.

$u = 0\,\mathrm{m\,s^{-1}}$, $s = 0.6\,\mathrm{m}$, $a = 3\,\mathrm{m\,s^{-2}}$

$$v^2 = u^2 + 2as$$

$$= 0^2 + 2 \times 3 \times 0.6$$

$$= 3.6$$

$$v = 1.90\,\mathrm{m\,s^{-1}}$$

When the string becomes slack, C will be slowed down by the resistive force but there will be no tension force pulling it. Hence C will decelerate and come to rest.

$$a\,\mathrm{m\,s^{-2}}$$
$$\longrightarrow\!\!\!\!\gg$$

$$22\,\mathrm{N} \longleftarrow\!\!\!\!\!\!\bullet\ \mathrm{C}$$

The new resultant force for particle C is $-22\,\mathrm{N}$.

Substituting into $F = ma$ gives $-22 = 4a$, so $a = -5.5\,\mathrm{m\,s^{-2}}$.

$u = 1.90\,\mathrm{m\,s^{-1}}$, $v = 0\,\mathrm{m}$, $a = -5.5\,\mathrm{m\,s^{-2}}$

$$v^2 = u^2 + 2as$$

$$s = \frac{v^2 - u^2}{2a}$$

$$= \frac{0^2 - 1.90^2}{2 \times -5.5}$$

$$= 0.327\,\mathrm{m}$$

> The initial velocity when the string is slack is the final velocity when D hits the floor.

Hence the total distance travelled by C is $0.6 + 0.327 = 0.927\,\mathrm{m}$.

Particle C is $1.5 - 0.927 = 0.573\,\mathrm{m}$ from the pulley when it comes to rest.

Exercise 16.5A

M **1** Particles J and K are connected by a light inextensible string which passes over a smooth pulley. J has a mass of 8 kg and K has a mass of 3 kg.

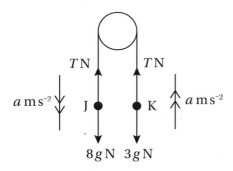

When the system is released from rest, find:

a the acceleration of J

b the tension in the string.

M **2** Particles G and H, of masses $3m$ and $2m$, are connected by a light inextensible string with G at rest on a smooth table. The string passes over a smooth fixed pulley at the edge of the table with H hanging freely beneath the pulley.

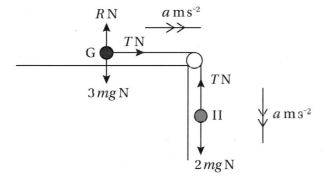

The system is released from rest.

a Find the force exerted on G by the table in terms of m and g.

b Find the acceleration in terms of g.

c Find the tension in terms of m and g.

M **3** Two particles, of masses 3 kg and 4 kg respectively, are connected by an inextensible string via a smooth fixed pulley, both hanging vertically at the same height. The system is released from rest.

a Find the acceleration in terms of g.

b Find the tension in terms of g.

c Find the difference in height between the particles after 1.4 s.

(M) **4** Particles P_1 and P_2 are connected by a light inextensible cord passing over a smooth pulley. P_1 has a mass of 4 kg and P_2 is heavier than P_1. When the system is released from rest, the acceleration of each particle is equal to $\frac{1}{5}g$ m s^{-2}.

    a Find the mass of P_2.

    b Find the force exerted on the pulley.

(M) **5** Particles A and B, of masses 3 kg and 7.5 kg, are connected by a light inextensible wire that passes over a smooth pulley. A is at rest on a smooth table 5 m from the pulley whilst B is hanging freely 2 m above the ground.

    When the system is released, how long does it take A to reach the pulley?

(PF) (M) **6** Two particles, of masses x kg and y kg, are connected by a light inextensible string passing over a smooth fixed pulley. Both particles are hanging freely.

    Show that when the system is released from rest, the tension is given by $\dfrac{2xyg}{x+y}$.

(M) **7** A ball of mass 500 g is suspended vertically from a smooth fixed pulley at the edge of a table. The ball is connected via a light inextensible string and the pulley to a flowerpot of mass 300 g on the table. The flowerpot is subject to a resistance to motion of 2 N. The ball is released and the string snaps after 0.2 s.

    Assuming that the flowerpot does not reach the pulley, find the total time the flowerpot is in motion.

(PS) (M) **8** Two rocks, P and Q, are connected by an inextensible rope passing over a smooth pulley. Q is 3 kg heavier than P.

    Given that the tension in the rope is 35.91 N, find the mass of each rock and the acceleration of the system.

> **Stop and think**
>
> In **question 1**, J is heavier than K so J moves down whilst K moves up. However, in **question 2**, H is lighter than G and yet is still able to pull it. How is this possible?

SUMMARY OF KEY POINTS

› You should know and be able to apply Newton's three laws of motion:

  › Newton's first law states that a particle will remain at rest or continue to move at a constant velocity in a straight line unless acted upon by a resultant force.

  › Newton's second law states that the resultant force F acting upon a particle is proportional to the mass m of the particle and the acceleration a of the particle such that $F = ma$.

  › Newton's third law states that every action has an equal and opposite reaction.

› Using SI units, 1 newton = $1 \, \text{kg m s}^{-2}$.

› There are five types of force you need to be aware of:

  › The normal reaction acts perpendicular to a surface.

  › Tension is the force in a taut string and acts away from an object.

  › Thrust (or compression) is the force in an inflexible rod and acts towards an object.

  › Resistive forces oppose motion.

  › Weight ($W = mg$) acts vertically downwards. Usually g is taken as $9.8 \, \text{m s}^{-2}$ but values of $9.81 \, \text{m s}^{-2}$ or $10 \, \text{m s}^{-2}$ may be used as well.

› The resultant force should be found in the direction of motion.

› Objects are modelled as particles so their mass is assumed to be concentrated at a single point.

› String is modelled as light and inextensible:

  › The string is modelled as light so that the tension is the same for both parts of the string.

  › The string is modelled as inextensible so that the acceleration is the same for both particles.

› Pulleys are modelled as smooth.

› Problems involving lifts often require you to consider the reaction force acting on a person in the lift as well as considering the lift as a whole.

› Problems involving connected particles usually require the solution of simultaneous equations.

› If two particles are connected by a string, then the tensions are equal and opposite.

› Particles connected via a pulley will be travelling in different directions.

EXAM-STYLE QUESTIONS 16

Answers page 564

1 A particle of mass 0.56 kg moves under the action of a single force **F** N. The acceleration of the particle is $(-14\mathbf{i} + 48\mathbf{j}) \, \text{m s}^{-2}$.

  a Which of these is the angle between the acceleration and the vector **i**?

  A 16°   **B** 74°

  C 106°   **D** 164°   **[2 marks]**

  b Find the magnitude of the force **F**.   **[3 marks]**

(M) **2** A particle A of mass 4 kg rests on a smooth horizontal table. Particle A is attached to one end of a light inextensible string which passes over a smooth pulley fixed at the edge of the table. The other end of the string is attached to a particle B of mass 3 kg which hangs freely below the pulley 1.3 m above the ground. The system is released from rest with the string taut. In the subsequent motion, particle A does not reach the pulley before B reaches the ground.

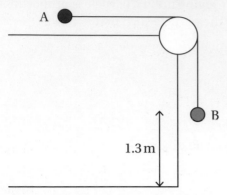

a Find the tension in the string before B reaches the ground. **[5 marks]**

b Find the time taken by B to reach the ground. **[3 marks]**

(CM) (PS) **3** A student is set the following question:

'Two particles are attached to the ends of a light inextensible string which passes over a smooth fixed pulley. The particles are held in position with the string taut and the hanging parts of the string vertical. The particles are then released from rest. The particles have masses of 6 kg and m kg, where $m < 6$, and the initial acceleration of each particle has a magnitude of $\frac{2}{3}g$ m s^{-2}.' Find:

a the tension in the string immediately after the particles are released

b the value of m.

The student's solution is shown below (continuing on to the next page):

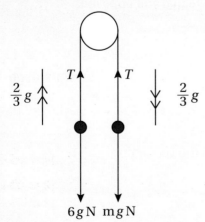

a Left:

$$T - 6g = 6 \times \frac{2}{3}g = 4g$$

$$T = 10g = 98\,\text{N}$$

b Right:

$$mg - T = \frac{2}{3}mg$$

$$mg - 98 = \frac{2}{3}mg$$

$$\frac{1}{3}mg = 98$$

$$mg = 294$$

$$m = 294 \div 9.8 = 30\,\text{kg}$$

Assess the student's solution and make corrections where necessary.　　　　**[7 marks]**

(M) **4** The forces $\mathbf{F}_1$ and $\mathbf{F}_2$ are given by $\mathbf{F}_1 = \begin{bmatrix} -8 \\ 3 \end{bmatrix}\,\text{N}$ and $\mathbf{F}_2 = \begin{bmatrix} -11 \\ -6 \end{bmatrix}\,\text{N}$.

Find values for a and b such that:

a $a\mathbf{F}_1 + b\mathbf{F}_2 = \begin{bmatrix} 10 \\ 3 \end{bmatrix}\,\text{N}$　　　　**[3 marks]**

b $a\mathbf{F}_1 + b\mathbf{F}_2$ is parallel to the force $\begin{bmatrix} -2 \\ 3 \end{bmatrix}\,\text{N}$.　　　　**[4 marks]**

(M) **5** Two particles, A and B, are connected by a light inextensible vertical cable with A above B. The mass of A is 500 g and the mass of B is 200 g. Initially the system is at rest. A vertical force of X N is applied to A which raises the particles 24 m in 4 s. Find the magnitude of X.　　　　**[5 marks]**

(M) (CM) **6** A man uses the lift in a hotel. The lift starts from rest on the ground floor and moves vertically. The lift initially accelerates with a constant acceleration of $0.8\,\text{m s}^{-2}$ for 15 s. It then decelerates for 10 s before coming to rest.

The mass of the man is 90 kg and the mass of the lift is 160 kg. The lift is pulled up by means of a vertical cable attached to the top of the lift. By modelling the cable as light and inextensible, find:

a the tension in the cable when the lift is accelerating　　　　**[3 marks]**

b the magnitude of the force exerted by the lift on the man when the lift is decelerating.　　　　**[3 marks]**

c Explain how you have used the assumption that the cable is light.　　　　**[1 mark]**

(M) **7** A particle is acted upon by three forces $\mathbf{F}_1$, $\mathbf{F}_2$ and $\mathbf{F}_3$, given by:

　　　$\mathbf{F}_1 = (q\mathbf{i} - 4q\mathbf{j})\,\text{N}$, where q is a positive constant

　　　$\mathbf{F}_2 = (7\mathbf{i} + 3\mathbf{j})\,\text{N}$

　　　$\mathbf{F}_3 = (-2\mathbf{i} - 17\mathbf{j})\,\text{N}$

The resultant of the three forces, $\mathbf{R}$, is parallel to $(\mathbf{i} - \mathbf{j})$.

a Find the value of q.　　　　**[4 marks]**

b Find the magnitude of $\mathbf{R}$, correct to 2 d.p.　　　　**[2 marks]**

The force $\mathbf{F}_1$ is removed.

c Find the angle the new resultant force makes the vector $\mathbf{i}$, correct to the nearest degree.　　　　**[3 marks]**

Ⓜ **8** A particle P of mass 7 kg is projected vertically upwards from a point A with speed 22 m s^{-1}. The point A is 12.5 m above horizontal ground. The particle P moves freely under gravity until it reaches the ground with speed V m s^{-1}.

a Find the value of V. **[3 marks]**

The ground is soft and, after P reaches the ground, P sinks vertically downwards into the ground before coming to rest. The ground exerts a constant resistive force of magnitude 3000 N on P.

b Find the vertical distance that P sinks into the ground before coming to rest. **[4 marks]**

Ⓜ **9** Two particles P_1 and P_2, of mass 4.5 kg and 5.5 kg respectively and initially at rest, are joined by a light horizontal rod as shown in the diagram. A constant force of magnitude 7 N is applied to P_2 and the system moves under the action of this force for 10 s. During the motion, the resistance to the motion of P_1 has a constant magnitude of 2 N and the resistance to the motion of P_2 has a constant magnitude of 3 N.

Find:

a the acceleration of the particles as the system moves under the action of the 7 N force **[3 marks]**

b the speed of the particles after 10 s **[2 marks]**

c the tension in the rod as the system moves under the action of the 7 N force. **[3 marks]**

After 10 s, the 7 N force is removed and the system decelerates to rest.
The resistances to motion are unchanged. Find:

d the distance moved by P_1 as the system decelerates **[4 marks]**

e the thrust in the rod as the system decelerates. **[3 marks]**

Ⓜ **10** A particle Q of mass 5 kg moves under the action of a single force **F** N. The initial velocity of Q is $\begin{bmatrix} -9 \\ 7 \end{bmatrix}$ m s^{-1} and 4 s later its velocity is $\begin{bmatrix} 7 \\ -5 \end{bmatrix}$ m s^{-1}.

a Find the angle between the direction of motion of Q after 4 s and the vector **j**. **[3 marks]**

b Find the magnitude |**F**|. **[5 marks]**

c Find the time at which Q is moving parallel to the vector $\begin{bmatrix} 2 \\ 1 \end{bmatrix}$. **[5 marks]**

Ⓜ ⓒⓜ **11** A car of mass 750 kg is pulling a trailer of mass 250 kg along a straight horizontal road using a tow bar. The resistances to motion of the car and the trailer are 300 N and 100 N respectively. The engine of the car is exerting a constant driving force of 1.5 kN. The car and trailer are modelled as particles.

a Find the acceleration of the car and trailer. **[3 marks]**

b Find the magnitude of the tension in the tow bar. **[3 marks]**

c How have you used the assumption that the car and trailer can be modelled as particles? **[1 mark]**

The car is moving along the road when the driver sees a hazard ahead. He reduces the force produced by the engine to zero and applies the brakes. The brakes produce a force on the car of magnitude F N, and both the car and trailer decelerate.

d Given that the resistances to motion are unchanged and the magnitude of the thrust in the tow bar is 125 N, find the value of F. **[7 marks]**

12 Two particles, A and B, are such that A is heavier and has a mass of 2.2 kg. The particles are connected by a light inextensible string which passes over a smooth, fixed pulley. Initially particle A is 3.75 m above a horizontal surface. The particles are released from rest with the string taut and the hanging parts of the string vertical. After particle A has been descending for 2.5 s, it strikes the ground. Particle A reaches the ground before particle B has reached the pulley.

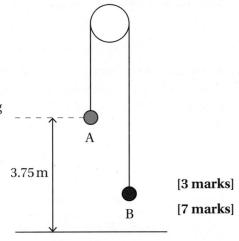

a Show that the acceleration of A as it descends is $1.2\,\mathrm{m\,s^{-2}}$. **[3 marks]**

b Show that the mass of particle B is 1.72 kg. **[7 marks]**

c State how you have used the information that the string is light. **[1 mark]**

When particle A strikes the ground, it does not rebound and the string becomes slack. Particle B then moves freely under gravity, without reaching the pulley, until the string becomes taut again.

d Find the time between the instant when particle A strikes the ground and the instant when the string becomes taut again. **[6 marks]**

13 A car which has run out of petrol is being towed by a breakdown truck along a straight horizontal road. The truck has mass 1600 kg and the car has mass 900 kg. The truck is connected to the car by a horizontal rope which is modelled as light and inextensible. The truck's engine provides a constant driving force of 2500 N. The resistances to motion of the truck and the car are modelled as constant and of magnitude 500 N and 300 N respectively. Find:

a the acceleration of the truck **[3 marks]**

b the tension in the rope. **[3 marks]**

c How have you used the modelling assumption that the rope is inextensible? **[1 mark]**

When the breakdown truck is moving at $25\,\mathrm{m\,s^{-1}}$, the rope breaks. The engine of the truck provides the same driving force as before and the magnitude of the resistance to the motion of the truck is assumed not to have changed.

d Show that the truck reaches a speed of $37\,\mathrm{m\,s^{-1}}$ approximately 8 seconds earlier than it would have done if the rope had not broken. **[7 marks]**

EXAM-STYLE EXTENSION QUESTIONS

Chapter 1 Algebra and functions 1: Manipulating algebraic expressions

(PF) 1   **a**   Show that $8^{\frac{1}{3}} = \pm 2$ is not true.   **[2 marks]**

      b   Find the value(s) of $(-8)^{-\frac{2}{3}}$.   **[3 marks]**

      c   Simplify $\left(-8y^{\frac{1}{4}}\right)^{-\frac{2}{3}}$.   **[4 marks]**

2   **a**   Find the remainder when $x^3 - 6x^2 + 3x + 10$ is divided by:

      i $x - 1$ **[2 marks]**   **ii** $x + 1$ **[1 mark]**

      b   Hence, or otherwise, find all the solutions to the equation $x^3 - 6x^2 + 3x + 10 = 0$.   **[4 marks]**

3   **a**   Find the first three terms of the expansion in ascending powers of x of $(1+x)(2-5x)^3$.   **[4 marks]**

      b   Hence, or otherwise, find an approximation of $(1.01)(1.95^3)$.   **[3 marks]**

(PF) 4   Show that $\dfrac{a+\sqrt{b}}{c-\sqrt{b}}$ can be written as $\dfrac{ac+b}{c^2-b} + \dfrac{\sqrt{b}(a+c)}{c^2-b}$.   **[5 marks]**

(PS) 5   Simplify $\dfrac{18x^2 - 71x + 28}{4x^3 - 24x^2 + 29x + 21}$.   **[6 marks]**

(PS) 6   **a**   Find the first three terms, in ascending powers of x, of the binomial expansion of $(1+px)^{10}$ where p is a constant.   **[4 marks]**

      b   The first 3 terms are 1, $30x$ and $27qx^2$ where q is a constant. Find the values of p and q.   **[2 marks]**

      c   Hence, or otherwise, find an approximation of 1.03^{10}.   **[2 marks]**

      d   Explain why your answer in **part c** is only an approximation.   **[1 mark]**

7   **a**   Fully factorise $6x^4 - 47x^3 + 81x^2 \div 7x - 15$.   **[5 marks]**

      b   Hence, or otherwise, solve $6x^4 - 47x^3 + 81x^2 - 7x - 15 = 0$.   **[2 marks]**

(PS) 8   **a**   Use the binomial expansion to expand $\left(3-\sqrt{x}\right)^4$.   **[4 marks]**

      b   Using your expansion from **part a**, write down the expansion of $\left(\dfrac{3}{2} - \dfrac{\sqrt{x}}{2}\right)^4$.   **[2 marks]**

(PS) 9   Simplify $\dfrac{2x^3 + x^2 - 12x + 9}{2x^4 + x^3 - 8x^2 - x + 6}$.   **[8 marks]**

(PF) 10   Show that $\dfrac{4x}{1-\sqrt{x}} + \dfrac{3\sqrt{x}+1}{\sqrt{x}}$ can be written as $\dfrac{x+3}{1-x} + \dfrac{\sqrt{x}\left(4x^2 - x + 1\right)}{x(1-x)}$.   **[6 marks]**

Chapter 2 Algebra and functions 2: Equations and inequalities

1   **a**   Find the set of values of x for which $7 < 2x + 7 < 17$ where x is an integer.   **[3 marks]**

      b   **i**   Solve $2x^2 + 12x + 7 = 0$ giving any solutions as exact values.   **[3 marks]**

          ii   Hence, or otherwise, find the set of values for which $2x^2 + 12x + 7 < 0$.   **[2 marks]**

2   Solve the simultaneous equations $x + y = 1$ and $x^2 + 3y = 7$.   **[4 marks]**

3 **a** Solve the inequality $5x - 3 > \dfrac{9}{x}$. **[5 marks]**

 b Hence, or otherwise, solve the inequality $5x - 3 < \dfrac{9}{x}$. **[2 marks]**

(CM) **4** Consider the equation $y = ax^2 + bx + c$.

 a Express b in the discriminant in terms of a and c where y has two distinct real roots. **[2 marks]**

 b Express b in the discriminant in terms of a and c where y has two equal real roots. **[2 marks]**

 c Express b in the discriminant in terms of a and c where y has no real roots. **[2 marks]**

5 Consider the function $f(x) = 4x^2 - 15x + 14$.

 a Solve $f(x) = 0$. **[3 marks]**

 b Find the coordinates of the turning point of $f(x)$ by completing the square. **[3 marks]**

 c Sketch the graph of $f(x)$. **[3 marks]**

(PS) **6** Find the range of possible values of b for which $2x^2 + bx - 9 = 0$ has:

 a no real roots **[2 marks]**

 b two equal roots **[2 marks]**

 c two distinct real roots. **[2 marks]**

(PS) **7** **a** Solve the simultaneous equations $(x-3)^2 + (y+2)^2 = 25$ and $x + y = 7$. **[5 marks]**

 b Hence, or otherwise, explain whether $x + y = 7$ is a tangent to the curve $(x-3)^2 + (y+2)^2 = 0$. **[2 marks]**

(PS) **8** Derive the quadratic formula for the equation $x^2\sqrt{a} + x\sqrt{b} + \sqrt{c} = 0$ by completing the square. **[8 marks]**

(CM) **9** **a** Solve the simultaneous equations $2x - 3y = 3$, $z + 9y = 0$ and $z^2 - x - 7$. Clearly state the solutions. **[8 marks]**

 b Explain the relationship between the number of equations and the number of variables, for all values to be found simultaneously. **[2 marks]**

(M) (PS) **10** The curved surface area of a cone is given by the formula $\pi r l$ where r is the radius of the base of the cone and l is the slant length from the base of the cone to the point. The slant length of a cone is 25 units. If the curved surface area of the cone cannot exceed 450 units^2 and the area of the base of the cone must be greater than 100 units^2, find the range of values for the radius, r, to 2 decimal places. **[7 marks]**

Chapter 3 Algebra and functions 3: Sketching curves

1 **a** Sketch the graph of the curve with equation $y = x^2 - 2x - 3$. State the coordinates of the points of intersection of the curve with the axes. **[3 marks]**

 b Find the coordinates of the point(s) of intersection of the curve with equation $y = x^2 - 2x + 3$ and the line $y = x + 2$ giving your answer in exact form. **[5 marks]**

2 The figure shows a sketch of the curve with equation $y = f(x)$.

The curve crosses the x-axis at the points $(1, 0)$ and $(2, 0)$ and the turning point is $(1.5, 1.25)$. The graph crosses the y-axis at $(0, -10)$.

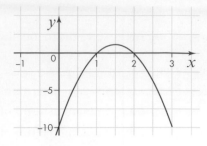

On separate diagrams sketch the following functions:

a i $f(-x)$ **[2 marks]**

ii $3f(x)$ **[2 marks]**

iii $f(x - 1)$ **[2 marks]**

On each diagram show clearly the coordinates of any intercept points with the x-axis and the coordinates of the turning point.

b Find the transformation that would result in the curve crossing the x-axis at the point $(-1, 0)$ and the turning point being at $(-0.5, 1.25)$. The graph would also go through the origin. **[3 marks]**

 3 **a** The surface area, S, of a sphere is directly proportional to the square of its radius, r. If $S = 36\pi \text{ cm}^2$ when $r = 3 \text{ cm}$, work out the surface area of a sphere with a radius of 7 cm. Sketch the graph of the relationship between S and r.

b Use your graph to find:

i S when $r = 4 \text{ cm}$ **[2 marks]** **ii** r when $S = 288\pi \text{ cm}^2$ **[2 marks]**

 4 $x \propto y^2$ and $z \propto \sqrt[3]{x}$

a Show that z is directly proportional to $y^{\frac{2}{3}}$. **[3 marks]**

b If $y = 8$ when $z = 21$ find the exact value of z when $y = 5$. **[3 marks]**

 5 The figure shows a sketch of the curve with equation $y = f(x)$.

State the transformation for $f(x)$ to match each of the following graphs. Justify your choice of transformation in each case.

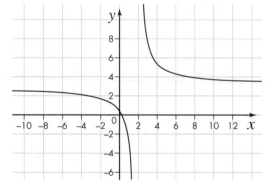

a

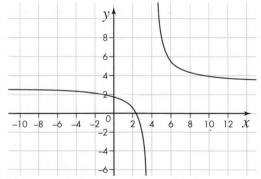

[2 marks]

b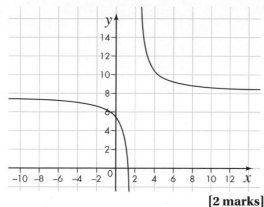

c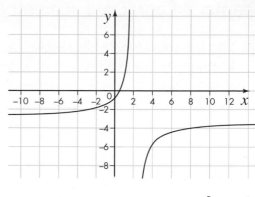

[2 marks]   [2 marks]

(PS) 6 **a** Sketch the curve given by the function $f(x) = \dfrac{1}{(x-2)^2} + 3$.

The function f(x) is transformed using single, simple transformations. For each of the following equations, work out the corresponding transformation and sketch the curve.

b i $\dfrac{1}{(2x-2)^2} + 3$ **[2 marks]**   **ii** $\dfrac{1}{(x+1)^2} + 3$ **[2 marks]**   **iii** $\dfrac{-1}{(x-2)^2} - 3$ **[2 marks]**

(PS) 7 **a** Sketch the graphs of $f(x) = (x-1)(x+2)^2$ and $g(x) = \dfrac{1}{x}$ on the same pair of axes.
Clearly highlight any asymptotes or points of intersection.
Write down the number of points of intersection.   **[3 marks]**

b On a separate pair of axes, sketch the graphs of –f(x) and g(x) – 8.
Clearly highlight any asymptotes or points of intersection.
Do the transformed graphs intersect? If so, how many times?   **[3 marks]**

(CM) 8 **a** Find the quartic equation for the following graph.
Justify your choice of equation.   **[3 marks]**

b Sketch the graph of $f(x) = (x+3)^2(2x+3)^2$ clearly
showing the coordinates of any intercepts with
the x-axis.   **[3 marks]**

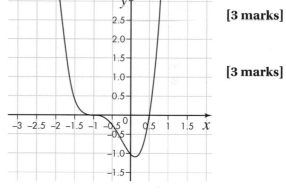

(PS) 9 **a** Given that $f(x) = 3^{x+2}$, on the same pair of axes sketch the graphs of f(3x) and 3f(x).   **[4 marks]**

b If originally $f(x) = 3^x$, what transformations would have been required to
achieve the same results as **part a** above?   **[2 marks]**

c For each transformation, in **part b** above, find the equation of the transformed function.   **[2 marks]**

(CM) 10 Given that f(x) = tan$x$ for –270° ≤ x ≤ 270°, sketch, on different pairs of axes, the following transformations:

a f(–x)   **[2 marks]**   **b** –f(x)   **[2 marks]**
c f(2x)   **[2 marks]**   **d** 2f(x)   **[2 marks]**
e Explain the reason(s) why there are similarities between graphs **a** and **b** but not graphs **c** and **d**.
                                                                                                                                        [2 marks]

Chapter 4 Coordinate geometry 1: Equations of straight lines

1 Line l_1 has equation $2y - 3x - 4 = 0$.

   a Find the gradient of line l_1.   **[2 marks]**

   A second line l_2 is parallel to line l_1 and passes through the point $(4, 2)$.

   b Find the equation of line l_2, give your answer in the form $ax + by + c = 0$
where a, b and c are integers.   **[3 marks]**

2 A line, l, passes through the points $(-4, 2)$ and $(2, 1)$.

   Find an equation for line l in the form $ax + by + c = 0$.   **[4 marks]**

(PS) (CM) (M) 3 Sarah is entering a sunflower growing competition. She records the height of the
sunflower each day. She starts recording with day 0 and a height of 0 cm. She records
the height of her sunflower on day 40 as 50 cm. She decides to use a straight line model
to predict the height y cm of her sunflower after x days.

   a Find a straight line equation that Sarah would use for this model.   **[1 mark]**

   b What is the meaning of the gradient of the straight line equation?   **[1 mark]**

   c Use the model to estimate how long it will take for the sunflower to reach 3.1 metres.   **[2 marks]**

   d Comment on the validity of your answer to **part c**.   **[1 mark]**

4 A line l_1 has equation $y - 3x - 2 = 0$.

   a Find the gradient of line l_1.   **[2 marks]**

   A second line l_2 is perpendicular to line l_1 and passes through the point $(3, 4)$.

   b Find the equation of line l_2, give your answer in the form $ax + by + c = 0$.   **[4 marks]**

(PF) (CM) 5 Line l_1 has equation $2y + x - 4 = 0$ and line l_2 has equation $y - 2x + 1 = 0$.

   a Show that line l_1 is perpendicular to line l_2.   **[4 marks]**

   b Find the coordinates of the point of intersection of line l_1 and line l_2.   **[3 marks]**

(PF) (PS) 6 The points with coordinates $(-3, 3)$, $(4, 4)$ and $(3, 6)$ are the vertices of a triangle.
Show that the triangle is a right-angled triangle.   **[4 marks]**

(PS) (CM) (M) 7 Plumber A uses the formula $y = 15x + 30$ to work out the cost y (£) for a job that takes x hours.

   Plumber B has a call-out charge of £15 and an hourly rate of £18.

   a Write a formula that plumber B could use to work out the cost y (£) for
a job that takes x hours.   **[1 mark]**

   b Jenni needs a plumber to complete a job that she expects to take a total of $3\frac{1}{2}$ hours.
Which plumber would be cheaper for her to use?   **[2 marks]**

   c Sanjit needs a plumber for a job. Both plumbers quote the same total
cost for the job. How long is the job expected to take?   **[2 marks]**

(PF) (CM) (PS) 8 Show that the points with coordinates $(-3, -2)$, $(1, -1)$ and $(6, \frac{1}{4})$ are in a straight line.   **[3 marks]**

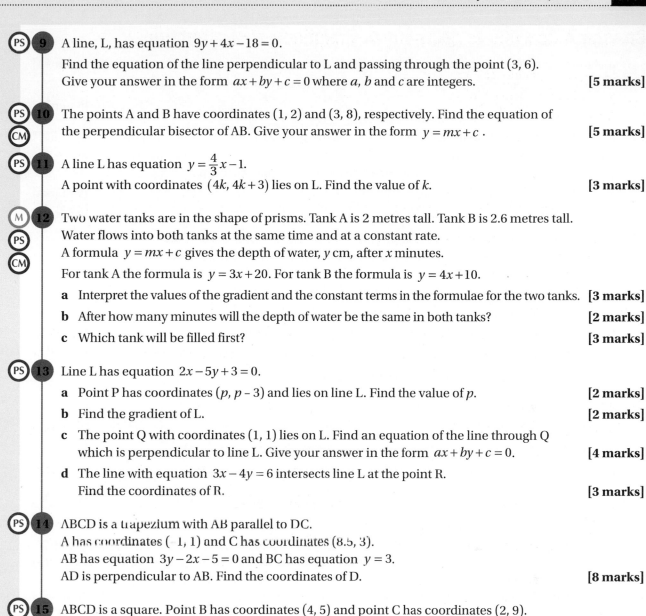

(PS) **9** A line, L, has equation $9y + 4x - 18 = 0$.

Find the equation of the line perpendicular to L and passing through the point (3, 6).
Give your answer in the form $ax + by + c = 0$ where a, b and c are integers. **[5 marks]**

(PS) (CM) **10** The points A and B have coordinates (1, 2) and (3, 8), respectively. Find the equation of the perpendicular bisector of AB. Give your answer in the form $y = mx + c$. **[5 marks]**

(PS) **11** A line L has equation $y = \frac{4}{3}x - 1$.
A point with coordinates $(4k, 4k + 3)$ lies on L. Find the value of k. **[3 marks]**

(M) (PS) (CM) **12** Two water tanks are in the shape of prisms. Tank A is 2 metres tall. Tank B is 2.6 metres tall.
Water flows into both tanks at the same time and at a constant rate.
A formula $y = mx + c$ gives the depth of water, y cm, after x minutes.

For tank A the formula is $y = 3x + 20$. For tank B the formula is $y = 4x + 10$.

a Interpret the values of the gradient and the constant terms in the formulae for the two tanks. **[3 marks]**

b After how many minutes will the depth of water be the same in both tanks? **[2 marks]**

c Which tank will be filled first? **[3 marks]**

(PS) **13** Line L has equation $2x - 5y + 3 = 0$.

a Point P has coordinates $(p, p - 3)$ and lies on line L. Find the value of p. **[2 marks]**

b Find the gradient of L. **[2 marks]**

c The point Q with coordinates (1, 1) lies on L. Find an equation of the line through Q which is perpendicular to line L. Give your answer in the form $ax + by + c = 0$. **[4 marks]**

d The line with equation $3x - 4y = 6$ intersects line L at the point R.
Find the coordinates of R. **[3 marks]**

(PS) **14** ABCD is a trapezium with AB parallel to DC.
A has coordinates (-1, 1) and C has coordinates (8.5, 3).
AB has equation $3y - 2x - 5 = 0$ and BC has equation $y = 3$.
AD is perpendicular to AB. Find the coordinates of D. **[8 marks]**

(PS) **15** ABCD is a square. Point B has coordinates (4, 5) and point C has coordinates (2, 9).
AD has equation $y = -2x + 3$.
Work out the coordinates of points A and D. **[10 marks]**

Chapter 5 Coordinate geometry 2: Circles

(PF) **1** The circle C has the equation $x^2 + y^2 - 8x + 20y + 35 = 0$.

a Find the coordinates of the centre of C. **[3 marks]**

b Show that the radius of C is 9. **[2 marks]**

(PF) **2** A circle has centre (3, –8) and a radius of 13.

a Verify that the point A(–9, –3) lies on the circumference. **[2 marks]**

b Find the equation of the tangent to the circle at A. Give your answer in the form $ax + by + c = 0$, where a, b and c are integers. **[4 marks]**

3 Line L_1 passes through the points $(7, -3)$ and $(9, -5)$.

   a Find the equation of the line L_1 in the form $ax + by = c$, where a, b and c are integers. **[3 marks]**

   Circle C has the equation $(x + 3)^2 + (y - 2)^2 = 17$. L_1 intersects C at the points G and H.

   b Find the midpoint of GH. **[6 marks]**

(M) **4** A circular piazza is drawn on a coordinate grid. Each square on the grid represents $1\,\text{m}^2$. The entrance and exit of the piazza are at opposite ends of a diameter of the piazza with coordinates $(-3, 4)$ and $(5, 10)$ respectively. The entrance and exit are joined by a straight yellow path.

   a Find the length of the yellow path. **[2 marks]**

   b Find the equation of the circle in the form $x^2 + y^2 + ax + by + c = 0$. **[4 marks]**

   At the exit, there is a hedge running perpendicular to the yellow path.

   c Find the equation of the hedge in the form $ax + by = c$, where a, b and c are integers. **[3 marks]**

5 The y-axis is a tangent of the circle C. The centre of C is the point X$(5, 3)$. C intersects the x-axis at the points A and B. The x-coordinate of B is greater than the x-coordinate of A.

   a Find the coordinates of the points A and B. **[4 marks]**

   b Find the area of the triangle ABX. **[2 marks]**

(PS) **6** The circle C has centre O$(-2, 5)$. The point A$(4, -7)$ lies on C.

   a Find the equation of the circle. Write the answer in the form $(x - a)^2 + (y - b)^2 = c$. **[3 marks]**

   Line L is the perpendicular bisector of the line segment AO. Points X and Y lie on C and L.

   b Find the distance XY, giving the answer in the form $a\sqrt{15}$. **[4 marks]**

(PF) **7** The points Y$(-1, -5)$ and Z$(11, 1)$ lie on the circumference of circle C.

   a Show that the centre of the circle lies on the straight line L with equation $2x + y = 8$. **[5 marks]**

   Given that X$(-4, 6)$ also lies on the circumference of C,

   b find the coordinates of the centre of C. **[6 marks]**

   c Hence, or otherwise, find the equation of the circle in the form $(x - a)^2 + (y - b)^2 = r^2$. **[3 marks]**

(PF) **8** A circle has the equation $(x + 4)^2 + (y - 9)^2 = 40$.

   a Show that the line L with equation $y = 3x + 1$ is a tangent to the circle. **[5 marks]**

   b Find the equation of the other tangent to the circle which is parallel to L. **[4 marks]**

9 The equation of circle C is $(x + 2)^2 + (y - 11)^2 = 49$.

   a State the centre and radius of C. **[2 marks]**

   The point P has coordinates $(-3, 24)$. T is a point on the circumference of C.

   Given that PT is a tangent of C,

   b find the length of PT. **[4 marks]**

10 The circle C has the equation $x^2 + y^2 - 14x - 26y + 138 = 0$.

   Find the equations of all the tangents to C of the form $y = \frac{1}{2}x + c$. **[10 marks]**

Chapter 6 Trigonometry

1 Solve the equation $\tan^2 x° = 2$, $0 \leqslant x < 360$. **[3 marks]**

2 ABC is a triangular piece of land.
AB = 15 m, BC = 20 m and angle BAC = 85°.
Calculate the area of the triangle. **[3 marks]**

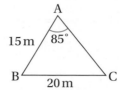

3 $\sin\theta = \dfrac{1}{\sqrt{3}}$ and $90° < \theta < 180°$.
Find the exact value of:

a $\cos\theta$ **[1 mark]** **b** $\tan\theta$ **[1 mark]**

(M) 4 A walker travels 2.5 km on a bearing of 035° from A to B

Then she walks 4.6 km on a bearing of 172° from B to C.

Finally she walks 3.7 km east from C to D.

Find the distance in a straight line from A to D. **[5 marks]**

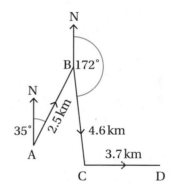

(M) 5 Here is an equation: $4\sin^2 x + 3\cos x = 4$

Here is a student's solution.

$$4\sin^2 x + 3\cos x = 4$$
$$4(1 - \cos^2 x) + 3\cos x = 4$$
$$-4\cos^2 x + 3\cos x = 0$$
$$4\cos^2 x = 3\cos x$$
$$4\cos x = 3$$
$$\cos x = \frac{4}{3}$$

This has no solution because the cosine cannot be greater than 1.

The student's solution contains two mistakes.

a Identify the two errors. **[1 mark]** **b** Solve the equation. **[2 marks]**

6 The equation of this graph is $y = \cos(ax + b)°$.
Find suitable values of the constants a and b. **[3 marks]**

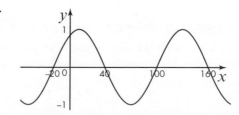

PS **7** P and Q are two points on the graph of $y = \tan kx$ and both have the same y-coordinate.

The length of PQ is 10. Find all the possible value of k. **[3 marks]**

8 Find where the graphs of $y = 5\sin 2x°$ and $y = 8\cos 2x°$ cross, when $0 < x < 360$. **[2 marks]**

PS **9** Two discs A and B have radius 5 cm and 10 cm respectively.

Point P on disc A and point Q on disc B are initially together, as shown in the diagram.

When disc A rotates anticlockwise, disc B rotates clockwise, without slipping.

Disc A rotates 500° clockwise.

Calculate the distance apart of P and Q now. **[4 marks]**

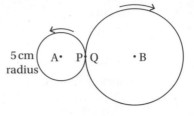

10 ABCD is a quadrilateral.

AB = 21 cm, AD = 24 cm, BC = 17 cm, CD = 28 cm

Angle $A = 70°$

Calculate the other angles. **[5 marks]**

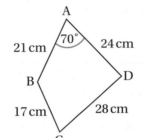

PS **11** ABC is a triangle. AB = 15 cm, BC = 20 cm and the area is 120 cm².

Find the length of AC. **[4 marks]**

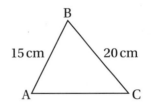

12 Solve the equation $\tan^2 x° = \tan x° + 12$, $0 \leqslant x < 360$. **[3 marks]**

PS **13** PQRS is a quadrilateral.

PQ = 8 cm, RS = 12 cm, QR = x cm, PS = $2x$ cm

Angle P = angle $R = 60°$

Find the value of x. **[5 marks]**

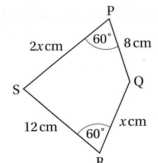

14 Solve the equation $\sin x° = 2\cos^2 x°$, $0 \leqslant x < 360$. **[3 marks]**

15 Solve the equation $5\sin x° + 9 = 12\cos^2 x$, $0 \leqslant x < 360$. **[4 marks]**

Chapter 7 Exponentials and logarithms

(PF) 1 Prove that:

a $\log_{n^2} a = \frac{1}{2}\log_n a$ **[2 marks]** **b** $\log_a b \times \log_b c = \log_a c$ **[2 marks]**

2 The equation of a curve is $y = ka^x$ where k and a are constants.
Two points on the curve are (1.5, 172) and (2.7, 131). Find the values of k and a. **[4 marks]**

3 The equation of a curve is $y = 1.4^{0.5x+2}$.
Write this in the form $y = ce^{kx}$ where c and k are constants to be found. **[2 marks]**

4 $\log_5 2 = c$

Write in terms of c:

a $\log_5 40$ **[1 mark]** **b** $\log_5 0.08$ **[1 mark]**

5 The equation of a curve is $y = 4e^{1.2x}$.
Find the coordinates of the point where the gradient of the curve is 6. **[3 marks]**

6 Solve the equation $\log_8 x + \log_8 2x + \log_8 x^2 = 10$. **[3 marks]**

(CM) 7 A student is making x the subject of the equation $y = 4e^{2x+2}$.
(PF)

Here is the student's solution. The student has made two mistakes.

$$y = 4e^{2x+2}$$
$$y = 4e^{2x} + 4e^2$$
$$y - 4e^2 = 4e^{2x}$$
$$e^{8x} = y - 4e^2$$
$$8x = \ln(y - 4e^2)$$
$$x = \frac{1}{8}\ln(y - 4e^2)$$

a Indicate the two mistakes the student has made. **[1 mark]**

b Show that $x = \ln(0.5y^{-0.5}) = \ln\frac{y^{0.5}}{2} - 1$. **[3 marks]**

(M) 8 The mean distance (d million km) of three planets from the Sun and the time each takes to orbit the Sun (t years) are shown in this table.

Planet	Mars	Jupiter	Saturn
Distance, d (10^6 km)	227.9	778.3	1427
Time, t (years)	1.88	11.86	29.49

$d \propto t^n$ where n is a constant.

a Find an equation for d in terms of k. **[4 marks]**

b Deduce the mean distance of the Earth from the Sun. **[1 mark]**

9 Solve the equation $5^{2x} + 6 = 5^{x+1}$. **[3 marks]**

(M) 10 The mass of a drug in a body decreases with time. Initially there is 0.05 g.

After t hours, the mass is 0.05×097^t g.

Find the rate at which the mass is decreasing, in g/hour, when the mass has reduced to 0.01 g. **[4 marks]**

Ⓜ **11** £A is invested and the rate of interest is p% a year. If you assume the interest is added continuously, the value, £y, after t years is modelled by a curve with the equation $y = Ae^{kt}$.

  a Find an expression for k in terms of p. [2 marks]

  b If the rate of interest is p% a month, what is the formula for k in this case? [2 marks]

12 x and y are variables. This graph show four points on the graph of $\log_{10} y$ against x.

Find an equation for y in terms of x. [4 marks]

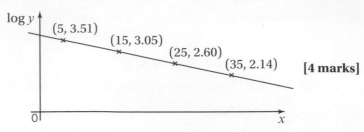

Ⓜ Ⓟⓢ **13** Plutonium -238 is radioactive and decays over time.

If a sample has a mass of m now, the mass in t years' time will be me^{-kt} where k is a constant. Plutonium-238 has a half-life of 88 years, which means that half the mass will have decayed in 88 years' time.

Work out how long it will be until the sample is one-tenth of its original mass. [4 marks]

Ⓜ **14** The population of a city is 8.24 million. 10 years ago the population was 6.85 million.

  a Assuming that the growth is exponential and continues at the same rate, find a formula to predict the population, p, in t years' time. [3 marks]

  b Why might the assumption in **part a** be unrealistic? [1 mark]

  c Further study suggests that the population will increase by 12% in the next 10 years. Modify your equation to take account of this finding. [2 marks]

15 Solve the equation $e^x + e^{-x} = 4$. [4 marks]

Chapter 8 Differentiation

Ⓜ **1** The speed, $v\,\text{m s}^{-1}$, of a vehicle after x seconds is given by the formula:

$$v = x^2(5 - x),\ 0 \leqslant x \leqslant 4$$

  a Find the times when the acceleration is zero. [2 marks]

  b Sketch a graph of the acceleration of the vehicle. [1 mark]

2 Differentiate the expression $4x\left(\sqrt{x^3} - 2\sqrt{x}\right)$ with respect to x [2 marks]

ⒸⓂ **3** Here is the equation of a curve: $y = 2x^3 - 3x^2 - 36x + 10$.
Find the stationary points on the curve and show whether each one is a maximum or a minimum point. [5 marks]

4 Find the gradient at each point where the curve $y = x(2x - 3)(x + 4)$ crosses the x-axis. [4 marks]

5 $y = \dfrac{x^2 + 3x}{\sqrt[3]{x}}$, $x > 0$

A student is trying to differentiate this and writes the following:

$$y = \dfrac{x^2 + 3x}{x^{\frac{1}{3}}}$$

$$\dfrac{dy}{dx} = \dfrac{2x + 3}{\frac{1}{3}x^{-\frac{2}{3}}}$$

$$\dfrac{dy}{dx} = 3x^{\frac{2}{3}}(2x + 3)$$

a What mistake has the student made? [1 mark]

b Show that the correct answer is $\dfrac{dy}{dx} = \dfrac{5x + 6}{3\sqrt[3]{x}}$. [3 marks]

6 The diagram shows a cuboid with a square cross-section.

The sides of the cuboid are x cm and h cm.

The volume of the cuboid is 500 cm^2.

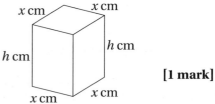

a Show that the surface area of the cuboid is $2x^2 + \dfrac{2000}{x} \text{ cm}^2$. [1 mark]

b Show that the surface area is a maximum when the shape is a cube. [3 marks]

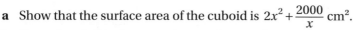 **7** $f(x) = \sqrt{2x}$

Show that $f''(x)\{f(x)\}^3 = -1$. [4 marks]

8 The equation of a curve is $y = x^3 - 3x^2 - 6x + 10$

The point P on the curve has an x-coordinate of 3.

Show that the tangent at P passes through the point $(7, 4)$. [4 marks]

 9 $y = 2x^2 + 3x + 1$

Find $\dfrac{dy}{dx}$ from first principles. [5 marks]

10 P is a point with positive coordinates on the curve $y = x^2 + 9$

The tangent to the curve at P passes through the origin.

Find the coordinates of the point where the normal to the curve at P crosses the x-axis. [5 marks]

 11 A card is a square of side 30 cm.

A square of side x cm is cut from each corner of the card.

The sides of the card are folded up to make a tray.

Find the maximum possible volume of the tray.

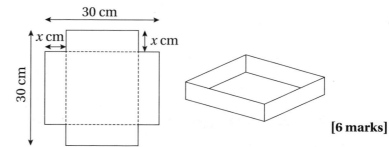

[6 marks]

12 The diagram shows a cone with a height of h cm and a base radius of r cm.

 $h + r = 60$

Show that the volume is a maximum when the ratio of the radius to the height is $2:1$.

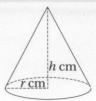

[5 marks]

13 The equation of a curve is $y = x^2 + \dfrac{a^2}{x^2}$, $a > 0$, $x \neq 0$.

Find the coordinates of any stationary points. [4 marks]

14 The equation of a curve is $y = \dfrac{36}{x^2}$.

Find the area of the triangle formed by the normal at $(6, 1)$ and the coordinate axes. [5 marks]

(PF) 15 **a** Show that $\dfrac{1}{x+a} - \dfrac{1}{x} = -\dfrac{a}{x(x+a)}$. [1 mark]

b Hence differentiate $\dfrac{1}{x}$ from first principles. [3 marks]

Chapter 9 Integration

1 Integrate $\dfrac{(x-3)^2}{\sqrt{x}}$ with respect to x. [3 marks]

(PS) 2 The equation of a curve is $y = \mathrm{f}(x)$, $x > 0$.

$$\mathrm{f}'(x) = \frac{1}{4} - \frac{9}{x^2}$$

The curve passes through $(4, 5)$ and $(9, k)$ Find the value of k. [4 marks]

3 Find $\displaystyle\int \left(2x + \frac{3}{x}\right)^2 \mathrm{d}x$ [3 marks]

4 The sketch shows the curve $y = x^2 + x - 6$

Find the area between the curve and the x-axis. [4 marks]

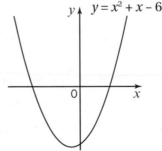

(PS) 5 $\dfrac{1}{2}x \times \dfrac{\mathrm{d}y}{\mathrm{d}x} = x^2 - \dfrac{1}{x^2}$ Find an expression for y in terms of x. [3 marks]

6 Find the area between the curve $y = \dfrac{1}{4}x^4$ and the line $y = 4$. [4 marks]

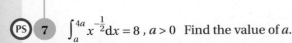

(PS) 7 $\displaystyle\int_a^{4a} x^{-\frac{1}{2}}\mathrm{d}x = 8$, $a > 0$ Find the value of a. [3 marks]

8 Find the area enclosed by the curve $y = x^2 - 3x$, the x-axis, and the lines $x = 2$ and $x = 4$. **[4 marks]**

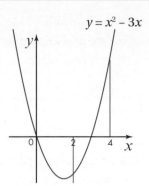

9 Find the area between the curve $y = \sqrt{4x}$ and the straight line $y = x$. **[4 marks]**

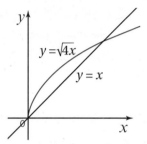

PF 10 The equation of this curve is $y = \dfrac{100}{x^2}$.

a Show that the area between the curve, the x-axis, $x = 5$ and $x = a$ is approximately 20 when a is large. **[3 marks]**

b State and prove a similar result if the equation of the curve is $y = \dfrac{100}{x^3}$. **[2 marks]**

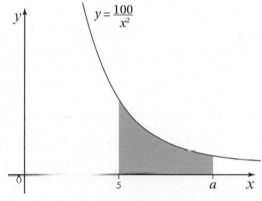

11 Find the area between the curves $y = \frac{1}{2}x^2$ and $y = \frac{1}{8}x^3$. **[4 marks]**

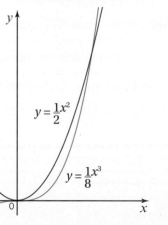

12 Integrate $\dfrac{x^2+1}{x^{\frac{1}{2}}} - \dfrac{x^2-1}{x^{\frac{3}{2}}}$ with respect to x. **[3 marks]**

13 Find the area between the curve $y = x(4 - x)$ and the straight line $x + y = 4$. **[4 marks]**

14 The equation of a curve is $y = x(x-3)(x+1)$.

Find the total area between the curve and the x-axis. **[5 marks]**

(PS) **15** The diagram show a quadratic curve.

The area below the x-axis is 108.

Find the equation of the curve. **[4 marks]**

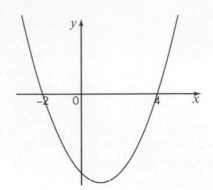

Chapter 10 Vectors

 1 The vector $\mathbf{a} = \begin{bmatrix} -4\sqrt{3} \\ 11 \end{bmatrix}$.

   a Find the magnitude of $\mathbf{a}$. **[2 marks]**

   b Find the angle between a and the vector $\mathbf{i}$. **[2 marks]**

(PS) **2** A port is situated at the position vector $(3\mathbf{i} + 10\mathbf{j})$.

(M) A yacht leaves the port and travels 8 km south-west to an island, then 12 km east to a dock.

   a Find the position vector of the dock. Give the answer in the form $(p\mathbf{i} + q\mathbf{j})$ km, where p and q are surds in the form $a + b\sqrt{2}$. **[3 marks]**

   b Show that the bearing of the port from the dock is 312°, correct to the nearest integer. **[3 marks]**

(PS) **3** JKLM is a parallelogram with position vectors $\mathbf{j} = (2\mathbf{i} - 9\mathbf{j})$, $\mathbf{k} = (12\mathbf{i} - 2\mathbf{j})$, $\mathbf{l} = (a\mathbf{i} + 3\mathbf{j})$ and $\mathbf{m} = (5\mathbf{i} + b\mathbf{j})$.

   a Find the values of a and b. **[4 marks]**

   b Find the size of the angle MJK. Write the angle correct to the nearest degree. **[3 marks]**

(PS) **4** WXY is a triangle with position vectors $\mathbf{w} = \begin{bmatrix} -4 \\ 3 \end{bmatrix}$ and $\mathbf{y} = \begin{bmatrix} 17 \\ 2 \end{bmatrix}$. The vector $\overrightarrow{WX}$ is given by $\begin{bmatrix} 15 \\ 9 \end{bmatrix}$.

(PF)   **a** Prove that WXY is a right-angled triangle. **[5 marks]**

   b Hence find the position vector of Z such that WXYZ is a rectangle. **[2 marks]**

   c Find the area of WXYZ. **[3 marks]**

(PS) **5** PQR is an isosceles triangle such that PR = RQ.

P has position vector $\mathbf{p} = 5\mathbf{i} - 20\mathbf{j}$. Q has position vector $\mathbf{q} = -\mathbf{i} + 16\mathbf{j}$.

R has position vector $\mathbf{r} = -4\mathbf{i} + a\mathbf{j}$.

Find the value of a. **[8 marks]**

(PS) **6** Points A, B, C and D have position vectors $13\mathbf{i} + 12\mathbf{j}$, $3\mathbf{i} + 16\mathbf{j}$, $-\mathbf{i} + 6\mathbf{j}$ and $9\mathbf{i} + 2\mathbf{j}$, respectively.

(CM)   **a** Prove that ABCD is a square. **[6 marks]**

(PF) Point E is the centre of ABCD.

   b Find the position vector of F such that AFBE is also a square. **[5 marks]**

PS **7** Vectors **m** and **n** are given by $\mathbf{m} = \begin{bmatrix} 3 \\ a \end{bmatrix}$ and $\mathbf{n} = \begin{bmatrix} -4 \\ 7 \end{bmatrix}$. The magnitude of **m** is $\sqrt{13}$.
Given that $a < 0$,

   a find the value of a [2 marks]

   b find the value of b such that the vector $b\mathbf{m} + 4\mathbf{n}$ is parallel to **i**. [3 marks]

PS **8** Point X has position vector $(-2\mathbf{i} - 4\mathbf{j})$. The vectors $\overrightarrow{XY}$ and $\overrightarrow{XZ}$ are given by $(7\mathbf{i} - 8\mathbf{j})$ and $(13\mathbf{i} - 13\mathbf{j})$, respectively. W is the midpoint of the line segment YZ.

   a Find the position vector of W. [4 marks]

   b Show that $\overrightarrow{WX}$ has a magnitude of $\frac{29}{2}$. [3 marks]

PS **PF** **9** The quadrilateral PQRS has position vectors $\mathbf{p} = (3\mathbf{i} + 14\mathbf{j})$, $\mathbf{q} = (4\mathbf{i} + 8\mathbf{j})$, $\mathbf{r} = (17\mathbf{i} + 4\mathbf{j})$ and $\mathbf{s} = (9\mathbf{i} + 15\mathbf{j})$.

   a Prove that $\overrightarrow{PQ}$ is perpendicular to $\overrightarrow{PS}$. [4 marks]

   b Prove that $\overrightarrow{QR} = \overrightarrow{SR}$. [4 marks]

   c Prove that PQRS is a kite. [4 marks]

PS **10** Parallelogram ABCD has vertices A$(-3, 1)$, B$(3, 4)$ and C$(2, -1)$.
Find the position vector of D. [4 marks]

PS **CM** **PF** **11** The points A, B, C have position vectors $\mathbf{a} = -5\mathbf{i}$, $\mathbf{b} = (-2\mathbf{i} + 3\mathbf{j})$ and $\mathbf{c} = (\mathbf{i} + 6\mathbf{j})$, respectively.
Use vector methods to show that A, B and C are collinear. [4 marks]

PS **12** Point A has position vector $\begin{bmatrix} 2 \\ 13 \end{bmatrix}$ and point B has position vector $\begin{bmatrix} 12 \\ 3 \end{bmatrix}$.
C divides AB in the ratio $1 : 4$.
Find the position vector of point C. [4 marks]

PS **CM** **PF** **13** The three distinct points A, B, and C have position vectors **a**, **b** and **c**, respectively, and are collinear.
Prove that $k\mathbf{a} + l\mathbf{b} + m\mathbf{c} = 0$ where $k + l + m = 0$ [4 marks]

PF **CM** **PS** **14** Use vector methods to prove that the diagonals of a parallelogram bisect each other.
(Hint: start with a parallelogram OABC.) [5 marks]

PS **15** Points A and B have position vectors **a** and **b**, respectively.
P is the midpoint of AB.
Q is a point on OA such that OQ : QA is $1 : 3$.
OP and BQ intersect at the point T.
Find the position vector of the point T. [5 marks]

Chapter 11 Proof

All of these questions involve (PS) (PF)

1 **a** Prove that the square of any odd number can be written as $4m + 1$. **[3 marks]**

    b Show that the same does not apply for even numbers. **[2 marks]**

2 For any consecutive integers a, b and c, where $a < b < c$, prove that $b^2 - ac = 1$. **[3 marks]**

3 The diagram shows a triangular prism.

    The surface area of the prism is $486\,\text{cm}^2$.

    Prove that $11x^2 + 15x - 243 = 0$. **[5 marks]**

(2x + 3) cm   (x + 3) cm   2x cm   4x cm

4 Prove that $x^2 - 10x + 26$ is always positive. **[3 marks]**

5 Given that $y = x^3$, prove that $\dfrac{\mathrm{d}y}{\mathrm{d}x} = 3x^2$ by using differentiation from first principles. **[4 marks]**

6 Given that $y = 2x^2 - 8x + 28$, prove that $y \geq 20$. **[4 marks]**

(CM) 7 Every day Henry goes up a flight of six steps.

    He can take the steps either one at a time, two at a time, or in a combination of one and two steps at a time. Show that there are exactly 13 different ways that Henry can go up the steps. **[3 marks]**

(CM) 8 A line, l, has equation $2y - x - 4 = 0$.

    Points A and B have coordinates $(-2, -1)$ and $(2, 1)$, respectively.

    Prove that AB and line l are parallel. **[5 marks]**

(CM) 9 Given that θ is acute, prove $\dfrac{\cos^2\theta}{1 - \sin\theta} = 1 + \sin\theta$. **[3 marks]**

(CM) 10 A circle has equation $x^2 - 2x + y^2 - 4y = 20$.

    Prove that the point $(5, 6)$ lies outside the circle. **[5 marks]**

(CM) 11 **a** Given that n is an odd number, prove that $n^2 - 1$ is a multiple of 8. **[4 marks]**

    b Show that this result is not true when n is an even number. **[2 marks]**

(CM) 12 Given that m is an integer, prove that $m^3 - m$ is a multiple of 6. **[3 marks]**

(CM) 13 Given that a, b, c and d are four consecutive integers with $a < b < c < d$, prove that $bd - ac = b + c$. **[4 marks]**

14 Given that a, b, c and d are four consecutive integers with $a < b < c < d$, prove that $bc - ad = 2$. **[3 marks]**

(CM) 15 A triangle has sides which can be written as $x^2 + 1$, $x^2 - 1$ and $2x$.

    Prove that the triangle is right angled. **[4 marks]**

(CM) 16 The difference between two numbers is 3.

    Prove that the difference between the squares of the numbers is a multiple of 3. **[4 marks]**

Chapter 12 Data presentation and interpretation

1 The coded mean of a local businesses profit (£x) per month is £24.

The code used was $y = \dfrac{(x - 50000)}{2500}$. Work out the uncoded mean monthly profit. **[2 marks]**

2 Jenson's parents think he doesn't spend enough time studying each day. Jenson decides to keep a record for the next 60 days of the amount of time he studies. The grouped frequency table shows his results.

Time (the nearest minute)	Frequency
15–19	1
20–24	4
25–29	3
30–34	10
35–39	21
40–44	19
45–50	2

a Calculate an estimate of the mean time he spends studying per day. **[3 marks]**

b Find the standard deviation. **[3 marks]**

c Use linear interpolation to find the median time he spends studying per day. **[4 marks]**

3 Two workers each make components for aircraft. The manager monitors how many hours it takes each worker to make the next 10 components. The times are given below.

   Worker 1: 3, 6, 5, 5, 7, 9, 5, 3, 6, 6
   Worker 2: 2, 5, 11, 4, 7, 9, 2, 12, 9, 8

a Find the median for worker 1. **[1 mark]**

b Find the interquartile range for worker 1. **[3 marks]**

c Draw box plots for worker 1 and worker 2. **[2 marks]**

d Make a statement comparing the workers.
   Which worker would be better to employ, and why? **[3 marks]**

e An outlier is defined as any item that is more than 1.5 × interquartile range above the upper quartile or more than 1.5 × interquartile range below the lower quartile.
   Does either worker have any outliers in their times? Show your method clearly. **[4 marks]**

4 A supermarket compares two brands of biscuits. The masses, q grams, of 50 biscuits from brand Q are summarised by:

   $\sum q = 640\,\text{g}$ $\sum q^2 = 9341\,\text{g}$

The mean mass of 20 biscuits from brand R was 13.7 g and the standard deviation 1.3 g.

a Find the mean mass and the standard deviation for brand Q. **[4 marks]**

b Compare the brands Q and R. **[2 marks]**

c Find the mean and the standard deviation for all 70 biscuits. **[4 marks]**

5 Explain what is meant by:

a a census **[1 mark]**

b a sample. **[1 mark]**

A company that produces tyres places a unique code on the outside wall of every tyre. For quality control of the tread wear, the company selects a random sample of 45 tyres.

c Describe how the company could select the sample. **[1 mark]**

d Give a reason why the company should take a sample and not the population. **[1 mark]**

6 The weights, in kg, of 280 pumpkins grown on a local farm are given in the table.

Weight (kg)	Frequency
$0 < w \leqslant 1$	9
$1 < w \leqslant 2$	19
$2 < w \leqslant 2.5$	38
$2.5 < w \leqslant 3$	99
$3 < w \leqslant 3.5$	101
$4 < w \leqslant 6$	14

  a Calculate an estimate of the mean weight of a pumpkin grown on the farm. **[2 marks]**

  b Calculate an estimate of the standard deviation of the pumpkins grown on the farm. **[3 marks]**

  c Use linear interpolation to estimate the median pumpkin weight. **[2 marks]**

7 The times, measured to the nearest second, that 115 students in a year group took to send an identical text message are summarised in the table.

Time (seconds)	Frequency
1–10	12
11–20	19
21–25	47
26–30	27
31–40	8
41–60	2

A histogram was drawn to represent these data. The 1–10 group was represented by a bar of width 2 and height 3.

  a Find the width and height of the 26–30 group. **[3 marks]**

  b Use linear interpolation to estimate the median of this distribution. **[2 marks]**

  c Calculate estimates for the mean and the standard deviation. **[3 marks]**

8 A primary school class measured the lengths (to the nearest cm) of 79 pencils. Their results are summarised in the table.

Length	3–8	9–10	11–12	13	14–15	16–20
Frequency	11	14	24	10	10	10
Midpoint		9.5	11.5	13	14.5	18

The teacher drew a histogram to represent the data. The 11–12 group was shown by a block of width 4 cm and height 24 cm.

  a Find the width and height of the block for the 16–20 group. **[3 marks]**

  b Use linear interpolation to estimate the median length of a pencil. **[2 marks]**

  c Write down the midpoint of the 3–8 group. **[1 mark]**

  d Estimate the mean and the standard deviation of the length of a pencil. **[5 marks]**

9 A keen amateur gardener planted 25 Tom Thumb tomato plants. She recorded the number of ripe tomatoes produced by each plant. These data are displayed in the stem and leaf diagram.

```
1| 5 means 15 tomatoes.

0 | 6  8  8  9  9                    (5)
1 | 0  0  2  3  5  7  7  8          (8)
2 | 1  1  3  4  6                    (5)
3 | 0  0  1  6                       (4)
4 | 0  7                             (2)
5 |                                  (0)
6 | 8                                (1)
```

  a Determine the median number of tomatoes per plant. **[1 mark]**

  b Calculate the upper quartile and the lower quartile. **[3 marks]**

An outlier is defined as any item which is more than 1.5 × interquartile range above the upper quartile or more than 1.5 × interquartile range below the lower quartile.

  c Illustrate the data in a box plot indicating clearly any outliers. **[7 marks]**

10 A butcher cuts up steaks and then sells them at a nominal weight of 500 g.
A sample of ten steaks is taken and their weights, in grams, are shown below.

501.1, 493.6, 500.9, 472.1, 512.4, 504.9, 501.6, 496.0, 481.7, 503.5

  a Find the mean and variance of the weights. [3 marks]

  b For the next sample of ten steaks, $\sum x = 4897.3$ and $\sum x^2 = 2\,399\,040.4$.
     Find the mean and variance for this sample. [3 marks]

  c Compare the two samples. [2 marks]

(PS) 11 The following data set shows the marks scored
on the Y12 mock exam in maths.

Marks scored	Frequency
0–19	4
20–39	34
40–59	45
60–79	87
80–99	52
100–119	21
120–140	7

  a Draw a cumulative frequency graph to
display this data set. [3 marks]

  b Find the median and the 90th percentile. [3 marks]

  c The school wishes that 70% of students
to pass. What should the pass mark be? [3 marks]

12 The frequency distribution for the masses (to the nearest kg) for 75 objects is shown in the table below

Mass (kg)	3–7	8–9	10–18	19–20	21–30
Frequency	8	13	37	11	6

  a Represent the data set as a histogram. [3 marks]

  b Find an estimate for the median. [4 marks]

13 Data collected by a university has been coded in the following way:

$x = t + 75$

$$y = \frac{(v + 250)}{0.05}$$

If the regression line of y on x is given by $y = 1.8x + 72.1$ find the regression line of v on t. [3 marks]

Chapter 13 Probability and statistical distributions

1 In a sixth-form canteen 70% of the students eat chips at lunchtime. 25% of the students
eat burger and chips, and 9% of the students do not eat either burger or chips.

  a What percentage of the students eat burger or chips but not both? [2 marks]

  b Given that a randomly selected student eats burger, what is the probability of the
student eating chips as well? [2 marks]

2 Mark goes to the gym. He does either weights or cardio. He then either goes for a swim or into the
sauna. The probability that he does weights is $\frac{3}{5}$, and the probability that he goes for a swim given that
he has done weights is $\frac{4}{7}$. The probability that he goes into the sauna given that he has done cardio is $\frac{5}{}$.

  a Find the probability that he does weights or goes into the sauna but not both. [3 marks]

  b Find the probability that he does cardio. [3 marks]

  c Find the probability that he has done weights given that he is in the sauna. [3 marks]

3   A new game is released for the Christmas fair. An unbiased spinner with sides labelled 1, 3, 3, 5, 5, 5, 7, 9 is spun twice.

Let X be the random variable 'sum of the two scores on the spinner'.

   a   Find the probability distribution of X.   **[3 marks]**

   b   Find $P(X < 10)$.   **[2 marks]**

4   The discrete random variable X has the probability function:

$$P(X = x) = \frac{x}{21} \quad \text{for } x = 1, 2, 3, 4, 5, 6$$

   a   Draw up a table showing the cumulative distribution function, $F(x)$.   **[2 marks]**

   b   Find $P(X \leqslant 4)$.   **[2 marks]**

   c   Find $P(2 < X \leqslant 3)$.   **[3 marks]**

5   The probability of a packet of crisps containing a code for a free gift is 0.12.

Find the probability that in a random sample of 50 packets there are:

   a   fewer than 3 codes   **[2 marks]**

   b   more than 5 codes   **[2 marks]**

   c   exactly 4 codes.   **[2 marks]**

The wholesalers sell the crisps in boxes of 50. Molly buys 12 boxes of them.

   d   What is the expected number of packets of crisps with codes for free gifts?   **[3 marks]**

6   A company producing bags carries out a survey on the number of their bags with defects. It is found that the probability of a bag having poor stitching is 0.04 and that a bag with poor stitching has a probability of 0.82 of breaking. The probability of a bag with good stitching breaking is 0.13.

   a   Draw a tree diagram to represent this information.   **[3 marks]**

   b   Find the probability that a randomly chosen bag has exactly one of the two defects, poor stitching or breaking.   **[3 marks]**

The company also finds that the reflective markings on the bags wear off with a probability of 0.11 and that this defect is independent of poor stitching or breaking. A bag is chosen at random.

   c   Find the probability that the bag has none of the three defects.   **[2 marks]**

   d   Find the probability that the bag has exactly one of these three defects.   **[4 marks]**

7   A cake stall has two red velvet sponges, one chocolate orange and one vanilla sponge left. Two cakes are selected at random from the stall.

   a   Draw a tree diagram to show all the possible outcomes and associated probabilities, stating each probability clearly.   **[3 marks]**

   b   Find the probability that the cakes selected are not red velvet.   **[2 marks]**

8   Two bags contain beads that are identical apart from colour. Bag A has three red and two blue beads. Bag B has four red and three blue beads. A bead is randomly selected from bag A and placed in bag B. A bead is now randomly selected from bag B.

   a   Illustrate this situation in a tree diagram.   **[3 marks]**

   b   Calculate the probability that the bead selected from bag B is red.   **[3 marks]**

   c   Given that the selected bead is red, calculate the probability that it came from bag A.   **[3 marks]**

9 The discrete random variable X has probability distribution given by:

x	0	1	2	3
P($X = x$)	0.25	a	0.45	2a

Determine the value of a. **[2 marks]**

10 J and M are events with P(J') = 0.4, P(M) = 0.49 and P($J \cup M$) = 0.9.

a Find P(J). **[1 mark]**

b Find P($J \cap M$). **[2 marks]**

c Draw a Venn diagram and annotate it to show all the probabilities. **[3 marks]**

d Find ($J' \cap M'$). **[2 marks]**

e Are J and M independent? Explain how you know. **[2 marks]**

11 The cumulative distribution function for a discrete random variable, X, is given by:

$$F(x) = \frac{x}{15} \quad \text{for } x = 1, 2, ..., 15$$
$$F(x) = 1 \quad \text{for } x > 15$$
$$F(x) = 0 \quad \text{for } x < 1$$

Find the following:

a P($X < 4$) **[1 mark]**   **b** P($2 < X \leqslant 11$) **[2 marks]**   **c** P($X > 13$) **[2 marks]**

12 The number of aces, X, served can be by a tennis player during a set modelled by a continuous probability distribution.

x	0	1	2	3	4	5
P($X = x$)	0.01	0.1	0.23	0.73	0.88	1

Calculate the probability of each variable. **[2 marks]**

13 The probability of getting heads on a biased coin is 0.8. The coin is flipped 7 times and the outcome is modelled using a binomial distribution.

a How would you write this mathematically? **[1 mark]**

b Find P($X \geqslant 5$) where X is the probability of getting heads when the coin is flipped. **[3 marks]**

14 Eggs are packed in boxes of 12. The probability that an egg is broken is 0.41

Find the probability in a random box of eggs:

a Exactly one egg is broken **[2 marks]**

b Less than 3 eggs are broken **[3 marks]**

c 5 or more eggs are broken **[3 marks]**

15 Seven unbiased dice are rolled. Find the probabilities that

a There will be exactly three 1s **[1 mark]**

b There will be a number that appears four times **[4 marks]**

16 At a school fayre you have a one in three chance of winning a game by getting a hoop over a bottle. A Y12 pays to have 10 attempts.

a What is the probability she gets exactly two wins? **[2 marks]**

b What is the probability she gets no more than two wins? **[3 marks]**

c What is the most likely number of wins? **[3 marks]**

d What is the probability she achieves this? **[3 marks]**

Chapter 14 Statistical sampling and hypothesis testing

1 A student claims to be able to identify his favourite brand of coffee. Cups of five different brands of coffee are made, and he gets 10 attempts at guessing which one is his favourite. He is correct on 5 out of the 10 guesses.

  a Write down the null and alternative hypotheses. **[3 marks]**

  b Carry out a hypothesis test at the 5% significance level and state your findings clearly. **[6 marks]**

2 A school increases its prices of hot dinners. Last year 57% of students felt that school dinners were good value for money. Mia is on the school council and believes that fewer students now think they are getting value for money. She asks a random sample of 40 students and finds that only 15 think they are getting value for money. Test Mia's claim at the 1% significance level. **[7 marks]**

3 It is reported that only 16% of ten-year-old girls believe in Santa Claus. A group of 100 ten-year-old boys are to be tested to see if the same proportion of boys believe in Santa Claus at age 10. Find the critical region for a test at the 10% significance level. **[7 marks]**

4 During a lesson, a sixth form student monitors how many times her teacher uses a particular word. In the 60-minute class, the teacher used the word 38 times. During the next lesson the same student records the number of times the same word is used. This time it was used 43 times.

Has the teacher increased the number of times he uses this word?
Test at the 5% significance level. **[7 marks]**

5 Laura is trying to be Head Girl. She claims to have 55% of the school backing her. The opposing student thinks she is over estimating this and asks a random sample of students from the school at break time. Of the 37 students she asks, only 16 say they are backing her.

What are the results of this test at the 10% significance level? **[7 marks]**

6 A test statistic has a distribution B$(35, p)$.

Given H$_0$: $p = 0.2$ and H$_1$: $p \neq 0.2$, find the critical region for the test statistic such that the probability in each tail is as close to possible to 5%. **[5 marks]**

7 On average, 63% of people pass their driving theory test first time after they have used a mobile app. People think this is too low, and the company making the app has decided to update it. It is found that, since the update, 15 out of 18 people passed the test first time. Test at the 5% significance level whether the app has improved. **[7 marks]**

8 A machine making lenses for spectacles has been producing 18% mis-shapen lenses. After being serviced, a sample of 40 lenses is taken and 3 are mis-shapen.

Is there evidence at the 5% significance level that the machine is now producing fewer mis-shapen lenses and hence costing the company less in wastages? **[7 marks]**

9 A sports scientist claims that a new exercise programme is 75% effective at increasing lung capacity. In a trial, a sample of 26 people were tested and it was found that 14 had actually increased their lung capacity.

  a Design a suitable statistical test to see whether there is evidence to support the sports scientist's claim. **[3 marks]**

  b Carry out the test at the 5% level of significance, stating your conclusions clearly. **[7 marks]**

  c State any assumptions you have made and why these are important. **[2 marks]**

(10) In January, a survey is given to all sixth form students about their New Year resolutions. 23% of the students want to give up takeaways. The sixth formers wonder if this is just in their school or whether it is 23% of all young people in the area. A random sample of 50 young people contains 6 who are also trying to give up takeaways.

    a Write down suitable hypotheses. **[2 marks]**

    b What is the critical region for the suitable hypotheses at the 10% significance level? **[3 marks]**

    c Carry out the test, stating your conclusions clearly. **[5 marks]**

Chapter 15 Kinematics

(M) (1) A rocket is launched vertically from the ground at $53.9\,\text{m}\,\text{s}^{-1}$.
Find the length of time for which the rocket is more than $88.2\,\text{m}$ above the ground. **[5 marks]**

(PS) (2) (M) A car accelerates uniformly from rest for $30\,\text{s}$, before maintaining a constant speed of $26\,\text{m}\,\text{s}^{-1}$ for $T\,\text{s}$. The car then decelerates uniformly for $20\,\text{s}$ until it stops.
Given that the total distance travelled by the car is $910\,\text{m}$, find the value of T. **[4 marks]**

(3) A particle moves such that after t s its displacement, $r = (24t^2 - 12t - t^4 + 15)$ m.
Find the position of the particle when it is moving at its maximum velocity. **[8 marks]**

(PS) (4) (M) An athlete is competing in the $100\,\text{m}$ against one other athlete. She starts from rest with a uniform acceleration of $0.8\,\text{m}\,\text{s}^{-2}$. Her opponent has a $2\,\text{s}$ head start and accelerates uniformly at $0.6\,\text{m}\,\text{s}^{-2}$.

    a How long does it take the athlete to overtake her opponent?
Give the answer correct to 1 decimal place. **[7 marks]**

    b Show that the athlete still has $33.1\,\text{m}$ left to run when she overtakes her opponent. **[2 marks]**

(M) (5) A ball is projected vertically upwards at $U\,\text{m}\,\text{s}^{-1}$ from the top of a building $24\,\text{m}$ above the ground. The ball reaches a maximum height above the ground of $64\,\text{m}$.

    a Find the value of U. **[3 marks]**

    b Find the speed of the ball when it is $10\,\text{m}$ above the ground. **[3 marks]**

    c Find the time taken for the ball to reach the ground. **[4 marks]**

(M) (6) Two motorists drive between two sets of traffic lights.
The first motorist accelerates steadily from rest for $T\,\text{s}$ to a speed of $16\,\text{m}\,\text{s}^{-1}$, and then maintains this constant speed for $16\,\text{s}$ before decelerating uniformly for $9\,\text{s}$.
The second motorist accelerates uniformly from rest at $2\,\text{m}\,\text{s}^{-2}$ for $(T+5)\,\text{s}$, and then immediately decelerates uniformly until she stops.
Given that both motorists start and finish at the same time and cover the same distance:

    a sketch a speed–time graph for this information **[3 marks]**

    b find the value of T **[6 marks]**

    c find the distance between the traffic lights. **[2 marks]**

(PS) (7) A particle moves along a straight line PQRS with a constant deceleration.

(PF) It passes through P and then 4 s later it passes through Q, where PQ is 96 m.
14 s later, the particle reaches S, where QS is 147 m.

a Show that the particle comes to instantaneous rest at S. **[7 marks]**

Given that PR is 240 m,

b find the number of seconds that elapse after passing P before the particle passes
through R for the second time. **[4 marks]**

(PF) (8) Initially a particle is displaced 3 m from the origin with a velocity of 30 m s^{-1}.

(M) The acceleration of the particle at t s is given by $(24t - 42)$ m s^{-2}.

a Find an expression for the displacement after t s. **[5 marks]**

b Show that the particle is at instantaneous rest twice during its journey. **[2 marks]**

c Show that the particle travels $4\frac{1}{2}$ m during the third second. **[3 marks]**

(M) (9) A car is travelling at 24 m s^{-1} along a straight road with a speed limit of 40 mph.

(PF) Given that 1 mile is approximately 1.6 km,

a show that the car is travelling faster than the speed limit. **[2 marks]**

The car passes a police car monitoring traffic from a layby. 12 s later, the police pursue
the speeding car, accelerating uniformly in 6 s to V m s^{-1} and then maintaining this velocity
until they catch up with the speeding car, T s after it passed them.
Given that the police travel 1.344 km at V m s^{-1}, find:

b the value of T **[8 marks]**

c the value of V. **[2 marks]**

(10) A particle has an initial velocity of U m s^{-1}, where $U < 7$. It decelerates uniformly at $\frac{1}{10}$ m s^{-2}
until it reaches a velocity of $(3U - 18)$ m s^{-1}. It then moves at this velocity for a further 10 s.
The total distance travelled by the particle is 200 m.
For the motion of the particle, sketch:

a i a speed–time graph **[2 marks]** **ii** an acceleration–time graph. **[2 marks]**

b Find the value of U. **[7 marks]**

(11) The velocity of a particle at time t s is shown in
the diagram.
For the first 8 s, the velocity is given by $kt(16 - t)$ m s^{-1},
where k is a constant.
The particle then decelerates uniformly until it comes to
rest after X s.
The distance travelled during the first 8 s is 256 m.

a Find the value of k. **[4 marks]**

b Find the maximum velocity of the particle. **[2 marks]**

Given that the distance travelled by the particle whilst decelerating is the same as whilst
it was accelerating,

c show that X is $\frac{56}{3}$. **[2 marks]**

Chapter 16 Forces

(M) **1** A parcel of weight 83.3 N is pulled along a rough horizontal floor by a horizontal cord. The tension in the cord is a constant 40 N and the parcel experiences a constant resistance to its motion of 23 N.

Given that the parcel is initially at rest, find the velocity of the parcel after it has been pulled 25 m. **[7 marks]**

(PF) **2** A particle P of mass 7 kg is moving under the action of a constant force **F**.
P has a velocity of $(-3\mathbf{i} + 5\mathbf{j})$ m s^{-1} at time $t = 0$ s.
The displacement of P at $t = 6$ s is given by $\mathbf{r} = (18\mathbf{i} + 3\mathbf{j})$ m.

a Find the distance of P from the origin at $t = 10$ s. **[5 marks]**

b Show that the magnitude of **F** is 17.5 N. **[3 marks]**

c Find the value of t when P is moving parallel to **j**. **[4 marks]**

3 Three forces act at a single point. $F_1 = (4a\mathbf{i} - 3a\mathbf{j})$ N, $F_2 = (3\mathbf{i} - 7\mathbf{j})$ N and $F_3 = (b\mathbf{i} + 2\mathbf{j})$ N, $a > 0$.
Given that the magnitude of F_1 is 15 N,

a find the value of a. **[2 marks]**

Given that the resultant force **R** is $(6\mathbf{i} + c\mathbf{j})$ N,

b find the value of b **[2 marks]**

c find the value of c. **[2 marks]**

d Find the angle that **R** makes with **i**. **[3 marks]**

(M) **4** A particle P, of mass $\frac{1}{3}$ kg, is acted upon by a single force F for 10 s.

The initial velocity of the particle is $\begin{bmatrix} -48 \\ 311 \end{bmatrix}$ m s^{-1} and the velocity of the particle after 10 s is $\begin{bmatrix} 168 \\ 116 \end{bmatrix}$ m s^{-1}.
Find:

a the magnitude of F **[6 marks]**

b the bearing of F, correct to the nearest degree **[2 marks]**

c the distance travelled by P during the 10 s. **[4 marks]**

(M) **5** Particles A and B are connected by a rope which passes over a smooth pulley such that both halves of the rope are vertical. The rope is assumed to be light and inextensible. Particle B is twice the mass of particle A.

a How have you used the assumption that the rope is light? **[1 mark]**

b How long does it take particle B to descend 0.588 m? **[6 marks]**

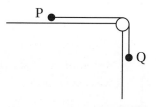

(M)
(PS) **6** Particles P and Q are connected by a light inextensible string via a smooth pulley. The pulley is at the edge of a rough horizontal table. Particle P, of mass 4 kg, is on the table, 6 m from the pulley. Particle Q, of mass 6 kg, is suspended from the pulley, 3 m above the ground. Initially both particles are at rest. When the system is released, P experiences a resistive force of $2.5g$ N as it moves across the table.

a Find the velocity of Q when it reaches the ground. **[6 marks]**

b Determine whether P reaches the pulley. **[4 marks]**

(M) **7** A man of mass 75 kg is standing in a lift of mass M kg. The lift is accelerating upwards at $0.6 \, \text{m s}^{-2}$.

When a boy of mass 25 kg enters the lift, the acceleration is reduced by $0.1 \, \text{m s}^{-2}$ and the tension in the cable is 215 N more than before.

Find the value of M. **[8 marks]**

(M) **8** A particle B is suspended 2 m above the ground from a particle A by a light inextensible vertical string. Particle A is 4 kg heavier than particle B. A vertical force of 113 N is applied to A so that both particles accelerate upwards. From rest, it takes the particles 4 s to reach a velocity of $6 \, \text{m s}^{-1}$.

a Find the mass of particle A. **[6 marks]**

When the particles are travelling at $6 \, \text{m s}^{-1}$, the string snaps but the 113 N vertical force is still applied to A.

b Find the new acceleration of particle A. **[2 marks]**

c Find the time after the string snaps until particle B hits the ground. **[4 marks]**

(M) (PS) **9** A rock of mass 12 kg is dropped from a height of 30 m onto soft muddy ground.

When the rock hits the ground, it sinks into the mud before coming to rest. The mud exerts a constant resistance to motion of 2500 N on the rock.

a Find the distance that the rock sinks into the mud. **[6 marks]**

b Find the total time from the moment the rock is dropped until it comes to rest. **[3 marks]**

(M) (PS) **10** A van of mass 500 kg pulls a trailer of mass 900 kg along a straight horizontal road by a horizontal metal cable. The vehicles experience resistive forces of 420 N and 600 N, respectively. The driving force exerted by the engine of the van is 1.86 kN. Starting from rest, the vehicles travel 67.5 m before the metal cable snaps. Find the further distance travelled by the trailer before it comes to rest. **[10 marks]**

(PF) (M) **11** Two particles are connected by a light inextensible string which passes over a smooth pulley at the edge of a rough horizontal table. One particle is at rest on the table. The other particle is suspended from the string vertically below the pulley. Both particles are the same mass, M kg. The resistive force acting upon the particle on the table is F N. When the system is released from rest, the particles accelerate at $a \, \text{m s}^{-2}$.

Prove that $a < \frac{1}{2}g$. **[6 marks]**

(PF) (M) **12** A car of mass M kg tows a caravan of mass $1.5M$ kg along a straight horizontal road. The car and caravan each experience a resistance to motion which is proportional to their mass. Starting from rest, the car travels 64 m in 8 s at a uniform acceleration, at which point the car and caravan become uncoupled. The car continues to exert the same driving force and experiences the same resistance to motion and now accelerates at $a \, \text{m s}^{-2}$.

The resistance to motion experienced by the car is F N.

a Show that $F = \dfrac{2M(a-5)}{3}$. **[8 marks]**

b Find the tension whilst the car and caravan were coupled in terms of M and a. **[3 marks]**

Answers

Short answers are given in this book, with full worked solutions available to teachers by emailing education@harpercollins.co.uk

1 Algebra and functions 1: Manipulating algebraic expressions

Prior knowledge 1   page 1

1 a $3a + 3b$

   b $4c - cd$

   c $e^2f - e^2g + e^3h$

   d $5i + ij + 7k - 14j$

   e $-5l^2 - l$

   f $2m^2 - mn - 2m - 2n$

2 a $x^2 + 3x + 2$

   b $x^2 + x - 12$

   c $x^2 - 5x + 6$

   d $x^2 - 6x + 9$

   e $2x^2 + 5x + 2$

   f $x^3 + 6x^2 + 11x + 6$

3 a a^{11}

   b a

   c b^4

   d c^{-6}

   e $16d^4$

   f ef^7

4 a $\sqrt{6}$

   b $2\sqrt{30}$

   c $6\sqrt{2}$

   d $3\sqrt{2}$

   e $\frac{\sqrt{5}}{5}$

   f 2

Exercise 1.1A   page 6

1 a $4x^3 + 12x^2 - 8x$

   b $x^3 + x^2y - x^3y$

2 a $x^3 + x^2y - x - y$

   b $6x - x^2 - x^3$

3 a $x^3 + 3x^2 - 4$

   b $6 + 4x - 6x^2 - 4x^3$

4 $(2x-1)(x+2) - (1-x)(3x^2+6) = 2x^2 + 3x - 2$
$$- (3x^2 + 6 - 3x^3\,6x)$$

$$= 2x^2 + 3x - 2$$
$$- 3x^2 - 6 + 3x^3 + 6x$$
$$= 3x^3 - x^2 + 9x - 8$$

5 $(2n)^2 + (2n+2)^2 = 4n^2 + 4n^2 + 8n + 4$
$$= 8n^2 + 8n + 4$$
$$= 4(2n^2 + 2n + 1)$$

Since 4 is a multiple of 2, $4(2n^2 + 2n + 1)$ must be an even number.

6 $(x+3)(2x^2 + x - 2) - (x+1)(3 - 2x - x^2)$
$$= x(2x^2 + x - 2) + 3(2x^2 + x - 2)$$
$$- x(3 - 2x - x^2) + 1(3 - 2x - x^2)$$

Should be:

$$= x(2x^2 + x - 2) + 3(2x^2 + x - 2)$$
$$- x(3 - 2x - x^2) - 1(3 - 2x - x^2)$$
$$= 2x^2 + x^2 - 2x + 3x^2 + 3x - 6 - 3x + 2x^2 + x^3 + 3 - 2x - x^2$$

Should be:

$$= 2x^3 + x^2 - 2x + 6x^2 + 3x - 6 - 3x + 2x^2 + x^3 - 3 + 2x + x^2$$
$$= 3x^2 - 6x + 3$$

Should be:

$$= 3x^3 + 10x^2 - 9$$

7 Let $p = 2n + 1$ where n is an integer. Consequently p is an odd number.
$$p^2 = (2n+1)^2$$
$$= 4n^2 + 4n + 1$$
$$= 4(n^2 + n) + 1$$

Since $4(n^2 + n)$ is even, $4(n^2 + n) + 1$ is odd.

8 $2p(2n+1)(2m+1) = 2p(4mn + 2n + 2m + 2)$
$$= 8mnp + 4pn + 4mp + 4p$$
$$= 2(4mnp + 2pn + 2mp + 2p),$$
an even number

Exercise 1.2A   page 8

1 a $x^3 + 6x^2 + 6x + 8$

   b $16x^4 + 32x^3 + 24x^2 + 8x + 1$

2 a $32 - 80x + 80x^2 - 40x^3 + 10x^4 - x^5$

   b $\frac{x^4}{16} + \frac{3x^3}{2} + \frac{27x^2}{2} + 54x + 81$

3 $3x^5 - 10x^4 + 10x^3 - 5x + 2$

4 1080

5 $x^9 - 6x^8 + 12x^7 - \frac{13}{2}x^6 - 6x^5 + 6x^4 + \frac{3x^3}{4} - \frac{3x^2}{2} + \frac{1}{8}$

6 $c = \frac{11}{5}$ or $c = -3$

Exercise 1.2B   page 11

1 a 10   **b** 15   **c** 8
2 a 1   **b** 3   **c** 3   **d** 1
They are the coefficients for a cubic expansion from Pascal's triangle.
3 $^5C_0, {}^5C_1, {}^5C_2, {}^5C_3, {}^5C_4, {}^5C_5$
4 All the rows prior to the index required have to be written in Pascal's triangle: a laborious and error-prone exercise.
5 $^{16}C_9 = \dfrac{16!}{7!9!}$
$= 11\,440$

Exercise 1.3A   page 14

1 a $16 + 32x + 24x^2 + 8x^3 + x^4$
   b $27 + 135x + 225x^2 + 125x^3$
2 a $270x^3$
   b $10\,240x^3$
3 a $8 + 6x + \dfrac{3x^2}{2} + \dfrac{x^3}{8}$
   b $81x^4 - 648x^3y + 1944x^2y^2 - 2592xy^3 + 1296y^4$
4 a $90x^3$
   b $4320x^3$
5 $b = 5$
6 $243 - \dfrac{1053x}{5} + \dfrac{1728x^2}{25} - \dfrac{1368x^3}{125} + \dfrac{528x^4}{625} - \dfrac{16x^5}{625}$
7 $\dfrac{259x^3}{2}$
8 $729 - 3645x + \dfrac{30375x^2}{4} - \dfrac{16875x^3}{2}$

$2.75^6 \approx 432$

Exercise 1.4A   page 16

1 a $2(x + 3)$ **b** $3(x^3 + 3)$
   c $4x(2 - x^3)$
   d $xy + 2xy - xz + 3yz$ doesn't factorise.
2 a $(x - 3)(x - 4)$ **b** $(2x + 1)(x + 2)$
   c $(x + 4)(x - 3)$ **d** $(x - 7)(x + 7)$
3 $7x(2x^2 - 8)$
   Should be:
   $= 7x(2x - 8)$
4 a $4xy(1 - 2y)$ **b** $(2x - 3)(3x - 2)$ **c** $xy^2(x + 2y)$
   d $6x(3x - 2)$ **e** $(9xy + 7z)(9xy - 7z)$
   f $x(2x^2 - 7x + 10)$ the quadratic cannot be factorised.
5 $(4x - 2)(x + 2)$ or $(2x - 1)(2x + 4)$
6 $(7x + 2y)(2x - 3y)$
7 $(6x + 1)(x + 15)$ $(2x + 1)(x + 15)$
   $(6x + 3)(x + 5)$ $(2x + 3)(3x + 5)$

$(3x + 1)(2x + 15)$ $(2x + 15)(3x + 1)$
$(3x + 3)(2x + 5)$ $(2x + 5)(3x + 3)$
$(6x + 15)(x + 1)$ $(3x + 15)(2x + 1)$
$(6x + 5)(x + 3)$ $(3x + 5)(2x + 3)$
8 $p = \dfrac{2qr}{r^2 + 6r - q}$
9 $(x - 3)^2 + (y + 5)^2 = 16$

Exercise 1.5A   page 20

1 a 2

   b $\dfrac{x + 2}{x + 3}$
2 $f(1) = 1^3 + 4(1^2) + 1 - 6 = 0$ so $(x - 1)$ is a factor of $f(x)$.
   $(x - 1)(x + 2)(x + 3)$
3 a $\dfrac{x + 2}{x + 5}$

   b $\dfrac{x - 2}{x + 3}$
4 a $(x - 1)(3x - 1)(x + 2)$
   b $(x + 1)(x + 3)(2x - 1)(x - 2)$
   c $(x - 4)(x + 4)(x + 3)(x - 3)$
5 a $2x^2 + 17x + 48 + \dfrac{60}{(x - 1)}$
   b $x^2 + 8x + 27 + \dfrac{60}{(x - 2)}$
   c $x^2 + 1 - \dfrac{10}{(x + 4)}$
6 a 61
   b −90
   c −1080
7 a $x^3 + 2x^2 + 4x - 10 + \dfrac{61}{(x - 2)}$
   b $8x^3 - 6x^2 - 47x + 84 - \dfrac{90}{(x + 1)}$
8 $(x + 1)(x - 2)(x - 3)^2$

Exercise 1.6A   page 23

1 a $2a^9$ **b** $2b^4$ **c** $3c^3$
   d $27d^6$ **e** e^{-6} **f** f
2 a 4   **b** ± 5 **c** $\dfrac{1}{3}$
3 a $6a^4$ **b** $2b^5$ **c** $\dfrac{1}{2}c^{-4}$ or $\dfrac{1}{2c^4}$
   d $d^{\frac{17}{12}}$ **e** $90e^3$ **f** $\dfrac{1}{3f^2g^2}$
4 a $\dfrac{1}{16}$ **b** $\dfrac{1}{36}$ **c** ± 4

5 $3x^6 \times 2x^{-4}$

  $= 5x^{10}$

  Should be:

  $= 27x^6 \times 2x^{-4}$

  $= 54x^2$

6 $8^{\frac{8}{3}}$

7 $4\pi x^{-\frac{2}{3}}$

8 a $\dfrac{9}{2a^3}$ b $\dfrac{1}{216b^{\frac{1}{3}}}$ c $32c^{\frac{7}{2}}$

9 $\dfrac{9}{2}\pi x^{\frac{3}{2}}$

10 $\sqrt[3]{y} = x$

   $\sqrt[5]{z} = x$

   $y^{\frac{1}{3}} = z^{\frac{1}{5}}$

   $z = y^{\frac{5}{3}}$

Exercise 1.7A   page 25

1 a $3a^2$ b $5b$

2 a $3\sqrt{e} + e$ b $8\sqrt{f} - 6f$

3 a $7c^2$ b $3d\sqrt{3d}$

4 a $g\sqrt{h} + h\sqrt{gh}$ b $j\sqrt{k} - k\sqrt{2j}$

5 $b - a^2$

6 $(\sqrt{b} - a)^2 = (\sqrt{b} - a)(\sqrt{b} - a)$

     $= b - a\sqrt{b} - a\sqrt{b} + a^2$

     $= b - 2a\sqrt{b} + a^2$

7 $u - b$

8 $(\sqrt{ab} + \sqrt{a})^2 = (\sqrt{ab} + \sqrt{a})(\sqrt{ab} + \sqrt{a})$

     $= ab + a\sqrt{b} + a\sqrt{b} + a$

     $= ab + 2a\sqrt{b} + a$

9 $(\sqrt{a} - \sqrt{ab})(\sqrt{a} + \sqrt{ab})$

10 $(b - \sqrt{ab})(b + \sqrt{ab})$

Exercise 1.8A   page 27

1 a $\dfrac{2\sqrt{3}}{3}$

  b $8 - 4\sqrt{3}$

  c $\dfrac{15x + 5x\sqrt{2}}{7}$

2 a $\dfrac{6\sqrt{2} + 4}{7}$

  b $\dfrac{12x + 4x\sqrt{7} + 15 + 5\sqrt{7}}{2}$

  c $\dfrac{5x\sqrt{2} - 2x}{23}$

3 a $\dfrac{16 - 6\sqrt{7}}{2}$

  b $\dfrac{28 + 10\sqrt{3}}{22}$

  c $\dfrac{y^2 - y\sqrt{x} - xy + x\sqrt{x}}{y^2 - x}$

4 a $2^{\frac{1}{2}}$ b $2^{-\frac{3}{2}}$ c $2^{\frac{1}{2}}$

5 a $\dfrac{2x\sqrt{x} + x^2}{4 - x}$

  b $\dfrac{y^2 + 2y\sqrt{x} + x}{y^2 - x}$

  c $\dfrac{3\sqrt{x} - \sqrt{xy} - 6 + 2\sqrt{y}}{y - 9}$

6 a $\dfrac{3x\sqrt{2} + 2y}{6}$

  b $\dfrac{2\sqrt{x} + 2x + \sqrt{y} - x\sqrt{y}}{2 - 2x}$

  c $\dfrac{6x\sqrt{x} - 8x - 8\sqrt{x}}{4x - x^2}$

Exam-style questions 1   page 29

1 a x^3 b $3x$

2 a $\dfrac{y\sqrt{x}}{x}$

  b $\dfrac{6 + 3\sqrt{x}}{4 - x}$

3 a $f(3) = 54 + 9 - 75 + 12 = 0$ so $(x - 3)$ is a factor of $f(x)$.

  b $f(-1) = -2 + 1 + 25 + 12 = 57$ so $(x + 1)$ is not a factor of $f(x)$.

4 2^8

5 $^{10}C_3 = 120$

6 $(x - 7)(x + 2)(x - 1)$

7 $5^{\frac{3}{2}}$

8 $\left(a + \sqrt{b}\right)^2 + \left(a - \sqrt{b}\right)^2$

  $= \left(a + \sqrt{b}\right)\left(a + \sqrt{b}\right) + \left(a - \sqrt{b}\right)\left(a - \sqrt{b}\right)$

  $= a^2 + 2a\sqrt{b} + b + a^2 - 2a\sqrt{b} + b$

  $= 2a^2 + 2b$

  which is rational if a and b are rational.

9 a $\dfrac{x - 2\sqrt{x} + 1}{x - 1}$

  b $\dfrac{6\sqrt{x} + 2x\sqrt{2}}{9 - 2x}$

10 $(x - 1)(x + 2)^2(x - 3)$

11 a The index on the bracket should be 5 and the contents of the bracket should be $\left(1 - \frac{3x}{4}\right)$.

b The coefficient of $x < 0$ (it is -12).

12 $2x^2 + 25x + 156 + \frac{792}{(x-5)}$

13 $x = \pm 64$

14 a $a = 3$

b $81 - 216x + 216x^2 - 96x^3 + 16x^4$

c 61.4656

15 $\frac{2x - 3}{7x - 1}$

16 One number is 1 too high. As the numbers are symmetrical, the 16 should be a 15.

17 $\left(a + \sqrt{b}\right) \times \left(a - \sqrt{b}\right)^3 = \left(a + \sqrt{b}\right)\left(a - \sqrt{b}\right)\left(a - \sqrt{b}\right)\left(a - \sqrt{b}\right)$

$= \left(a^2 - b\right)\left(a^2 - 2a\sqrt{b} + b\right)$

$= a^4 - 2a^3\sqrt{b} + a^2 b - a^2 b + 2ab\sqrt{b} - b^2$

$= a^4 - 2a^3\sqrt{b} + 2ab\sqrt{b} - b^2$

which is irrational unless b is a square number.

18 a $\frac{6x\sqrt{x} - 3x - 9\sqrt{x}}{4x^2 - 9x}$

b $\frac{3x^2\sqrt{x} + 2x^2 - x\sqrt{x} + x + \sqrt{x}}{x^2 - x}$

19 $(x - 1)(x + 3)(2x - 3)(2x + 3)$

20 $2x^4 - 23x^3 + 106x^2 - 227x + 552 - \frac{1080}{(x+2)}$

21 a $1 + 3\sqrt{x} + 3x + x\sqrt{x}$

b $8 - 24\sqrt{x} + 24x - 8x\sqrt{x}$

22 $\frac{49x^2 - 49x + 12}{36x^2 + 66x + 30}$

23 a $1 + 3x + \frac{15x^2}{4} + \frac{5x^3}{2} + \dots$

b $1.05^6 \approx 1.34$

c It is only an approximation because the expansion isn't complete.

24 This is not a possible correct expression for the volume of the cuboid.

Possible alternative: $2x^3 + 3x^2 - 11x - 6$

2 Algebra and functions 2: Equations and inequalities

Prior knowledge   page 32

1 $x = -\frac{1}{2}$ or $x = 3$

2 $x = 3$ or -4

3 $b^2 - 4ac = (-8)^2 - (4)(2)(-13) = 168$ so two real roots.
$x = 5.240$ or -1.24

4 $x = 2$, $y = -3$

5 $x = 2 + \sqrt{3}$ or $2 - \sqrt{3}$

6 $x < 4$

7

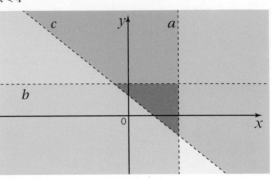

Exercise 2.1A   page 34

1 a Intercepts and turning point at $(0, 0)$.

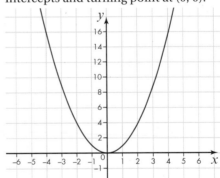

b y-intercept and turning point at $(0, 4)$.

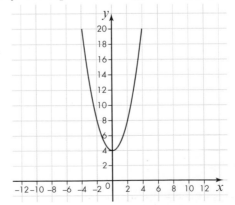

c x-intercepts at $(-4, 0)$ and $(1, 0)$, y-intercept at $(0, -4)$ and turning point at $\left(-\frac{3}{2}, -\frac{25}{4}\right)$.

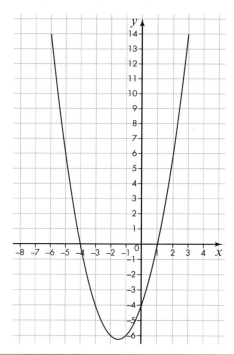

b Line of symmetry at $x = -1.75$.

x-intercepts at $(-3, 0)$ and $(-\frac{1}{2}, 0)$, y-intercept at $(0, 3)$ and turning point at $(-1.75, -3.1)$.

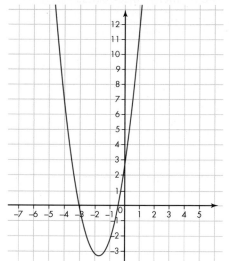

c Line of symmetry at $x = -1.5$.

x-intercepts at $(-4, 0)$ and $(1, 0)$, y-intercept at $(0, 4)$ and turning point at $(-1.5, 6.3)$.

2

x	-3	-2	-1	0	1	2	3	4	5
x^2	9	4	1	0	1	4	9	16	25
$-2x$	6	4	2	0	-2	-4	-6	-8	-10
-8	-8	-8	-8	-8	-8	-8	-8	-8	-8
y	7	0	-5	-8	-9	-8	-5	0	7

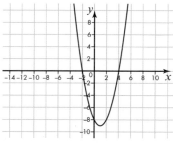

3 **a** Line of symmetry at $x = -1$, y-intercept at $(0, 1)$, x-intercept and turning point at $(-1, 0)$.

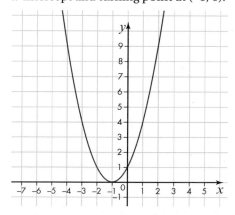

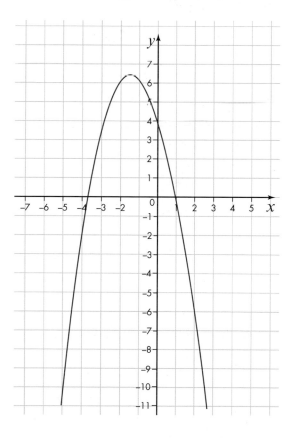

4 Line of symmetry at $x = 1.2$ (approx.).

x-intercepts at $(-0.7, 0)$ and $(3, 0)$, y-intercept at $(0, -6)$ and turning point at approx. $(1.2, -10)$.

$(3x + 2)(x - 3) = 0$

Expand the brackets.

$$y = 3x^2 - 7x - 6$$

This equation has the approximate x-intercepts and the y-intercept at the turning point when $x = 1.2$, $y = -10.08$ in the equation.

5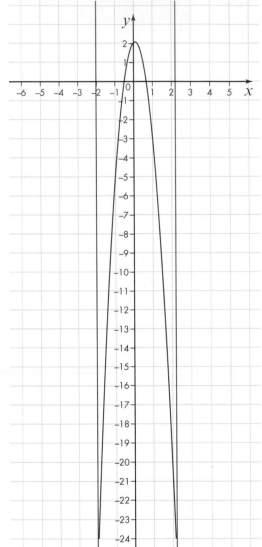

$x = -2$ and $x = 2$

Exercise 2.2A   **page 36**

1 **a** -48

no real roots.

   b 1

two distinct real roots.

   c 0

two equal real roots.

2 $b^2 - 4ac = (-5)^2 - (4)(2)(3)$

$= 25 - 24$

$= 1$

$b^2 - 4ac > 0$ so has two **distinct** real roots.

3 **a** 29

two distinct real roots.

   b -80

no real roots.

   c 0

two equal real roots.

4 $k^2 - 3k + 2 = 0$ or $-k^2 + 3k - 2 = 0$

For $k = 1$, $x = -2$ and for $k = 2$, $x = -3$.

5 **a** 52

two distinct real roots.

   b 0

two equal real roots.

   c -119

no real roots.

6 **a** $c < -\frac{25}{8}$

   b $c = -\frac{25}{8}$

   c $c > -\frac{25}{8}$

Exercise 2.3A   **page 38**

1 **a** $(x + 2)^2 - 4$

   b $2[(x - 2)^2 - 4]$

   c $(x + 4)^2 - 9$

2 $x^2 - 10x + 11 = (x - 5)^2 - 25 + 11$

$= (x - 5)^2 - 14$

3 **a** $(x - 4)^2 - 21$

   b $\left(x + \frac{3}{2}\right)^2 - \frac{37}{4}$

   c $2\left[\left(x + \frac{3}{4}\right)^2 + \frac{63}{16}\right]$

4 $y = \left(x - \frac{5}{2}\right)^2 + \frac{3}{4}$

$x = \frac{5}{2}, y = \frac{3}{4}$

5 $2x^2 + 8x - 5$

6 a $5\left[\left(x+\dfrac{1}{20}\right)^2-\dfrac{241}{400}\right]$

b $5\left(x^2-\dfrac{18}{5}\right)$ but can go no further.

c $-3\left[\left(x+\dfrac{7}{6}\right)^2-\dfrac{109}{36}\right]$

7 $\quad 2x^2-3x+11=2\left(x^2-\dfrac{3x}{2}+\dfrac{11}{2}\right)$

$$=2\left[\left(x-\dfrac{3}{4}\right)^2-\dfrac{9}{16}+\dfrac{11}{2}\right]$$

$$=2\left[\left(x-\dfrac{3}{4}\right)^2+\dfrac{79}{16}\right]$$

$$2\left[\left(x-\dfrac{3}{2}\right)^2+\dfrac{79}{16}\right]>0$$

Consequently $b^2-4ac<0$ because the curve of this equation will not intercept the x-axis and so there are no real roots.

8 $\quad (1-2x)[(x+4)^2-1]=(1-2x)(x^2+8x+16-1)$

$$=(1-2x)(x^2+8x+15)$$

$$=x^2+8x+15-2x^3-16x^2-30x$$

$$=15-22x-15x^2-2x^3$$

Exercise 2.4A page **41**

1 a $x=3$ or -3

b $x=-1$ or -2

c i Cannot be factorised.

ii, iii, iv $\quad x=\dfrac{5\pm\sqrt{53}}{2}$

d $x=0$ or 6

2 Factorisation:
- can only be used on equations that factorise
- sometimes spotting factors can be difficult
- can solve: $x^2+3x+2=0$
- cannot solve: $x^2-5x-7=0$

Quadratic formula:
- can be used to solve any equation with real roots including ones that don't factorise
- cumbersome and consequently easy to make a mistake
- can solve: $x^2-5x-7=0$
- cannot solve: $x^2-5x+7=0$

Completing the square:
- can be used to solve any equation with real roots including ones that don't factorise

- can be cumbersome manipulations if b is odd and $a>0$; consequently easy to make a mistake
- can solve: $x^2-5x-7=0$
- cannot solve: $x^2-5x+7=0$

Calculator:
- easy if you know how to use your calculator
- does not help you to understand the underlying methods
- can solve: $x^2+3x+2=0$ and $x^2-5x-7=0$
- cannot solve: $x^2-5x+7=0$

3 a $x=4\pm\sqrt{13}$

Completed the square. The equation was already in form of a completed square.

b $x=\dfrac{5\pm\sqrt{113}}{4}$

Does not factorise so used the quadratic formula. Alternatively, could have completed the square but chose not to as $a>1$.

c $x=\dfrac{1}{3}$ or -5

Equation factorises so easy to do this.

d $x=\dfrac{1}{2}$ or $\dfrac{7}{2}$

Equation factorises so easy to do this.

4 a $b^2-4ac=16-(4)(3)(-11)=148$

148 not a square number so does not factorise.

b $x=\dfrac{-2\pm\sqrt{37}}{3}$

5 $x=1$ or 7

6 $\qquad ax^2+bx+c=0$

$$a\left(x^2+\dfrac{bx}{a}+\dfrac{c}{a}\right)=0$$

$$\left(x+\dfrac{b}{2a}\right)^2-\dfrac{b^2}{4a^2}+\dfrac{c}{a}=0$$

$$\left(x+\dfrac{b}{2a}\right)^2=\dfrac{b^2}{4a^2}-\dfrac{c}{a}$$

$$\left(x+\dfrac{b}{2a}\right)^2=\dfrac{b^2}{4a^2}-\dfrac{4ac}{4a^2}$$

$$\left(x+\dfrac{b}{2a}\right)^2=\dfrac{b^2-4ac}{4a^2}$$

$$x+\dfrac{b}{2a}=\dfrac{\pm\sqrt{b^2-4ac}}{2a}$$

$$x=\dfrac{-b\pm\sqrt{b^2-4ac}}{2a}$$

Exercise 2.5A   page 43

1 **a** $x = 1$ and $y = -6$

   b $x = 4$ and $y = 5$

   c $y = -2$ and $x = 9$

2 Add ① and ③.

$$23x = 138$$
$$x = 6$$

Substitute x into ②.

$$18 - y = 22$$
$$y = -4$$

3 If there are n unknowns then you need n distinct equations involving the n unknowns.

4 $2x + y = -1$ ①

$3x - 2y = -12$ ②

Subtract ② from ①.

$$x - 3y = -11$$

As neither the coefficient of x nor y is the same in both equations, when one equation is subtracted from the other an unknown is not eliminated.

5 $x = \frac{2}{3}$ and $y = -\frac{1}{4}$

6 46 years old.

7 $x = -5$

$z = -2$

$y = 4$

Exercise 2.5B   page 45

1 **a** $x = 1$ and $y = -6$

   b $x = 4$ and $y = 5$

   c $y = -2$ and $x = 9$

2 $5x + 6y = 6$ ①

$3x - y = 22$ ②

Rearrange ②.

$$3x - 22 = y$$

Substitute into ①.

$$5x + 6(3x - 22) = 6$$

Expand and simplify.

$$23x - 132 = 6$$
$$x = 6$$

Substitute x into ②.

$$18 - 22 = y$$

Solve.

$$y = -4$$

3 $(6, -5)$.

494

4 $y = -\frac{1}{4}$ and $x = \frac{2}{3}$

5 $2c + 7b = 170$ ①

$3c + 8b = 250$ ②

From ①.

$$c = \frac{170 - 7b}{2}$$

Substitute into ②.

$$3\left(\frac{170 - 7b}{2}\right) + 8b = 250$$

$$b = 2\text{p and } c = 78\text{p}$$

The simultaneous equations only work if the cost of a small bag of sweets is 78p and the cost of a bar of chocolate is 2p. Although two equations can be formed and solved simultaneously from the given figures, 2p is too little for a bar of chocolate so the student must be correct: one of the calculations is wrong.

6 $x = -5$, $z = -2$ and $y = 4$

Exercise 2.6A   page 47

1 **a** $x = \frac{7}{2}$ or -4

    $y = -\frac{1}{2}$ or 7

   b $x = \frac{2}{3}$ or 6

    $y = -\frac{56}{3}$ or -8

   c $x = -5$ or 1

    $y = -20$ or 4

2 **a** Substitution or elimination: for elimination the second equation would need to be multiplied by 2 so that y can subsequently be eliminated.

   b Substitution only: neither addition nor subtraction of the equations will eliminate a variable.

   c Substitution only: neither addition nor subtraction of the equations will eliminate a variable.

3 **a** $x = -\frac{13}{4}$ or 3

    $y = -\frac{57}{8}$ or -4

   b $x = \pm 2$ $y = -1$ or 3

   c $y = -5 \pm \sqrt{165}$ and $x = 5 \pm \sqrt{165}$

4 $x = -\frac{4}{3}$ or 2

$y = -\frac{47}{9}$ or -3

So the coordinates of the points of intersection of this line and this curve are $(-\frac{4}{3}, -\frac{47}{9})$ and $(2, -3)$.

5 $(5.135, 6.135)$ and $(-6.135, -5.135)$.

6 They intersect once at $(-5, 5)$.

Exercise 2.7A   page 51

1 **a** $x > 2$

    b $x \leq 7$

    c $x < -2$

    d $x \geq \dfrac{8}{7}$

2 **a**

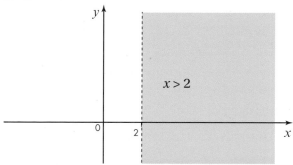

    b

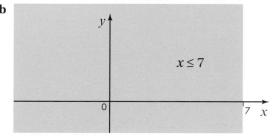

    c

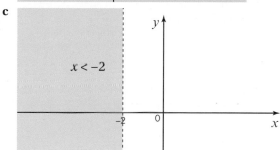

    d

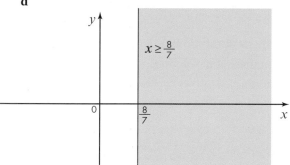

3 $8.8 \leq r \leq 10$ miles

4 **a** $\{1, 2, 3, 4\}$

    b $\{-10, -9, -8, -7, -6, -5, -4, -3, -2, -1, 0, 1, 2,$
       $3, 4, 5, 6, 7\}$

    c No solutions

    d $\{0, 1, 2, 3, 4, 5\}$

5 **a** x cannot be both less than -2 and greater than 3 in this way.

      Can be rewritten as $x < -2$ **or** $x > 3$

    b x cannot be both less than -2 or greater than 3 in this way.

      Can be rewritten as $x < -2$ **or** $x > 3$

    c x cannot be both less than -2 and greater than 3.

6 $y \leq 5, y \geq 2x - 1$ and $y \leq 5 - x$

7 **a** $x \leq 3$ **b** $x \geq 3$

    c $x \leq -3$ **d** $x \geq -3$

8 $417 < j < 480$

Exercise 2.8A   page 55

1 **a** $x < -4$ or $x > 1$

    b $2 \leq x \leq 4$

    c $x < -3$ or $x > 3$

    d $0 < x < 6$

2 **a** $-3 < x < -\dfrac{1}{2}$

    b $-4 < x < 1$

    c $x < -2$ or $x > 2$

    d $0 < x < \dfrac{5}{3}$

3 **a** $\{-5, -4, -3, -2, -1, 0\}$

    b {all integers less than 2.5}

    c $\{-2, -1, 0, 1\}$

    d {all integers less than 0.25}

4 So $8.9 \leq r \leq 10$ miles

5 **a** $\{2, 3, 4, 5\}$

    b $\{-1, 0, 1, 2, 3\}$

    c Inequality true for all values of x.

    d {All integers except $-2, -1, 0$ }

6 **a** $\{-3, -2, -1, 0\}$

    b $\{-5, -4, -3, -2, -1, 0\}$

    c $\{-5\}$

    d {all integers excluding $-6, -5, -4, 1$}

Exam-style questions 2   page 56

1

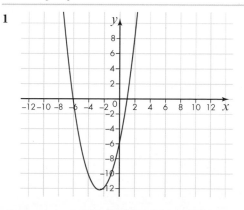

Line of symmetry at $x = -2.5$

x-intercepts at $(-6, 0)$ and $(1, 0)$, y-intercept at $(0, -6)$ and turning point at $(-2.5, -12.3)$ (approx.).

2 $b^2 - 4ac = 9 - (4)(2)(-7) = 65 > 0$ so two distinct real roots

3 $x = -1.479$ or $x = 2.479$

$x < -1.479$ or $x > 2.479$

4 $x = 5$ and $y = -3$

5 $x \leqslant \dfrac{17}{4}$

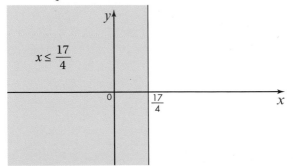

6 $x \geqslant 10$ or $x \leqslant -1$

7 x-intercepts at $(-\frac{1}{2}, 0)$ and $(2, 0)$ and y-intercept at $(2, 0)$; $a < 0$

$y = (-2x - 1)(x - 2)$

$y = -2x^2 + 3x + 2$

Satisfies the intercepts and $a < 0$.

8 The statement '$2x^2 + 3x + 13 \geqslant 0$ for all values of x' is correct. The discriminant is < 0, consequently the equation has no real roots and so it does not intercept the x-axis. As $a > 0$ all of the curve will be above the x-axis.

9 $(\frac{7}{8}, -\frac{81}{16})$

10 $y = -\frac{1}{2}$ and $x = \frac{3}{2}$

So the coordinates of the point of intersection of these two lines are $(\frac{3}{2}, -\frac{1}{2})$.

11 $(\frac{27}{5}, \frac{23}{5})$ and $(3, 1)$

12 a $x \geqslant \frac{11}{13}$

$x < -\frac{20}{33}$

{All integers except 0}

b No solution (as no overlap of inequalities).

13 $x < -\frac{3}{2}$ or $x > \frac{1}{3}$

14 a So $x > \frac{5}{3}$, i.e. {all integers greater than $\frac{5}{3}$}

b $x < -7$ or $x > -3$ {all integers less than -7 or greater than -3}

15 $x < -2.07$ or $x > 2.737$

16 a $a < -\dfrac{25}{24}$

b $a = -\dfrac{25}{24}$

c $24a > 25$

$a > -\dfrac{25}{24}$

17 $a^2 x^2 + 3bx + \sqrt{c} = 0$

$a^2\left(x^2 + \dfrac{3bx}{a^2} + \dfrac{\sqrt{c}}{a^2}\right) = 0$

$\left(x + \dfrac{3b}{2a^2}\right)^2 - \dfrac{9b^2}{4a^4} + \dfrac{\sqrt{c}}{a^2} = 0$

$\left(x + \dfrac{3b}{2a^2}\right)^2 = \dfrac{9b^2}{4a^4} - \dfrac{\sqrt{c}}{a^2}$

$\left(x + \dfrac{3b}{2a^2}\right)^2 = \dfrac{9b^2}{4a^4} - \dfrac{4a^2\sqrt{c}}{4a^4}$

$\left(x + \dfrac{3b}{2a^2}\right)^2 = \dfrac{9b^2 - 4a^2\sqrt{c}}{4a^4}$

$x + \dfrac{3b}{2a^2} = \dfrac{\pm\sqrt{9b^2 - 4a^2\sqrt{c}}}{2a^2}$

$x = \dfrac{-3b \pm \sqrt{9b^2 - 4a^2\sqrt{c}}}{2a^2}$

18 width $= 2$

length $= 7$

19 $4.4 < r \leqslant 6.4$

20 One set of solutions is $y = \frac{1}{3}$, $x = -2$, $z = \frac{1}{7}$

Another set of solutions is $y = 2$, $x = \frac{1}{2}$, $z = -\frac{4}{7}$

3 Algebra and functions 3: Sketching curves

Prior knowledge   **p 59**

1   **a** $(x-7)(x+4)=0$ so $x=-4$ or 7

    b $(x+7)(x-7)=0$ so $x=-7$ or 7

    c $(x-3)^2=0$ so $x=3$

    d $x(5x-1)=0$ so $x=0$ or $\frac{1}{5}$

    e $(3x-5)(3x+5)=0$ so $x=-\frac{5}{3}$ or $\frac{5}{3}$

    f $(3x-5)(2x-1)=0$ so $x=\frac{5}{3}$ or $\frac{1}{2}$

2   **a** $x=4\pm\sqrt{17}$

    b $x=1\pm\sqrt{4}$

       $x=-1$ or 3

    c $x=-3\pm\sqrt{7}$

3   **a** $(4,-17)$, minimum

    b $(1,-4)$, minimum

    c $(-3,-7)$, minimum

4   **a**   0 so one real root.

    b -56 so no real roots.

    c   53 so two real roots.

Exercise 3.1A   **p 62**

1   **a**

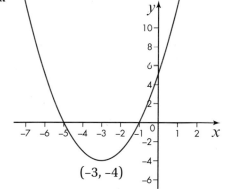

$(-3,-4)$

    b

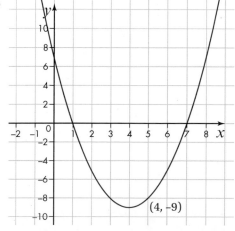

$(4,-9)$

2 x^2-4x+3

3   **a**

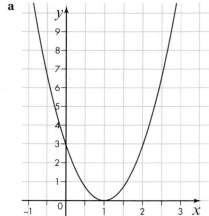

    b

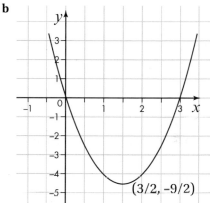

$(3/2, -9/2)$

4 $x^2-4x-21=(x-7)(x+3)$

     So should cut x-axis at $(-3, 0)$ and $(7, 0)$.

     $a>0$ so graph should be U-shaped.

     When $x=0$, $y=-21$ so y-intercept at $(0, -21)$.

     $x^2-4x-21=(x-2)^2-25$

     So turning point at $(2, -25)$.

5   **a**

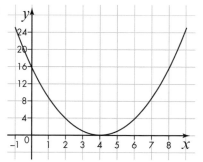

b

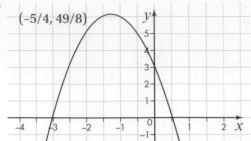

(−5/4, 49/8)

6 a 4 m   **b** 3 s

  c $y = 3 + 2x - x^2$

Exercise 3.2A   **p 64**

1 a

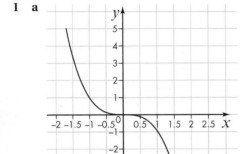

b

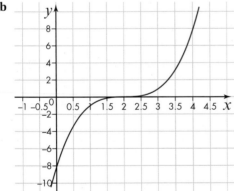

2 a

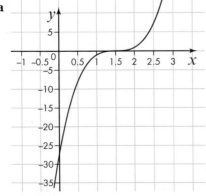

b

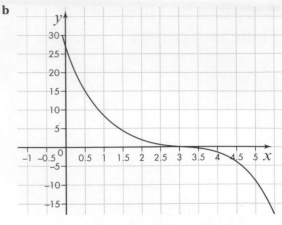

3 The sketch is *not* correct.

Point of inflection should be at (−3, 0) and curve should be a reflection in the *y*-axis of the curve shown.

4 The graphs are reflections of each other in the line $x = 2$.

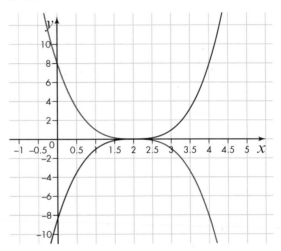

5 C $y = (3 - 4x)^3$ only

Exercise 3.2B   **p 66**

1 a

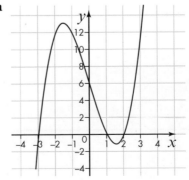

b

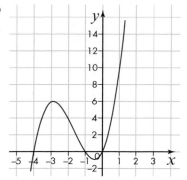

2 a

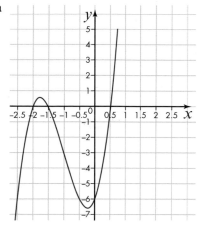

b

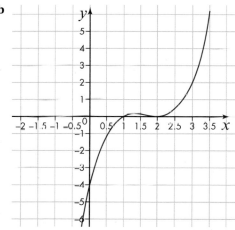

3 a $x(x-3)(x+1)$

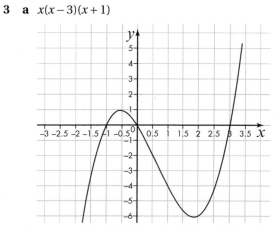

b $x^2(x-7)$

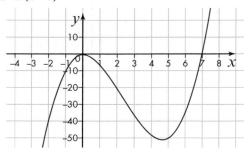

4 a

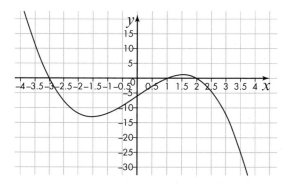

b

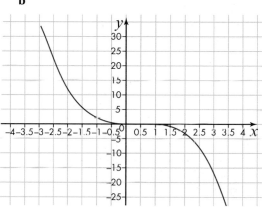

5

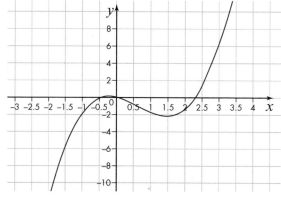

499

Exercise 3.3A   p 69

1 a

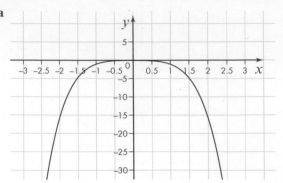

b

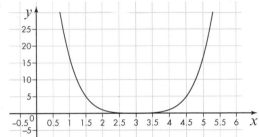

2 a

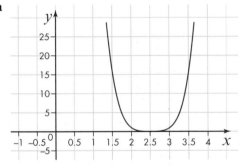

b

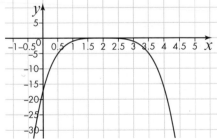

3 The curve is the wrong shape: it should *not* be a parabola.

The curve should be a U-shape sitting on the x-axis at (3, 0) and intercepting the y-axis at (0, 81).

4 They do not generate the same curve because they are different functions.

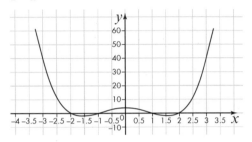

5 B $y = (4 - 5x)^4$ and **C** $y = (5x - 4)^4$

Exercise 3.3B   p 72

1 a

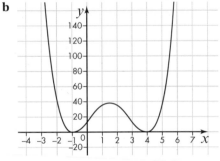

b

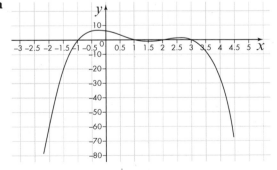

2 a

b

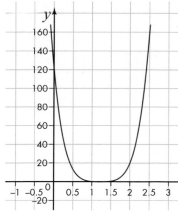

3 $x(x-1)(x+2)(x+3)$

4 **a**

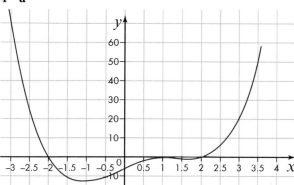

b

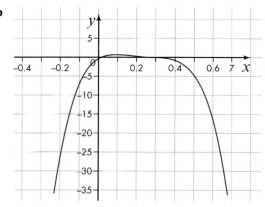

5 $(x-3)^2(x+3)^2$

Exercise 3.4A **p 74**

1 **a**

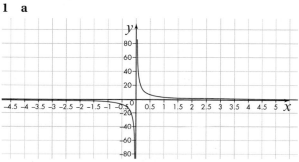

b

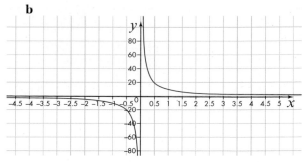

c In both graphs the asymptotes are in the same locations. The key difference is that the curves of the graph of $y = \dfrac{5}{x}$ are much closer to the origin than those of $y = \dfrac{10}{x}$.

2 **a** y-intercept at $(0, 1)$; asymptotes at $x = -2$ and $y = 0$.

b y-intercept at $(0, 0.75)$; asymptotes at $x = 4$ and $y = 0$.

3 **a** $y = \dfrac{1}{(x+1)}$

b $y = \dfrac{-2}{(x+1)}$

4 **a** iv   **b** ii

c iv   **d** i

5 **a**

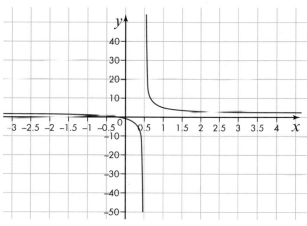

b

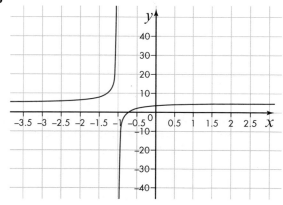

6 No. The asymptotes are in the correct locations at $x = \frac{2}{3}$ and $y = 2$. However, the graph drawn is for a reciprocal with a *positive* numerator whereas the given equation has a negative numerator.

Exercise 3.4B   p 78

1 a

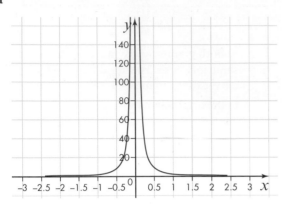

b

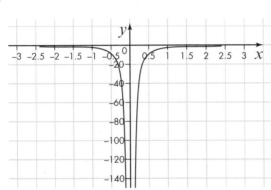

c In both graphs the asymptotes are in the same locations. The key differences are (**i**) that the curves of the graph of $y = \frac{3}{x^2}$ are much closer to the origin than those of $y = \frac{-6}{x^2}$ and (**ii**) the graph of $y = \frac{-6}{x^2}$ is a reflection in the *x*-axis.

2 a Asymptotes at $x = -2$ and $y = 0$; *y*–intercept at $(0, 0.25)$.

b Asymptotes at $x = -1$ and $y = 0$; *y*-intercept at $(0, -5)$.

3 a $y = \dfrac{1}{(x-3)^2} + 2$ **b** $y = \dfrac{1}{(x-4)^2} - 3$

4 a iii   **b** i

   c ii   **d** i

5 a

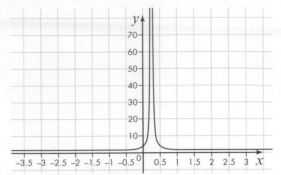

b

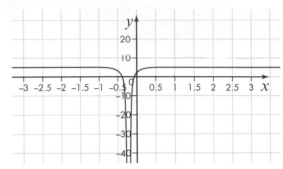

6 Yes, the graph is for the given function. Correct shape with asymptotes at $x = \frac{3}{2}$ and $y = 2$; *y*-intercept at $(0, 2\frac{7}{9})$.

Exercise 3.5A   p 82

1 There are three points of intersection.

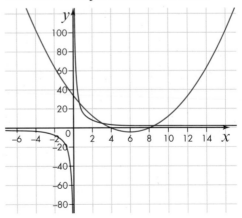

2 There are no solutions to $\dfrac{1}{x^2} = -x^2$ because the graphs do not intersect.

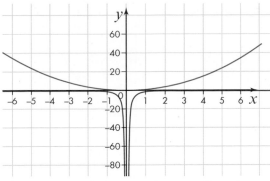

3 There are two solutions to $x^4 = x(x-3)(x-4)$ because the graphs intersect twice.

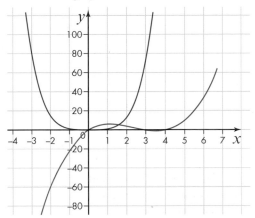

4 There are two solutions to $\dfrac{(x+1)(2x-1)^3}{(3-x)^2} = 1$

because the graphs intersect twice.

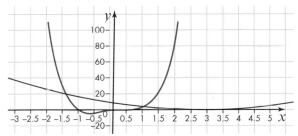

5 There are no solutions to $(x+4)^4 = -2$ because the graphs do not intersect.

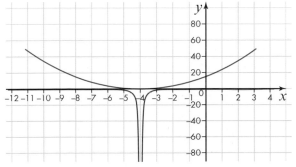

6

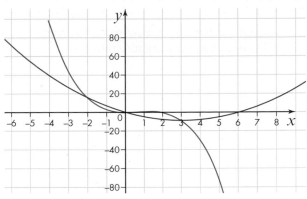

$$x(x-6) = x^2(2-x)$$
$$x = -2, 0 \text{ or } 3$$
$$y = 16, 0 \text{ or } -9$$

7 There are two points of intersection.

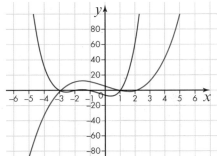

8 There are two points of intersection.

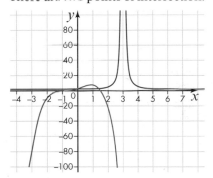

Exercise 3.6A   **p 85**

1 a 54

   b 5

2

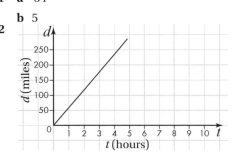

$t = 4.5$, $d = 270$ miles.

3 14.49 g

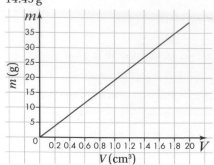

0.78 cm^{-3}

4 a C, iii **b** A, ii **c** B, i

5

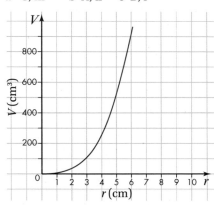

a 268 cm^3.

b $r = 6$ cm.

c Substituting $r = 4$ into the equation gives $V = 268$.

Likewise substituting $V = 288\pi$ into the equation gives $r = 6$.

6 a $y = 5\sqrt{x}$

b 20

c 100

Exercise 3.7A **p 89**

1 a one solution.

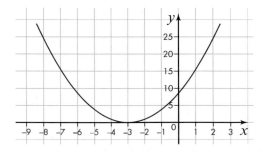

b no solutions.

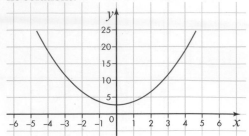

c one solution.

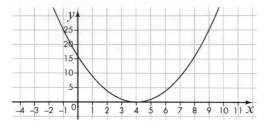

d one solution.

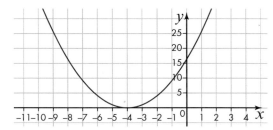

2 a asymptotes at $x = -2$ and $y = 0$.

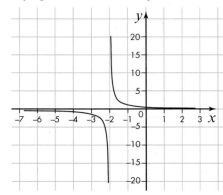

b asymptotes at $x = 0$ and $y = 2$.

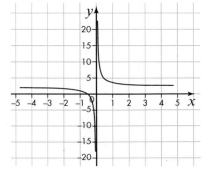

c asymptotes at $x = 1$ and $y = 0$.

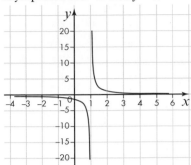

d asymptotes at $x = 0$ and $y = -1$.

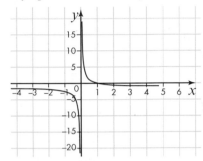

3 $f(x - 2)$: one solution.

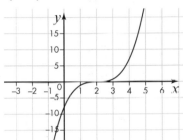

$f(x) + 2$: one solution.

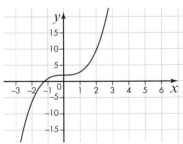

4 a asymptotes at $x = -2$ and $y = 0$.

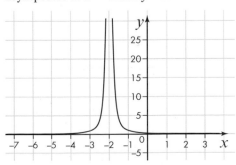

b $f(x + 2) = \dfrac{1}{(x + 2)^2}$

c $(0, \frac{1}{4})$

5 $f(x + 1)$: P at $(-1, 1)$; asymptote at $y = 0$.

$f(x) + 1$: P at $(0, 2)$; asymptote at $y = 1$.

$f(x + 1): = f(x) + 1$ has one solution (one point of intersection).

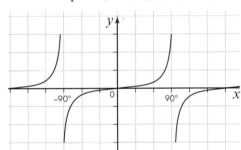

6 a, b asymptotes at $x = -90$ and 90.

Axis intercepts at $(-180, 0)$, $(0, 0)$ and $(180, 0)$.

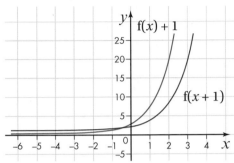

c, d asymptotes at $x = -90$ and 90.

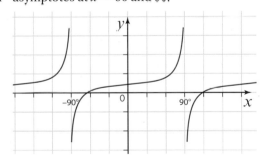

7 a $f(x + 2) = (x + 3)(2x + 3)^3$: two solutions.

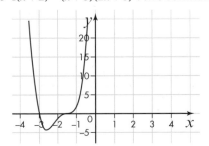

b $f(x) + 2 = (x + 1)(2x - 1)^3 + 2$: two solutions.

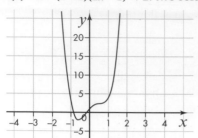

c $f(x - 1) = x(2x - 3)^3$: two solutions.

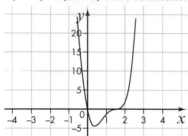

d $f(x) - 1 = (x + 1)(2x - 1)^3 - 1$: two solutions.

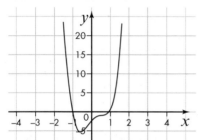

8 **a, b** $f(x + 2) = 2^{x+3}$: no solutions.

$f(x) + 2 = 2^{x+1} + 2$: no solutions.

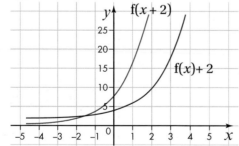

c If $f(x) = 2^x$ then $f(x + 3) = 2^{x+3}$
and $f(x + 1) + 2 = 2^{x+1} + 2$

9 $f(x - 1) = (x - 1)^4 - 5(x - 1)^3 + 5(x - 1)^2 + 5(x - 1) - 6$
$\qquad = x^4 - 2x^3 + x^2 - 2x^3 + 4x^2 - 2x + x^2 - 2x$
$\qquad + 1 - 5x^3 - 5x + 15x^2 - 10x + 5$
$\qquad + 5x^2 - 10x + 5 + 5x - 5 - 6$
$\qquad = x^4 - 9x^3 + 26x^2 - 24x$

Exercise 3.8A **p 95**

1 **a** one solution.

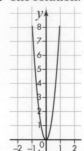

b one solution.

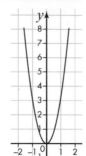

c one solution.

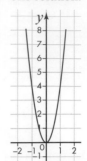

d one solution.

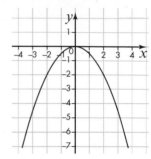

2 **a** asymptotes at $x = 0$ and $y = 0$.

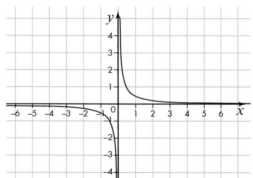

b asymptotes at $x = 0$ and $y = 0$.

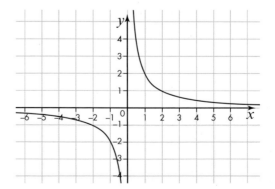

c asymptotes at $x = 0$ and $y = 0$.

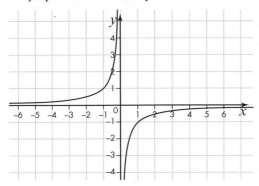

d asymptotes at $x = 0$ and $y = 0$.

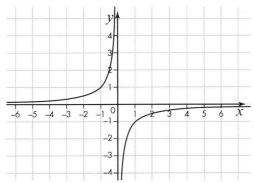

3 $f(-2x)$: $P(1, 0)$, $R(-\frac{1}{2}, 0)$, $Q(0, -2)$.

$f(-2x) = 0$ has two solutions.

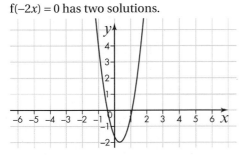

$-2f(x)$: $P(-2, 0)$, $R(1, 0)$, $Q(0, 4)$.

$-2f(x) = 0$ has two solutions.

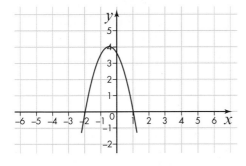

4 **a** asymptotes at $x = -1$ and $y = 0$.

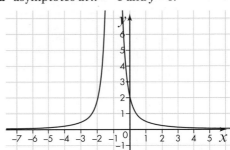

b $2f(x) = \dfrac{2}{(x+1)^2}$

c $(0, 2)$

5 $f(-x)$ and $-f(x)$: in both cases asymptote at $y = 0$.

For $f(-x)$, P is at $(0, 1)$.

For $-f(x)$, P is at $(0, -1)$.

$f(-x) = -f(x)$: has no solutions.

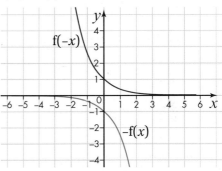

6 **a, b** asymptotes at $x = -135°$, $-45°$, $45°$ and $135°$.

Axis intercepts at $(-180°, 0)$, $(-90°, 0)$, $(0, 0)$, $(90°, 0)$ and $(180°, 0)$.

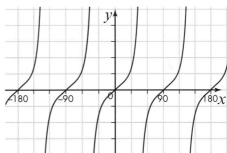

c, d asymptotes at $x = -90°$ and $90°$.

Axis intercepts at $(-180°, 0)$, $(0, 0)$ and $(180°, 0)$.

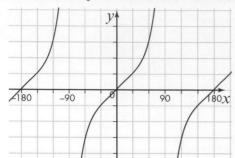

7 a $f(2x) = (2x + 1)(4x - 1)^3$: two solutions.

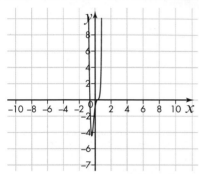

b $2f(x) = 2(x + 1)(2x - 1)^3$: two solutions.

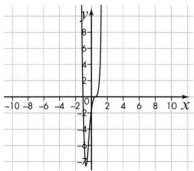

c $f\left(\dfrac{x}{2}\right) = \left(\dfrac{x}{2} + 1\right)(x - 1)^3$: two solutions.

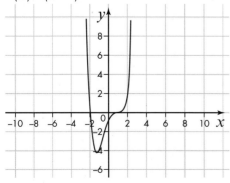

d $\dfrac{1}{2} f(x) = \dfrac{1}{2}(x + 1)(2x - 1)^3$: two solutions.

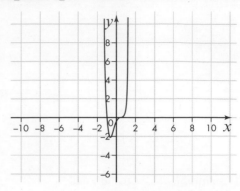

8 a, b, d $f(2x) = 2^{2x+1}$: no solutions. $2f(x) = (2)(2^{x+1})$: no solutions.

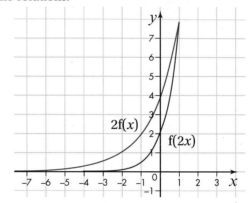

c $f(2x + 1) = 2^{2x+1}$ and $2f(x + 1) = (2)(2^{x+1})$

9 $f\left(\dfrac{x}{3}\right) = \left(\dfrac{x}{3}\right)^4 - 5\left(\dfrac{x}{3}\right)^3 + 5\left(\dfrac{x}{3}\right)^2 + 5\left(\dfrac{x}{3}\right) - 6$

$\quad = \dfrac{x^4}{81} - \dfrac{5x^3}{27} + \dfrac{5x^2}{9} + \dfrac{5x}{3} - 6$

Exam-style questions 3   p 98

1 a $f(x + 2)$ **b** $f(-x)$

    c $f(x) + 1$ **d** $f(x)$

2 The number of points the cable will touch the rollercoaster in the given interval is 2

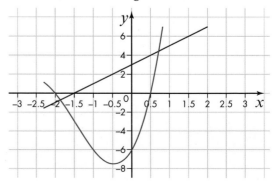

3

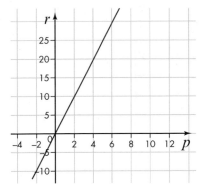

a $r = 145$
b $p = 50$

4 a

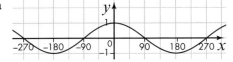

b

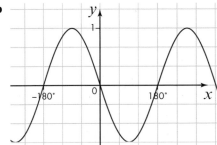

c

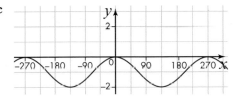

d

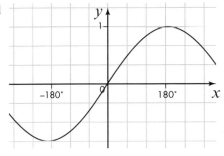

5

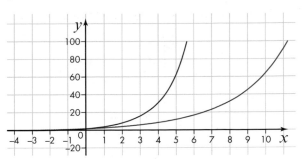

The rabbit and mice populations are only the same when they are both first introduced to the island.

6 $l = 7, A = 49\pi$

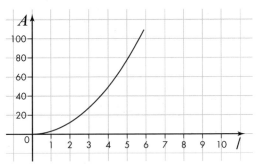

The shape is a circle.

7 a Asymptotes at $x = 2$ and $y = 3$.

There are two points of intersection.

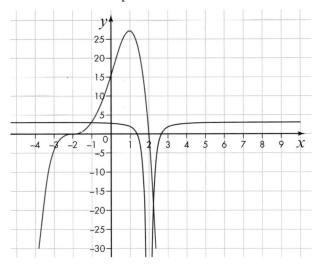

b Asymptotes at $x = 4$ and $y = 3$.

There are two points of intersection.

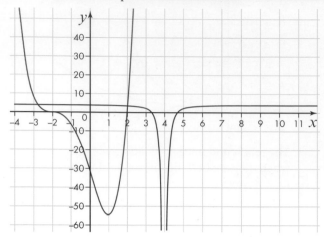

8 a f(x) + 2   **b** −f(x)

c f(x − 2)   **d** f(−x)

9 a

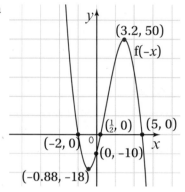

b

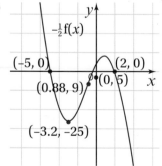

c

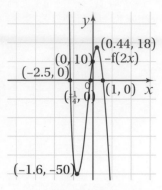

d

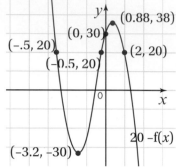

10 a

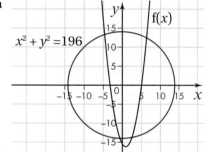

b i f(x) + a where a > 2

ii f(x − a) where a > 9 or a < −11

4 Coordinate geometry 1: Equations of straight lines

Prior knowledge   **p 101**

1   $(0, 3)$, $(-\frac{3}{2}, 0)$

2   **a**   A: 0, 2

     B: 2, 3

     C: $-\frac{1}{2}$, 3

     D: −2, 3

     E: 0, 3

     F: −2, $\frac{3}{2}$

  b   A, E

     D, F

  c   B, C

3   $y = 16 - 8x$

4   $y = 2x - 2$

Exercise 4.1A   **p 103**

1   **c**   because the right-hand side is not zero

  d   because the right-hand side is not zero

  e   because c is not an integer

  f   because a, b and c are not integers.

2   **a** $5x - y + 4 = 0$ where $a = 5$, $b = -1$ and $c = 4$

  b $2x + y - 3 = 0$ where $a = 2$, $b = 1$ and $c = -3$

3   **a** $x - 3y - 21 = 0$ where $a = 1$, $b = -3$ and $c = -21$

  b $2x + 5y - 30 = 0$ where $a = 2$, $b = 5$ and $c = -30$

  c $8x - 6y + 21 = 0$ where $a = 8$, $b = -6$ and $c - 21$

4   $y = 4x + \frac{3}{2}$

  Gradient $= 4$ and y intercept $= \frac{3}{2}$; coordinates $(0, \frac{3}{2})$

5   Equation written down incorrectly. It should be $y = 3 - \frac{5}{2}x$.

  Not all expressions in the equation have been multiplied by 2. It should be $2y = 6 - 5x$.

  Although the values are wrong this step is actually correct. With the correct values it should be $5x + 2y = 6$.

  Although the values are wrong this step is actually correct. With the correct values it should be $5x + 2y - 6 = 0$.

6   $3x - 5y - 12 = 0$ where $a = 3$, $b = -5$ and $c = -12$

7   coordinates of the x intercept are $\left(\frac{2}{3}, 0\right)$.

  coordinates of the y intercept are $\left(0, \frac{2}{5}\right)$.

Exercise 4.2A   **p 105**

1   **a** $y = 2x - 6$

  b $y = 3x + 3$

  c $y = 2x - 2$

  d $y = -5x + 13$

2   **a** $y = -4x - 13$

  b $y = -x$

3   $2x + y - 3 = 0$

4   $3x - y + 3 = 0$

5   $y = 3x + 3$

6   $(3, 7)$.

7   The line does not go through the origin.

Exercise 4.3A   **p 108**

1   **a**   1

  b   3

  c   −3

2   **a** $\frac{2}{5}$

  b   −3

3   Should be $m = \dfrac{y_2 - y_1}{x_2 - x_1}$

  y coordinate incorrect in (x_2, y_2). It should be $(5, -8)$.

  Numerator and denominator confused. Should be:

  $m = \dfrac{-8 - 2}{5 - 1}$

  $m = \dfrac{-10}{4}$

  $m = \dfrac{-5}{2}$

4   All except b

5   $m = \dfrac{y_2 - y_1}{x_2 - x_1}$

  $(x_1, y_1) = \left(5, \frac{1}{2}\right)$

  $(x_2, y_2) = (2, 2)$

  $m = \dfrac{2 - \frac{1}{2}}{2 - 5}$

  $m = \dfrac{\frac{3}{2}}{-3}$

  $m = -\frac{1}{2}$

  $m = \dfrac{y_2 - y_1}{x_2 - x_1}$

  $(x_1, y_1) = \left(11, -\frac{5}{2}\right)$

  $(x_2, y_2) = (2, 2)$

$$m = \frac{2 - -\frac{5}{2}}{2 - 11}$$

$$m = \frac{\frac{9}{2}}{-9}$$

$$m = -\frac{1}{2}$$

The gradient, m, is the same between both of the pairs of points, so we can conclude that all the points lie on a straight line.

6 The descent is the steepest part because $\frac{1}{16} > \frac{1}{20}$.

7 −4

Exercise 4.4A p 111

1 a $y = x + 1$

   b $y = 3x - 6$

2 a $y = -3x$

   b $y = -\frac{5}{2}x - \frac{23}{2}$

3 $\frac{y - y_1}{y_2 - y_1} = \frac{x - x_1}{x_2 - x_1}$

   Let $(x_1, y_1) = (1, 2)$

   Let $(x_2, y_2) = (5, -8)$

   y_1 and y_2 and x_1 and x_2 confused. The next line should be:

   $\frac{y - 2}{-8 - 2} = \frac{x - 1}{5 - 1}$

   With this correction, the remainder of the calculation should be:

   $$\frac{y - 2}{-10} = \frac{x - 1}{4}$$
   $$4(y - 2) = -10(x - 1)$$
   $$4y - 8 = -10x + 10$$
   $$4y = -10x + 18$$
   $$\text{So } y = \frac{-5x + 9}{2}$$

4 $8x - y + 6 = 0$

5 $7x + y - 14 = 0$

   Gradient $= -7$ and y-intercept $(0, 14)$

6 Line A:

   $y = \frac{23}{3} - \frac{1}{3}x$

   Line B:

   $y = 1 - x$

   Line B is steeper because $1 > \frac{1}{3}$

7 a $y = -4x + \frac{7}{3}$

   b $y = \frac{63}{130}x - \frac{22}{65}$

8 $21x + 25y - 35 = 0$

9 Line A:

   $y = 4x - 7$

   Line B:

   $y = 4 - 7x$

   point of intersection is $(1, -3)$.

10 $y = -\frac{4}{3}x + \frac{22}{3}$

   When $x = 1$, $y = 6$ so this line does not pass through the point $(1, 2)$.

Exercise 4.5A p 115

1 a the lines are parallel.

   b the lines are parallel.

2 Line A:

   $(x_1, y_1) = (1, 7)$

   $(x_2, y_2) = (3, 11)$

   $$m = \frac{y_2 - y_1}{x_2 - x_1}$$

   $$m = \frac{11 - 7}{3 - 1}$$

   $$m = 2$$

   Line B:

   $(x_1, y_1) = (2, -3)$

   $(x_2, y_2) = (5, 3)$

   $$m = \frac{y_2 - y_1}{x_2 - x_1}$$

   $$m = \frac{3 - -3}{5 - 2}$$

   $$m = 2$$

   $m = 2$ for both, so the lines are parallel.

3 $y = 7x - 21$

4 a the lines are not parallel.

   b the lines are parallel.

5 Line A:

   $(x_1, y_1) = (-1, 3)$

   $(x_2, y_2) = (3, -1)$

   $$m = \frac{y_2 - y_1}{x_2 - x_1}$$

   $$m = \frac{-1 - 3}{3 - -1}$$

   $$m = -1$$

Line B:

$(x_1, y_1) = (-2, 2)$

$(x_2, y_2) = (-3, 3)$

$$m = \frac{y_2 - y_1}{x_2 - x_1}$$

$$m = \frac{3 - 2}{-3 - -2}$$

$$m = -1$$

$m = -1$ for both, so the lines are parallel.

6   $y = 4x + 3$

7   **a**   the lines are parallel.
    b   the lines are not parallel.

8   Arrange the equations in the form $y = mx + c$:

$2x - y + 6 = 0$, $y = 2x + 6$, $m = 2$

$2x + 4y - 44 = 0$, $4y = 44 - 2x$, $y = 11 - \frac{1}{2}x$, $m = -\frac{1}{2}$

$2x - y - 4 = 0$, $y = 2x - 4$, $m = 2$

$2x + 4y - 24 = 0$, $4y = 24 - 2x$, $y = 6 - \frac{1}{2}x$, $m = -\frac{1}{2}$

So $2x - y + 6 = 0$ and $2x - y - 4 = 0$ are parallel lines as $m = 2$ for both lines.

And $2x + 4y - 44 = 0$ and $2x + 4y - 24 = 0$ are parallel lines as $m = -\frac{1}{2}$ for both lines.

Exercise 4.5B   p 117

1   **a**   the lines are not perpendicular.

    b   the lines are perpendicular.

2   $m_1 = -\frac{1}{2}$ for the given line but the gradient of a line perpendicular to this will be $m_2 = 2$.

Let $(x_1, y_1) = (1, 0)$

$$y - y_1 = m(x - x_1)$$

(x_1, y_1) incorrectly substituted. Correcting this and using correct value of m_2 gives:

$y - 0 = 2x - 2$

So $y = 2x - 2$

3   **a** $m_1 = 3$ and $m_2 = -\frac{1}{3}$ so $m_1 m_2 = -1$ so the lines are perpendicular.

    b $m_1 = 3$ and $m_2 = 12$ so $m_1 m_2 \neq -1$ so the lines are not perpendicular.

4   Line A:

$(x_1, y_1) = (1, 1)$

$(x_2, y_2) = (3, 9)$

$$m = \frac{y_2 - y_1}{x_2 - x_1}$$

$$m = \frac{9 - 1}{3 - 1}$$

$$m = 4$$

Line B:

$(x_1, y_1) = (4, 0)$

$(x_2, y_2) = (8, -1)$

$$m = \frac{y_2 - y_1}{x_2 - x_1}$$

$$m = \frac{-1 - 0}{8 - 4}$$

$$m = -\frac{1}{4}$$

$m_1 = 4$ and $m_2 = -\frac{1}{4}$ so $m_1 m_2 = -1$ so the lines are perpendicular.

5   $2x - y + 6 = 0$, $y = 2x + 6$, $m = 2$

$2x + 4y - 44 = 0$, $4y = 44 - 2x$, $y = 11 - \frac{1}{2}x$, $m = -\frac{1}{2}$

$2x - y - 4 = 0$, $y = 2x - 4$, $m = 2$

$2x + 4y - 24 = 0$, $4y = 24 - 2x$, $y = 6 - \frac{1}{2}x$, $m = -\frac{1}{2}$

So $2x - y + 6 = 0$ and $2x + 4y - 44 = 0$ are perpendicular lines as $m_1 m_2 = -1$.

And $2x - y + 6 = 0$ and $2x + 4y - 24 = 0$ are perpendicular lines as $m_1 m_2 = -1$.

And $2x - y - 4 = 0$ and $2x + 4y - 44 = 0$ are perpendicular lines as $m_1 m_2 = -1$.

And $2x - y - 4 = 0$ and $2x + 4y - 24 = 0$ are perpendicular lines as $m_1 m_2 = -1$.

6   $y = \frac{5}{2} - \frac{1}{4}x$

7   **a** $m_1 = \frac{5}{3}$ and $m_2 = -\frac{3}{5}$ so $m_1 m_2 = -1$ so the lines are perpendicular.

    b $m_1 = \frac{9}{10}$ and $m_2 = \frac{10}{9}$ so $m_1 m_2 \neq -1$ so the lines are not perpendicular.

8   A is parallel to C, B is parallel to D, A is perpendicular to B and D, and C is perpendicular to B and D.

So the quadrilateral can only be a square or a rectangle.

Exercise 4.6A   p 122

1   $y = \frac{8}{5}x$

2   **a**   False: $m_1 = 2$ and $m_2 = -2$ so $m_1 m_2 \neq -1$ so the lines are not perpendicular.

    b   False: $c_1 = 6$ and $c_2 = 12$ so the lines do not share the same y intercept.

    c   False: negative gradient means the object is travelling back to the start.

    d   True: speed is the gradient and the gradients are the same (the difference in the signs just indicates direction) so they are travelling at the same speed.

e False: $m_1 \neq m_2$ so the lines are not parallel.

f False: positive gradient means the object is travelling away from the start.

3 $y = 180x + 50$

$y = 120x + 67.50$

m is the cost per year to run the printer and is £180 in Anna's case and £120 in Bhavini's case.

c is the cost to buy the printer and is £50 in Anna's case and £67.50 in Bhavini's case.

The cost of one ink cartridge for Anna's printer is £180 ÷ 4 = £45.

The cost of one ink cartridge for Bhavini's printer is £120 ÷ 4 = £30.

Nothing is significant about the x intercepts.

4 $y = 5x$

$y = 3x + 10$

$y = 2x + 15$

Taxi firm A is the cheapest on a 4-mile journey.

Taxi firm C is the cheapest on a 7-mile journey.

5 $y = \frac{75}{43}x + \frac{1}{172}$

Exam-style questions 4   p 124

1 $y = 3x + 3$

2 **a** Line A:

$m = -2$

Line B:

$m = \frac{1}{2}$

b Line A is steeper because $2 > \frac{1}{2}$.

3 $y = -x + 5$

4 **a** runner A will have run 6 miles.

b Runner A:

$y = 2x + 3$

Runner B:

$y = 3x$

c Runner B runs faster. Gradient = speed and $3 > 2$.

5 **a** $y = -\frac{1}{2}x + 10$

b y intercept = (0, 10)

x intercept = (20, 0)

c 100 units^2

6 Line AB:

$(x_1, y_1) = (-2, 8)$

$(x_2, y_2) = (4, 5)$

$m = \frac{y_2 - y_1}{x_2 - x_1}$

$m = \frac{5 - 8}{4 - -2}$

$m_1 = -\frac{1}{2}$

Line BC:

$(x_1, y_1) = (-2, 8)$

$(x_2, y_2) = (-10, -8)$

$m = \frac{y_2 - y_1}{x_2 - x_1}$

$m = \frac{-8 - 8}{-10 - -2}$

$m_2 = 2$

$m_1 m_2 = \left(-\frac{1}{2}\right)(2) = -1$

So AB and BC are perpendicular to each other and ABC is a right-angled triangle.

7 Pay as you go:

$y = 2x + 5$

4 rides: $x = 4$ so $y = £13$

17 rides: $x = 17$ so $y = £39$

For other option (book of tickets):

Cost of 4 rides = £5 + £10 = £15

so pay as you go option is cheaper.

Cost of 17 rides = £5 + (3 × £10) = £35

so book of tickets option is cheaper.

8 $y = -2x - 5$

9 $y = \dfrac{2\sqrt{6} + 2}{10}x + 3$

10 $y = \frac{5}{3}x + \frac{617}{21}$

11 **a** Line A:

$y = \frac{9}{5}x + 32$

Line B:

$y = \frac{5}{9}(x - 32)$

The two equations are the inverses of each other.

b (−40, −40).

c The transformation linking the two lines is a reflection in the line $y = x$.

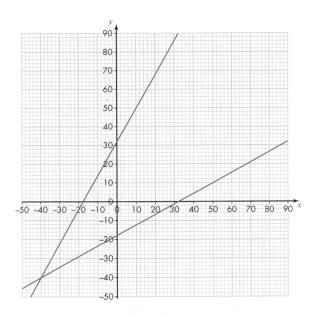

d The point of intersection is the only point where the temperature in degrees Celsius and the temperature in degrees Fahrenheit have the same numerical value, i.e. at $(-40, -40)$.

5 Coordinate geometry 2: Circles

Prior knowledge   p 127

1 $-\dfrac{2}{5}$

2 $x - 3y + 19 = 0$

3 $(9, 4)$.

4 $(x - 7)^2 - 26$

Exercise 5.1A   p 131

1 **a** Centre $= (-5, 8)$, radius $= 6$

   b Centre $= (19, 33)$, radius $= 20$

   c Centre $= (0, -4)$, radius $= 3\sqrt{5}$

   d Centre $= (-3, -10)$, radius $= 2\sqrt{7}$

2 **a** $(x + 5)^2 + (y - 9)^2 = 49$ and $x^2 + y^2 + 10x - 18y + 57 = 0$

   b $(x + 11)^2 + (y + 1)^2 = 169$ and $x^2 + y^2 + 22x + 2y - 47 = 0$

   c $(x - 3)^2 + y^2 = 48$ and $x^2 + y^2 - 6x - 39 = 0$

   d $(x - 14)^2 + (y - 6)^2 = 44$ and $x^2 + y^2 - 28x - 12y + 188 = 0$

3 **a** $(x - 5)^2 + (y + 3)^2 = 16$

   b $(x + 4)^2 + (y - 2)^2 = 9$

   c $(x + 1)^2 + (y - 4)^2 = 20$

4 **a** Centre $= (9, -7)$, radius $= 12$

   b Centre $= (-4, 0)$, radius $= 5$

   c Centre $= (-5, -9)$, radius $= 3\sqrt{3}$

   d Centre $= (-15, 3)$, radius $= \dfrac{9}{2}$

5 $(x + 4)^2 + (y - 7)^2 = 41$.

6 $(x - 9)^2 + (y - 2)^2 = 81$

7 The circle cuts the x-axis at $(5, 0)$ and $(-3, 0)$.

8 The centre is 5 units from the origin.

Exercise 5.1B   p 133

1 55

2 **a** $(8, 1)$

   b 10

   c $(x - 8)^2 + (y - 1)^2 = 100$.

3 $(x - 9)^2 + (y - 3)^2 = 169$.

4 **a** The student has correctly worked out that $AB^2 = 1369$, but the equation of the circle needs the square of the radius, not the square of the diameter. The diameter is $\sqrt{1369} = 37$, so the radius is $\dfrac{37}{2}$.

   b $(x + 0.5)^2 + (y - 17)^2 = (\dfrac{37}{2})^2$

5 **a** $x = 1$

     $y = -11$

   b $4\sqrt{5}$

   c $(x - 5)^2 + (y + 3)^2 = 80$

   d $(-3 - 5)^2 + (1 + 3)^2 = (-8)^2 + (4)^2 = 80$

   Since both sides of the equation are satisfied, H lies on C.

6 **a** $p = -5$

     $q = 0$

   b $(x + 2)^2 + (y - 7)^2 = 58$.

7 The equation of the circle can be rewritten as $(x + 3)^2 + (y - 8)^2 = 125$, so $(-3, 8)$ is the centre of the circle.

Substitute $y = 2x + 14$ into $x^2 + y^2 + 6x - 16y = 52$.

$x^2 + (2x + 14)^2 + 6x - 16(2x + 14) = 52$

Expand and simplify.

$x^2 + 4x^2 + 56x + 196 + 6x - 32x - 224 = 52$

$$5x^2 + 30x - 80 = 0$$
$$x^2 + 6x - 16 = 0$$

Solve.

$$(x + 8)(x - 2) = 0$$
$$x = -8 \text{ or } 2$$

M and N are the two intersection points, it doesn't matter which is which.

When $x = -8$, $y = 2 \times -8 + 14 = -2$, so M $= (-8, -2)$

When $x = 2$, $y = 2 \times 2 + 14 = 18$, so N $= (2, 18)$

The midpoint of MN $= \left(\frac{-8+2}{2}, \frac{-2+18}{2} \right) = (-3, 8)$

Since the midpoint of MN is the centre of the circle, MN is a diameter of the circle.

Exercise 5.2A   p 137

1 a $d^2 = (x_1 - x_2)^2 + (y_1 - y_2)^2$

$DE^2 = (-1 - 1)^2 + (11 - 5)^2$

$DE^2 = (-2)^2 + (6)^2 = 40$

$EF^2 = (1 - 13)^2 + (5 - 9)^2$

$EF^2 = (-12)^2 + (-4)^2 = 160$

$DF^2 = (-1 - 13)^2 + (11 - 9)^2$

$DF^2 = (-14)^2 + (2)^2 = 200$

Since $40 + 160 = 200$, $DE^2 + EF^2 = DF^2$.

b Gradient of DE $= \frac{11 - 5}{-1 - 1} = -3$

Gradient of EF $= \frac{9 - 5}{13 - 1} = \frac{1}{3}$

Since $m_1 \times m_2 = -1$, DE and EF are perpendicular.

c $(x - 6)^2 + (y - 10)^2 = 50$.

2 a $PQ^2 = (-5 - 11)^2 + (-4 - 8)^2 = (-16)^2 + (-12)^2 = 400$

$PR^2 = (-5 - 13)^2 + (-4 - 2)^2 = (-18)^2 + (-6)^2 = 360$

$QR^2 = (11 - 13)^2 + (8 - 2)^2 = (-2)^2 + (6)^2 = 40$

Since $PR^2 + QR^2 = PQ^2$, Pythagoras' theorem is satisfied and the triangle PQR is right-angled.

b $(100\pi - 60)$ m^2

c £204

3 $\frac{149}{26}$

$\frac{17}{2}$

4 a 10

b 4

c Gradient of AB $= -\frac{1}{3}$

Gradient of AC $= 1$

Gradient of BC $= -1$

d $m_1 \times m_2 = -1$ so AC and BC are perpendicular.

Since the angle in a semicircle is a right angle, AB is a diameter of the circle.

5 a Gradient of $L_1 = \frac{6 - 1}{5 - (-10)} = \frac{1}{3}$

Gradient of $L_2 = \frac{4 - (-2)}{9 - 11} = -3$

$m_1 \times m_2 = -1$ so L_1 and L_2 are perpendicular.

b $x^2 + y^2 - 16x - 4y + 43 = 0$

6 a $(x - 8)^2 + (y - 4)^2 = 50$.

b 10 or 34

Exercise 5.3A   p 142

1 a $(8, 1)$

b 4

c $-\frac{1}{4}$

d $4y + x - 12 = 0$.

e When $x = 4$, $y - 1 = -\frac{1}{4}(4 - 8)$

$y = 1 - 1 + 2 = 2$ as required.

f $(x - 4)^2 + (y - 2)^2 = 34$.

2 $(x - 9)^2 + (y - 8)^2 = 37$.

3 a $(2, -5)$.

b $(x - 8)^2 + (y - 2)^2 = 85$.

4 28 m

5 a $(-3, -5)$

b 16

c 88

6 a $3x + 2y = 7$

b $2x = 3y + 9$

d 90°.

e AC is a diameter.

7 a $(x - 15)^2 + (y - 4)^2 = 25$.

b $(x + 2)^2 + (y - 8)^2 = 25$.

c $(x - 10)^2 + (y - 6)^2 = 85$.

d $(x + 8)^2 + (y + 12)^2 = 130$.

8 a $(2, -3)$ and $(6, 5)$.

b $4\sqrt{5}$

Exercise 5.4A   p 146

1 $3x + 2y = 37$

2 a $2^2 + (-2 - 1)^2 = 4 + 9 = 13$

Since both sides of the equation agree, T lies on the circle.

b Centre of circle $= (0, 1)$

Gradient of radius $= \frac{1 - (-2)}{0 - 2} = -\frac{3}{2}$

Gradient of the tangent $= \frac{2}{3}$

The equation of the tangent is given by

$y + 2 = \frac{2}{3}(x - 2)$

$3y + 6 = 2x - 4$

Hence $2x = 3y + 10$

3 a $(3, 6)$

b 8

c 9 m

4 a $(7, \frac{19}{4})$

b $\frac{75}{4}$

5 7

6 a 63

b Use $d^2 = (x_1 - x_2)^2 + (y_1 - y_2)^2$

For PT:

$PT^2 = (10 - 31)^2 + (63 - 35)^2$

$PT^2 = (-21)^2 + (28)^2 = 1225$

For PX:

$PX^2 = (10 - 15)^2 + (63 - 23)^2$

$PX^2 = (-5)^2 + (40)^2 = 1625$

TX is the radius, so $TX^2 = 400$

Hence $PT^2 + TX^2 = PX^2$ because $1225 + 400 = 1625$.

c 350

7 a $(x - 5)^2 + (y - 11)^2 = 8$

b $(x + 8)^2 + (y - 23)^2 = 176.4$

Exercise 5.4B   p 150

1 a Substitute $y = x - 10$ into $x^2 + y^2 = 50$.

$x^2 + (x - 10)^2 = 50$

$x^2 + x^2 - 20x + 100 = 50$

$2x^2 - 20x + 50 = 0$

$x^2 - 10x + 25 = 0$

$(x - 5)^2 - 0$

The equation has repeated roots so $y = x - 10$ is a tangent of the circle.

b Substitute $x = 7y - 50$ into $x^2 + y^2 = 50$.

$(7y - 50)^2 + y^2 = 50$

$49y^2 - 700y + 2500 + y^2 = 50$

$50y^2 - 700y + 2450 = 0$

$y^2 - 14y + 49 = 0$

$(y - 7)^2 = 0$

The equation has repeated roots so $7y = x + 50$ is a tangent of the circle.

2 Substitute $y = 4x - 5$ into $x^2 + y^2 + 10x - 18y + 38 = 0$.

$x^2 + (4x - 5)^2 + 10x - 18(4x - 5) + 38 = 0$

$x^2 + 16x^2 - 40x + 25 + 10x - 72x + 90 + 38 = 0$

$17x^2 - 102x + 153 = 0$

$x^2 - 6x + 9 = 0$

$(x - 3)^2 = 0$

The equation has repeated roots so $y = 4x - 5$ is a tangent of the circle.

3 8 or -32

4 $y = -\frac{1}{2}x + 6$ and $y = -\frac{1}{2}x + 26$.

5 $6x + 17y - 408 = 0$

$3x + 2y - 48 = 0$

Exam-style questions 5   p 151

1 a The equation of the circle is given by $(x - a)^2 + (y - b)^2 = r^2$, so centre = (9, –8). Gemma has copied the signs from the equation but needed to change them both.

b Olivia has stated the length of the radius, not the diameter.

Radius = $\sqrt{484} = 22$

Diameter = $22 \times 2 = 44$

2 $x^2 + y^2 - 14x + 6y - 6 = 0$

3 (–6, 0) (–4, 0)

(0, 2) (0, 12).

4 Dilgusha has not halved the coefficients of x and y when completing the square.

She has added 14^2 rather than subtracting.

When square-rooting -182, which shouldn't have been possible in this situation, she has square-rooted $+182$ and made it negative.

Correct solution:

Complete the square on x and y.

$(x + 3)^2 - 9 + (y - 7)^2 - 49 + 22 = 0$

$(x + 3)^2 + (y - 7)^2 = 36$

Centre = (–3, 7)

Radius = $\sqrt{36} = 6$

5 a 150 cm

b $(x - 60)^2 + (y - 130)^2 = 22\,500$

c $3y - 4x + 600 = 0$

d 72 m

e Unlikely athlete will remain at same point whilst turning.

Hammer's motion will not be horizontal because it will be affected by gravity.

6 a $(x - 11)^2 + (y - 4)^2 = 40$.

b $x - 3y + 21 = 0$

7 23

8 a $(x - 7)^2 + (y - 2)^2 = 676$.

b $12x + 5y - 263 = 0$.

c $26\sqrt{3}$

9 a $(x + 4)^2 + (y - 11)^2 = 50$

b A is (–9, 16) and B is (1, 6).

c $(-\frac{13}{2}, \frac{37}{2})$.

10 a Equation is given by $(x-6)^2 + (y-3)^2 = 9$

The centre is (6, 3). The radius is 3.

Method 1:

Similar triangles PTX and PAO:

$\dfrac{XP}{OP} = \dfrac{TX}{OA}$

Let length of $XP = a$

$\dfrac{a}{\sqrt{6^2 + (3+a)^2}} = \dfrac{3}{6}$

$2a = \sqrt{36 + (3+a)^2}$

$4a^2 = 36 + (3+a)^2$

$4a^2 = 36 + 9 + 6a + a^2$

$3a^2 - 6a - 45 = 0$

$a^2 - 2a - 15 = 0$

$(a-5)(a+3) = 0$

$a = 5$ (can't be –3)

b 18

11 a Substitute $y = mx$ into the equation of the circle C.

$(x-5)^2 + (mx-3)^2 = 2$

$x^2 - 10x + 25 + m^2x^2 - 6mx + 9 = 2$

$(1 + m^2)x^2 + (-10 - 6m)x + 32 = 0$

Given that $y = mx$ is a tangent to the circle C, $b^2 - 4ac = 0$:

$(-10 - 6m)^2 - 4 \times (1 + m^2) \times 32 = 0$

$100 + 120m + 36m^2 - 128 - 128m^2 = 0$

$92m^2 - 120m + 28 = 0$

$23m^2 - 30m + 7 = 0$

b $4\sqrt{2}$

12 130 units^2

6 Trigonometry

Prior knowledge   p 155

1 17.2 cm

2 53.1°

3 $x = \frac{1}{2}$ or –1

Exercise 6.1A   p 157

1 100°

2 a 168°   **b** 156°

   c 144°   **d** 132°

   e 108°

3 $\cos 143° = -0.7986$

4 99°

5 a 70.7°

   b 109.3°

6 a

angle	0	30°	60°	90°	120°	150°	180°
sine	0	0.5	0.866	1	0.866	0.5	0
cosine	1	0.866	0.5	0	–0.5	–0.866	–1

b

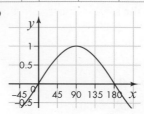

c Reflection symmetry in the line $x = 90°$

d

e Rotation symmetry of order 2, centre (90°, 0)

f 45°

g 135°

7 a 41.3°

   138.7°

b There is no solution in the given range.

Exercise 6.2A   p 164

1 a 28.2 cm

   b 271.9 cm^2

2 a Area $= \dfrac{1}{2} \times 35 \times 40 \times \sin 70° = 658$ cm^2

   b $\dfrac{BC}{\sin 70°} = \dfrac{40}{\sin 60°}$

   $\Rightarrow BC = \dfrac{40 \sin 70°}{\sin 60°} = 43.4$ cm

3 $X = 69.5°$

   $Y = 59.2°$

   $Z = 51.3°$

4 a 29.5 cm

   b 17.2 cm

5 70.5° or 109.5°

6 86.1°

7 141°

8 a 7.63 m

9 a i 38.7° 141.3°

ii 23.6°

iii no solution

b $\sin Y = \dfrac{20}{YZ} \Rightarrow YZ = \dfrac{20}{\sin Y}$

YZ will have its smallest possible value when sin Y has its largest value. The largest value is 1 (when Y = 90°) so the smallest possible value of YZ is 20 cm.

Exercise 6.3A p 168

1 a 0.5 **b** −0.5 **c** −0.5

  d −0.5 **e** 0.5

2 a −0.966 **b** 0.966 **c** 0.966

  d −0.966 **e** 0.966

3 a 0.940 **b** 0.940 **c** 0.940

  d 0.940 **e** −0.940

4 a 71.8° and 108.2°

  b 200.5° and 339.5°.

  c 234.3° and = 305.7°.

  d No solution

5 a 104.5°.

   255.5°.

  b 84.3° and 275.7

  c 90° and 270°

  d 180°

6 a 77.3°, 282.7°, 437.3°, 642.7°

  b 192.7°, 347.3°, 552.7°, 707.3°

  c 0°, 180°, 360°, 540°, 720°

  d 90°, 450°

7 a 0°, 360°, −360°

  b −30°, −150°, 210°, 330°

  c 45.6°, 314.4°, −45.6°, −314.6°

  d 270°, −90°

8 45°, 225°, −135° and −315°.

9 a Translation of $\begin{bmatrix} 270 \\ 0 \end{bmatrix}$ or translation of

$\begin{bmatrix} -450 \\ 0 \end{bmatrix}$ are two possible answers.

  b Rotation of 180° about (135, 0) is one possible answer.

Exercise 6.4A p 171

1 −0.577

2 0.973

3 125° and −55°

4 a 2.9° and 182.9°

  b 153.4° and 333.4°

  c 78.7° and 258.7°

  d 91.1° and 271.1°

5 a 0, ±180°, ±360°, ±540°, etc.

  b 45° ± any multiple of 180°

  c −45° ± any multiple of 180°

6 76.0° and 256.0°

7 149.0° and 329.0°.

8 a For example, a translation of $\begin{bmatrix} 180 \\ 0 \end{bmatrix}$ or

  $\begin{bmatrix} 360 \\ 0 \end{bmatrix}$ or $\begin{bmatrix} -180 \\ 0 \end{bmatrix}$

  b For example, a rotation of 180° about (0, 0), (90, 0), (−90, 0) or (270, 0).

9 a $\tan\theta = \dfrac{\sin\theta}{\cos\theta}$. If 0° < θ < 90° then 0 < cos θ < 1. Dividing sin θ by a number less than 1 gives an answer that is greater than sin θ so tan θ > sin θ.

  b

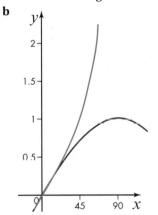

  tan x is always above sin x.

Exercise 6.5A p 174

1 a 56.6°

  b 28.3°

  c 46.6°

  d 23.3°

2 30°, 150°, 210°, 330°

3 83.1° or 156.9°

4 29.0° or 331.0°

5 28.2°, 118.2°, 208.2°, 298.2°

6 6.4° and 11.6°

7 a 37.8° or 172.2°

8 a 7.2°

   b −4.2°, 0.6° or 3°

   c The number of cycles is $360 \div 7.2 = 50$ and there are two solutions in each cycle. There will be $50 \times 2 = 100$ solutions.

Exercise 6.6A   p 176

1 Calculator operation

2 a ± 0.733

   b ± 0.927

3 a ± 0.898

   b ± 2.041

4 One is a reflection of the other in the line $y = \tfrac{1}{2}$.

   This ensures that corresponding values always add to 1.

5 90° 270°

   48.2° 311.8°

6 0°, 180°, 360° 270°

7 60° 300°

   120° 240°

8 19.5°, 160.5°

   30°, 150°

9 194.5°, 345.5°

10 a two solutions

    b 38.2°

Exam-style questions 6   p 177

1 a 30° or 150°

   b 25°

     85°

2 20.7 cm

3 a

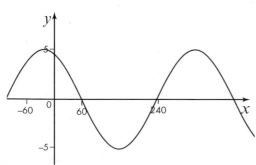

b 113.1° or 186.9°

4 The shortest side is opposite the smallest angle, which is 50°.

   If the shortest side is x:

$$\frac{x}{\sin 50°} = \frac{35}{\sin 68°} \Rightarrow x = \frac{35 \sin 50°}{\sin 68°} = 28.9 \,\text{cm}$$

5 −104.0°

6 $\dfrac{\sin X}{24.9} = \dfrac{\sin 50°}{19.5} \Rightarrow \sin X = \dfrac{24.9 \sin 50°}{19.5} = 0.9782$

   $X = 78°$ or $180 - 78 = 102°$. There are two possible values for the angle. Carla and Larry have each found one.

7 0°, 60°, 180°, 300° and 360°.

8 132 cm²

9 70.5° or 289.5°

   120° or 240°

10 $x = 0, 60, 180$ or 300.

11 8.25 m.

12 $\pm \dfrac{\sqrt{39}}{3}$

13 a

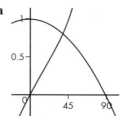

   b 38.2° and 141.8°

14 a When $x = 0$, $0.35 + 0.2 \cos(100x + 60°) = 0.35 + 0.1 = 0.45$

    After a short time, say 0.1 s the height is $0.35 + 0.2 \cos 70° = 0.42 \,\text{m}$, which is less than 0.45 m.

   b When $0.35 + 0.2 \cos(100x + 60°) = 0.5$ then $0.2 \cos(100x + 60°) = 0.15$ and so $\cos(100x + 60°) = 0.75$ Therefore $100x + 60 = 41.4$ or $360 - 41.4$ or $360 + 41.4 \ldots$

    Therefore $x = -0.19$ or 2.59 or 3.41 … The negative value is not appropriate. The formula shows that the height is greater than 0.5 if x is between 2.59 and 3.41 (For example, if $x = 3$ the height is 0.55 m) The time in each revolution is $3.41 - 2.59 = 0.82$ or about 0.8 s.

7 Exponentials and logarithms

Prior knowledge

page 180

1 **a** 2^4 **b** 2^{-5} **c** $2^{\frac{1}{2}}$

  d $2^{\frac{3}{2}}$ **e** 2^0

2 **a** n^5 **b** n^{-1}

  c n^6 **d** $n^{\frac{5}{6}}$

3 **a**

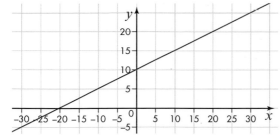

  b

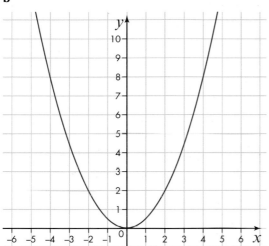

  c

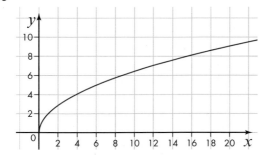

4 $y = -0.4x + 20$

5 3

6 £25 958.40

Exercise 7.1A

page 183

1 **a**

x	−3	−2	−1	0	1	2	3
$y = 1.5^x$	0.30	0.44	0.67	1	1.5	2.25	3.38

  b

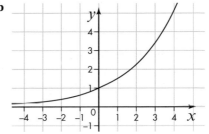

  c $x \approx 2.7$

2 **a**

x	−3	−2	−1	0	1	2	3
$y = 0.6^x$	4.63	2.78	1.67	1	0.6	0.36	0.22

  b

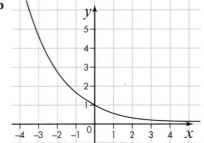

  c $x \approx 1.4$

  d $x \approx -2.7$

3 **a**

Year	2020	2030	2040	2050
Predicted population	50 000	54 000	58 320	62 986

  b $y = 50 \times 1.08^x$

  c

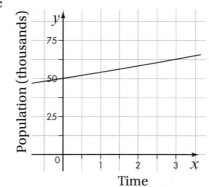

d In 2110, $x = 9$

Formula gives $y = 50 \times 1.08^9 = 99.95$

This is double 50.

e Birth rate, death rate and net immigration and emigration rates can all change over time.

4 a £405

b Multiply the original price by 0.9 for each year.

c

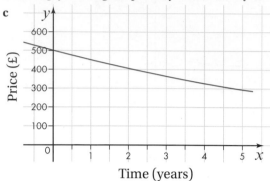

d No. After 5 years, price = $500 \times 0.9^5 = 295$, which is more than half of 500.

5 a

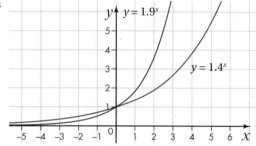

b When $x > 0$ **c** When $x < 0$

6 3.5

0.5

7 $K = 500$

$a = 1.3$

8 a

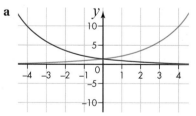

b A reflection in the y-axis.

c If n is a positive integer, a reflection in the y-axis maps the graph of $y = n^x$ onto the graph of $y = \left(\frac{1}{n}\right)^x$

9 a

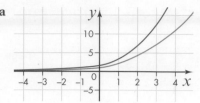

b a translation of $\begin{bmatrix} -1 \\ 0 \end{bmatrix}$

c A stretch, factor 2 from the x-axis which is $y = 2 \times 2^x$

Exercise 7.2A   page 186

1 a 3   **b** 6

   c −3   **d** −6

2 a $\frac{1}{2}$ **b** 1

   c $\frac{3}{2}$ **d** $-\frac{1}{2}$

3 a 1.3010   **b** −0.6990

   c 0.6990   **d** −0.1505

4 a 1.9031   **b** 0.1761

   c 0.9823   **d** −0.1761

5 a $2k$ **b** $2 + k$

   c $-k$ **d** $1 - \frac{1}{2}k$

6 a $k + 2h$ **b** $\frac{1}{2}k - h$

   c $2 + 2k - 3h$ **d** $-k - h$

7 a 100   **b** 50   **c** 10^{100}

8 a $2 \log 2$ **b** $1 - \log 2$ **c** $\log 2 + \log 3$

   d $\log 7$ cannot be found in this way.

   e $3 \log 2$ **f** $2 \log 3$

   g $\log 11$ cannot be found in this way.

   h $2 \log 2 + \log 3$

Exercise 7.2B   page 188

1 a 6   **b** 3

   c 5   **d** 4

2 a $\frac{1}{2}$ **b** $\frac{3}{2}$

   c −2   **d** $\frac{1}{4}$

3 $2\log_a 4 = \log_a 16$

$\log_a 8 - \log_a 2 = \log_a 4$

$\log_a \frac{1}{2} - \log_a \frac{1}{6} = \log_a 3$

$3\log_a 2 = \log_a 8$

4 a 7 **b** −5

c 4.5 **d** $\frac{8}{3}$

5 a $1+c$ **b** $-c$

c $3c$ **d** $\frac{1}{2}c$

6 a $n=6$ **b** $n=16$

c n can be any positive integer.

d $n=9$

7 $a = n^{3.5}$

$b = n^{4.5} = n^{1+3.5} = n \times n^{3.5} = na$

8 a i $b = a^x$ **ii** $a = b^y$

b $a = b^y$

Substitute for b: $a = (a^x)^y = a^{xy}$

This means that $xy = 1$ and the result follows.

Exercise 7.3A page 190

1 a 2.183 **b** 3.726 **c** 0.721

2 a 1.322 **b** 3.887 **c** 5.674

3 a 2.249 **b** 3.173 **c** 6.586

4 a 4.7 years **b** Just over 8 years

5 a 1.246 **b** 0.4921 **c** 0.942

6 a $y = 10 \times 1.5^t$ **b** 33.75 m²

c 5.1 weeks **d** 11.4 weeks

e Changes in temperature and light conditions can affect the growth rate.

Growth will not continue when the pond is fully covered.

7 a 1.7325 million

1.82 million (to 3 s.f.)

b 1.65×1.05^t

c 8.5 years

d The number cannot increase beyond the capacity of the airport.

Changes in travel patterns could make the model incorrect.

8 a Just over 13 years

b 10 years

c The size of an atom puts a limit on how small components can be made.

Exercise 7.4A page 194

1 a $\log y = \log 250 + 2\log x$

b 2

c $(0, \log 250)$

2 $\log V = \log \frac{4}{3}\pi + 3\log r$

The graph is a straight line with a gradient of 3.

3 a $\log F = \log Gm_1m_2 - 2\log r$

b −2

c Intercept is $(0, \log Gm_1m_2)$.

4 a $\log P = \log Ac^t = \log A + t\log c$

This is a straight line graph with a gradient of $\log c$.

b $c = 1.03$ $A = 93.3$

c 3%

d Changes in birth rates, life expectancy and immigration rates or emigration rates will change the annual rate of growth.

5 a 0.5

b the intercept on the y-axis is $\log a$.

c $n = 0.5$ $a = 31.6$ **d** 316

6 a 2

b 40 000 to 2 s.f.

7 a

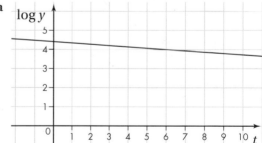

b $A = 25\,000$ to 2 s.f.

$c = 0.86$ to 2 s.f.

8 $y = \frac{100}{x^{\frac{2}{3}}}$ or $x^{\frac{2}{3}}y = 100$

Exercise 7.5A   **page 200**

1   **a**   7.39   **b**   1.65   **c**   0.368

    d   6.57   **e**   0.0498

2   **a**   **i**   2.70481   **ii**   2.71815   **iii**   2.71828

    b   The answers increase and approach the value of e as n increases.

    c   The value of the expression approaches e.

3   **a**

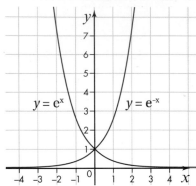

    b   Reflection in the y-axis

4   **a** $4e^x$ **b** $-e^{-x}$

    c $0.5e^{0.5x}$ **d** $30e^{3x}$

5   **a**   1   **b** e^2 **c**   2

6   Gradient $= 3e^{3x}$

    a   3   **b** $3e^6$ **c**   6

7   **a** $y = e^x + 1$ is C.

        $y = e^{x+1}$ is A.

        $y = (e + 1)^x$ is B.

    b $y = e^x + 1$ is C and a translation of $\begin{bmatrix} 0 \\ 1 \end{bmatrix}$

        $y = e^{x+1}$ is A and a translation of $\begin{bmatrix} -1 \\ 0 \end{bmatrix}$

    c   A is $y = e^{x+1} = e \times e^x$ so it is a stretch of $y = e^x$ by a scale factor of e.

8   **a** $(0, e^2)$

    b   A translation of $\begin{bmatrix} 2 \\ 0 \end{bmatrix}$ will map $y = e^{x+4}$ onto $y = e^{(x-2)+4}$, which is $y = e^{x+2}$

9   **a**   2.718055…

    b   the value of the sum is a bit less than e.

    c   Adding this gives 2.718253… which is closer to e.

    d   It suggests that adding more terms will get even closer to e.

10   **a**

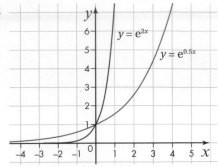

    b   a stretch of scale factor 4 from the y-axis.

11 $x = \dfrac{e}{2}$ or $x = -e$

Exercise 7.6A   **page 202**

1   **a**   7.5   **b**   12

    c   −1.5   **d**   3.75

2   **a**   7.60   **b**   3.51

    c   4.91   **d**   ±1.18

3   **a**   109   **b**   7.78

    c   0.00916   **d**   0.809

4 $\ln\dfrac{1}{a} = \ln 1 - \ln a = -\ln a$

5   **a** $y = \ln 4x = \ln 4 + \ln x$

        The graph is a translation of $y = \ln x$ by $\begin{bmatrix} 0 \\ \ln 4 \end{bmatrix}$.

    b

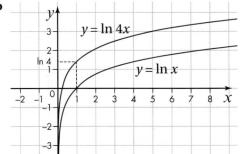

6   **a**   1.61   **b**   0.134

    c   1.16   **d**   0.925

7   **a**

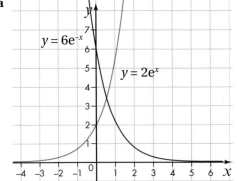

b (0.549, 3.46)

8 **a** (ln 2, 2)

 b 2

 c $\frac{1}{2}$

 d The gradient of $y = \ln x$ at $(x, \ln x)$ is $\frac{1}{x}$.

9 **a** $y^2 + y = 6$

 b $y = -3$ or 2

 c $e^x = -3$ has no solution

  If $e^x = 2$, then $x = 0.693$ to 3 s.f.

10 $x = -0.693$:

 $x = 1.39$ to 3 s.f.

11 **a** $\ln(e + e^2) = \ln(e(1 + e)) = \ln e + \ln(1 + e)$

  $= 1 + \ln(1 + e)$

 b $\ln(e^2 - e^4) = \ln(e^2(1 - e^2) = \ln e^2(1 + e)(1 - e)$

  $= \ln e^2 + \ln(1 + e) + \ln(1 - e)$

  $= 2 + \ln(1 + e) + \ln(1 - e)$

Exercise 7.7A   **page 206**

1 **a** £5000

 b **i** £5986   **ii** £12 298   **iii** £30 248

2 **a** £2000 **b** 3.2 years **c** 25%

 d 25%   **e** 25%   **f** They are the same.

3 **a** When $t = 5$, $x = Ne^{-0.75} = 0.472N$

 b If the model is valid, it will fall by just over 50% again, to less than 25% of the current value.

4 **a**

Number of years from now	0	10	20	30
Population	50	61.07	74.59	91.11

 b In each case the percentage increase is 22.1%.

5 $A = 40$

 $k = 0.049$

6 **a**

Days	0	2	5	10	15
Mass remaining (mg)	5.0	4.4	3.7	2.7	2.0

b

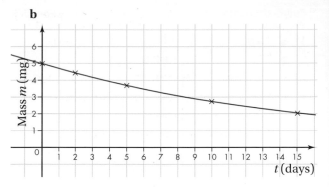

 c 11.6 days   **d** 11.55 days   **e** 23.10 days

7 **a** 3.73 mg

 b

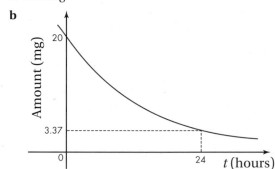

 c 12.36 mg

 d

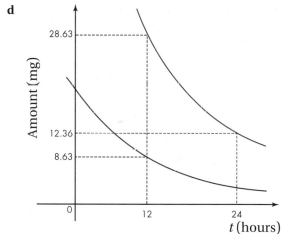

Exam-style questions 7   **page 209**

1 a $x \approx 3.4$

    b

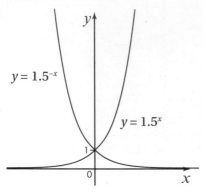

$y = 1.5^{-x}$

$y = 1.5^{x}$

2 7.12

3 a $(0, e^3)$ **b** 7.389   **c** 3.21

4 a $c = 2.5$ **b** −1.25

5 a $\dfrac{1}{2} + k$ **b** $2k$

6 $t = 2(\ln 10r - 1)$

7 a £10 000   **b** 35.0%   **c** 2.53%

8 a When $t = 0$, $y = 375$

        $375 = Ae^0 = A$

    b 0.002 31

    c 595

        This is not very close to the actual value of 540.

9 a 2000 to 3 s.f.

    b 2.82 to 3 s.f.

    c The population may not have enough food or space to keep growing at the same rate.

10 1.25 to 3 s.f.

11 a $x = 0.693$ to 3 s.f.

     b $x = 1.386$ to 3 d.p.

12 Let $y = \log_a b$

This can be written as $a^y = b$

Take the yth root: $a = b^{\frac{1}{y}}$

Therefore $\log_b a = \dfrac{1}{y}$ and so $\log_a b = \dfrac{1}{\log_b a}$

13 $\log_{10}\left(8\sqrt{5}\right) = \log_{10} 8 + \log_{10} \sqrt{5} = \log_{10}\left(2^3\right) + \dfrac{1}{2}\log_{10} 5$

$\qquad = 3\log_{10} 2 + \dfrac{1}{2}\log_{10}\dfrac{10}{2}$

$\qquad = 3\log_{10} 2 + \dfrac{1}{2}\left(\log_{10} 10 - \log_{10} 2\right)$

$\qquad = \dfrac{1}{2}\left(1 + 5\log_{10} 2\right)$

14 a $A = 200$

        $k = -0.060$

    b $y = 200e^{-0.060t}$

        When $t = 10$, $y = 200e^{-0.60} = 110$ and this is less than the recorded value.

    c $k = -0.040$ and $A = 181$, hence $y = 181e^{-0.040t}$

    d When t is large, the temperature approaches 0. This will only be the case if the temperature of the surroundings is 0 °C.

15 $n = -1.5$

      $a = 10^{6.6} \approx 4\,000\,000$

16 $x = \dfrac{\log 6.828}{\log 2} = 2.772$ or $\dfrac{\log 1.172}{\log 2} = 0.228$

8 Differentiation

Prior knowledge   **page 212**

1 a 7   **b** −0.3

   c $\frac{1}{2}$ **d** −2

2 a $2x^2 - 8x$ **b** $x^3 + 3x^2$

   c $x^2 - x - 20$

   d $9x^2 + 6x + 1$

3 a x^{-3} **b** $x^{\frac{1}{2}}$ **c** $x^{-\frac{1}{2}}$

   d $x^{\frac{1}{3}}$ **e** $x^{\frac{3}{2}}$

Exercise 8.1A   **page 214**

1 a 250 m min⁻¹

   b −2.5 m s⁻²

   c 25 l min⁻¹

   d 0.7 l min⁻¹

   e 4 percentage points per hour

2 a 5 m s⁻²   **b** about 10 m s⁻²

   c 3.3 m s⁻²

3 a i 5 m   **ii** 20 m   **iii** 45 m

   b 20 m s⁻¹

   c i 10 m s⁻¹   **ii** 35 m s⁻¹

4 a 20 m s⁻¹ upwards and 5 m s⁻¹ downwards

   b 5 m s⁻¹ upwards and 20 m s⁻¹ downwards

   c about 10 m s⁻¹

5 a 1.3 m s⁻¹   **b** 0.5 m s⁻¹

Exercise 8.2A   **page 219**

1 $\frac{dy}{dx} = 2x$

   a $y = 16$ and $\frac{dy}{dx} = 8$ **b** $y = 6.25$ and $\frac{dy}{dx} = 5$

   c $y = 1.44$ and $\frac{dy}{dx} = -2.4$ **d** $y = 1225$ and $\frac{dy}{dx} = 70$

2 a 3   **b** −1

   c 26   **d** −7.6

3 14 and −14

4 a and **b**

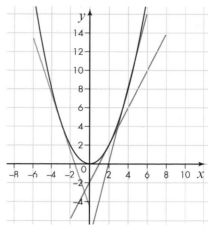

   c about 2, −3 and 4

   d $\frac{dy}{dx} = x$

   e Tangents at other points should confirm this.

5 In both cases $k = 0.5 \times 100 = 50$.

   At (10, 50) the gradient is 10 and at (−10, 50) the gradient is −10.

6 a

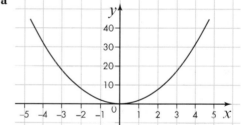

   b $\frac{dy}{dx} = 4x$.

Exercise 8.2B   **page 221**

1 a 3   **b** $6x$ **c** $6x + 4$

2 a $2x + 4$ **b** $2x - 4$ **c** $4 - 2x$

3 a $2x + 6$

   b $2x - 2$

   c $8x + 12$

4 a $\frac{dy}{dx} = 2x + 1$

   b i (0, −6)   **ii** 1

   c 5 and −5

d

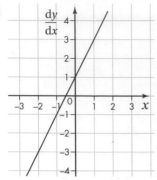

5 a $-2x$ **b** 2 and -6

c

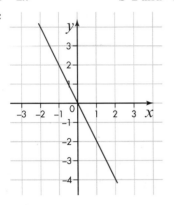

6 a $2x - 6$

b -4 and 16

c

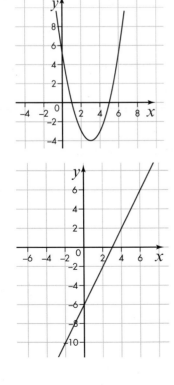

7 a 0

  b 120

8 a 10

  b 35

9 10

10 $\frac{1}{2}$ or 0.5

Exercise 8.3A   **page 224**

1 a 6

  b 6.1

  c 6.05

  d 6.01

  e As Q gets closer to P, the gradient of the chord gets closer to the gradient of the tangent.

2 $\frac{dy}{dx} = 2x$ so the gradient at P is $2 \times 2 = 4$
The gradient of PQ $= \dfrac{4 - 3.61}{2 - 1.9} = 3.9$, which is a little smaller than 4.

3 a $f(x + \delta x) - f(x) = 2(x + \delta x)^2 - 2x^2$
$$= 2x^2 + 4x\,\delta x + 2(\delta x)^2 - 2x^2$$
$$= 4x\,\delta x + 2(\delta x)^2$$

  b The gradient of the chord is
$$\frac{f(x + \delta x) - f(x)}{\delta x} = \frac{4x\,\delta x + 2(\delta x)^2}{\delta x} = 4x + 2\delta x$$

  c $f'(x) = \lim\limits_{\delta x \to 0} \dfrac{f(x + \delta x) - f(x)}{\delta x} = \lim\limits_{\delta x \to 0}(4x + 2\delta x) = 4x$

4 a

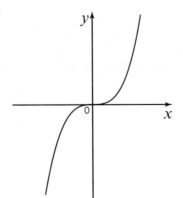

  b $3x^2\delta x + 3x(\delta x)^2 + (\delta x)^3$

  c $\dfrac{f(x + \delta x) - f(x)}{\delta x} = \dfrac{3x^2\delta x + 3x(\delta x)^2 + (\delta x)^3}{\delta x}$
$$= 3x^2 + 3x\delta x + (\delta x)^2$$

  d As $\delta x \to 0$, $3x\delta x + (\delta x)^2 \to 0$ and so $\lim\limits_{\delta x \to 0}\dfrac{\delta y}{\delta x} = 3x^2$

e $3x^2$

f The gradient is never negative.

Exercise 8.4A page 226

1 a $6x^2$ **b** $2x^3$

  c $0.5x^4$ **d** $150x^2$

2 a $3x^2 + 8x - 8$ **b** $6x^2 - 10x + 6$

3 a $4x^3 + 16x$ **b** $5x^4 - 30x^2 + 2$

4 a $3x^2 - 4x$

  b i 4

    ii 7

  c Visual check

5 a $2x^3 - 4x + 1$

  b i 3

    ii −1

    iii 9

  c Visual check

6 a $6x^2 + 10x$

  b $3x^2 - 16x + 16$

  c $3x^2 + 2x + 1$

7 a $3x^2 + 6x$

  b i 9

    ii 0

    iii 9

8 8

  −4

  8

9 a $\dfrac{dy}{dx} = 0.012x^2 - 0.6x + 7.5$

  When $x = 10$, $\dfrac{dy}{dx} = 1.2 - 6 + 7.5 = 2.7$, which is the correct value.

  b $0.3\ \text{m s}^{-1}$.

Exercise 8.5A page 229

1 a $-\dfrac{2}{x^3}$

  b $-\dfrac{6}{x^4}$

  c $\dfrac{1}{3}x^{-\frac{2}{3}}$ **d** $10x^{\frac{3}{2}}$

2 a $5x^{-\frac{1}{2}}$ or $\dfrac{5}{\sqrt{x}}$ **b** $-\dfrac{50}{x^2}$

  c $20x + \dfrac{20}{x^3}$

3 a $-\dfrac{24}{x^2}$

  b i $-\dfrac{2}{3}$ **ii** $-\dfrac{3}{2}$

    iii −6   **iv** $-\dfrac{1}{24}$

  c If $x \neq 0$, then x^2 is positive and the gradient $-\dfrac{24}{x^2}$ is negative.

4 a i $\dfrac{1}{4}$

    ii $\dfrac{1}{6}$

    iii $\dfrac{1}{20}$

  b $x = 1$ and the coordinates are (1, 1).

  c $x = \frac{1}{4}$ and the coordinates are $(\frac{1}{4}, \frac{1}{2})$.

5 $\dfrac{4}{3}$

  $\dfrac{1}{3}$

6 a $y = \dfrac{1}{2}x + 2x^{-1}$

  $\dfrac{dy}{dx} = \dfrac{1}{2} - 2x^{-2} = \dfrac{1}{2} - \dfrac{2}{x^2}$

  When $= 2$, $\dfrac{dy}{dx} = \dfrac{1}{2} - \dfrac{2}{2^2} = \dfrac{1}{2} - \dfrac{1}{2} = 0$

  When $= -2$, $\dfrac{dy}{dx} = \dfrac{1}{2} - \dfrac{2}{(-2)^2} = \dfrac{1}{2} - \dfrac{1}{2} = 0$

  b $-7\frac{1}{2}$

    $\frac{3}{8}$

  c $\dfrac{dy}{dx} = \dfrac{1}{2} - \dfrac{2}{x^2}$

  If x is large, $\dfrac{2}{x^2}$ is a small positive number and the gradient is close to $\frac{1}{2}$.

  The larger x is, the closer the gradient is to $\frac{1}{2}$.

7 a $-5x^{-2} = -\dfrac{5}{x^2}$

  b i $-\dfrac{5}{4}$

    ii −0.05

  c (1, 7) and (−1, −3).

Exercise 8.5B page 232

1 a $\dfrac{dV}{dr} = 4\pi r^2$

  b 36π

2 a $20 - 10t$

b and c

i $20\ \text{m s}^{-1}$.

ii $0\ \text{m s}^{-1}$.

iii $-10\ \text{m s}^{-1}$.

The minus sign means the stone is returning to the ground.

3 a $\dfrac{3}{2}$

b $\dfrac{9}{2}$

4 a Volume $= \pi r^2 h$

$$\pi r^2 h = 1000$$

$$h = \frac{1000}{\pi r^2}$$

b $-\dfrac{2000}{\pi r^3}$

c 2.950 to 3 s.f.

Exercise 8.6A page 236

1 a $(6, -56)$.

b $\dfrac{d^2 y}{dx^2} = 2$ which is positive when $x = 6$ (in fact it is always positive) so the point is a minimum point.

c

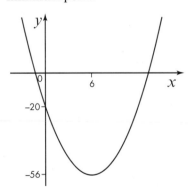

2 a $(1.5, 9.5)$

b $\dfrac{d^2 y}{dx^2} = -4$ which is negative when $x = 1.5$ (in fact it is always negative) so the point is a maximum point.

c

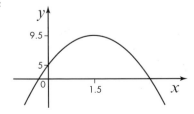

3 a and b

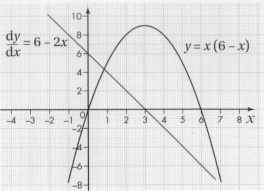

$\dfrac{dy}{dx} = 6 - 2x$ $\quad y = x(6 - x)$

c Where the graph of $\dfrac{dy}{dx}$ crosses the x-axis gives the x-coordinate of a turning point.

d

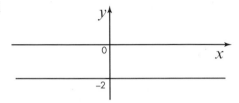

4 a $(1, 13)$ and $(2, 12)$.

b $(1, 13)$ is a maximum point because $f''(x)$ is negative at $x = 1$.

$(2, 12)$ is a minimum point because $f''(x)$ is positive at $x = 2$.

c

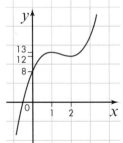

5 a $(10, -1400)$ and $(-6, 648)$

b $(10, -1400)$ is a minimum point because $\dfrac{d^2 y}{dx^2}$ is positive for $x = 10$.

$(-6, 648)$ is a maximum point because $\dfrac{d^2 y}{dx^2}$ is negative for $x = -6$.

6 $(0, 0)$ is a maximum point.

$(1, -1)$ is a minimum point.

$(-1, -1)$ is a minimum point.

7 $(\sqrt{40}, \sqrt{10})$

8 $28\ \text{m s}^{-1}$

9 a 26.45 m

b Air resistance will reduce the maximum height of the ball.

10 a Volume $= x^2 h = 1000$

Rearrange: $h = \dfrac{1000}{x^2}$

b The total surface area is

$2x^2 + 4xh = 2x^2 + 4x \times \dfrac{1000}{x} = 2x^2 + \dfrac{4000}{x^2}$

c $x = 10$

1 a

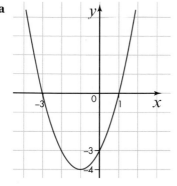

b $\dfrac{dy}{dx} = 2x + 2$

c $y = 8x - 12$

d $x + 8y = 99$

e $y + 4x = -12$

$4y = x + 3.$

2 a When $x = -1$, $y = (-1)^3 - 4 \times (-1)^2 = -5$ so $(-1, -5)$ is on the curve.

b $\dfrac{dy}{dx} = 3x^2 - 8x$

c $y = 11x + 6$

d $11y + x = -56$

3 $24y = x + 287$

4 a $x + 3y = 76$

b $5y = 3x + 300$

5 a $y + 4x = 14$

b 24.5

6 a $3y = x + 4$

b $6y = x + 32$

7 a

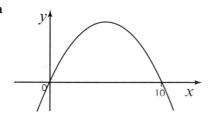

b $19\frac{1}{8}$ or 19.125

1 a $2 - \dfrac{8}{x^2}$

b 1.5

2

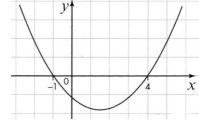

3 a You cannot find the derivative of a product by finding the derivative of each term separately and multiplying the results.

b $\dfrac{dy}{dx} = 6x - 1$

4 $y = x^{\frac{1}{2}}(x + 4) = x^{\frac{3}{2}} + 4x^{\frac{1}{2}}$

$\dfrac{dy}{dx} = \dfrac{3}{2}x^{\frac{1}{2}} + 2x^{-\frac{1}{2}}$

When $x = 9$, $\dfrac{dy}{dx} = \dfrac{3}{2} \times 9^{\frac{1}{2}} + 2 \times 9^{-\frac{1}{2}}$

$= \dfrac{9}{2} + \dfrac{2}{3} = \dfrac{31}{6} = 5\frac{1}{6}$

5 a $4x^3 - 16x$

b $(0, 0)$, $(2, -16)$ and $(-2, -16)$

c

6 $(3, -32)$ and $(-1, 8)$

7 a $\dfrac{dy}{dx} = 0.4x - 0.03x^2$

When $x = 10$, $\dfrac{dy}{dx} = 4 - 3 = 1$

Speed is 1 m s^{-1} upwards.

b At the highest point, $\dfrac{dy}{dx} = 0$

$0.4x - 0.03x^2 = 0$

$x(0.4 - 0.03x) = 0$

Either $x = 0$, which is when the drone starts, or $0.4 - 0.03x = 0$

$x = \dfrac{0.4}{0.03} = 13\frac{1}{3}$

This is $\frac{2}{3}$ of 20 seconds, showing the ascent takes twice as long as the descent.

8 a $2x - 250x^{-2}$

b At a stationary point, $2x - 250x^{-2} = 0$.

$2x^3 = 250$

$x^3 = 125$

$x = 5$

$\dfrac{d^2y}{dx^2} = 2 + 500x^{-3}$

When $x = 5$, $\dfrac{d^2y}{dx^2} = 2 + 500 \times 5^{-3} = 6 > 0$ so this is a minimum point.

When $x = 5$, $y = 5^2 + \frac{250}{5} = 75$

The minimum value is 75.

9 $\left(\sqrt{10}, 2\sqrt{10}\right)$.

10 a $A = x(240 - 2x)$.

b $A = 240x - 2x^2$

$\dfrac{dA}{dx} = 240 - 4x$

For maximum area, $\dfrac{dA}{dx} = 240 - 4x = 0$.

$x = 60$

$\dfrac{d^2A}{dx^2} = -4 < 0$ so this will be a maximum.

Maximum area $= 60 \times 120 = 7200$ m²

11 a $f(x + \delta x) - f(x) = 2(x + \delta x)^2 - 2x^2$

$= 2(x^2 + 2x\,\delta x + (\delta x)^2) - 2x^2$

$= 2x^2 + 4x\,\delta x + 2(\delta x^2 - 2x)^2 = 4x\,\delta x + 2(\delta x)^2$

b $\dfrac{f(x + \delta x) - f(x)}{\delta x} = 4x + 2\delta x$

$f'(x) = \lim_{\delta x \to 0} \dfrac{f(x + \delta x) - f(x)}{\delta x} = \lim_{\delta x \to 0} \left(4x + 2\delta x\right) = 4x$

12 $y = 8x - 9$

13 a Volume of a cone, $V = \frac{1}{3}\pi r^2 h$

$h = 60 - r$ so $V = \frac{1}{3}\pi r^2(60 - r)$

b $V = \frac{1}{3}\pi\left(60r^2 - r^3\right)$

$\dfrac{dV}{dr} = \frac{1}{3}\pi\left(120r - 3r^2\right)$

When the volume is a maximum, $\dfrac{dv}{dr} = 0$.

$\frac{1}{3}\pi\left(120r - 3r^2\right) = 0$

$120r - 3r^2 = 0$

Divide by 3 and factorise.

$r(40 - r) = 0$

$r = 0$ or 40

$\dfrac{d^2V}{dr^2} = \frac{1}{3}\pi\,(120 - 6r)$

When $r = 40$, $\dfrac{d^2V}{dr^2} = \frac{1}{3}\pi(120 - 240) < 0$ so this is a maximum point.

c $\dfrac{32000\pi}{3}$

14 240.25

15 The area of the triangle is $\frac{1}{2} \times 2a \times \dfrac{40}{a} = 40$ and is independent of a.

16 $3x^2 - 4x$

9 Integration

Prior knowledge
page 244

1 $4x^{-\frac{1}{2}}$

2

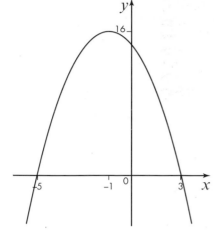

3 a $12x^2 - 6$

b $\dfrac{dy}{dx} = \dfrac{1}{2} - 2x^{-2}$

c $\dfrac{dy}{dx} = 2x^{-\frac{2}{3}}$

Exercise 9.1A
page 246

1 a $3x^2 + c$ **b** $2x^2 + 2x + c$

  c $\dfrac{7}{2}x^2 - 5x + c$ **d** $3x - 2x^2 + c$

2 a $\dfrac{2}{5}x^5 + c$ **b** $-\dfrac{2}{3}x^{-3} + c$

  c $\dfrac{5}{4}x^4 + c$ **d** $\dfrac{5}{-2}x^{-2} + c$

3 a $2x^3 - 2x^2 + c$ **b** $\dfrac{2}{5}x^5 - \dfrac{5}{2}x^2 + 10x + c$

  c $4x^2 - \dfrac{10}{3}x^3 + c$

4 a $\dfrac{2}{3}x^{\frac{3}{2}} + c$ **b** $\dfrac{4}{5}x^{\frac{5}{2}} + c$

  c $6x^{\frac{1}{2}} + c$ **d** $20\sqrt{x} + c$

5 a $\dfrac{1}{3}x^3 - 2x^2 + c$

  b $\dfrac{1}{4}x^4 - \dfrac{8}{3}x^3 + c$

  c $-5x^{-2} + c$ or $-\dfrac{5}{x^2} + c$

6 a $\dfrac{8}{3}x^{\frac{3}{2}} + c$

  b $\dfrac{4}{3}x^{\frac{3}{2}} + c$

  c $8\sqrt{x} + c$

7 $\dfrac{1}{2}x^4 - x^2 + c$

8 a You cannot integrate a product by integrating each term separately and multiplying the results.

  b $\dfrac{2}{3}x^3 - \dfrac{1}{2}x^2 - x + c$

9 a $4x + x^{-2} + c$

  b $= \dfrac{4}{3}x^{\frac{3}{2}} + 16x^{\frac{1}{2}} + c$

  c $-4x^{-1} - 4x^{-\frac{1}{2}} + c$

10 The derivatives of f(x) and g(x) are the same so the derivative of f(x) − g(x) is 0.

This means that f(x) − g(x) is $\int 0 \, dx = $ a constant.

Exercise 9.1B
page 247

1 a $y = 2x^2 - 2x + c$

  b $y = 2x^2 - 2x + 3$

2 a $y = \dfrac{2}{3}x^{\frac{3}{2}} + c$

  b $y = \dfrac{2}{3}x^{\frac{3}{2}} + 7$

3 a $y = 0.2x^2 + 3x$

  b $y = 0.2x^2 + 3x + 5$

  c $y = 0.2x^2 + 3x - 20$

4 $y = x - \dfrac{10}{x} - 1$

5 a If $x \neq 0$, then x^2 is positive. This means that $\dfrac{2}{x^2}$ is always positive.

  b $y = -\dfrac{2}{x} + 5$

6 a $y = x^3 - 3x + 2$

  b The turning points are where $3x^2 - 3 = 0$.

    $x^2 = 1$

    $x = 1$ or -1

    There are 2 turning points.

    When $x = 1$, $y = 1 - 3 + 2 = 0$ so one point is (1, 0).

    When $x = -1$, $y = -1 + 3 + 2 = 4$ so the other point is (−1, 4).

7 **a** $y = 6.5 - \dfrac{10}{x^2}$

b

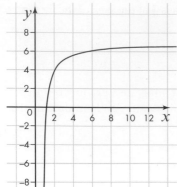

8 **a** $y = 0.006x^5 - 0.15x^4 + x^3 + 100$

b After 10 seconds, $x = 10$ and

$y = 0.006 \times 10^5 - 0.15 \times 10^4 + 10^3 + 100$
$= 600 - 1500 + 1000 + 100 = 200$

The distance from A is 200 m.

9 $(0, 0)$, $(2, 0)$ and $(4, 0)$

10 **a** $y = 12x^{\frac{1}{3}} + 6$

b 38.6 cm to 3 s.f.

Exercise 9.2A   page 255

1 **a** $\dfrac{7}{9}$

   b $\dfrac{56}{9}$

   c $\dfrac{28}{9}$

2 **a** 6.2

   b 600

   c 8

3 **a** 60

   b 22

   c 28

4 **a** $9\dfrac{1}{3}$

   b 40

   c 32.10 to 4 s.f.

5 **a** $\dfrac{4}{3}$

   b $\dfrac{4}{3}$

   c $\dfrac{4}{3}$

6 **a** **i** $16\dfrac{1}{4}$

   ii $16\dfrac{1}{4}$

   iii 0

b

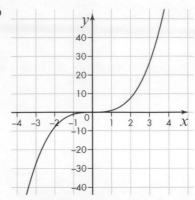

The integral from –2 to 2 in **part iii** above is zero because the graph is symmetrical and the areas above and below the x-axis are the same.

7 Shaded area $= \displaystyle\int_0^9 x^{\frac{1}{2}}\,dx = \left[\dfrac{2}{3}x^{\frac{3}{2}}\right]_0^9 = \left[\dfrac{2}{3}\times 27\right] - [0] = 18$

Area of OABC $= 3 \times 9 = 27$

18 is $\dfrac{2}{3}$ of 27

8 9.28

9 **a** 36

   b 36

   c

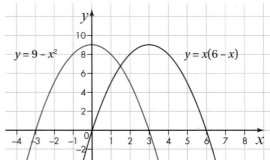

One graph is a translation of the other and the areas are the same.

10 **a** **i** $\dfrac{1}{3}$

   ii $\dfrac{1}{4}$

   iii $\dfrac{1}{7}$

   b $\displaystyle\int_0^1 x^n\,dx = \left[\dfrac{1}{n+1}x^{n+1}\right]_0^1 = \dfrac{1}{n+1}$ when n is a positive integer.

11 **a** 33.75 m

   b 30.25 m

12 $26\dfrac{2}{3}$

13 a 4

b 7

c $10 - \dfrac{30}{n}$

d $\dfrac{30}{n}$ is always positive for any positive value of n so $10 - \dfrac{30}{n}$ is always less than 10.

14 a (1, 3) and (4, 0)

b $4\frac{1}{2}$

Exam-style questions 9   **page 258**

1 $2x^4 - 4x^3 + c$

2 $116\frac{2}{3}$ m

3 $-5x^{-1} - 2x^{-\frac{1}{2}} + c$

4 $y = \dfrac{16}{x} + 2$

5 a

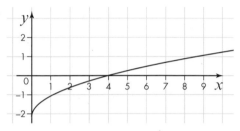

b 0

c The area below the x-axis between 0 and 4 is the same as the area above it between 4 and 9.

6 a $y = \int 0.6x - 10 \, dx = 0.3x^2 - 10x + c$

When $x = 0$, $y = 100$ so $c = 100$

$y = 0.3x^2 - 10x + 100$

When $x = 9$, $y = 0.3 \times 9^2 - 10 \times 9 + 100 = 34.3$

b When $x = 20$, $\dfrac{dy}{dx} = 0.6 \times 20 - 10 = 2$

This is positive and implies that the volume is increasing. This cannot be the case. If there is a leak, $\dfrac{dy}{dx}$ will be negative.

7 a You cannot integrate a fraction by integrating the numerator and the denominator separately and dividing one result by the other.

b $\frac{5}{8}x^2 + 2x^{-1} + c$

8 $y = x^2 + 6x + 9$.

9 $k = 3$

10 Area beneath the curve between A and B is

$$\int_{-a}^{a} x^2 \, dx = \left[\frac{1}{3}x^3\right]_{-a}^{a} = \left[\frac{1}{3}a^3\right] - \left[-\frac{1}{3}a^3\right] = \frac{2}{3}a^3$$

The area of rectangle ABCD $= 2a \times a^2 = 2a^3$

The area between AB and the curve is $2a^3 - \frac{2}{3}a^3 = \frac{4}{3}a^3$ and this is $\frac{2}{3}$ of the area of the rectangle.

11 $20\frac{5}{6}$.

12 $6\frac{2}{3}$

10 Vectors

Prior knowledge   **page 261**

1 a 13

b $11\sqrt{5}$

2 $32.0°$

3 25.6 cm

Exercise 10.1A   **page 265**

1 a $29.7°$

b $145.0°$

c $140.2°$

d $-66.0°$

2 a i 25

074°

ii 37

199°

iii 218

123°

b i $8\sqrt{5}$

297°

ii $5\sqrt{2}$

082°

iii $2\sqrt{5}$

153°

3 a Both are correct. If a is positive, then the bearing is 045° but if a is negative then the bearing is 225°.

b $\begin{bmatrix} -a \\ a \end{bmatrix}$ ($a > 0$; same number but different signs)

4 16 possible pairs

5 a i $\frac{1}{5}(-4\mathbf{i} + 3\mathbf{j})$

ii $\frac{1}{145}(143\mathbf{i} - 24\mathbf{j})$

b i $-20\mathbf{i} + 48\mathbf{j}$

    ii $189\mathbf{i} + 48\mathbf{j}$

6 $(28.8\mathbf{i} - 3.4\mathbf{j})\,\text{km}\,\text{h}^{-1}$

7 a $\begin{bmatrix} 7 \\ -7 \end{bmatrix}$

  b $\begin{bmatrix} \sqrt{3} \\ 1 \end{bmatrix}$

  c $\begin{bmatrix} -7 \\ -7\sqrt{3} \end{bmatrix}$

8 a $21.3\,\text{km}$

  b $((14 + \tfrac{9}{2}\sqrt{2}\,)\mathbf{i} - \tfrac{9}{2}\sqrt{2}\,\mathbf{j})\,\text{km}$

  c $107°$

Exercise 10.2A   **page 268**

1 a i $\begin{bmatrix} 2 \\ 12 \end{bmatrix}$ **ii** $\begin{bmatrix} 7 \\ -4 \end{bmatrix}$ **iii** $\begin{bmatrix} 13 \\ -4 \end{bmatrix}$

  b i $\begin{bmatrix} -4 \\ -2 \end{bmatrix}$ **ii** $\begin{bmatrix} 0 \\ 10 \end{bmatrix}$ **iii** $\begin{bmatrix} -2 \\ -11 \end{bmatrix}$

  c i $\begin{bmatrix} 10 \\ 20 \end{bmatrix}$ **ii** $\begin{bmatrix} 11 \\ 4.5 \end{bmatrix}$ **iii** $\begin{bmatrix} -20 \\ 12 \end{bmatrix}$

  d i $\begin{bmatrix} 3 \\ -11 \end{bmatrix}$ **ii** $\begin{bmatrix} 1 \\ 18 \end{bmatrix}$ **iii** $\begin{bmatrix} 29 \\ -4.25 \end{bmatrix}$

2 a i $\begin{bmatrix} 3 \\ 4 \end{bmatrix}$

  ii It is the vector which joins the end of $\begin{bmatrix} 4 \\ 1 \end{bmatrix}$ to the start of $\begin{bmatrix} -1 \\ 3 \end{bmatrix}$.

  b i $\begin{bmatrix} 5 \\ -2 \end{bmatrix}$

  ii

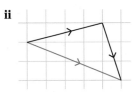

  iii $-\mathbf{b}$ is the opposite vector, so is parallel to (but in the opposite direction to **b**). $\begin{bmatrix} 5 \\ -2 \end{bmatrix}$ is the vector which joins the end of $\begin{bmatrix} 4 \\ 1 \end{bmatrix}$ to the start of $-\begin{bmatrix} -1 \\ 3 \end{bmatrix}$.

c i $\begin{bmatrix} 8 \\ 2 \end{bmatrix}$ **ii**

  iii It is the vector twice as long as **a** and parallel to **a**.

3 a $q = -\tfrac{1}{2}$

    $p = 6$

  b $q = -2$

    $p = 9$

  c $q = 2$

    $p = 3$

4 a $p = 2$

  b $q = -3$

5 a $b = -1$

    $a = 7$

  b $b = 6$

    $a = -18$

  c Any pair of values for which $a = -5b$

Exercise 10.3A   **page 273**

1 $\overrightarrow{ST} = \overrightarrow{SR} + \overrightarrow{RT} = -4\mathbf{a} + 6\mathbf{b} = 2(-2\mathbf{a} + 3\mathbf{b})$

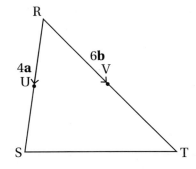

$\overrightarrow{RU} = 2\mathbf{a}$

$\overrightarrow{RV} = 3\mathbf{b}$

$\overrightarrow{UV} = \overrightarrow{UR} + \overrightarrow{RV} = -2\mathbf{a} + 3\mathbf{b}$

Hence $\overrightarrow{ST} = 2\overrightarrow{UV}$.

If two vectors have a common factor then they are parallel.

2 a i $6\mathbf{a} + 6\mathbf{c}$ **ii** $-6\mathbf{a} + 6\mathbf{c}$

  iii $6\mathbf{a} + 4\mathbf{c}$ **iv** $3\mathbf{a} - 6\mathbf{c}$

  v $3\mathbf{a} + 3\mathbf{c}$ **vi** $-3\mathbf{a} - \mathbf{c}$

  vii $-3\mathbf{a} + 2\mathbf{c}$ **viii** $3\mathbf{c}$

  b The vectors are parallel.

3 $\frac{3}{13}(15\mathbf{a}+8\mathbf{b})$

4 **a** $\overrightarrow{EF}=\overrightarrow{ED}+\overrightarrow{DF}=-12\mathbf{a}+9\mathbf{b}=3(-4\mathbf{a}+3\mathbf{b})$

$\overrightarrow{MN}=\overrightarrow{MD}+\overrightarrow{DN}=-4\mathbf{a}+3\mathbf{b}$

Since $\overrightarrow{EF}$ is parallel to $\overrightarrow{MN}$, FEMN is a trapezium.

b 64 units^2.

5 $10\mathbf{i}-\mathbf{j}$

6 $\frac{3}{2}(\mathbf{a}+\mathbf{b})$

7 **a** $\overrightarrow{OM}=\frac{2}{3}\overrightarrow{OA}=\frac{2}{3}\mathbf{a}$

$\overrightarrow{MB}=\overrightarrow{MO}+\overrightarrow{OB}=-\frac{2}{3}\mathbf{a}+\mathbf{b}$

$\overrightarrow{MX}=\lambda(-\frac{2}{3}\mathbf{a}+\mathbf{b})$

$\overrightarrow{OX}=\overrightarrow{OM}+\overrightarrow{MX}=\frac{2}{3}\mathbf{a}+\lambda(-\frac{2}{3}\mathbf{a}+\mathbf{b})$

$\qquad=(\frac{2}{3}-\frac{2}{3}\lambda)\mathbf{a}+\lambda\mathbf{b}$

$\overrightarrow{ON}=\frac{2}{3}\overrightarrow{OB}=\frac{2}{3}\mathbf{b}$

$\overrightarrow{NA}=\overrightarrow{NO}+\overrightarrow{OA}=-\frac{2}{3}\mathbf{b}+\mathbf{a}$

$\overrightarrow{NX}=\mu(-\frac{2}{3}\mathbf{b}+\mathbf{a})$

$\overrightarrow{OX}=\overrightarrow{ON}+\overrightarrow{NX}=\frac{2}{3}\mathbf{b}+\mu(-\frac{2}{3}\mathbf{b}+\mathbf{a})$

$\qquad=(\frac{2}{3}-\frac{2}{3}\mu)\mathbf{b}+\mu\mathbf{a}$

Coefficients of **a**: $\frac{2}{3}-\frac{2}{3}\lambda=\mu$, so $3\mu+2\lambda=2$ ①

Coefficients of **b**: $\frac{2}{3}-\frac{2}{3}\mu=\lambda$, so $2\mu+3\lambda=2$ ②

Put ① = ②.

$3\mu+2\lambda=2\mu+3\lambda$

$\mu=\lambda$

Substitute for μ in ②.

$2\lambda+3\lambda=2$

$5\lambda=2$

$\lambda=\frac{2}{5},\ \mu=\frac{2}{5}$

Hence $\overrightarrow{OX}=(\frac{2}{3}-\frac{2}{3}\times\frac{2}{5})\mathbf{b}+\frac{2}{5}\mathbf{a}=\frac{2}{5}\mathbf{a}+\frac{2}{5}\mathbf{b}$

$\qquad=\frac{2}{5}(\mathbf{a}+\mathbf{b})$

b $\overrightarrow{OM}=\frac{p}{p+q}\overrightarrow{OA}=\frac{p}{p+q}\mathbf{a}$

$\overrightarrow{MB}=\overrightarrow{MO}+\overrightarrow{OB}=-\frac{p}{p+q}\mathbf{a}+\mathbf{b}$

$\overrightarrow{MX}=\lambda(-\frac{p}{p+q}\mathbf{a}+\mathbf{b})$

$\overrightarrow{OX}=\overrightarrow{OM}+\overrightarrow{MX}=\frac{p}{p+q}\mathbf{a}+\lambda(-\frac{p}{p+q}\mathbf{a}+\mathbf{b})$

$\qquad=(\frac{p}{p+q}-\frac{p}{p+q}\lambda)\mathbf{a}+\lambda\mathbf{b}$

$\overrightarrow{ON}=\frac{p}{p+q}\overrightarrow{OB}=\frac{p}{p+q}\mathbf{b}$

$\overrightarrow{NA}=\overrightarrow{NO}+\overrightarrow{OA}=-\frac{p}{p+q}\mathbf{b}+\mathbf{a}$

$\overrightarrow{NX}=\mu(-\frac{p}{p+q}\mathbf{b}+\mathbf{a})$

$\overrightarrow{OX}=\overrightarrow{ON}+\overrightarrow{NX}=\frac{p}{p+q}\mathbf{b}+\mu(-\frac{p}{p+q}\mathbf{b}+\mathbf{a})$

$=(\frac{p}{p+q}-\frac{p}{p+q}\mu)\mathbf{b}+\mu\mathbf{a}$

Coefficients of **a**: $\frac{p}{p+q}-\frac{p}{p+q}\lambda=\mu$, so $\mu+\frac{p}{p+q}\lambda$

$\qquad\qquad\qquad\qquad=\frac{p}{p+q}$ ①

Coefficients of **b**: $\frac{p}{p+q}-\frac{p}{p+q}\mu=\lambda$, so $\frac{p}{p+q}\mu+\lambda$

$\qquad\qquad\qquad\qquad=\frac{p}{p+q}$ ②

Put ① = ②:

$\mu+\frac{p}{p+q}\lambda=\frac{p}{p+q}\mu+\lambda$

$\frac{\mu(p+q)+p\lambda}{p+q}=\frac{p\mu+\lambda(p+q)}{p+q}$

$\mu(p+q)+p\lambda=p\mu+\lambda(p+q)$

$p\mu+q\mu+p\lambda=p\mu+p\lambda+q\lambda$

$q\mu=q\lambda$

$\mu=\lambda$

Substitute for μ in ①.

$\lambda+\frac{p}{p+q}\lambda=\frac{p}{p+q}$

$\frac{\lambda(p+q)+p\lambda}{p+q}\lambda=\frac{p}{p+q}$

$\lambda(p+q)+p\lambda=p$

$p\lambda+q\lambda+p\lambda=p$

$\lambda(2p+q)=p$

$\lambda=\frac{p}{2p+q}$

Hence $\overrightarrow{OX}=\left(\frac{p}{p+q}-\frac{p}{p+q}\times\frac{p}{2p+q}\right)\mathbf{a}+\frac{p}{2p+q}\mathbf{b}$

$=\left(\frac{p}{p+q}-\frac{p^2}{(p+q)(2p+q)}\right)\mathbf{a}+\frac{p}{2p+q}\mathbf{b}$

$=\left(\frac{p(2p+q)}{(p+q)(2p+q)}-\frac{p^2}{(p+q)(2p+q)}\right)\mathbf{a}+\frac{p}{2p+q}\mathbf{b}$

$=\left(\frac{p(2p+q)-p^2}{(p+q)(2p+q)}\right)\mathbf{a}+\frac{p}{2p+q}\mathbf{b}$

$=\left(\frac{2p^2+pq-p^2}{(p+q)(2p+q)}\right)\mathbf{a}+\frac{p}{2p+q}\mathbf{b}$

$=\left(\frac{p^2+pq}{(p+q)(2p+q)}\right)\mathbf{a}+\frac{p}{2p+q}\mathbf{b}$

$=\left(\frac{p(p+q)}{(p+q)(2p+q)}\right)\mathbf{a}+\frac{p}{2p+q}\mathbf{b}$

$=\left(\frac{p}{2p+q}\right)\mathbf{a}+\frac{p}{2p+q}\mathbf{b}$

$=\frac{p}{2p+q}(\mathbf{a}+\mathbf{b})$

8 **a** $\overrightarrow{JK}$ and $\overrightarrow{LM}$ are parallel.

b $6\mathbf{a}-6\mathbf{b}$

$6\mathbf{a}-18\mathbf{b}$

Exercise 10.3B   **page 277**

1 Since OABC is a parallelogram, $\overrightarrow{OA}=\overrightarrow{CB}=\mathbf{a}$ and $\overrightarrow{OC}=\overrightarrow{AB}=\mathbf{c}$.

a $\overrightarrow{OB}=\overrightarrow{OC}+\overrightarrow{CB}=\mathbf{c}+\mathbf{a}$

$\overrightarrow{DE}=\overrightarrow{DA}+\overrightarrow{AE}=\frac{1}{2}\overrightarrow{OA}+\frac{1}{2}\overrightarrow{AB}=\frac{1}{2}\mathbf{c}+\frac{1}{2}\mathbf{a}$

$\qquad\qquad\qquad\qquad\qquad\qquad=\frac{1}{2}(\mathbf{c}+\mathbf{a})$

So $\overrightarrow{DE} = \frac{1}{2}\overrightarrow{OB}$

If two vectors have a common factor then they are parallel.

b $\overrightarrow{CF} = \overrightarrow{CO} + \overrightarrow{OF} = \overrightarrow{CO} + 2\overrightarrow{OA} = -\mathbf{c} + 2\mathbf{a}$

$\overrightarrow{CE} = \overrightarrow{CB} + \overrightarrow{BE} = \overrightarrow{CB} + \frac{1}{2}\overrightarrow{BA} = \mathbf{a} + \frac{1}{2}(-\mathbf{c})$
$\qquad = -\frac{1}{2}\mathbf{c} + \mathbf{a} = \frac{1}{2}(-\mathbf{c} + 2\mathbf{a})$

So $\overrightarrow{CE} = \frac{1}{2}\overrightarrow{CF}$

If two vectors have a common factor then they are parallel.

If two vectors are parallel and include a common point, all three points are collinear.

2 $\overrightarrow{TU} = 3\mathbf{a} - 2\mathbf{c}$

$\overrightarrow{AX} = 4\mathbf{a}$

$\overrightarrow{TX} = 3\mathbf{a} - 6\mathbf{c} + 6\mathbf{a} = 9\mathbf{a} - 6\mathbf{c} = 3(3\mathbf{a} - 2\mathbf{c})$

$\overrightarrow{TU}$ and $\overrightarrow{TX}$ have a common factor, so are parallel.

They also share a common point, so T, U and X are collinear.

3 $\overrightarrow{BC} = \overrightarrow{BA} + \overrightarrow{AD} + \overrightarrow{DC} = -15\mathbf{a} - 14\mathbf{b} + 5\mathbf{a}$
$\qquad\qquad\qquad = -10\mathbf{a} - 14\mathbf{b} = 2(-5\mathbf{a} - 7\mathbf{b})$

$\overrightarrow{BE} = \overrightarrow{BA} + \overrightarrow{AE} = \overrightarrow{BA} + 1.5\overrightarrow{AD}$
$\qquad = -15\mathbf{a} + 1.5(-14\mathbf{b}) = -15\mathbf{a} - 21\mathbf{b} = 3(-5\mathbf{a} - 7\mathbf{b})$

If two vectors have a common factor then they are parallel.

If two vectors are parallel and include a common point, all three points are colinear.

4 $12\mathbf{a} + 16\mathbf{b}$

5 $6\mathbf{p} + 5\mathbf{r}$

Exercise 10.4A page 281

1 a i Supermarket and church

ii Supermarket and museum

iii Church and hospital

b $(\mathbf{i} + 11\mathbf{j})$ km

c church and museum.

2 a A $= 4\mathbf{i} - 2\mathbf{j}$, B $= -\mathbf{i} + 3\mathbf{j}$, C $= -3\mathbf{i} - 3\mathbf{j}$

b $\overrightarrow{AB} = -5\mathbf{i} + 5\mathbf{j}$,

$\overrightarrow{AC} = -7\mathbf{i} - \mathbf{j}$

$\overrightarrow{BC} = -2\mathbf{i} - 6\mathbf{j}$

c $AB = 5\sqrt{2}$

$AC = 5\sqrt{2}$

$BC = 2\sqrt{10}$

d ABC is isosceles because two sides are exactly the same length.

3 a $3\sqrt{13}$

b $q = 10$ or -14

4 a $\overrightarrow{AB} = \mathbf{b} - \mathbf{a} = (8\mathbf{i} + 4\mathbf{j}) - (-\mathbf{i} - 4\mathbf{j}) = 9\mathbf{i} + 8\mathbf{j}$

$\overrightarrow{DC} = \mathbf{c} - \mathbf{d} = (3\mathbf{i} + 6\mathbf{j}) - (-6\mathbf{i} - 2\mathbf{j}) = 9\mathbf{i} + 8\mathbf{j}$

$\overrightarrow{BC} = \mathbf{c} - \mathbf{b} = (3\mathbf{i} + 6\mathbf{j}) - (8\mathbf{i} + 4\mathbf{j}) = -5\mathbf{i} + 2\mathbf{j}$

$\overrightarrow{AD} = \mathbf{d} - \mathbf{a} = (-6\mathbf{i} - 2\mathbf{j}) - (-\mathbf{i} - 4\mathbf{j}) = -5\mathbf{i} + 2\mathbf{j}$

This confirms that the shape has two pairs of parallel sides.

Also, since AB $\neq$ BC and the gradients of AB and BC are not perpendicular, the shape is a parallelogram.

b $\overrightarrow{CD} = \mathbf{d} - \mathbf{c} = (-6\mathbf{i} - 2\mathbf{j}) - (3\mathbf{i} + 6\mathbf{j}) = -9\mathbf{i} - 8\mathbf{j}$

$\overrightarrow{FE} = \mathbf{e} - \mathbf{f} = (-5\mathbf{i} - 14\mathbf{j}) - (4\mathbf{i} - 6\mathbf{j}) = -9\mathbf{i} - 8\mathbf{j}$

$\overrightarrow{DE} = \mathbf{e} - \mathbf{d} = (-5\mathbf{i} - 14\mathbf{j}) - (-6\mathbf{i} - 2\mathbf{j}) = \mathbf{i} - 12\mathbf{j}$

$\overrightarrow{CF} = \mathbf{f} - \mathbf{c} = (4\mathbf{i} - 6\mathbf{j}) - (3\mathbf{i} + 6\mathbf{j}) = \mathbf{i} - 12\mathbf{j}$

This confirms that the shape has two pairs of parallel sides.

Also, since CD $=$ DE (because $9^2 + 8^2 = 1^2 + 12^2$), and the gradients of CD and DE are not perpendicular, the shape is a rhombus.

c $\overrightarrow{DF} = \mathbf{f} - \mathbf{d} = (4\mathbf{i} - 6\mathbf{j}) - (-6\mathbf{i} - 2\mathbf{j}) = 10\mathbf{i} - 4\mathbf{j}$

$\overrightarrow{CB} = \mathbf{b} - \mathbf{c} = (8\mathbf{i} + 4\mathbf{j}) - (3\mathbf{i} + 6\mathbf{j}) = 5\mathbf{i} - 2\mathbf{j}$

$\overrightarrow{FB} = \mathbf{b} - \mathbf{f} = (8\mathbf{i} + 4\mathbf{j}) - (4\mathbf{i} - 6\mathbf{j}) = 4\mathbf{i} + 10\mathbf{j}$

$\overrightarrow{DC} = \mathbf{c} - \mathbf{d} = (3\mathbf{i} + 6\mathbf{j}) - (-6\mathbf{i} - 2\mathbf{j}) = 9\mathbf{i} + 8\mathbf{j}$

There are no identical sides, but DF $=$ 2CB.

This confirms that there is one pair of parallel sides. Hence the shape is a trapezium.

5 a $\overrightarrow{OS} = 4\mathbf{i} - 7\mathbf{j}$

b $5\mathbf{i}$

c 100

6 $\overrightarrow{JK} = \mathbf{k} - \mathbf{j} = (9\mathbf{i} - 4\mathbf{j}) - (-3\mathbf{i} + \mathbf{j}) = 12\mathbf{i} - 5\mathbf{j}$

$\overrightarrow{JL} = \mathbf{l} - \mathbf{j} = (3\mathbf{i} + 5\mathbf{j}) - (-3\mathbf{i} + \mathbf{j}) = 6\mathbf{i} + 4\mathbf{j}$

$\overrightarrow{KL} = \mathbf{l} - \mathbf{k} = (3\mathbf{i} + 5\mathbf{j}) - (9\mathbf{i} - 4\mathbf{j}) = -6\mathbf{i} + 9\mathbf{j}$

Magnitude of $\overrightarrow{JK} = \sqrt{12^2 + (-5)^2} = 13$

Magnitude of $\overrightarrow{JL} = \sqrt{6^2 + 4^2} = \sqrt{52} = 2\sqrt{13}$

Magnitude of $\overrightarrow{KL} = \sqrt{(-6)^2 + 9^2} = \sqrt{117} = 3\sqrt{13}$

Since $(3\sqrt{13})^2 + (2\sqrt{13})^2 = 13^2$, the sides of the triangle obey Pythagoras' theorem.

Hence the triangle is right-angled.

page 283

Exam-style questions 10

1 a $23.7\,\text{m}\,\text{s}^{-1}$

  b $332°$

2 $6.1\,\text{km}$

3 a $p = 4$

  b $22.6°$

4 a $\overrightarrow{OP} = \frac{2}{5}(12\mathbf{a} + 7\mathbf{c})$

  b $\overrightarrow{OB} = \overrightarrow{OC} + \overrightarrow{CB} = 7\mathbf{c} + 12\mathbf{a}$

   $\overrightarrow{OP} = \frac{2}{5}\overrightarrow{OB}$

   $\overrightarrow{OP}$ and $\overrightarrow{OB}$ have a common factor, so are parallel.

   They also share a common point, so O, P and B are collinear.

  c $2 : 3$.

5 a i $\begin{bmatrix} 5 - 2k \\ -8 \end{bmatrix}$

   ii $\sqrt{(5 - 2k)^2 + (-8)^2}$

  b For magnitude $> 2\sqrt{17}$, $k < \frac{3}{2}$ or $k > \frac{7}{2}$.

6 $p = 11$ and $q = 9$

7 a 70

  b $(-4\mathbf{i} - 6\mathbf{j})$

8 $19\mathbf{d} - 10\mathbf{b}$

9 a $3\sqrt{10}$

  b $\overrightarrow{CD} = \mathbf{d} - \mathbf{c} = (-5\mathbf{i} - 6\mathbf{j}) - (-4\mathbf{i} + \mathbf{j}) = (-\mathbf{i} - 7\mathbf{j})$

   $\overrightarrow{FE} = \mathbf{e} - \mathbf{f} = (4\mathbf{i} - 3\mathbf{j}) - (5\mathbf{i} + 4\mathbf{j}) = (-\mathbf{i} - 7\mathbf{j})$

  c $(-2\mathbf{i} - 5\mathbf{j})$

  d $XC^2 = (-2 - (-4))^2 + (-5 - 1)^2 = 40$

   $XD^2 = (-2 - (-5))^2 + (-5 - (-6))^2 = 10$

   $CD^2 = (-4 - (-5))^2 + (1 - (-6))^2 = 50$

   Since $XD^2 + XC^2 = CD^2$, XC is perpendicular to XD.

  e 60

  f $\overrightarrow{CG} = \mathbf{g} - \mathbf{c} = (2\mathbf{i} + 3\mathbf{j}) - (-4\mathbf{i} + \mathbf{j}) = (6\mathbf{i} + 2\mathbf{j}) = 2(3\mathbf{i} + \mathbf{j})$

   Since $\overrightarrow{CG}$ and $\overrightarrow{DX}$ have a common factor, they are parallel and DXGC is a trapezium.

11 Proof

1 Let $2n + 1$ be an odd number. Then $2n + 3$ and $2n + 5$ are consecutive odd numbers.

$2n + 1 + 2n + 3 + 2n + 5 = 6n + 9$

$\qquad = 3(2n + 3)$

$\qquad$ which is a multiple of 3.

2 Let $2n$ be an even number and $2n + 1$ be the next number.

$2n + 2n + 1 = 4n + 1$ which is an odd number.

3 Let $2n$ be an even number and $2n + 1$ be the next number.

$(2n + 1)^2 - (2n)^2 = 4n^2 + 4n + 1 - 4n^2$

$\qquad = 4n + 1$

$\qquad = (2n) + (2n + 1)$

4 Let $2n$ be an even number and $2n + 2$ be the next even number.

$(2n)^2 + (2n + 2)^2 = 4n^2 + 4n^2 + 8n + 4$

$\qquad = 8n^2 + 8n + 4$

$\qquad = 4(2n^2 + 2n + 1)$

$\qquad$ which is a multiple of 4.

5 $(2n - 1)^2 - (2n + 1)^2 = 4n^2 - 4n + 1 - (4n^2 + 4n + 1)$

$\qquad = -8n$

$\qquad$ which is a multiple of 8.

1 $n^3 - n = n(n^2 - 1) = n(n - 1)(n + 1)$

$(n - 1)$, n and $(n + 1)$ are three consecutive integers.

When $n = 2$, the integers are 1, 2 and 3.

When $n > 2$, the integers are three consecutive positive integers greater than 1, 2 and 3.

2 The shape is a rectangle with a rectangular part removed.

The area of the rectangle is

$(3x - 1)(2x + 5) = 6x^2 + 13x - 5$

The removed part has an area of

$((3x - 1) - (x + 3))((2x + 5) - x)$

$= (2x - 4)(x + 5)$

$= 2x^2 + 6x - 20$

The area of the shape is

$6x^2 + 13x - 5 - (2x^2 + 6x - 20) = 4x^2 + 7x + 15$

Hence $4x^2 + 7x + 15 = 201$

$\qquad 4x^2 + 7x - 186 = 0$

$\qquad (4x + 31)(x - 6) = 0$

$\qquad\qquad\qquad x = 6$

Perimeter $= 2((3x - 1) + (2x + 5))$

$\qquad\qquad = 2((3(6) - 1) + (2(6) + 5)) = 2(17 + 17)$

$\qquad\qquad = 68\,\text{cm}$

3 $n^2 - 6n + 10 = (n - 3)^2 - 9 + 10$

$\qquad\qquad\quad = (n - 3)^2 + 1$

$\qquad\qquad$ which is positive for all values of n.

4 Every even number can be expressed as $2n$.

$(2n)^2 = 4n^2$ which is a multiple of 4.

Every odd number can be expressed as $2n + 1$.

$(2n + 1)^2 = 4n^2 + 4n + 1$

$\qquad\qquad = 4(n^2 + n) + 1$

$\qquad\qquad$ which is one more than a multiple of 4.

5 Join the four corners of the cyclic quadrilateral to the centre of the circle. This will create four isosceles triangles. Label the base angles of the four isosceles triangles as shown in the diagram.

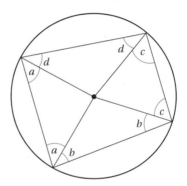

$a + a + b + b + c + c + d + d = 360°$

$\qquad\qquad 2a + 2b + 2c + 2d = 360°$

$\qquad\qquad\quad a + b + c + d = 180°$

One pair of opposite angles is $(a + d)$ and $(b + c)$.

The other pair of opposite angles is $(a + b)$ and $(c + d)$.

Hence the opposite angles of a cyclic quadrilateral have a sum of 180°.

6 Let ABCDEF be a seating arrangement with A to F sitting in alphabetical order and with F next to A.

Start with A and B next to each other.

For ABC, E must come next. Hence there are two possibilities starting ABC: ABCEDF and ABCEFD.

For ABD, C must not come next.

For ABDE, C must come next. Hence there is one possibility starting ABDE: ABDECF.

For ABDF, there are two possibilities: ABDFCE and ABDFEC.

For ABE, C must come next. D cannot follow C. Hence there is one possibility starting ABE: ABECFD.

For ABFC, E must come next. Hence there is one possibility starting ABFC: ABFCED.

For ABFD, C must not come next. Hence there is one possibility starting ABFD: ABFDEC.

For ABFE, C must come next but would have to be followed by D, so this is not a possibility.

Therefore there are eight possible seating arrangements.

7 Area of large square $= a^2$

Height of one triangle $= \sqrt{a^2 - b^2}$

Area of one triangle $= \frac{b}{2}(\sqrt{a^2 - b^2})$

Length of small square $= \sqrt{a^2 - b^2} - b$

Area of small square $= (\sqrt{a^2 - b^2} - b)(\sqrt{a^2 - b^2} - b)$

Area of four triangles plus small square

$= \frac{4b}{2}(\sqrt{a^2 - b^2}) + (\sqrt{a^2 - b^2} - b)(\sqrt{a^2 - b^2} - b)$

$= 2b(\sqrt{a^2 - b^2}) + a^2 - b^2 - 2b\sqrt{a^2 - b^2} + b^2$

$= a^2$

8 The flaw is between the following two stages:

Factorising $a(b - a) = (b + a)(b - a)$

Dividing leaves $a = b + a$

If $b = a$ then $b - a = 0$ and you cannot divide by zero.

Exercise 11.2A   **page 293**

1 Case 1, ABC: A is next to B so this is allowed.

Case 2, ACB: A is not next to B so this is not allowed.

Case 3, BAC: A is next to B so this is allowed.

Case 4, BCA: A is not next to B so this is not allowed.

Case 5, CAB: A is next to B so this is allowed.

Case 6, CBA: A is next to B so this is allowed.

Therefore there are four possible seating arrangements.

2 A prime number is an integer which has two and only two distinct factors. 1 is not a prime number.

$n = \{20, 21, 22, 23, 24, 25, 26, 27, 28, 29, 30\}$

Case 1, $n = 20$: the factors of 20 are 1, 2, 4, 5, 10, 20 so 20 is not prime.

Case 2, $n = 21$: the factors of 21 are 1, 3, 7, 21 so 21 is not prime.

Case 3, $n = 22$: the factors of 22 are 1, 2, 11, 22 so 22 is not prime.

Case 4, $n = 23$: the factors of 23 are 1, 23 so 23 is prime.

Case 5, $n = 24$: the factors of 24 are 1, 2, 3, 4, 6, 8, 12, 24 so 24 is not prime.

Case 6, $n = 25$: the factors of 25 are 1, 5, 25 so 25 is not prime.

Case 7, $n = 26$: the factors of 26 are 1, 2, 13, 26 so 26 is not prime.

Case 8, $n = 27$: the factors of 27 are 1, 3, 9, 27 so 27 is not prime.

Case 9, $n = 28$: the factors of 24 are 1, 2, 4, 7, 14, 28 so 28 is not prime.

Case 10, $n = 29$: the factors of 24 are 1, 29 so 29 is prime.

Case 11, $n = 30$: the factors of 24 are 1, 2, 3, 5, 6, 10, 15 so 30 is not prime.

Consequently there are two prime numbers, 23 and 29, between 20 and 30 inclusive.

3 $n = \{1, 4, 9, 16, 25, 36, 49, 64, 81, 100\}$

Consequently there are no square numbers below 100 ending in 7.

4 $n = \{1, 2, 3, 4, 5, 6, 7, 8, 9\}$

Case 1, $n = 1$: $(1)^2 + 3(1) + 19 = 23$, which is prime.

Case 2, $n = 2$: $(2)^2 + 3(2) + 19 = 29$, which is prime.

Case 3, $n = 3$: $(3)^2 + 3(3) + 19 = 37$, which is prime.

Case 4, $n = 4$: $(4)^2 + 3(4) + 19 = 47$, which is prime.

Case 5, $n = 5$: $(5)^2 + 3(5) + 19 = 59$, which is prime.

Case 6, $n = 6$: $(6)^2 + 3(6) + 19 = 73$, which is prime.

Case 7, $n = 7$: $(7)^2 + 3(7) + 19 = 89$, which is prime.

Case 8, $n = 8$: $(8)^2 + 3(8) + 19 = 107$, which is prime.

Case 9, $n = 9$: $(9)^2 + 3(9) + 19 = 127$, which is prime.

Consequently, the expression $n^2 + 3n + 19$ is prime for all positive integer values of n below 10.

5 $x = \{1, 3, 5\}$ and $y = \{1, 3, 5\}$

Case 1: $x + y = 1 + 1 = 2 = 2 \times 1$

Case 2: $x + y = 1 + 3 = 4 = 2 \times 2$

Case 3: $x + y = 1 + 5 = 6 = 2 \times 3$

Case 4: $x + y = 3 + 1 = 4 = 2 \times 2$

Case 5: $x + y = 3 + 3 = 6 = 2 \times 3$

Case 6: $x + y = 3 + 5 = 8 = 2 \times 4$

Case 7: $x + y = 5 + 1 = 6 = 2 \times 3$

Case 8: $x + y = 5 + 3 = 8 = 2 \times 4$

Case 9: $x + y = 5 + 5 = 10 = 2 \times 5$

So if x and y are odd positive integers less than 7 then their sum is always divisible by 2.

6 Case 1, 1, 1 and 9: $1 + 1 < 9$, so two sides have a sum shorter than the third.

Case 2, 1, 2 and 8: $1 + 2 < 8$, so two sides have a sum shorter than the third.

Case 3, 1, 3 and 7: $1 + 3 < 7$, so two sides have a sum shorter than the third.

Case 4, 1, 4 and 6: $1 + 4 < 6$, so two sides have a sum shorter than the third.

Case 5, 1, 5 and 5: $1 + 5 > 5$ and $5 + 5 > 1$, so any two sides sum to more than the third.

Case 6, 2, 2 and 7: $2 + 2 < 7$, so two sides have a sum shorter than the third.

Case 7, 2, 3 and 6: $2 + 3 < 6$, so two sides have a sum shorter than the third.

Case 8, 2, 4 and 5: $2 + 4 > 5$, $2 + 5 > 4$ and $4 + 5 > 2$, so any two sides sum to more than the third.

There are two distinct triangles: 1, 5, 5 and 2, 4, 5.

7 Case 1, 1, 1, 1 and 7: $1 + 1 + 1 < 7$, so three sides have a sum shorter than the fourth.

Case 2, 1, 1, 2 and 6: $1 + 1 + 2 < 6$, so three sides have a sum shorter than the fourth.

Case 3, 1, 1, 3 and 5: $1 + 1 + 3 = 5$, so three sides have a sum equal to the fourth.

Case 4, 1, 1, 4 and 4: $1 + 1 + 4 > 4$ and $1 + 4 + 4 > 1$, so any three sides have a sum greater than the fourth.

Case 5, 1, 2, 2 and 5: $1 + 2 + 2 = 5$, so three sides have a sum equal to the fourth.

Case 6, 1, 2, 3 and 4: $1 + 2 + 3 > 4$, $1 + 2 + 4 > 3$, $1 + 3 + 4 > 2$ and $2 + 3 + 4 > 1$, so any three sides have a sum greater than the fourth.

Case 7, 1, 3, 3 and 3: $1 + 3 + 3 > 3$ and $3 + 3 + 3 > 1$, so any three sides have a sum greater than the fourth.

Case 8, 2, 2, 2 and 4: $2 + 2 + 2 > 4$ and $2 + 2 + 4 > 2$, so any three sides have a sum greater than the fourth.

Case 9, 2, 2, 3 and 3: $2 + 2 + 3 > 3$ and $2 + 3 + 3 > 2$, so any three sides have a sum greater than the fourth.

Consequently, there are five quadrilaterals with sides of integer length that have a perimeter of 10.

8 Note that any tetronimo must fit into a 1 by 4 rectangle or a 2 by 3 rectangle (neither 2 by 5 or 3 by 3 would be possible).

Case 1: There is only one possibility in a 1 by 4 rectangle: four in a row.

For cases 2 to 16, consider 2 by 3 rectangles with two squares removed.

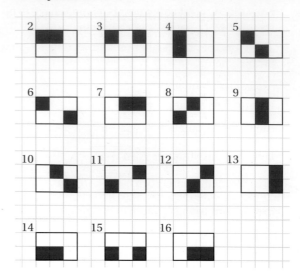

Case 2: L shape

Case 3: T shape

Case 4: 2 by 2 square

Case 5: Squares are not all joined by edges

Case 6: S shape

Case 7: L shape

Case 8: Squares are not all joined by edges

Case 9: Squares are not all joined by edges

Case 10: Squares are not all joined by edges

Case 11: S shape

Case 12: Squares are not all joined by edges

Case 13: Square

Case 14: L shape

Case 15: T shape

Case 16: L shape

Consequently, there are five possible tetronimos.

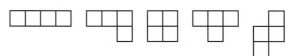

9 The first team could have scored 0, 1 or 2 before half-time.

The second team could have scored 0 or 1 before half-time.

The half-time score could be:

0–0, 0–1, 1–0, 1–1, 2–0, 2–1

Consequently there are six possible half-time scores for a match which finishes 2–1.

10 $n = \{1, 2, 3, 4, 5, 6, 7, 8, 9\}$

Case 1: $n^6 - n = 1 - 1 = 0 = 2 \times 0$

Case 2: $n^6 - n = 64 - 2 = 62 = 2 \times 31$

Case 3: $n^6 - n = 729 - 3 = 726 = 2 \times 363$

Case 4: $n^6 - n = 4096 - 4 = 4092 = 2 \times 2046$

Case 5: $n^6 - n = 15\,625 - 5 = 15\,620 = 2 \times 7810$

Case 6: $n^6 - n = 46\,656 - 6 = 46\,650 = 2 \times 23\,325$

Case 7: $n^6 - n = 117\,649 - 7 = 117\,642 = 2 \times 58\,821$

Case 8: $n^6 - n = 262\,144 - 8 = 262\,136 = 2 \times 131\,068$

Case 9: $n^6 - n = 531\,441 - 9 = 531\,432 = 2 \times 265\,716$

So if n is a positive integer less than 10 then $n^6 - n$ is a multiple of 2.

Exercise 11.3A   page 296

1 a A rectangle with sides of 10 cm and 26 cm is not a square.

b −26 is not positive.

c When −26 is doubled, the answer is smaller (−52).

d −10 cubed is not positive.

e 2 is a prime number but is not odd.

f $\log(1 + 1) = \log 2$ but $\log 1 + \log 1 = 0$

g $(1 + 1)^2 = 4$ but $1^2 + 1^2 = 2$

2 $2^3 < 3^2$. 3 is the only counter example.

3 When $x = -4$, $x^2 + 8x + 15 = (-4)^2 + 8(-4) + 15 = -1$

4 ABCED has Carlos not next to Dupika.

5 Any prime number multiplied by 2 will be even.

6 If $x + y = 1 + 3 = 4$, which is even, x and y don't have to be even.

7 2 is a prime number that is not odd.

8 If $x = \sqrt{2}$ and $y = 2\sqrt{2}$ then $xy = \sqrt{2} \times 2\sqrt{2} = 4$ which is rational.

9 If $x = -3$ and $y = 3$ then $x^2 = y^2$ but $x \neq y$.

Exam-style questions 11   page 298

1 $(3n + 5)^2 - (3n - 5)^2 = 9n^2 + 30n + 25 - 9n^2 + 30n - 25$

$$= 60n$$

which is a multiple of 12.

2 The 12 ways are: SSUM, SSMU, SUSM, SMSU, SUMS, SMUS, USSM, MSSU, USMS, MSUS, UMSS and MUSS.

3 $x = \{2, 4, 6\}$ and $y = \{2, 4, 6\}$

Case 1: $x - y = 2 - 2 = 0$

Case 2: $x - y = 2 - 4 = -2$

Case 3: $x - y = 2 - 6 = -4$

Case 4: $x - y = 4 - 2 = 2$

Case 5: $x - y = 4 - 4 = 0$

Case 6: $x - y = 4 - 6 = -2$

Case 7: $x - y = 6 - 2 = 4$

Case 8: $x - y = 6 - 4 = 2$

Case 9: $x - y = 6 - 6 = 0$

All of these differences are divisible by 2.

4 When n is odd, the integers can be written as $a - \left(\frac{n-1}{2}\right)$, $a - \left(\frac{n-1}{2} - 1\right)$, ... $a - 2$, $a - 1$, a, $a + 1$, $a + 2$, ..., $a + \left(\frac{n-1}{2} - 1\right)$, $a + \left(\frac{n-1}{2}\right)$, where a is the middle value (or the mean of the first and last terms) and is an integer. These have a sum of na, which is a multiple of n.

When n is even, the integers can be written as $a - \left(\frac{n-1}{2}\right)$, $a - \left(\frac{n-1}{2} - 1\right)$, ... $a - 2$, $a - 1$, a, $a + 1$, $a + 2$, ..., $a + \left(\frac{n-1}{2} - 1\right)$, $a + \left(\frac{n-1}{2}\right)$, where a is the middle value (or the mean of the first and last terms) but is not an integer. These also have a sum of na, but this is not a multiple of n, because a is not an integer.

5 The three-digit cube numbers are:

$5^3 = 125$ $6^3 = 216$ $7^3 = 343$ $8^3 = 512$ $9^3 = 729$

343 is the only palindromic three-digit cube number.

6 When $\theta = \frac{\pi}{2}$, $\sin 2\left(\frac{\pi}{2}\right) = 0$ whereas $2\sin\left(\frac{\pi}{2}\right) = 2$.

7 When $n = 5$, $n^2 - n + 1 = 5^2 - 5 + 1 = 21$ which is not prime.

8 Area of large square − area of four triangles = area of small square

$$(a + b)^2 - 2ab = c^2$$
$$a^2 + 2ab + b^2 - 2ab = c^2$$
$$a^2 + b^2 = c^2$$

9 If $x = -4$ and $y = 3$, then $x < y$ but $x^2 > y^2$.

12 Data presentation and interpretation

Prior knowledge   page 300

1 a The data is quantitative and discrete

   b Mean = 32.88, median = 35, mode = 41 and range = 35

   c

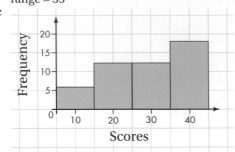

   The results show a higher frequency or modal group 40–49.

Exercise 12.1A   page 306

1 a continuous

   b discrete

   c discrete

   d continuous

   e continuous

2 a qualitative

   b quantitative

   c qualitative

   d quantitative

   e quantitative

   f qualitative

   g quantitative

3 Various ideas such as height, weight, texts sent, music tracks owned, etc.

4 a 9

   b 9

   c 13.05

   d The mean takes into account all the visits made so it is the best measure to use.

5 a 3.8

   b No effect, as the total remains the same.

6 Median = 5

   Mode = 5

   Mean = 4.96

   The best average to represent the data is the mean, as it uses all the data.

7 19.3 cm

8

Height (cm)	Frequency	fx
$100 < h \leqslant 120$	5	550
$120 < h \leqslant 140$	4	520
$140 < h \leqslant 160$	12	1800
$160 < h \leqslant 180$	13	2210
$180 < h \leqslant 200$	8	1520
Total	42	6600

Estimate of the mean = $\frac{6600}{42} = 157.14$

Exercise 12.2A   page 311

1 a Mean = 8, standard deviation = 4

   b Mean = 70, standard deviation = 14.14

   c Mean = 13, standard deviation = 3.63

2 a Mean = 5, standard deviation = 2.19

   b Mean = 16.5, standard deviation = 3.07

3 a Mean = 6.06, standard deviation = 1.12

   b All three averages are similar

4 Mean = 9

   Standard deviation = 0.82

5 Mean = 6.8

   Variance = 32.5

6 Mean = 0.6

   Standard deviation = 2.62

7 a Town route:

      Mean = 20

      Standard deviation = 4.60

      Country route:

      Mean = 20

      Standard deviation = 1.41

   b The country route is better as the average times are the same but the country route is more consistent.

8 a Mean = 2.075

Standard deviation = 0.185

b No ball bearing is 2 standard deviations either side of the mean, so consistent results.

Exercise 12.2B page 317

1 Mean of x: 53.71

Standard deviation: 3.21

2 a Mean = 4.31

Standard deviation = 2.29

b Mean = 3.831

Standard deviation = 7.1

3 a Mean = 67.07

Standard deviation = 9.97

b Mean = 19.48

Standard deviation = 5.54

4 Mean = 4.9 ounces

Standard deviation = 0.201 ounces

Exercise 12.3A page 323

1 Median = 48th value; this lies in the '3 in family' group.

2 a

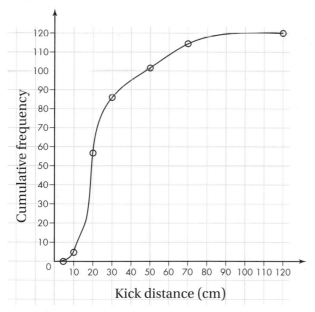

Kick distance (cm)

b 18.1

c 29.02

3 a

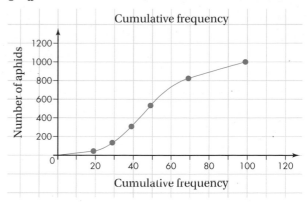

b 47.56

c 50.55

4 a Jenson: Q_0 3, Q_1 12, Q_2 31, Q_3 55, Q_4 66

Molly: Q_0 12, Q_1 17, Q_2 34, Q_3 57, Q_4 98

b Medians are similar but Molly's is slightly higher, suggesting she scores more runs. Larger range for Molly, suggesting Jenson is more consistent. Jenson has a smaller IQR suggesting again that Jenson is more consistent.

5 a

Length (cm)	Frequency	Upper class	CF
17.5–17.9	15	17.95	15
18.0–18.4	27	18.45	42
18.5–18.9	18	18.95	60
19.0–19.9	12	19.95	72
20.0–24.9	15	24.95	87
25.0–29.9	4	29.95	91

b

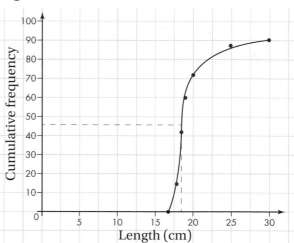

Length (cm)

c between 17 and 19 cm.

d 18.56

e Comparison should mention the fact that they are similar but interpolation is more accurate.

f Mean = 19.49 which is higher than the median

6 a 60

b Q_1 46, Q_2 56, Q_3 64

c

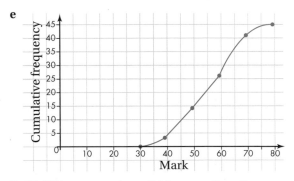

d no outliers

e

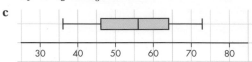

f 28; This measure is useful because it omits extreme values.

7 a A: True

B: True

C: False

D: False

b Class A is more consistent as it has a smaller range so less extremes. Class B has a smaller IQR so the middle 50% of the data is more consistent.

8 a i 63 minutes

ii 120 minutes

iii 34

iv 42

b 25

Exercise 12.3B page 327

1 a 3 and 25.5

b Width = 9 and height = $\dfrac{10}{63}$

c 15.05

d 15.04

2 a

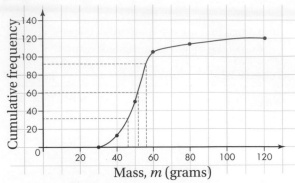

i about 51 g

ii about 45 g and 59 g

b

Mass, m (grams)	FD
$30 \leqslant m < 40$	1.3
$40 \leqslant m < 50$	3.7
$50 \leqslant m < 60$	5.6
$60 \leqslant m < 80$	0.4
$80 \leqslant m < 120$	0.15

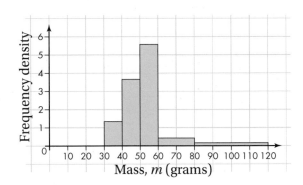

c $Q_3 + 1.5 \times IQR = 59 + 1.5(14) = 80 \, g$

3 a Width = 1.6 cm

b Height 2.3 cm

4 a Continuous data

b

Distance (km)	Number of people	FD
40–49	67	$\dfrac{67}{10} = 6.7$
50–59	124	$\dfrac{124}{10} = 12.4$
60–64	4023	$\dfrac{4023}{5} = 804.6$
64–69	2981	$\dfrac{2981}{5} = 596.2$
70–84	89	$\dfrac{89}{14} = 5.9$
85–149	75	$\dfrac{75}{65} = 1.2$

5 **a** Width = 3 cm

Height = 0.41 cm

b 9.14 kg

c Mean = 10.7 kg

Standard deviation = 6.84 kg

6 **a**

t	Frequency	Frequency density
5–10	10	2
10–14	15	3.75
14–18	22	5.5
18–25	21	3
25–40	18	1.2

b 33 people

c 19.1 minutes

d 8.1 minutes

e $Q_1 = 13.2$

$Q_2 = 17.3$

$Q_3 = 24$

7 **a** Width= 3.5 cm

Height = 4.8 cm

b Median = £272

c Mean = £306.6

Standard deviation = £125.9

Exercise 12.3C page 332

1 **a** and **c**

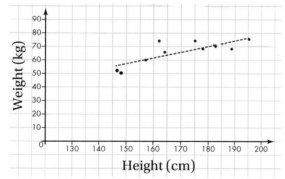

b Positive correlation: as height increases, so does weight.

d 78 ± 1 kg

2 **a** An example could be 'Weight and strength'

b An example could be 'Alcohol consumption and judgment'

c An example could be 'Tests scores in French and geography'

d An example could be 'Amount of money spent on food and death through getting stuck in bedding'

3

4 **a**

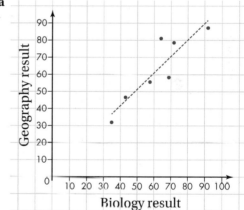

5 **a** and **c**

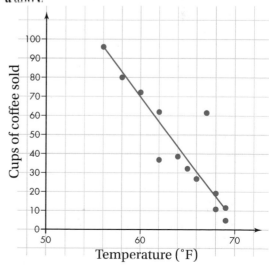

b Negative correlation: as temperature increases, coffees sold decreases

d 82 cups

Exercise 12.3D   **page 337**

1 a a is the y intercept, b is the gradient.

   b 4034 g

   c Extrapolation, hence unreliable, as the data does not extend to 60 cm.

2 100.4 (extrapolation, therefore not an accurate result)

3 $y = 3.41$ kg
Extrapolation, so not reliable.

4 a

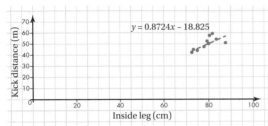

   b $y = 0.87x - 18.8$

   c About 86 cm

   d 42.1 m; this is extrapolation, so unreliable.

5 a Positive correlation

   b about 60 grams

   c Interpolation

   d For every 1 gram consumed in 2001, only 0.64 was consumed in 2014. The constant is 5.7408 grams.

   e 457.85

   f Extrapolation – not in data set, cannot say for sure.

6 a No outliers

   b

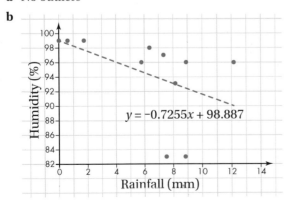

   c Negative correlation

d $y = -0.7255x + 98.88$

e 92.1%

Exam-style questions 12   **page 339**

1 a

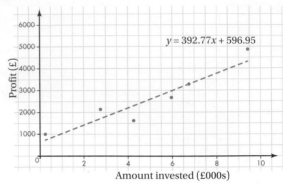

   b For every £1000 invested, a return of £393 profit can be expected.

   c £1971.65

2 Salaries below £10 500 and above £46 500 would be outliers.

3 a 321.5 kg

   b 290.1 kg

4 a Q_1 121, Q_2 187, Q_3 260

   b $Q_3 + 1.5 \times$ IQR $= 260 + 1.5(139) = 468.5$ and $121 - 1.5(139) = -87.5$, so no outliers

5 a 20.8 cm

   b

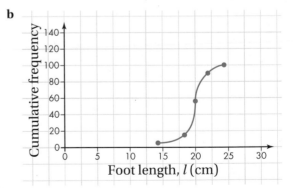

About 20 cm

   c 3.8%

   d Mean $= 20.8$
Standard deviation $= 1.96$

   e Mean is best as no obvious outliers, and it uses all the available data.

6 a Width = 12.5 cm

b Height = 9.7 cm

c 29.4 = median

21.8 = lower quartile

38.9 = upper quartile

IQR = 17.1

d Median is best due to the extreme values in the data set.

7 a

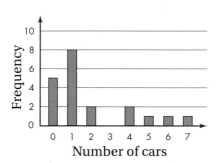

b Median is between the 10th and 11th values so is 1 car.

c Mean = $\frac{38}{20}$ = 1.9

$S_{xx} = \sum fx^2 - n\bar{x}^2 = 158 - 20 \times 1.9^2 = 85.8$

$\sigma = \frac{\sqrt{85.8}}{20} = 2.07$

d 1.9 + 2(2.07) = 6.04

The manufacturer should be concerned if a car is returned 7 or more times.

8 a Median = 19

Upper quartile = 22

Lower quartile = 15

Interquartile range = 7

Minimum value = 8, maximum value = 31

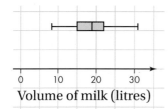

Volume of milk (litres)

b Outlier = UQ + 1.5(IQR)

= 22 + 1.5(7) = 32.5, therefore no outliers

Outlier = LQ – 1.5(IQR)

= 15 – 1.5(7) = 4.5, therefore no outliers

c Mean = 18.6 litres and median = 19 litres; these values are consistent as there are no outliers in the data set.

13 Probability and statistical distributions

Prior knowledge page 342

1 a

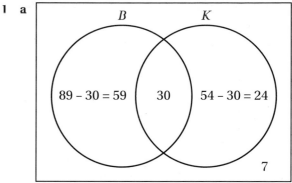

b 7 members

2 a $\frac{1}{3}$

b $\frac{4}{15}$

c $\frac{3}{5}$

Exercise 13.1A page 350

1 a $\frac{46}{200}$ = 0.23

b 0.43

2 a $\frac{1}{2}$ or 0.5

b 0.9

c $\frac{19}{100}$ or 0.19

3 0.8

4 a 0.8

b 0.1

5 a P(S or T) = 0.62

b P(S' and T') = 0.38

6 a 0.7

b 0.5

c 0.65

7 a

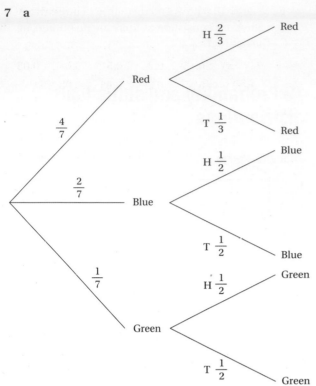

b P(head) = $\frac{8}{21} + \frac{1}{7} + \frac{1}{14} = \frac{25}{42}$ or 0.595 (3 s.f.)

Exercise 13.2A   **page 354**

1 a

x	P($X=x$)	x	P($X=x$)	x	P($X=x$)
2	$\frac{1}{36}$	6	$\frac{5}{36}$	10	$\frac{1}{12}$
3	$\frac{1}{18}$	7	$\frac{1}{6}$	11	$\frac{1}{18}$
4	$\frac{1}{12}$	8	$\frac{5}{36}$	12	$\frac{1}{36}$
5	$\frac{1}{9}$	9	$\frac{1}{9}$		

b

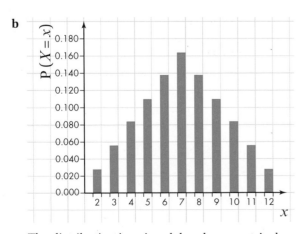

The distribution is unimodal and symmetrical.

2 a

a	P($A=a$)	a	P($A=a$)	a	P($A=a$)
1	$\frac{1}{36}$	8	$\frac{1}{18}$	18	$\frac{1}{18}$
2	$\frac{1}{18}$	9	$\frac{1}{36}$	20	$\frac{1}{18}$
3	$\frac{1}{18}$	10	$\frac{1}{18}$	24	$\frac{1}{18}$
4	$\frac{1}{12}$	12	$\frac{1}{9}$	25	$\frac{1}{36}$
5	$\frac{1}{18}$	15	$\frac{1}{18}$	30	$\frac{1}{18}$
6	$\frac{1}{9}$	16	$\frac{1}{36}$	36	$\frac{1}{36}$

b

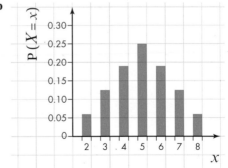

No distinguishable shape

3 a

x	2	3	4	5	6	7	8
P($X=x$)	$\frac{1}{16}$	$\frac{1}{8}$	$\frac{3}{16}$	$\frac{1}{4}$	$\frac{3}{16}$	$\frac{1}{8}$	$\frac{1}{16}$

b

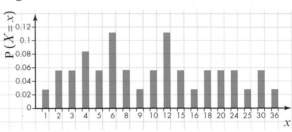

One mode and symmetrical

c $\frac{9}{16}$ or 0.5625

4 a $k = 0.1$

b

b	1	2	3	4
P($B=b$)	0.1	0.2	0.3	0.4

c

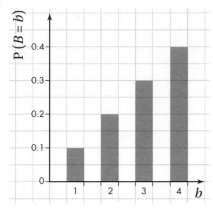

Negatively skewed distribution

5 a $k = \frac{1}{6}$

b $\frac{1}{2}$ or 0.5

c $\frac{5}{6}$

d $\frac{2}{3}$

6 a

x	1	2	3	4	5
$P(X=x)$	$1k$	$4k$	$9k$	$16k$	$25k$
$P(X=x)$	$\frac{1}{55}$	$\frac{4}{55}$	$\frac{9}{55}$	$\frac{16}{55}$	$\frac{25}{55}$

$k = \frac{1}{55}$

b

x	1	2	3	4	5
$P(X-x)$	$\frac{k}{1}$	$\frac{k}{2}$	$\frac{k}{3}$	$\frac{k}{4}$	$\frac{k}{5}$
$P(X=x)$	0.438...	0.219...	0.146...	0.109...	0.088...

$k = \frac{60}{137}$

c

x	1	2	3	4	5
$P(X=x)$	$\frac{1}{k}$	$\frac{2}{k}$	$\frac{3}{k}$	$\frac{4}{k}$	$\frac{5}{k}$
$P(X=x)$	$\frac{1}{15}$	$\frac{2}{15}$	$\frac{3}{15}$	$\frac{4}{15}$	$\frac{5}{15}$

$k = 15$

1 a

X	1	2	3	4	5
$P(X=x)$	0.15	0.2	0.5	0.1	0.05
$F(x)$	0.15	0.35	0.85	0.95	1

b

X	1	2	4	8
$P(X=x)$	0.25	0.25	0.25	0.25
$F(x)$	0.25	0.5	0.75	1

c

X	−3	−1	1	3
$P(X=x)$	0.8	0.1	0.02	0.08
$F(x)$	0.8	0.9	0.92	1

d

X	1	2	3	4	5
$P(X=x)$	0.2	$2a$	0.4	a	0.1
$F(x)$	0.2	$2a+0.2$	$2a+0.6$	$3a+0.6$	$3a+0.7$

2 a

X	1	2	3	4	5
$P(X=x)$	$\frac{1}{5}$	$\frac{1}{5}$	$\frac{1}{5}$	$\frac{1}{5}$	$\frac{1}{5}$
$F(x)$	$\frac{1}{5}$	$\frac{2}{5}$	$\frac{3}{5}$	$\frac{4}{5}$	1

b $P(X<3) = \frac{2}{5}$

c $\frac{3}{5}$ or 0.6

3 a

x	2	4	6	8
$P(X=x)$	0.3	0.3	0.2	0.2

b

x	−5	0	5	10
$P(X=x)$	0.05	0.35	0.1	0.5

c

x	2	5	6	22
$P(X=x)$	0.1	0.4	0.25	0.25

4

x	1	2	3	4
$P(X=x)$	0.32	0.23	0.46	0.22

$a = 0.55$

5

r	0	1	2
$P(R=r)$	$\frac{1}{8}$	$\frac{3}{8}$	$\frac{1}{2}$

6 a

s	1	2	4	6
$P(S=s)$	$\frac{1}{4}$	$\frac{1}{4}$	$\frac{1}{4}$	$\frac{1}{4}$

b Uniform probability distribution

c Unimodal, symmetrical

d P(both prime) = $\frac{1}{4} \times \frac{5}{6} = \frac{5}{24}$

7 a

x	1	2	3	4	5	6
$P(X=x)$	$\frac{1}{36}$	$\frac{3}{36}$	$\frac{5}{36}$	$\frac{7}{36}$	$\frac{9}{36}$	$\frac{11}{36}$

b $\frac{1}{9}$

8 a Substituting the values of x into the probability function,

$1225k = 1$

$k = \frac{1}{1225}$, as required

b $\frac{1072}{1225}$

9 $a = 0.1$, $b = 0.4$, $c = 0.2$ and $d = 0.8$

10 a Since all probabilities sum to 1,

$\frac{2}{18} + \frac{2k}{18} + \frac{7}{18} + \frac{k}{18} = 1$

Therefore $3k = 9$

$k = 3$, as required

b

x	2	3	4	6
$F(x)$	$\frac{2}{18}$	$\frac{8}{18}$	$\frac{15}{18}$	1

Exercise 13.3A   page 364

1 a i 0.302

ii 0.026

iii 0.000

b i 0.004

ii 0.275

iii 0.005

c i 0.000

ii 0.969

iii 0.465

d i 0.913

ii 0.010

iii 0.683

e i 1.000

ii 0.120

iii 0.781

2 0.104

3 $P(X=6) = {}^{10}C_6 \times (p)^6 \times (1-p)^4$

4 a 0.842 (3 s.f.)

b 0.0143 (3 s.f.)

5 a 0.002 (3 s.f.)

b 0.709 (3 s.f.)

6 a 0.250 (3 s.f.)

b 0.474 (3 s.f.)

7 a 0.319 (3 s.f.)

b 0.978 (3 s.f.)

Exam-style questions 13   page 365

1 a

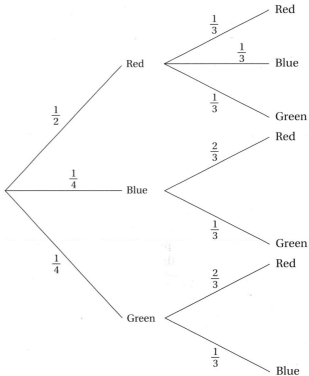

b $\frac{1}{6}$

2 a $\frac{7}{15}$

b $\frac{2}{15}$

c $\frac{3}{100}$

3 a

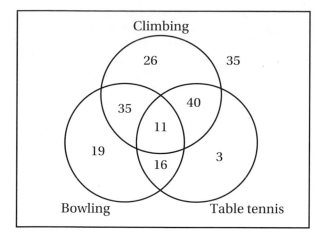

b 0.189 (3 s.f.)

c 0.103 (3 s.f.).

4 a 0.0125

b 0.6625

c i 0.375

ii 0.625

5 a 0.100 928

b 0.899 072

c 0.399 816

d 0.8274

6 a 0.969 (3 s.f.)

b 0.140 (3 s.f.)

c 0.891 (3 s f)

7 a 0.301 (3 s.f.)

b 0.139 (3 s.f.)

c 0.139 (3 s.f.)

8 a Since each person passing her is independent of each other and there is a fixed probability of taking a leaflet with a set number of people passing, the problem can be modelled as a binomial distribution.

b 0.416 (3 d.p.)

9 a 0.129 (3 s.f.)

b 0.000
(3 s.f.) (actual answer is 1×10^{-15})

10 a

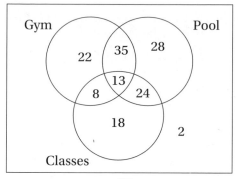

b P(gym only) = $\dfrac{22}{150}$

c P(no gym) = $1 - \dfrac{78}{150} = \dfrac{72}{150}$

d P(gym or pool but no classes) = $\dfrac{22 + 35 + 28}{150}$

$= \dfrac{85}{150}$

e P(given used the gym also used the pool and classes) = $\dfrac{56}{78}$

11

T	1	2	3	4	5
$P(T = t)$	$k(4 - 1)$ $= 3k$	$k(4 - 2)$ $= 2k$	$k(4 - 3)$ $= k$	$k(4 - 4)$ $= 0$	$k(4 - 5)$ $= -1k$
	$\dfrac{3}{5}$	$\dfrac{2}{5}$	$\dfrac{1}{5}$	0	$-\dfrac{1}{5}$

$3k + 2k + k - k = 5k$

$5k = 1$

$k - \dfrac{1}{5}$

14 Statistical sampling and hypothesis testing

Prior knowledge   **page 369**

1 Every member of the population has the same chance of being included in the sample.

2 Every student would be numbered, 1 to 100, and then using random number tables or random numbers on your calculator, you pick numbers until you have a sample of 30.

3 Everyone has the same chance of being selected but the sample doesn't take the proportions of individual strata into account.

Exercise 14.1A page 373

1   a   A sampling unit is an individual member of the population that can be sampled.

     b   A sampling frame is the collection of all of the sampling units.

     c   Sampling distribution means the type of distribution of a sample.

2   a   So everyone in the entire target population has an equal chance of being selected and each selection is independent of each other.

     b   A fair proportion but not all of the cricket club supporters.

     c   The sampling units are the cricket club supporters.

3   a   A census is a sampling method where every member of the population.

     b   Since the sample is taken from a large population of 500, this sample of 5 will be adequately representative.

     c   By giving each individual vacuum cleaner a number, from which a random sample of 5 can be taken.

     d   The sampling units are the vacuum cleaners.

4   a   The population is all teenagers, not just this sample of 242.

     b   The sampling distribution is the probability distribution of this sample of 242.

5   Number all of the staff, 1 to 50. The staff members with the following numbers would form the selection: 13, 15, 41, 46, 10

6   a   The responses obtained may not be representative as these responses may come from school students of certain group(s). Bias could also derive from not obtaining many responses.

     b   Taking a census approach would collect the data from all the students at the school, which will accurately represent the population of these school students.

7   a   Taking a random sample randomly selects a sample of 10 students, whereas a stratified sample approach is designed to be more representative of the whole population as the sample is aimed to represent all types of groups within the population.

     b   Taking a random sample randomly selects a sample of 10 students, whereas the quota sampling approach splits the population into distinct groups where a sample is taken. This quota sampling approach is generally more representative of the population than a random sample.

     c   Systematic sampling involves choosing subjects in a systematic (orderly or logical) way from the target population, whereas opportunity sampling is done as a matter of convenience by selecting people who are available.

Exercise 14.2A page 376

1   a $H_0 : p = 0.55$

     b $H_1 : p < 0.55$

     c   1-tail test

2   a $H_0 : p = 0.68$

     b $H_1 : p > 0.68$

     c   1-tail test

3   a $H_0 : p = 0.65$

     b $H_1 : p > 0.65$

     c   1-tail test

Exercise 14.2B page 385

1    $X \leqslant 26$

     $X \geqslant 38$

2    $X \geqslant 5$

3    $X \leqslant 0$

4    $X \leqslant 4$

5   Mistake 1:

Using a calculator in binomial probability mode, select $n = 50$ and $p = 0.1$.

Select $x = 7$, then select $x = 8$. This should state 'Select $x = 6$, then select $x = 7$.'

$P(X \geqslant 7) = 1 - 0.7702 = 0.2298 > \alpha$

$P(X \geqslant 8) = 1 - 0.8779 = 0.1221 > \alpha$

$P(X \geqslant 9) = 1 - 0.9421 = 0.0579 < \alpha$

So the critical region is $P(X \geqslant 8)$.

Mistake 2:

$P(X \geq 7) = 1 - 0.7702 = 0.2298 > \alpha$

$P(X \geq 8) = 1 - 0.8779 = 0.1221 > \alpha$

$P(X \geq 9) = 1 - 0.9421 = 0.0579 < \alpha$

Incorrect change in significance level, this should be 5%.

$P(X \geq 9) = 1 - 0.9421 = 0.0579 > \alpha$

$P(X \geq 10) = 1 - 0.9755 = 0.0245 < \alpha$

Combined, the critical regions are $P(X \geq 10)$ and $P(X \leq 1)$.

6 The alternate hypothesis states that the probability should be greater than 0.35; this solution looks at the lower tail.

Using $n = 30$, $p = 0.35$, try x values on your calculator systematically to look for the change.

$P(X \geq 16) = 1 - 0.9699 = 0.0301 > \alpha$

$P(X \geq 17) = 1 - 0.9876 = 0.0124 > \alpha$

$P(X \geq 18) = 1 - 0.9955 = 0.0045 < \alpha$

So the critical region is $P(X \geq 18)$.

7 Mistake 1:

The value of x has been inputted as the value in the bracket, not the value less than itself.

$P(X \geq 7) = 1 - 0.7858 = 0.2142 > \alpha$

$P(X \geq 8) = 1 - 0.8982 = 0.1018 > \alpha$

$P(X \geq 9) = 1 - 0.9591 = 0.0409 < \alpha$

So the critical region is $P(X \geq 9)$.

Mistake 2:

Incorrect significance level used, should be 5% as a two-tailed test.

$P(X \leq 3) = 0.2252 > \alpha$

$P(X \leq 2) = 0.0913 > \alpha$

$P(X \leq 1) = 0.0243 < \alpha$

So the critical region is $P(X \leq 1)$.

Combined, the critical regions are $P(X \geq 9)$ and $P(X \leq 1)$.

8 a 0.176 (3 s.f.)

   b 0.238 (3 s.f.)

   c 0.752 (3 s.f.)

   d 0.249 (3 s.f.)

9 a 0.0574 (3 s.f.)

   b 0.930 (3 s.f.)

10 a 0.051 (3 s.f.)   b 0.1672   c 0.761 (3 s.f.)

11 a 0.4   b 0.382 (3 s.f.)

12 the test value is $X = 3$, is not in the critical region (α), there is not enough evidence to suggest that the doctor's claim is correct.

13 James' claim is correct – that he has improved in his game.

14 there is enough evidence to suggest that the proportion of red sweets delivered has changed.

Exam-style questions 14   **page 388**

1 a A – quota   b D – random

   c C – stratified

2 Year 7 = 38 students

   Year 8 = 44 students

   Year 9 = 40 students

   Year 10 = 42 students

   Year 11 = 36 students

3 there is not enough evidence to suggest that Stan's claim is incorrect

4 a The problem can be modelled as a binomial distribution such that $X \sim B(30, 0.40)$, where X is the number of clients who bought a protein shake.

   The test can be tested under the following hypotheses:

   $H_0 : p = 0.40$

   $H_1 : p \neq 0.40$

   The critical region α lies where $\alpha < 0.025$

   there has been no change in the proportion of customers buying a protein shake.

   b The problem can be modelled as a binomial distribution such that $X \sim B(100, 0.35)$, where X is the number of customers who bought fruit.

   The test can be tested under the following hypotheses:

   $H_0 : p = 0.35$

   $H_1 : p > 0.35$

   there is not enough evidence to suggest that the manager's claim is correct

5 a The critical region is the range of values which would lead you to reject the null hypothesis.

  b The significance level is the probability of rejecting the null hypothesis when it is in fact true.

6 The hypotheses are as follows:

$H_0 : p = 0.2$

$H_1 : p \neq 0.2$

there is *not* enough evidence to suggest that Stan's claim was correct.

7 a The population would be all the products of this particular variety.

  b A possible sampling frame would be a list or barcodes of all of those products in a particular time frame.

  c Using a random sample – student's own description of this process, using random number tables or another reliable process.

8 Let p be the probability that a dice lands on a 6.

$H_0: p = \dfrac{1}{6}$

$H_1: p > \dfrac{1}{6}$

Significance level = 10%

Let X be the number of times a 6 was recorded out of the 20 rolls.

$X \sim B(20, \frac{1}{6})$

$P(X \geqslant 6) = 1 - P(X \leqslant 5)$

$1 - 0.8982 = 0.1018$

$0.1018 > 10\%$ so accept null hypothesis (not enough evidence to suggest that the dice is biased towards 6).

9 a $P(X \geqslant 24) = 1 - P(X \leqslant 23)$

   $1 - 0.729 = 0.271$

  b $P(X \geqslant 24) = 1 - P(X \leqslant 23)$

   $1 - 0.993 = 0.007$

  c Let p be the probability that a child improves his or her spellings.

   $H_0: p = 90\%$ (company states 90% improvement)

   $H_1: p < 90\%$ (school claims the improvement rate is not that high)

   Significance level = 5%

  d Let X be the number of children in the class.

   $X \sim B(25, 0.90)$

   $P(X \leqslant 19) = 0.0334$

   $P(X \leqslant 20) = 0.09799$

   $P(X \leqslant 21) = 0.2364$

   $0.2364 > 0.05$

So, if the number of children is 20 or more, the null hypothesis will be rejected and evidence will suggest the alternate hypothesis is true.

15 Kinematics

1 a $v = 58$

   b $s = 1000$

2 $583\,\text{cm}^2$

3 a $T = 6$

   b $T = 4$ or -1.5

4 a $f'(x) = 12x - 7$

   b $\int f(x)\,dx = 2x^3 - \frac{7}{2}x^2 + 8x + c$

1 Overall distance travelled $= 50\,\text{m}$

   Displacement from original position $= -10\,\text{m}$

2 Overall distance $= 65\,\text{m}$

   Displacement $= -15\,\text{m}$

3 a $56\,\text{m}$

   b $40\,\text{m}$

   c $2\,\text{m}\,\text{s}^{-1}$

4 a $22\,\text{m}$

   b $0\,\text{m}$

1 a $550\,\text{m}$

   b $198\,\text{m}$

   c $5\,\text{m}\,\text{s}^{-2}$

   d $25\,\text{m}$

   e $t = 0$ or $8\,\text{s}$

2 a $26\,\text{m}\,\text{s}^{-1}$

   b $0.4\,\text{s}$

3 a $-20\,\text{m}\,\text{s}^{-1}$

   b $7.5\,\text{s}$

4 a $20\,\text{m}\,\text{s}^{-1}$

   b $400\,\text{m}$

5 a $a = -\frac{5}{6}\,\text{m}\,\text{s}^{-2}$

     $t = 60\,\text{s}$

   b Deceleration is constant, etc.

6 a $8\,\text{s}$

   b 2 minutes.

   c $6\,\text{s}$

7 $u = 8\,\text{m}\,\text{s}^{-1}$

   $v = 23\,\text{m}\,\text{s}^{-1}$

8 a $16\,\text{s}$

   b $64\,\text{s}$

1 a $2.8\,\text{m}\,\text{s}^{-2}$

   b $27\,\text{m}\,\text{s}^{-1}$

   c $10\,\text{s}$

2 a $30\,\text{m}$

   b $17.0\,\text{m}\,\text{s}^{-1}$

3 a $75\,\text{m}$ from A's starting point

   b After $5s$

   c For A: $30\,\text{m}\,\text{s}^{-1}$

      For B: $-10\,\text{m}\,\text{s}^{-1}$

4 $3\,\text{s}$, $10\,\text{s}$ and $15\,\text{s}$.

5 $150.9\,\text{m}$

6 a $4\,\text{m}\,\text{s}^{-2}$

   b $56\,\text{m}\,\text{s}^{-1}$

7 $270\,\text{m}$

1 $15.9\,\text{m}$

2 $0.495\,\text{s}$

3 $8.98\,\text{m}$

4 a $30.7\,\text{m}\,\text{s}^{-1}$

   b $3.13\,\text{s}$

5 a $98\,\text{m}$

   b $176.4\,\text{m}$

   c $4\,\text{s}$

   d $58.8\,\text{m}\,\text{s}^{-1}$

   e $8\,\text{s}$

6 a $5.10\,\text{s}$

   b $31.9\,\text{m}$

   c $1.28\,\text{s}$

   d Maximum height is an overestimate because of the effect of air resistance.

   e No difference

7 $1.88\,\text{s}$

8 $\frac{10}{7}\,\text{s}$

Exercise 15.3B page 407

1 a $u = 0\,\text{m s}^{-1}$, $a = g\,\text{m s}^{-2}$, $s = h\,\text{m}$, $t = t\,\text{s}$

$$s = ut + \tfrac{1}{2}at^2$$
$$h = 0 \times t + \tfrac{1}{2}gt^2$$
$$h = \tfrac{1}{2}gt^2$$
$$2h = gt^2$$
$$t^2 = \frac{2h}{g}$$
$$t = \sqrt{\frac{2h}{g}}$$

  b It would take 2 times as long.

2 23.0 m

3 The ball is travelling in the opposite direction to the direction in which it was launched when it hits the ground, so its velocity is negative. Sawda's answer is correct.

Katie's solution is more efficient but she needs to ensure she chooses the correct sign for v.

4 a $u = 70\,\text{m s}^{-1}$, $a = -10\,\text{m s}^{-2}$, $t = T\,\text{s}$, $s > 165\,\text{m}$

$$70T - 5T^2 > 165$$
$$0 > 5T^2 - 70T + 165$$
$$5T^2 - 70T + 165 < 0$$
$$T^2 - 14T + 33 < 0$$

  b 8 seconds.

5 1.52 s

6 42.0 m s^{-1}

7 a Minimum value of U:

    $= 28\,\text{m s}^{-1}$

    Maximum value of U:

    $= 28.4\,\text{m s}^{-1}$

  b 1.02 s

8 Student's own proof

Exercise 15.4A page 410

1

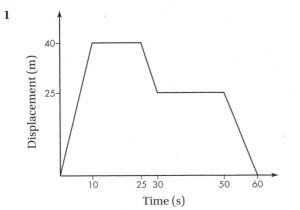

2 a

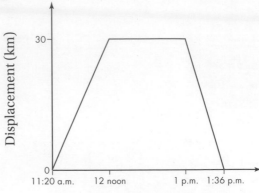

  b 136 minutes after 11:20 a.m. is 1:36 p.m.

3 a and **b**

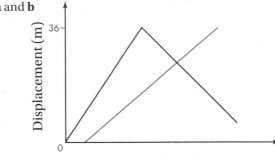

  c $t = 18\,\text{s}$

    $d = 24\,\text{m}$

4 a

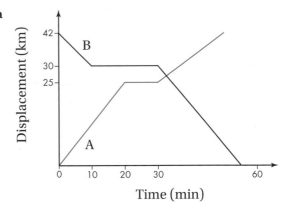

  b 5:17 pm

5 a

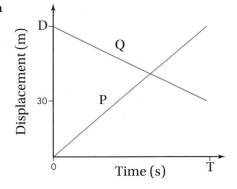

b $T = 40$

c 50 m

Exercise 15.4B   **page 414**

1 a $0.35\,\text{m}\,\text{s}^{-2}$

b 340 m

2 a 54 s

b

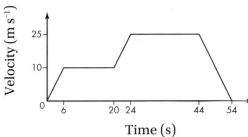

c 865 m

d $1.67\,\text{m}\,\text{s}^{-2}$

3 a

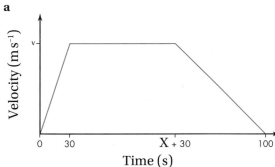

b 20 s

c $45\,\text{m}\,\text{s}^{-1}$

4 a

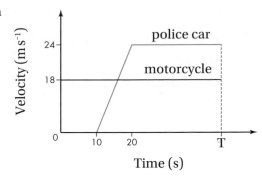

b 60 s

c 1080 m

5 a

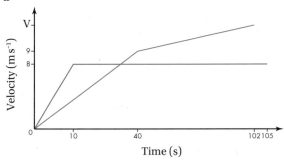

b 800 m

c $11\,\text{m}\,\text{s}^{-1}$

6 Stage 1: $u = 0\,\text{m}\,\text{s}^{-1}$, $t = T_1$ s, $v = V\,\text{m}\,\text{s}^{-1}$

Substitute into $s = \frac{1}{2}(u + v)t$.

$s = \frac{1}{2}(0 + V) \times T_1 = \frac{1}{2}VT_1$

Stage 2: $u = V\,\text{m}\,\text{s}^{-1}$, $t = T_2$ s

There is no acceleration, so substitute into $s = ut$.

$s = VT_2$

Stage 3: $u = V\,\text{m}\,\text{s}^{-1}$, $t = T_3$ s, $v = 0\,\text{m}\,\text{s}^{-1}$

Substitute into $s = \frac{1}{2}(u + v)t$.

$s = \frac{1}{2}(V + 0) \times T_3 = \frac{1}{2}VT_3$

The total distance is given by the sum of these distances.

Total distance $= \frac{1}{2}VT_1 + VT_2 + \frac{1}{2}VT_3$

       $= \frac{1}{2}V(T_1 + 2T_2 + T_3)$

Since $T = T_1 + T_2 + T_3$, total distance $= \frac{1}{2}V(T + T_2)$, which is the area of the trapezium with parallel sides of T and T_2 and a perpendicular height of V.

7 a

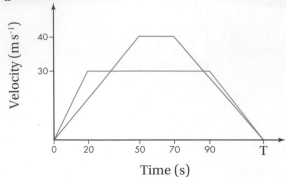

b 130 s

c 3000 m

d First Vehicle: $0.75\,\mathrm{m\,s^{-2}}$

Second Vehicle: $0.67\,\mathrm{m\,s^{-2}}$

8 a

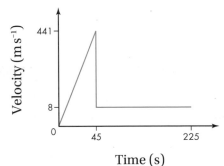

b

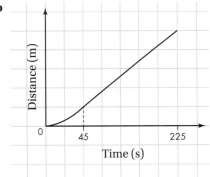

c

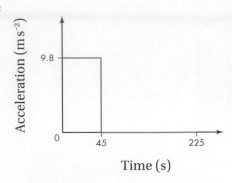

d 11 362.5 m

e Shorter. The skydiver will not descend in freefall at g due to air resistance.

9 a

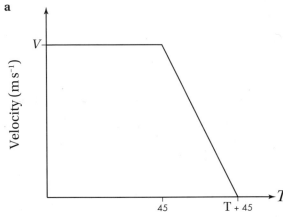

b $a = -1.6 = \dfrac{V}{T}$, so $V = 1.6T$

Area $= \dfrac{1}{2}(T + 45 + 45) \times V = 2300$

$\dfrac{1}{2}(T + 90) \times 1.6T = 2300$

$1.6T(T + 90) = 4600$

$1.6T^2 + 144T - 4600 = 0$

$T^2 + 90T - 2875 = 0$

c $40\,\mathrm{m\,s^{-1}}$

Exercise 15.5A   **page 417**

1 a $-3\,\mathrm{m\,s^{-1}}$

b $a = \dfrac{\mathrm{d}v}{\mathrm{d}t} = -8\,\mathrm{m\,s^{-2}}$

2 a $v = 15t^2 - 14t$

$a = 30t - 14$

b $v = 32\,\mathrm{m\,s^{-1}}$

c $a = 76\,\mathrm{m\,s^{-2}}$

3 a $t = \frac{1}{3}\,\text{s or } 5\,\text{s}$

b when $t < \frac{1}{3}$ s and when $t > 5$ s

c $t > \frac{8}{3}$ s

d $v = -16\,\text{m s}^{-1}$

$r = -20\,\text{m}$

4 $8\,\text{m s}^{-2}$

$-8\,\text{m s}^{-2}$

5 a $6\,\text{s and } 8\,\text{s}$

b $20.7\,\text{m}$

6 a Substitute $t = 4$: $r = 8\sqrt{4} - \frac{64}{4} = 8 \times 2 - 16 = 0$

b $v = 6\,\text{m s}^{-1}$

$a = -2\frac{1}{4}\,\text{m s}^{-2}$

7 a 17

b 12 m

17 m

20 m

c 257 m

7 a

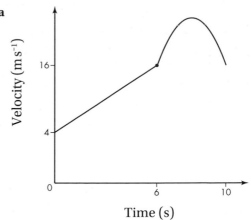

b $20\,\text{m s}^{-1}$

c i $2\,\text{m s}^{-2}$

ii $-2\,\text{m s}^{-2}$

d 45 m

117 m

8 927 m

Exercise 15.5B   **page 420**

1 a $r = \frac{1}{2}t^2 - 2t + 5$

b 21 m

c $10\,\text{m s}^{-1}$

2 a $v = 3t^2 - 20t + 31$

b $r = t^3 - 10t^2 + 31t - 30$

c $2\,\text{s}, 3\,\text{s and } 5\,\text{s}$

d $\frac{2}{3}\,\text{s and } 6\,\text{s}$

3 69 m

4 a $-3\,\text{m s}^{-1}$

b $4\,\text{s and } 6\,\text{s}$

c 228 m

5 a $\frac{46}{3}\,\text{m}$

b $\frac{121}{6}\,\text{m}$

c $\frac{46}{3}\,\text{m}$

6 a i $13.25\,\text{m s}^{-1}$

ii $21.6\,\text{m s}^{-1}$

b 129.3 m

Exam-style questions 15   **page 422**

1 a $4.5\,\text{m s}^{-2}$

b 156 m

2 a $42\,\text{m s}^{-1}$

b 8.57 s

3 a $10\,\text{m s}^{-1}$

b $-0.8\,\text{m s}^{-2}$

4 a

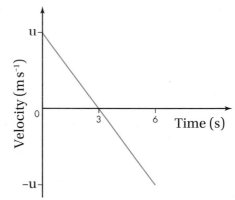

b $29.4\,\text{m s}^{-1}$

c Air resistance

5 6 m

6 a $24\,\text{m s}^{-1}$

b 72 m

c 12 s

7 a

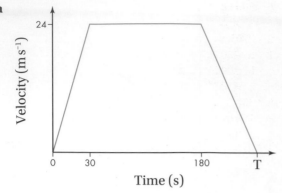

b 225 s

c 75 s.

d 51.2 m s^{-1}

8 8.70 m s^{-1}

9 a −2 m s^{-2}

b 241 m

10 a

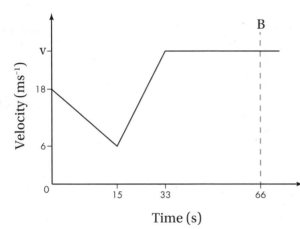

b 66 s

c 23 m s^{-1}

11 a $v = \frac{1}{2}t^2 - 6t + 10$

b 2 and 10 s

c 42.67 m

12 140 m

13 315 m

14 a $r = t^3 + kt^2$

$v = 3t^2 + 2kt$

$3 \times 6^2 + 2k \times 6 = 0$

$k = -9$

$a = 6t + 2k = 6 \times 13 + 2 \times -9 = 60 \, \text{m s}^{-2}$

b $r = t^2(t - 9)$

During first 6 s, $r = [6^2(6 - 9)] - [0] = -108$

From 6 s to 10 s, $r = [10^2(10 - 9)] - [6^2(6 - 9)]$
$= 100 - (-108) = 208$

Total distance $= 108 + 208 = 316$ m

16 Forces

Prior knowledge   page 426

1 a 2.8 m s^{-2}

b 18 m s^{-1}

2 7.62 N

113°

3 $x = 5$

$y = 15$

Exercise 16.1A   page 428

1 a $(-5\mathbf{i} + 13\mathbf{j})$ N

b $(8\mathbf{i} - 15\mathbf{j})$ N

c $(2.9\mathbf{i} - 1.8\mathbf{j})$ N

2 a $q = 3$

$p = -3$

b $p = -3.5$

$q = -4$

c $p = 5$

$q = -2$

3 a i 3.6 N

ii 123.7°

b i 9.4 N

ii −32°

c i 5.7 N

ii −135°

d i 11.4 N

ii 37.9°

4 a 254°

b $a = -8$

5 a 10

b 15

Exercise 16.2A   **page 431**

1 **a** 2.48 N

   b 1300 g

2 **a** $\begin{bmatrix} 4 \\ -6 \end{bmatrix} \text{m s}^{-2}$

   b 1.4 kg

3 Option **D** $(34\,\text{m s}^{-2})$

4 **a** $p = 13$

      $q = -17$

   b 26 N

   c $-22.6°$

   d $(16\mathbf{i} - 4\mathbf{j})\,\text{m s}^{-1}$

5 **a** $\begin{bmatrix} -1.5 \\ 2 \end{bmatrix} \text{m s}^{-2}$

   b $\begin{bmatrix} -1.05 \\ 1.4 \end{bmatrix} \text{N}$

   c 1.75 N

6 Liam's solution was correct. The vectors must be added or subtracted before the magnitude is calculated.

Exercise 16.2B   **page 437**

1 5880 N

2

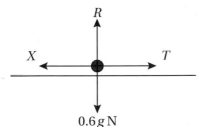

3 **a**

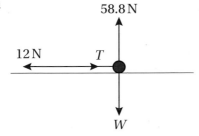

   b 12 N

   c 6 kg

4 600 kg

5 2530 N

6 270 kg

7 **a** 250 N

   b 260 N

   c 330 N

8 35 s

9 155 m

Exercise 16.3A   **page 440**

1 **a** 82.32 N

   b 281.25 N

2 0.029 s

3 **a** 3.53

   b 0.769

   c 13.56

4 **a** 474.75 N

   b 407.25 N

5 0.462 s

6 12.5 N

7 4.2 m

8 **a** 3.98 N

   b 3.32 m

   c $8.07\,\text{m s}^{-1}$

Exercise 16.4A   **page 446**

1 $6\,\text{m s}^{-2}$ 0.9 N

2 $T = \frac{33}{7}mg$

   $X = 11mg\,\text{N}$

3 2250 N

4 1260 N

5 2.89 s

6 1024 m

7 **a** 500 kg

   b 300 N

Exercise 16.4B   **page 448**

1 **a** 671 N

   b 451 N

   c 506 N

2 **a** 200 N

   b 1320 N

3 6.8 s

4 2.22 s

5 **a** **i** 80 kg

       ii $2.2\,\text{m s}^{-2}$

b i 56 kg

ii 72 kg

iii 2.7 m s^{-2}

Exercise 16.5A   page 453

1 a 4.45 m s^{-2}

b 42.8 N

2 a $R = 3mg$

b $\frac{2}{5}g$ m s^{-2}

c $\frac{6}{5}mg$ N

3 a $\frac{1}{7}g$ m s^{-2}

b $\frac{24}{7}g$ N

c 2.744 m.

4 a 6 kg

b 94.1 N

5 1.32 s

6 Let $x > y$

For the x kg particle: $xg - T = xa$

For the y kg particle: $T - yg = ya$

Add.

$xg - yg = (x + y)a$

Hence $a = \frac{g(x-y)}{x+y}$

Substitute for a in $T = ya + yg$.

$T = y\left(\frac{g(x-y)}{x+y}\right) + yg$

$T = \frac{yg(x-y) + yg(x+y)}{x+y}$

$T = \frac{xyg - gy^2 + xyg + gy^2}{x+y}$

$T = \frac{2xyg}{x+y}$ N

7 0.309 s

8 The mass of P is 2.7 kg and the mass of Q is 5.7 kg.

$a = 3.5$ m s^{-2}

Exam-style questions 16   page 455

1 a 106°

b 28 N

2 a 16.8 N

b 0.787 s

3 The student has made the mistake of misreading or assuming that $m > 6$ and therefore that the m kg particle will move downwards and the 6 kg particle upwards. The correct solution is given below.

For the 6 kg particle, resultant force = $(6g - T)$ N

$F = ma$

$6g - T = 6 \times \frac{2}{3}g = 4g$

$T = 2g = 19.6$ N

For the m kg particle, resultant force = $(T - mg)$ N

$F = ma$

$T - mg = m \times \frac{2}{3}g$

$2g - mg = \frac{2}{3}mg$

$2g = \frac{5}{3}mg$

$2 = \frac{5}{3}m$

$m = 1.2$ kg

4 a $b = -\frac{2}{3}$

$a = -\frac{1}{3}$

b Any values for a and b for which $a = -\frac{5}{2}b$, such as $a = 5$ and $b = -2$.

5 8.96 N

6 a 2650 N

b 774 N

c The tension is the same for both parts of the cable

7 a $q = -3$

b 2.83 N

c −70°

8 a 27 m s^{-1}

b 0.87 m

9 a 0.2 m s^{-2}

b 2 m s^{-1}

c 2.9 N

d 4 m

e 0.25 N

10 a 125.5°

b 25 N

c 2.3 s

11 a 1.1 m s^{-2}

b 375 N

c Assumed that mass is concentrated at a single point.

d 500 N

12 a $u = 0$ m s^{-1}, $s = 3.75$ m, $t = 2.5$ s

$s = ut + \frac{1}{2}at^2$

$3.75 = 0 \times 2.5 + \frac{1}{2} \times a \times 2.5^2$

$3.75 = 3.125a$

$a = 1.2$ m s^{-2}

b Resultant force for A = $(2.2g - T)$ N

$F = ma$

$2.2g - T = 2.2 \times 1.2 = 2.64$

Let mass of B = m kg

Resultant force for B = $(T - mg)$ N

$F = ma$

$T - mg = 1.2m$

Add.

$2.2g - mg = 1.2m + 2.64$

$2.2g - 2.64 = m(9.8 + 1.2)$

$18.92 = 11m$

$m = 1.72$ kg

c Assumed that acceleration is the same for both.

d 0.612 s

13 a 0.68 m s^{-2}

b 912 N

c Tensions are the same (equal and opposite).

d Resultant force for truck = (2500 − 500)

= 2000 N

$F = ma$

$2000 = 1600a$

$a = 1.25$ m s^{-2}

$u = 25$ m s^{-1}, $v = 37$ m s^{-1}

When broken, $a = 1.25$ m s^{-2}

$v = u + at$

$t = \dfrac{v - u}{a}$

$t = \dfrac{37 - 25}{1.25} = 9.6$ s

When not broken, $a = 0.68$ m s^{-2}

$v = u + at$

$t = \dfrac{v - u}{a}$

$t = \dfrac{37 - 25}{0.68} = 17.65$ s

Approximately an 8 s difference.

Chapter 1 Algebra and functions 1: Manipulating algebraic expressions   page 460

1 **a** $(+2) \times (+2) \times (+2) = 8$
$(-2) \times (-2) \times (-2) = -8$

  b $\dfrac{1}{4}$

  c $\dfrac{1}{4}y^{-\frac{1}{6}}$

2 **a** **i** 8

   ii 0

  b $x = -1, x = 2, x = 5$

3 **a** $8 - 52x + 90x^2$

  b 7.489

4 $\dfrac{a+\sqrt{b}}{c-\sqrt{b}} = \dfrac{(a+\sqrt{b})}{(c-\sqrt{b})}\dfrac{(c+\sqrt{b})}{(c+\sqrt{b})}$

  $= \dfrac{ac + a\sqrt{b} + c\sqrt{b} + b}{c^2 - b}$

  $= \dfrac{ac + b}{c^2 - b} + \dfrac{\sqrt{b}(a+c)}{c^2 - b}$

5 $\dfrac{9x - 4}{(2x+1)(x-3)}$

6 **a** Binomial expansion of $(1 + px)^{-10} = 1 + 10(px) + \ldots$
so the first three terms are:
$1 + 10px + 45px^{-2}$

  b $p = 3$ and $q = 5$

  c $1.03^{10} = 1.313521899442$

  d Expansion only done to term in x^3.

7 **a** $(3x - 1)(2x + 1)(x - 5)(x - 3)$

  b $x = \dfrac{1}{3}, -\dfrac{1}{2}, 5$ or 3

8 **a** $\left(3 - \sqrt{x}\right)^4 = 81 - 108x^{\frac{1}{2}} + 54x + x^2 - 12x^{\frac{3}{2}}$

  b $\dfrac{1}{16}\left(81 - 108x^{\frac{1}{2}} + 54x + x^2 - 12x^{\frac{3}{2}}\right)$

9 $\dfrac{x+3}{(x+1)(x+2)}$

10 $\dfrac{4x}{1-\sqrt{x}} + \dfrac{3\sqrt{x}+1}{\sqrt{x}} = \dfrac{4x}{(1-\sqrt{x})}\dfrac{(1+\sqrt{x})}{(1+\sqrt{x})} + \dfrac{3\sqrt{x}+1}{\sqrt{x}}\dfrac{\sqrt{x}}{\sqrt{x}}$

  $= \dfrac{4x + 4x\sqrt{x}}{1-x} + \dfrac{3x + \sqrt{x}}{x}$

  $= \dfrac{x(4x + 4x\sqrt{x})}{x(1-x)} + \dfrac{(1-x)(3x+\sqrt{x})}{x(1-x)}$

  $= \dfrac{4x^2 + 4x^2\sqrt{x} + 3x + \sqrt{x} - 3x^2 - x\sqrt{x}}{x(1-x)}$

  $= \dfrac{x+3}{1-x} + \dfrac{\sqrt{x}\left(4x^2 - x + 1\right)}{x(1-x)}$

Chapter 2 Algebra and functions 2: Equations and inequalities   page 460

1 **a** $0 < x < 5$

  b **i** $x = -3 \pm \sqrt{\dfrac{11}{2}}$

   ii $-3 - \sqrt{\dfrac{11}{2}} < x < -3 + \sqrt{\dfrac{11}{2}}$

2 $(4, -3)$ and $(-1, 2)$

3 **a** $x < -1.07$ or $x > 1.67$

  b $-1.07 < x < 1.67$

4 **a** $b < -\sqrt{4ac}$ or $b > \sqrt{4ac}$

  b $b = \pm\sqrt{4ac}$

  c $-\sqrt{4ac} < b < \sqrt{4ac}$

5 **a** $x = \dfrac{7}{4}$ or $x = 2$

  b $\left(\dfrac{15}{8}, \dfrac{-1}{16}\right)$

  c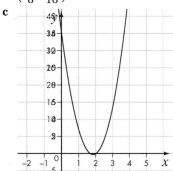

6 **a** No values for b

  b $b = \pm\sqrt{72}$

  c $b < -\sqrt{72}$ or $b > \sqrt{72}$

7 **a** $x = \dfrac{12 \pm \sqrt{14}}{2}, y = \dfrac{2 \mp \sqrt{14}}{2}$

  b As there are two pairs of solutions the linear equation must intersect the circle (twice).

8 $x^2\sqrt{a} + x\sqrt{b} + \sqrt{c} = 0$

  $\sqrt{a}\left(x^2 + \dfrac{\sqrt{b}}{\sqrt{a}}x + \dfrac{\sqrt{c}}{\sqrt{a}}\right) = 0$

  $\sqrt{a}\left(\left(x + \dfrac{\sqrt{b}}{2\sqrt{a}}\right)^2 - \dfrac{b}{4a} + \dfrac{\sqrt{c}}{\sqrt{a}}\right) = 0$

  $\left(x + \dfrac{\sqrt{b}}{2\sqrt{a}}\right)^2 = \dfrac{b}{4a} - \dfrac{4\sqrt{ac}}{4a}$

  $x + \dfrac{\sqrt{b}}{2\sqrt{a}} = \dfrac{\sqrt{b - 4\sqrt{ac}}}{2\sqrt{a}}$

  $x = -\dfrac{\sqrt{b}}{2\sqrt{a}} \pm \dfrac{\sqrt{b - 4\sqrt{ac}}}{2\sqrt{a}}$

  $= \dfrac{-\sqrt{b} \pm \sqrt{b - 4\sqrt{ac}}}{2\sqrt{a}}$

  $= \dfrac{\sqrt{a}(-\sqrt{b} \pm \sqrt{b - 4\sqrt{ac}})}{2a}$

9 a $x = 2, y = \frac{1}{3}, z = -3$

b Number of equations $\geq$ number of variables

10 $5.64 < r \leqslant 5.73$

Chapter 3 Algebra and Functions 3: Sketching curves page 461

1 a Points of intersection are $(-1, 0)$, $(3, 0)$ with the x-axis and $(0, -3)$ with the y-axis.

b $y = \dfrac{7 \pm \sqrt{29}}{2}$

2 a i

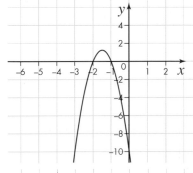

ii

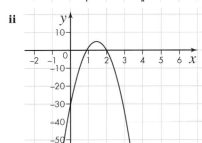

iii

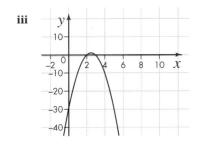

b f$(x + 2)$

3 a $S = 196\pi$ cm^2

b i $S = 64\pi$ cm^2

ii $r = +\sqrt{72}$ cm

4 a $x \propto y^2$ and $z \propto \sqrt[3]{x}$

$x = ky^2$

$z = K\sqrt[3]{x}$

Substituting gives:

$z = K\sqrt[3]{ky^2}$

$z \propto \sqrt[3]{y^2}$ or $z \propto y^{\frac{2}{3}}$

b $z = \dfrac{21(5)^{\frac{2}{3}}}{4}$

5 a f$(x - 2)$

b f$(x) + 5$

c $-$f(x)

6 a

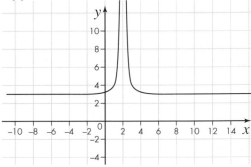

b i f$(2x)$

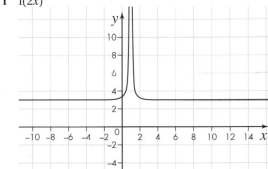

ii f$(x + 3)$

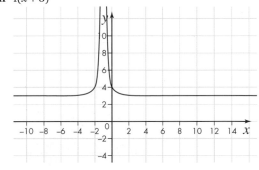

iii $-f(x)$

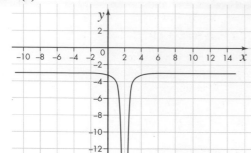

7 a Asymptotes at $x = 0$ and $y = 0$. Four points of intersection.

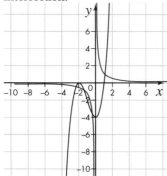

b Asymptotes at $x = 0$ and $y = -8$. Two points of intersection.

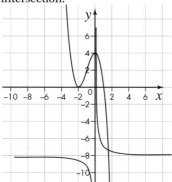

8 a $(x + 1)^3(2x - 1)$

b

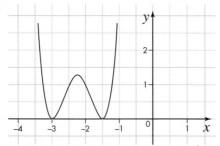

The graph will intercept the y-axis at $(0, 81)$

9 a

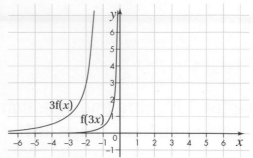

b $f(3x + 2)$ and $f(x + 3)$

c $f(x) = 3^{3x+2}$ and $f(x) = 3^{x+3}$

10 a $f(-x)$

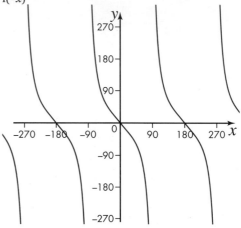

b $-f(x)$

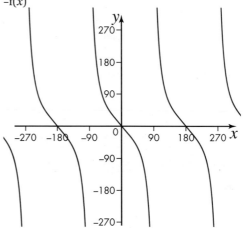

c f(2x)

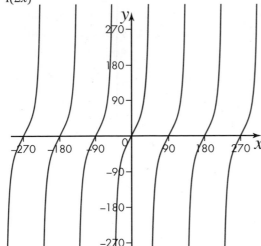

d 2f(x)

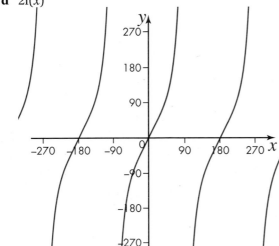

e f(−x) = −f(x) but f(2x) ≠ 2f(x)

Chapter 4 Coordinate geometry 1: Equations of straight lines page 464

1 **a** $\dfrac{3}{2}$

b $-3x + 2y + 8 = 0$

2 $x + 6y - 8 = 0$

3 **a** $y = 1.25x$

b The rate at which the sunflower grows in cm per day, which is 1.25 cm per day.

c 248 days

d A sunflower is unlikely to grow at a steady rate for this long, so a linear relationship is not valid.

4 **a** 3

b $x + 3y - 15 = 0$

5 **a** $2y + x - 4 = 0$

$$y = -\dfrac{1}{2}x + 2$$

Gradient is $\dfrac{-1}{2}$

$$y - 2x + 1 = 0$$

$$y = 2x - 1$$

Gradient is 2.

Product of gradients is $-\dfrac{1}{2} \times 2 = -1$

Therefore the lines are perpendicular.

b (1.2, 1.4)

6 Line joining (−3, 3) and (4, 4):

gradient $= \dfrac{4-3}{4-(-3)} = \dfrac{1}{7}$

Line joining (4, 4) and (3, 6):

gradient $= \dfrac{6-4}{3-4} = \dfrac{2}{-1} = -2$

Line joining (−3, 3) and (3, 6):

gradient $= \dfrac{6-3}{3-(-3)} = \dfrac{3}{6}$

$\dfrac{3}{6} \times (-2) = \dfrac{-6}{6} = -1$ so two of the sides are perpendicular.

7 **a** $y = 18x + 15$

b Plumber B

c 5 hours

8 Gradient of line joining (−3, −2) and (1, −1)

$$= \dfrac{\frac{1}{4} - 1 - (-2)}{1 - (-3)} = \dfrac{1}{4}$$

Gradient of line joining (−3, −2) and $\left(6, \dfrac{1}{4}\right)$

$$= \dfrac{\frac{1}{4} - (-2)}{6 - (-3)} = \dfrac{2\frac{1}{4}}{9} = \dfrac{\frac{9}{4}}{9} = \dfrac{1}{4}$$

As the two lines have a point in common and have the same gradient, they must be in a straight line.

9 $-9x + 4y + 3 = 0$

10 $y = -\dfrac{1}{3}x + \dfrac{17}{3}$

11 $k = 3$

12 **a** The gradient m gives the rate at which the depth changes in cm per minute. c gives the depth of water when x is 0, i.e. it gives the depth of water in the tank when the filling started.

b When the tanks have been filling for 10 minutes

c Tank A

13 **a** $p = 6$

　　b $\dfrac{2}{5}$

　　c $5x + 2y - 7 = 0$

　　d $(6, 3)$

14 $(1, -2)$

15 A $(0, 3)$ and D $(-2, 7)$

Chapter 5 Coordinate geometry 2: Circles　page 465

1 **a** $(4, -10)$

　　b $(x - 4)^2 - 16 + (y + 10)^2 - 100 + 35 = 0$
　　　　$(x - 4)^2 + (y + 10)^2 = 81$

　　Radius $= \sqrt{81} = 9$

2 **a** $\sqrt{(3 - (-9))^2 + ((-8) - (-3))^2} = \sqrt{12^2 + (-5)^2} = 13$

　　b $12x - 5y + 93 = 0$

3 **a** $x + y = 4$

　　b $\left(-\dfrac{1}{2}, \dfrac{9}{2}\right)$

4 **a** $10\,\text{m}$

　　b $x^2 + y^2 - 2x - 14y + 25 = 0$

　　c $4x + 3y = 50$

5 **a** A$(1, 0)$, B$(9, 0)$

　　b 12 units^2

6 **a** $(x + 2)^2 + (y - 5)^2 = 180$

　　b $6\sqrt{15}$

7 **a** Midpoint of YZ $= (5, -2)$
　　　Gradient of YZ $= \dfrac{1}{2}$
　　　Gradient of bisector $= -2$
　　　Equation of bisector is $\qquad y - (-2) = -2(x - 5)$
　　　　　　　　　　　　　　　$y + 2 = -2x + 10$
　　　　　　　　　　　　　　　$2x + y = 8$

　　b $(3, 2)$

　　c $(x - 3)^2 + (y - 2)^2 = 65$

8 **a**
$$(x + 4)^2 + (3x + 1 - 9)^2 = 40$$
$$(x + 4)^2 + (3x - 8)^2 = 40$$
$$x^2 + 8x + 16 + 9x^2 - 48x + 64 = 40$$
$$10x^2 - 40x + 40 = 0$$
$$x^2 - 4x + 4 = 0$$
$$(x - 2)^2 = 0$$

　　Repeated root, so line is a tangent.

　　b $y = 3x + 41$

9 **a** Centre $(-2, 11)$, radius 7

　　b 11

10 $y = \dfrac{1}{2}x - \dfrac{1}{2}$ and $y = \dfrac{1}{2}x + \dfrac{39}{2}$

Chapter 6 Trigonometry page 467

1 $x = 54.7$ or 234.7
　　$x = 125.3$ or 305.3

2 $C = 48.3°$
　　Area $= 109\,\text{m}^2$

3 **a** $\cos\theta = -\sqrt{\dfrac{2}{3}}$ 　　**b** $\tan\theta = -\dfrac{1}{\sqrt{2}}$

4

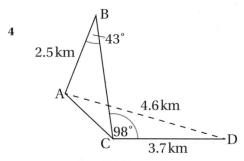

　AD $= 6.30\,\text{km}$ (3 s.f.)

5 **a** Not saying that $\cos x°$ could be 0 on the fifth line (the left-hand side should have been factorised in the third line); the fraction on the last line is upside down.

　　b $x = 90°$ or $270°$ or $41.4°$ or $318.6°$

6 $a = 3$ 　$b = -30$

7 $k = \dfrac{18}{n}$ where $n = 1, 2, 3, \dots$

8 $x = 29$ or 119 or 209 or 299

9 PQ $= 25.6$

10 ABD = 60.4°

CBD = 78.4°

Angle B = 138.8°

Angle D = 86.1°

11 B = 53.1°

AC = 16.3 cm

B = 126.9°

AC = 31.4 cm

12 76.0° or 256.0°

108.4° or 288.4°

13 $x = 5.87$

14 $x = 51.3°$

15 $x = 19.5$ or 160.5°

$x = 228.6$ or 311.4

Chapter 7 Exponentials and logarithms   page 469

1 **a** Let $\log_{n^2} a = x$. Then $a = \left(n^2\right)^x = n^{2x}$ so $\log_n a = 2x$ and $x = \frac{1}{2}\log_n a$.

b Let $\log_a b = y$ and $\log_b c = z$. Then $b = a^y$ and $c = b^z = \left(a^y\right)^z = a^{yz}$.

Hence $\log_a c = yz$ as required.

2 $a = 0.80$

$k = 240$

3 $y = 1.96e^{0.168x}$

4 **a** $1 + 3c$

b $= c - 2$

5 $y = 5$

$x = 0.186$

6 $x = 152.2$

7 **a** The second line, $y = 4e^{2x} + 4e^2$, should be

$y = 4e^{2x} \times 4e^2$

The fourth line is wrong because $4e^{2x} \neq e^{8x}$

b $y = 4e^{2x+2}$; $\ln y = \ln 4 + 2x + 2$; $\ln y = \ln 2^2 + 2x + 2$; $\ln y = 2\ln 2 + 2x + 2$

Divide by 2 and rearrange: $0.5\ln y - \ln 2 - 1 = x$

$x = \ln y^{0.5} - \ln 2 - 1$

$x = \ln\frac{y^{0.5}}{2} - 1$, as required.

8 **a** $d = 1496t^{0.667}$

b $d = 149.6$ km

9 0.431

10 3.046×10^{-3} g/hour.

11 **a** $k = \ln\left(1 + \frac{p}{100}\right)$

b $k = 12\ln\left(1 + \frac{p}{100}\right)$

12 $y = 5500(0.9)^x$

13 292 years.

14 **a** 8.24×1.0186^t.

b The rate of growth could change because of a change in the birth or death rates or because of emigration or immigration.

c 8.24×1.0114^t

15 $x = 1.317$ or -1.3170

Chapter 8 Differentiation page 470

1 **a** The acceleration is 0 initially, and again after $3\frac{1}{3}$ s.

b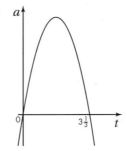

2 $10x^{\frac{3}{2}} - 12x^{\frac{1}{2}}$

3 $(3, -71)$ minimum point.

$(-2, 54)$ maximum point.

4 $(0, 0) -12, (1.5, 0) 16.5, (-4, 0) 44$

5 **a** The second line is incorrect: you cannot divide the numerator and denominator separately in that way.

b $y = \dfrac{x^2 + 3x}{x^{\frac{1}{3}}}$ so $y = x^{\frac{5}{3}} + 3x^{\frac{2}{3}}$

$\dfrac{dy}{dx} = \dfrac{5}{3}x^{\frac{2}{3}} + 2x^{-\frac{1}{3}} = \dfrac{\frac{5}{3}x + 2}{x^{\frac{1}{3}}} = \dfrac{5x + 6}{3\sqrt[3]{x}}$

6 **a** Volume $= x^2 h = 500$ so $h = \dfrac{500}{x^2}$

The surface area is

$A = 2x^2 + 4xh = 2x^2 + 4x\dfrac{500}{x^2} = 2x^2 + \dfrac{2000}{x}$.

b $\dfrac{dA}{dx} = 4x - \dfrac{2000}{x^2}$

At a minimum point $\dfrac{dA}{dx} = 0$

so $4x - \dfrac{2000}{x^2} = 0$ and $x^3 = 500$

Since this is the volume, it implies that h is the same as x so the shape is a cube.

7 $f(x) = \sqrt{2x} = \sqrt{2}x^{\frac{1}{2}}$

$f'(x) = \sqrt{2} \times \frac{1}{2}x^{-\frac{1}{2}} = \frac{1}{\sqrt{2}}x^{-\frac{1}{2}}$

$f''(x) = -\frac{1}{2} \times \frac{1}{\sqrt{2}}x^{-\frac{3}{2}}$

Now $\{f(x)\}^3 = \left\{\sqrt{2}x^{\frac{1}{2}}\right\}^3 = \left(\sqrt{2}\right)^3 x^{\frac{3}{2}}$

so $f''(x)\{f(x)\}^3 = -\frac{1}{2} \times \frac{1}{\sqrt{2}}x^{-\frac{3}{2}} \times \left(\sqrt{2}\right)^3 x^{\frac{3}{2}} = -1$

8 $\frac{dy}{dx} = 3x^2 - 6x - 6$

When $x = 3$, $y = 27 - 27 - 18 + 10 = -8$

and $\frac{dy}{dx} = 27 - 18 - 6 = 3$

The equation of the tangent is $y + 8 = 3(x - 3)$ or
$y = 3x - 17$. On the tangent, when $x = 7$, $y = 4$, which
shows that $(7, 4)$ is on the tangent.

9 $y = 2x^2 + 3x + 1$

$y + \delta y = 2(x + \delta x)^2 + 3(x + \delta x) + 1 =$

$\quad = 2x^2 + 4x\delta x + 2(\delta x)^2 + 3x + 3\delta x + 1$

$\delta y = (y + \delta y) - y = 4x\delta x + 2(\delta x)^2 + 3\delta x$

$\frac{\delta y}{\delta x} = \frac{4x\delta x + 2(\delta x)^2 + 3\delta x}{\delta x} = 4x + 2\delta x + 3$

Then $\frac{dy}{dx} = \lim_{\delta x \to 0} \frac{\delta y}{\delta x} = 4x + 3$.

10 $(111, 0)$

11 $2000\,\text{cm}^3$

12 Volume $= V = \frac{1}{3}\pi r^2 h$

If $h + r = 60$ then $h = 60 - r$ and $V = \frac{1}{3}\pi r^2(60 - r)$.

$V = \frac{1}{3}\pi\left(60r^2 - r^3\right)$

Therefore $\frac{dV}{dr} = \frac{1}{3}\pi\left(120r - 3r^2\right)$.

When the volume is a maximum, $\frac{dV}{dr} = 0$

so $\frac{1}{3}\pi(120r - 3r^2) \Rightarrow 40r - r^2 + 0 \Rightarrow r(40 - r) = 0$ and
so $r = 0$ or 40.

$r = 0$ gives a volume of 0, so $r = 40$ for a maximum
volume and then $h = 60 - r = 20$.

The ratio $r : h = 40 : 20 = 2 : 1$

$\frac{d^2V}{dr^2} = \frac{1}{3}\pi(120 - 6r)$, which is negative when $r = 40$,
confirming that the volume is a maximum.

13 $(\sqrt{a}, 2a)$ and $(-\sqrt{a}, 2a)$.

14 $48\frac{1}{6}$.

15 a $\frac{1}{x+a} - \frac{1}{x} = \frac{x - (x + a)}{(x + a)x} = \frac{x - x - a}{x(x + a)} = -\frac{a}{x(x + a)}$

b If $f(x) = \frac{1}{x}$, then $f(x + \delta x) = \frac{1}{x + \delta x}$ and

$f(x + \delta x) - f(x) = \frac{1}{x + \delta x} - \frac{1}{x} = -\frac{\delta x}{x(x + \delta x)}$

Then $\frac{\delta y}{\delta x} = \frac{f(x + \delta x) - f(x)}{\delta x} = -\frac{1}{x(x + \delta x)}$

and $\frac{dy}{dx} = \lim_{\delta x \to 0} \frac{\delta y}{\delta x} = -\frac{1}{x \times x} = -\frac{1}{x^2}$

Chapter 9 Integration   page 472

1 $\frac{2}{5}x^{\frac{5}{2}} - 4x^{\frac{3}{2}} + 18x^{\frac{1}{2}} + c$

2 5

3 $\frac{4}{3}x^3 + 12x - \frac{9}{x} + c$

4 $20\frac{5}{6}$

5 $x^2 + \frac{1}{x^2} + c$

6 $x = 2$ or -2
   12.8

7 $a = 16$

8 3

9 $2\frac{2}{3}$

10 a Area $= 20 - \frac{100}{a}$ This is always less than 20 but if a is
   large then $\frac{100}{a}$ is small. The larger a is, the closer
   the area is to 20.

b If the equation is $y = \frac{100}{x^3}$ then the area is $= 2 - \frac{100}{a}$
   This is always less than 2. The larger a is, the closer
   the area is to 2.

11 $2\frac{2}{3}$

12 $\frac{2}{5}x^{\frac{5}{2}} + 2x^{\frac{1}{2}} - \frac{2}{3}x^{\frac{3}{2}} - 2x^{-\frac{1}{2}} + c$

13 4.5

14 $11\frac{5}{6}$

15 $y = 3(x + 2)(x - 4)$.

Chapter 10 Vectors   page 474

1 a 13

b 122.2°.

2 a $(15 - 4\sqrt{2})\mathbf{i} + (10 - 4\sqrt{2})\mathbf{j}$

b The vector to return to port is $\begin{bmatrix} 4\sqrt{2} - 12 \\ 4\sqrt{2} \end{bmatrix}$

$\theta = \tan^{-1}\left[\frac{12 - 4\sqrt{2}}{4\sqrt{2}}\right] = 48°$

So the required bearing is $360° - 48° = 312°$.

3 a $a = 15, b = -4$

b 24°

4 a $|\overrightarrow{XY}|^2 = 136$
$|\overrightarrow{WY}|^2 = 442$
$|\overrightarrow{WX}|^2 = 306$
$306 + 136 = 442$
Pythagoras' theorem holds, so the triangle is right angled.

b $\begin{bmatrix} 2 \\ -7 \end{bmatrix}$

c 204

5 $a = -3$

6 a $\overrightarrow{AB} = \begin{bmatrix} 3 \\ 16 \end{bmatrix} \begin{bmatrix} 13 \\ 12 \end{bmatrix} = \begin{bmatrix} -10 \\ 4 \end{bmatrix}$

$\overrightarrow{BC} = \begin{bmatrix} -1 \\ 6 \end{bmatrix} - \begin{bmatrix} 3 \\ 16 \end{bmatrix} = \begin{bmatrix} -4 \\ -10 \end{bmatrix}$

$\overrightarrow{DC} = \begin{bmatrix} -1 \\ 6 \end{bmatrix} - \begin{bmatrix} 9 \\ 2 \end{bmatrix} = \begin{bmatrix} -10 \\ 4 \end{bmatrix}$

$\overrightarrow{AD} = \begin{bmatrix} 9 \\ 2 \end{bmatrix} - \begin{bmatrix} 13 \\ 12 \end{bmatrix} = \begin{bmatrix} -4 \\ -10 \end{bmatrix}$

$\overrightarrow{BC} = \overrightarrow{AD}$ and $\overrightarrow{AB} = \overrightarrow{DC}$, so the quadrilateral has two pairs of parallel and equal sides

Diagonal $\overrightarrow{AC} = \begin{bmatrix} -1 \\ 6 \end{bmatrix} - \begin{bmatrix} 13 \\ 12 \end{bmatrix} = \begin{bmatrix} -14 \\ -6 \end{bmatrix}$

$|\overrightarrow{AC}| = \sqrt{196 + 36} = \sqrt{232}$

$|\overrightarrow{AD}| = \sqrt{16 + 100}$

$|\overrightarrow{DC}| = \sqrt{16 + 100}$

As $116 + 116 = 232$, the triangle ABC satisfies Pythagoras' theorem so there is a right angle at vertex B.
The quadrilateral is a square.

b $f = 10i + 19j$

7 a $a = -2$.

b $b = 14$

8 a $\begin{bmatrix} 8 \\ -14.5 \end{bmatrix}$

b $|\overrightarrow{WX}| = \sqrt{10^2 + 10.5^2} = \sqrt{100 + 110.25}$
$= \sqrt{210.25} = 14.5 = \dfrac{29}{2}$

9 a $\overrightarrow{PQ} = q - p = \begin{bmatrix} 4 \\ 8 \end{bmatrix} - \begin{bmatrix} 3 \\ 14 \end{bmatrix} = \begin{bmatrix} 1 \\ -6 \end{bmatrix}$

$\overrightarrow{PS} = s - p = \begin{bmatrix} 9 \\ 15 \end{bmatrix} - \begin{bmatrix} 3 \\ 14 \end{bmatrix} = \begin{bmatrix} 6 \\ 1 \end{bmatrix}$

$\overrightarrow{PQ}$ makes an angle of $\tan^{-1}(\frac{1}{6})$ with the negative direction of the y-axis.

$\overrightarrow{PS}$ makes an angle of $\tan^{-1}(\frac{1}{6})$ with the positive direction of the x-axis.
They are perpendicular to each other.

b $\overrightarrow{QR} = r - q = \begin{bmatrix} 17 \\ 4 \end{bmatrix} - \begin{bmatrix} 4 \\ 8 \end{bmatrix} = \begin{bmatrix} 13 \\ -4 \end{bmatrix}$

$\overrightarrow{SR} = r - s = \begin{bmatrix} 17 \\ 4 \end{bmatrix} - \begin{bmatrix} 9 \\ 15 \end{bmatrix} = \begin{bmatrix} 8 \\ -11 \end{bmatrix}$

$|\overrightarrow{QR}| = \sqrt{13^2 + 4^2} = \sqrt{169 + 16} = \sqrt{185}$
$|\overrightarrow{SR}| = \sqrt{8^2 + 11^2} = \sqrt{64 + 121} = \sqrt{185}$

c $|\overrightarrow{PQ}| = \sqrt{1^2 + (-6)^2} = \sqrt{37}$
$|\overrightarrow{PS}| = \sqrt{6^2 + 1^2} = \sqrt{37}$
$|\overrightarrow{PQ}| = |\overrightarrow{PS}|$ and $|\overrightarrow{QR}| = |\overrightarrow{SR}|$ therefore the quadrilateral has two pairs of equal adjacent sides, and so is a kite.

10 $\begin{bmatrix} -4 \\ -4 \end{bmatrix}$

11 $\overrightarrow{AB} = b - a = \begin{bmatrix} -2 \\ 3 \end{bmatrix} - \begin{bmatrix} -5 \\ 0 \end{bmatrix} = \begin{bmatrix} 3 \\ 3 \end{bmatrix}$

$\overrightarrow{AC} = c - a = \begin{bmatrix} 1 \\ 6 \end{bmatrix} - \begin{bmatrix} -5 \\ 0 \end{bmatrix} = \begin{bmatrix} 6 \\ 6 \end{bmatrix} = 2\overrightarrow{AB}$

$\overrightarrow{AC}$ is a multiple of $\overrightarrow{AB}$ so they are parallel. Since they have a common point, they are collinear.

12 $\begin{bmatrix} 4 \\ 11 \end{bmatrix}$

13 $\overrightarrow{AB} = b - a$
$\overrightarrow{BC} = c - b$
$\overrightarrow{AC} = c - a$
A, B and C are collinear, therefore $b - a = s(c - b) = t(c - a)$ for some constants s and t.
$$sc - sb = tc - ta$$
$$ta - sb + sc - tc = 0$$
$$ta - sb + (s - t)c = 0$$
$$ka + lb + mc = 0$$
where $k + l + m = t + (-s) + (s - t) = t - s + s - t = 0$

14 In parallelogram OABC let a and c be the position vectors of A and C, respectively.
Let M be the midpoint of OB and let N be the midpoint of AC.

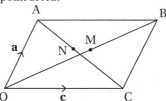

Position vector of N:
$\overrightarrow{AC} = c - a$
$\overrightarrow{AN} = \frac{1}{2}(c - a)$ as N is the midpoint of AC
$\overrightarrow{ON} = a + \frac{1}{2}(c - a) = \frac{1}{2}a + \frac{1}{2}c = \frac{1}{2}(a + c)$

Position vector of M:

$\overrightarrow{OB} = \mathbf{a} + \mathbf{c}$

$\overrightarrow{OM} = \frac{1}{2}(\mathbf{a} + \mathbf{c})$ as M is the midpoint of OB

Since the midpoints have the same position vector, they must be the same point. Hence the diagonals bisect each other.

15 $\overrightarrow{OT} = \frac{1}{5}\mathbf{a} + \frac{1}{5}\mathbf{b}$

Chapter 11 Proof   page 476

1 a Use $2n + 1$ to represent any odd number, and square it.

$(2n+1)^2 = 4n^2 + 4n + 1 = 4(n^2 + n) + 1 = 4m + 1$

(replace $n^2 + n$ with m).

   b Use a counter example, for example 2. Then $2^2 = 4$. This is a multiple of 4, not 1 more than a multiple of 4. (In fact, it can be proved that the square of any even number is a multiple of 4.)

2 Let the consecutive numbers be $(n - 1)$, n and $(n + 1)$.

$n^2 - (n-1)(n+1) = n^2 - (n^2 - 1) = n^2 - n^2 + 1 = 1$

3 Total surface area

= area of the triangular ends + area of the 3 rectangles

$= 2 \times \frac{1}{2}(2x)(x+3) + 4x(x+3) + 4x \times 2x + 4x(2x+3)$

$= 2x^2 + 6x + 4x^2 + 12x + 8x^2 + 8x^2 + 12x$

$= 22x^2 + 30x$

Given that the total surface area is 486 cm

$22x^2 + 30x = 486$

$11x^2 + 15x = 243$

$11x^2 + 15x - 243 = 0$, as required.

4 Complete the square:

$x^2 - 10x + 26 = (x-5)^2 + 1$

which is always positive.

5 Let δx be a small increase in x and let δy be the corresponding small increase in y.

$y = x^3$

$\Rightarrow y + \delta y = (x + \delta x)^3 = x^3 + 3x^2\delta x + 3x(\delta x)^2 + (\delta x)^3$

$\Rightarrow \quad \delta y = 3x^2\delta x + 3x(\delta x)^2 + (\delta x)^3$

$\Rightarrow \quad \frac{\delta y}{\delta x} = 3x^2 + 3x\delta x + (\delta x)^2$

and $\quad \frac{dy}{dx} = \lim_{\delta x \to 0} \frac{\delta y}{\delta x} = 3x^2$

6 Complete the square:

$y = 2x^2 - 8x + 28$

$= 2(x^2 - 4x + 14)$

$= 2((x-2)^2 + 10)$

$= 2(x-2)^2 + 20$

As $2(x-2)^2 \geqslant 0$, $2(x-2)^2 + 20 \geqslant 20$, i.e. $y \geqslant 20$

7 Use a sensible notation, such as 1 to indicate going up one step and 2 to indicate going up two steps.

Work systematically.

Taking all 6 steps one at a time:

1, 1, 1, 1, 1, 1

Taking one pair of steps as a 'two' and all the rest one at a time:

2, 1, 1, 1, 1

1, 2, 1, 1, 1

1, 1, 2, 1, 1

1, 1, 1, 2, 1

1, 1, 1, 1, 2

Taking two pairs of steps as a 'two' and all the rest one at a time:

2, 2, 1, 1

2, 1, 2, 1

2, 1, 1, 2

1, 2, 2, 1

1, 2, 1, 2

1, 1, 2, 2

Taking all the steps in twos:

2, 2, 2

This lists all 13 possible ways of going up the stairs one or two steps at a time.

8 $2y - x - 4 = 0$

$2y = x + 4$

$y = \frac{1}{2}x + 2$

So, the gradient of the line is $\frac{1}{2}$.

Gradient of AB $= \frac{y_2 - y_1}{x_2 - x_1} = \frac{1 - (-1)}{2 - (-2)} = \frac{2}{4} = \frac{1}{2}$.

As the line, l, and AB have the same gradient, they are either parallel or the same line.

Check the coordinates of one of the points.

If $(2, 1)$ lies on the line, then it will satisfy the equation of the line.

$2y - x - 4 = 2 \times 1 - 2 - 4 = 2 - 2 - 4 = -4$

As the coordinates of B do not satisfy the equation, AB is not the same line as l, and the two lines, AB and l are parallel.

9 Use $\sin^2\theta + \cos^2\theta = 1$.

$\frac{\cos^2\theta}{1 - \sin\theta} = \frac{1 - \sin^2\theta}{1 - \sin\theta} = \frac{(1 - \sin\theta)(1 + \sin\theta)}{1 - \sin\theta} = 1 + \sin\theta$

10 First find the centre and radius of the circle. To do this, complete the square twice.

$$x^2 - 2x + y^2 - 4y = 20$$
$$(x-1)^2 - 1 + (y-2)^2 - 4 = 20$$
$$(x-1)^2 + (y-2)^2 = 25$$

Circle centre is $(1, 2)$ and radius is 5 units.

If $(5, 6)$ is more than 5 units from the circle centre, then it lies outside the circle. Let d be the distance from the centre of the circle to $(5, 6)$.

$$d = \sqrt{(6-2)^2 + (5-1)^2} = \sqrt{16+16} = \sqrt{32}$$

$\sqrt{32} > 5$ therefore $(5, 6)$ lies outside the circle.

11 **a** Let $n = 2k + 1$.

Then $n^2 - 1 = (2k+1)^2 - 1 = 4k^2 + 4k + 1 - 1$
$$= 4k^2 + 4k = 4k(k+1)$$

As k and $k + 1$ are consecutive numbers, one of them must be even, i.e. a multiple of 2.

$4k(k+1)$ is a product of 4 and an even number so it must be a multiple of 8.

b Counter example, $n = 2$: $n^2 - 1 = 4 - 1 = 3$.
This is not a multiple of 8.

12 $m^3 - m = m(m^2 - 1) = m(m+1)(m-1)$
$$= (m-1)m(m+1)$$

This is the product of three consecutive numbers.

At least one of these numbers must be even and one must be a multiple of 3.

The product of an even number with a multiple of 3 is a multiple of 6, i.e. $(m-1)m(m+1)$ is a multiple of 6. Therefore $m^3 - m$ must also be a multiple of 6.

13 Let $a = x$, $b = x+1$, $c = x+2$, $d = x+3$.

$$bd - ac = (x+1)(x+3) - x(x+2)$$
$$= x^2 + 4x + 3 - (x^2 + 2x)$$
$$= x^2 + 4x + 3 - x^2 - 2x$$
$$= 2x + 3$$

$$b + c = x + 1 + x + 2$$
$$= 2x + 3$$

$bd - ac = b + c$, as required.

14 Let $a = x$, $b = x+1$, $c = x+2$, $d = x+3$

$$bc - ad = (x+1)(x+2) - x(x+3)$$
$$= (x^2 + 3x + 2) - (x^2 + 3x)$$
$$= x^2 + 3x + 2 - x^2 - 3x$$
$$= 2$$

$bc - ad = 2$, as required.

15 To prove the triangle is right angled, you need to show that the sides satisfy Pythagoras' theorem.

$$(2x)^2 = 4x^2$$
$$(x^2+1)^2 = x^4 + 2x^2 + 1$$
$$(x^2-1)^2 = x^4 - 2x^2 + 1$$

$$(x^4 - 2x^2 + 1) + (4x^2) = x^4 + 2x^2 + 1$$
$$(x^2-1)^2 + (2x)^2 = (x^2+1)^2$$

Pythagoras' theorem is satisfied and the hypotenuse is the side with length $x^2 + 1$.

16 Let the numbers be x and $x + 3$.

The difference between the squares of these numbers is $(x + 3)^2 - x^2 = x^2 + 6x + 9 - x^2$
$$= 6x + 9$$
$$= 3(2x + 3)$$

$3(2x + 3)$ is a multiple of 3, so the difference between the squares of the numbers is also a multiple of 3, as required.

Chapter 12 Data presentation and interpretation   page 477

1 $x = £110\,000$

2 **a** 36.3

b 6.47

c 37.5 minutes

3 **a** 5.5

b 1

c

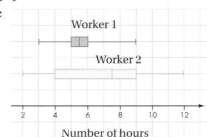

d Worker 1 has a lower median of 5.5 compared with 7.5 for worker 2. Worker 1 has an IQR of 1 compared with 5 for worker 2. Worker 1 is better to employ as they are faster and more consistent.

e

	Worker 1	Worker 2
LQ – 1.5 × IQR	5 – 1.5 × 1 = 3.5 so 3 is an outlier	4 – 1.5 × 5 = –3.5 no outliers
UQ + 1.5 × IQR	6 + 1.5 × 1 = 7.5 so 9 is an outlier	9 + 1.5 × 5 = 16.5 no outliers

4 **a** $\bar{x} = 12.8\,\text{g}$
$\sigma = 4.79\,\text{g}$

b Brand R has a more consistent mass and the mean is greater – this would be the better brand to buy.

c $\bar{x} = 13.06\,\text{g}$
$\sigma = 4.12\,\text{g}$

5 a Census – all members of the population are used.

b Sample – a selection of the population.

c Select 45 tyres using a random number generator based on the unique codes on the tyre wall, forming a random sample.

d A sample is quick, easy and cheap. A population is time consuming, expensive and unrealistic.

6 a $\bar{x} = 2.82\,\text{kg}$

b $\sigma = 0.801\,\text{kg}$

c $2.88\,\text{kg}$

7 a $h_2 = 13.5\,\text{cm}$
$w_2 = 1\,\text{cm}$

b $23.4\,\text{s}$

c $\bar{x} = 22.48\,\text{s}$
$\sigma = 8.51\,\text{s}$

8 a width $10\,\text{cm}$
height $4\,\text{cm}$

b $11.75\,\text{cm}$.

c $5.5\,\text{cm}$

d mean $\approx 11.7\,\text{cm}$
standard deviation $= 3.53\,\text{cm}$

9 a 18 tomatoes per plant

b $Q_3 = 30$, $Q_1 = 10$.

c

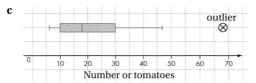

10 a $\bar{x} = 496.78\,\text{g}$
$\sigma^2 = 126.13\,\text{g}$

b $\bar{x} = 489.73\,\text{g}$
$\sigma^2 = 68.57\,\text{g}$

c Sample 1 has a higher mean than sample 2.
Sample 1 has a larger variance than sample 2.
Although sample 1 has a higher mean, there is less consistency than in sample 2.

11 a

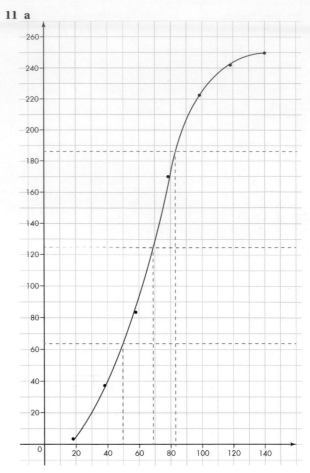

b Median about 70 marks
90th percentile about 107 marks

c 70% of 250 is 175, this is about 83 marks.
The pass mark should be 83 marks.

12 a

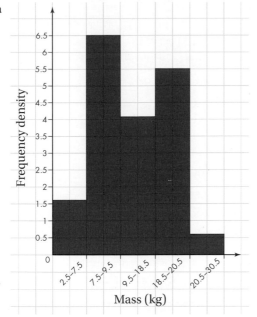

Mass (kg)	3–7	8–9	10–18	19–20	21–30
Frequency	8	13	37	11	6
Frequency density	1.6	6.5	4.1	5.5	0.6

b Median = 38th value

$$9.5 + \frac{17}{37} \times 10 = 14.1 \, \text{kg}$$

13 $\dfrac{(v + 250)}{0.05} = 1.8(t + 75) + 72.1$

$v + 250 = 0.05\,(1.8t + 135) + 0.05\,(72.1)$

$v + 250 = 0.09t + 6.75 + 3.605$

$v + 250 = 0.09t + 10.355$

$v = 0.09t - 239.645$

Chapter 13 Probability and statistical distributions   page 479

1 a $45 + 25 + 9 = 69$
$100 - 69 = 21$
$21 + 45 = 66\%$ eat burger only and 45% eat chips only so 66% eat burger or chips but not both.

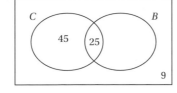

b $25 + 21 = 46\%$
$\dfrac{25}{46} = 54.3\%$

2 a

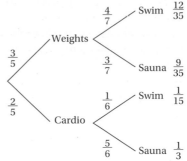

$\dfrac{12}{35} + \dfrac{1}{3} = \dfrac{71}{105}$

b $\dfrac{2}{5}$

c $P(\text{sauna}) = \dfrac{9}{35} + \dfrac{1}{3} = \dfrac{62}{105}$

$P(\text{weights} \mid \text{sauna}) = \dfrac{\frac{9}{35}}{\frac{62}{105}} = \dfrac{27}{62}$

3 a

X	2	4	6	8	10	12	14	16	18
$P(X = x)$	$\frac{1}{64}$	$\frac{4}{64}$	$\frac{10}{64}$	$\frac{14}{64}$	$\frac{15}{64}$	$\frac{10}{64}$	$\frac{7}{64}$	$\frac{2}{64}$	$\frac{1}{64}$

b $P(X < 10) = \dfrac{29}{64}$

4 a

X	1	2	3	4	5	6
$P(X = x)$	$\frac{1}{21}$	$\frac{2}{21}$	$\frac{3}{21}$	$\frac{4}{21}$	$\frac{5}{21}$	$\frac{6}{21}$
$F(x)$	$\frac{1}{21}$	$\frac{3}{21}$	$\frac{6}{21}$	$\frac{10}{21}$	$\frac{15}{21}$	$\frac{21}{21}$

b $P(X \leq 4) = \dfrac{10}{21}$

c $P(2 < X \leq 3) = \dfrac{3}{21}$

5 a $P(X < 3) = 0.05126$

b $P(X > 5) = 1 - P(X \leq 5)$
$= 1 - 0.4353 = 0.5647$

c $P(4) = 0.1334$

d $50 \times 12 = 600$ packets
$600 \times 0.12 = 72$

6 a

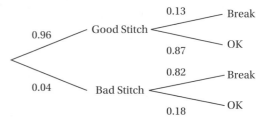

b $0.96 \times 0.13 + 0.04 \times 0.18 = 0.132$

c $0.96 \times 0.87 \times 0.89 = 0.743$

d $0.96 \times 0.87 \times 0.11 + 0.96 \times 0.13 \times 0.89$
$+ 0.04 \times 0.18 \times 0.89 = 0.209$

7 a

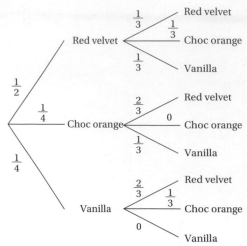

b $\frac{1}{4} \times \frac{1}{3} + \frac{1}{4} \times \frac{1}{3} = \frac{1}{6}$

8 a

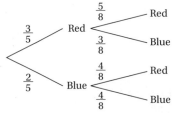

b
$$P(\text{red}) = P(\text{red}_A \text{ and red}_B) + P(\text{blue}_A \text{ and red}_B)$$
$$= \frac{3}{5} \times \frac{5}{8} + \frac{2}{5} \times \frac{4}{8} = \frac{23}{40}$$
$$P(\text{red}_A|\text{red}) = \frac{P(\text{red}_A \text{ and red}_B)}{P(\text{red})}$$
$$= \frac{\frac{3}{5} \times \frac{5}{8}}{\frac{23}{40}} = \frac{15}{23}$$

9 $\Sigma P(X = x) = 1 \Rightarrow 3a + 0.7 = 1 \Rightarrow a = 0.1$

10 a $P(J) = 0.6$

b $P(J \cap M) = 0.19$

c

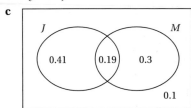

d $P(J' \cap M') = 0.1$

e If J and M are independent, $P(J \cap M) = P(J) \times P(M)$
0.19 ≠ 0.6 × 0.49 so not independent.

11 a $P(X < 4) = \frac{3}{15}$

b $P(2 < X \le 11) = \frac{9}{15} = \frac{3}{5}$

c $P(X > 13) = \frac{2}{15}$

12

X	0	1	2	3	4	5
P(X = x)	0.01	0.09	0.13	0.5	0.15	0.12

13 a $X \sim B(7, 0.8)$

b $P(X \ge 5) = 1 - P(X \le 4) = 1 - 0.148 = 0.852$

14 a $P(\text{Exactly } 1) = 0.0148$

b $P(<3 \text{ eggs}) = 0.0733$

c $P(\ge 5 \text{ eggs}) = 1 - 0.41 = 0.59$

15 a $P(3) = 0.0781$

b $6 \times P(4) = 0.0938$

16 $X \sim B(10, \frac{1}{3})$

a $P(\text{exactly } 2) = 0.195$

b $P(\le 2) = P(0) + P(1) + P(2) = 0.299$

c $np = 10 \times \frac{1}{3} = 3\frac{1}{3}$

d $P(3) = 0.26$, $P(4) = 0.228$ therefore 3 most likely

Chapter 14 Statistical sampling and hypothesis testing   page 482

1 a $H_0: p = \frac{1}{5}$

$H_1: p > \frac{1}{5}$ where p is the probability of selecting the favourite coffee brand.

b $P(X \ge 5) = 1 - P(X < 5)$
$$= 1 - 0.9672$$
$$= 0.0328$$

As this is below 5%, evidence suggests that the student can tell the difference, so reject the null hypothesis.

2 $H_0: p = 0.57$, $H_1: p < 0.57$
where p is the probability of good value for money.
$P(X \le 15) = 0.0102$
This is above 1% so evidence suggests that students still believe they are getting good value for money, so accept the null hypothesis.

3 $H_0: p = 0.16$, $H_1: p \ne 0.16$
where p is the probability of believing in Santa Claus.
Below 5%, $P(X \le 9)$
Above 5%, $P(X \ge 22)$

4 $H_0: p = \frac{38}{60}$

$H_1: p > \frac{38}{60}$

where p is the probability of saying the same word in an hour.
$P(X \ge 43) = 1 - P(X < 42)$
$$= 1 - 0.88728$$
$$= 0.11272$$

As this is above 5%, evidence suggests that the teacher has not increased his usage of the word, so accept the null hypothesis.

5 H_0: $p = 0.55$
H_1: $p < 0.55$
where p is the probability of her being backed.
$P(X \leq 16) = 0.1019$
As this is above 10%, evidence suggests that the student has not over estimated, so accept the null hypothesis.

6 H_0: $p = 0.2$
H_1: $p \neq 0.2$
Below 5%, $P(X \leq 2) = 0.019$
Above 5%, $P(X \geq 12) = 0.0344$

7 H_0: $p = 0.63$
H_1: $p > 0.63$
where p is the probability of passing first time.
$P(X \geq 15) = 1 - P(X < 15)$
$\qquad = 1 - 0.9439$
$\qquad = 0.0561$
As this is above 5%, evidence suggests that the app hasn't improved, so accept the null hypothesis.

8 H_0: $p = 0.18$
H_1: $p < 0.18$
where p is the probability of mis-shapen lenses.
$P(X \leq 3) = 0.0542$
As this is above 5%, evidence suggests that the service has not improved the machine, so accept the null hypothesis.

9 a H_0: $p = 0.75$
H_1: $p < 0.75$
where p is the probability of increasing lung capacity.

b $P(X \leq 14) = 0.00521$
As this is below 5%, evidence suggests that the programme does improve lung capacity, so reject the null hypothesis. Because $\frac{14}{26}$ is about 53% you do not need to look higher than 75%.

c That the people in the sample are only doing this programme, and are not undertaking any other form of exercise or taking performance-enhancing medication.

10 a H_0: $p = 0.23$
H_1: $p < 0.23$
where p is the probability of giving up takeaways.

b Below 10%, $P(X \leq 7) = 0.08444$

c $P(X \leq 6) = 0.0397$
As this is below 5% in the lower tail, evidence suggests that young people are trying to give up, so reject the null hypothesis.
Alternatively: 6 is below 7, which is the critical value, therefore you know that this must be in the critical region.

Chapter 15 Kinematics   page 483

1 7 s

2 $T = 10$

3 71 m.

4 a 12.9 s

b When $T = 6 + 4\sqrt{3}$, $s = 0.4 \times (6 + 4\sqrt{3})^2 = 66.9$ m
$100 - 66.9 = 33.1$ m

5 a $U = 28$

b $32.5\,\text{m s}^{-1}$

c $6.47\,\text{s}$

6 a

b $T = 7\,\text{s}$

c $384\,\text{m}$

7 a For PQ, $t = 4\,\text{s}$, $s = 96\,\text{m}$
$\quad s = ut + \frac{1}{2}at^2$
$\quad 96 = 4u + 8a$ ⬦①
For PS, $t = 18\,\text{s}$, $s = 243\,\text{m}$
$\quad 243 = 18u + 162a$
$\quad 27 = 2u + 18a$
$\quad 54 = 4u + 36a$ ⬦②
① − ②
$\quad -42 = 28a$
$\quad a = -1.5\,\text{m s}^{-2}$
$\quad u = 27\,\text{m s}^{-1}$
$\quad v = u + at$
$\quad v = 27 - 1.5 \times 18 = 0\,\text{m s}^{-1}$

b 20 seconds

8 a $r = 4t^3 - 21t^2 + 30t + 3$

b $\qquad v = 12t^2 - 42t + 30 = 0$
$\quad 2t^2 - 7t + 5 = 0$
$\quad (2t - 5)(t - 1) = 0$
$\qquad t = 1\,\text{s or } \frac{5}{2}\,\text{s}$

c When $t = 2$, $r = 4(2)^3 - 21(2)^2 + 30(2) + 3 = 11$
When $t = 2.5$, $r = 4(2.5)^3 - 21(2.5)^2 + 30(2.5) + 3 = 9.25$
When $t = 3$, $r = 4(3)^3 - 21(3)^2 + 30(3) + 3 = 12$
Distance $= (11 - 9.25) + (12 - 9.25) = 4\frac{1}{2}$ m

9 a $24\,\text{m s}^{-1} = 86\,400$ metres per hour $= 86.4\,\text{km h}^{-1}$
≈ 54 mph, which is faster than the speed limit of 40 mph.

b $T = 60$

c $V = 32$

10 a i $v\,(\mathrm{m\,s^{-1}})$

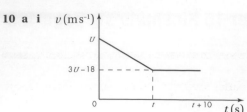

ii $a\,(\mathrm{m\,s^{-2}})$

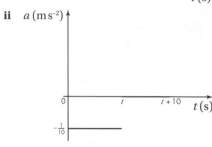

b $U = 6.25$

11 a $k = \frac{3}{4}$

b $48\,\mathrm{m\,s^{-1}}$

c $\frac{1}{2} \times 48 \times (X - 8) = 256$

$$X - 8 = \frac{32}{3}$$

$$X = \frac{56}{3}$$

Chapter 16 Forces   page 485

1 $10\,\mathrm{m\,s^{-1}}$

2 a $74.3\,\mathrm{m}$

b $\mathbf{F} = m\mathbf{a}$

$\mathbf{F} = 7(2\mathbf{i} - 1.5\mathbf{j}) = (14\mathbf{i} - 10.5\mathbf{j})\,\mathrm{N}$

Magnitude of $\mathbf{F} = \sqrt{14^2 + (-10.5)^2} = 17.5\,\mathrm{N}$

c $1.5\,\mathrm{s}$

3 a 3

b -9

c -14

d $66.8°$

4 a $9.7\,\mathrm{N}$

b $132°$

c $2220\,\mathrm{m}$

5 a Tensions are equal.

b $0.6\,\mathrm{s}$

6 a $1.43\,\mathrm{m\,s^{-1}}$

b P does not reach the pulley.

7 350

8 a $7\,\mathrm{kg}$

b $6.34\,\mathrm{m\,s^{-2}}$

c $2.41\,\mathrm{s}$

9 a $1.48\,\mathrm{m}$

b $2.60\,\mathrm{s}$

10 $60.8\,\mathrm{m}$

11 For the system $Mg - F = 2Ma$

$$F = Mg - 2Ma$$

For a rough table $F > 0$

Therefore $Mg - 2Ma > 0$

$$2a < g$$

$$a < \tfrac{1}{2}g$$

12 a Let the driving force exerted by the car be $P\,\mathrm{N}$.
The resistance to motion on the car is F so for the caravan it is $1.5F$.

Using $s = ut + \frac{1}{2}at^2$ with $u = 0$, $s = 64$, $t = 8$, we get

$$a = \frac{2s}{t^2}$$

$$= \frac{2 \times 64}{64}$$

$$a = \mathrm{m\,s^{-2}}$$

Resolving forces on the car and caravan together:

$$P - F - 1.5\,F = 2.5\,Ma$$

Hence $P = 5M + \frac{5}{2}\,F$

For the car when the caravan becomes uncoupled:

$$P - F = Ma$$

Therefore $(5M + \frac{5}{2}\,F) - F = Ma$

$$\frac{3}{2}\,F = Ma - 5M$$

$$F = \frac{2M(a - 5)}{3}$$

b $T = \dfrac{M(5a - 2)}{2}$

FORMULAE

Formulae that students are expected to know for AS-levels Mathematics are given below and should be learnt by heart.

Pure Mathematics

Quadratic Equations

$ax^2 + bx + c = 0$ has roots $\dfrac{-b \pm \sqrt{b^2 - 4ac}}{2a}$

Laws of Indices

$a^x a^y \equiv a^{x+y}$

$a^2 \div a^y \equiv a^{x-y}$

$(a^x)^y \equiv a^{xy}$

Laws of Logarithms

$x = a^n \Leftrightarrow n = \log_a x$ for $a > 0$ and $x > 0$

$\log_a x + \log_a y \equiv \log_a(xy)$

$\log_a x - \log_a y \equiv \log_a\left(\dfrac{x}{y}\right)$

$k\log_a x \equiv \log_a(x^k)$

Coordinate Geometry

A straight line graph, gradient m passing through (x_1, y_1) has equation $y - y_1 = m(x - x_1)$

Straight lines with gradients m_1 and m_2 are perpendicular when $m_1 m_2 = -1$

Trigonometry

In the triangle ABC

Sine rule: $\dfrac{a}{\sin A} = \dfrac{b}{\sin B} = \dfrac{c}{\sin C}$

Cosine rule: $a^2 = b^2 + c^2 - 2bc\cos A$

Area $= \dfrac{1}{2}ab\sin C$

$\cos^2 A + \sin^2 A \equiv 1$

Mensuration

Circumference and area of circle, radius r and diameter d:

$$C = 2\pi r = \pi d \qquad A = \pi r^2$$

Pythagoras' theorem:

In any right-angled triangle where a, b and c are the lengths of the sides and c is the hypotenuse, $c^2 = a^2 + b^2$

Area of a trapezium $= \dfrac{1}{2}(a + b)h$, where a and b are the lengths of the parallel sides and h is their perpendicular separation.

Volume of a prism = area of cross section × length

Calculus and Differential Equations

Differentiation

Function	Derivative
x^n	nx^{n-1}
e^{kx}	ke^{kx}
$f(x) + g(x)$	$f'(x) + g'(x)$

Integration

Function	Integral
x^n	$\dfrac{1}{n+1}x^{n+1} + c, \ n \neq -1$
$f'(x) + g'(x)$	$f(x) + g(x) + c$

Area under a curve $= \displaystyle\int_a^b y\,dx\ (y \geqslant 0)$

Vectors

$|x\mathbf{i} + y\mathbf{j} + z\mathbf{k}| = \sqrt{(x^2 + y^2 + z^2)}$

Statistics

The mean of a set of data: $\bar{x} = \dfrac{\sum x}{n} = \dfrac{\sum fx}{\sum f}$

Mechanics

Forces and Equilibrium

Weight = mass $\times g$

Friction: $F \leqslant \mu R$

Newton's second law in the form: $F = ma$

Kinematics

For motion in a straight line with variable acceleration:

$$v = \frac{dr}{dt} \qquad a = \frac{dv}{dt} = \frac{d^2r}{dt^2}$$

$$r = \int v \, dt \qquad v = \int a \, dt$$

GLOSSARY

1-tail test   A test of the region under one of the tails (sides) of a statistical distribution.

2-tail test   A test of the region under both of the tails (sides) of a statistical distribution.

acceleration   The rate of change of velocity with respect to time.

algebraic fraction   A fraction that includes algebraic terms.

alternative hypothesis   A hypothesis which the researcher tries to prove.

asymptote   A straight line that a curve approaches but never meets.

base of a logarithm   In the expression $\log_a x$, the base of the logarithm is a.

bearing   An angle measured clockwise from north.

bias   The tendency of a statistical measurement to over- or under-estimate the value in a population.

binomial   An expression containing exactly two terms, for example $(x + 2)$.

binomial distribution   The number of successes, x, in n repeated trials with a constant probability, p.

bivariate data   Data that has two variables.

box and whisker plot   A type of diagram used to display patterns of quantitative data, splitting the data set into quartiles.

categorical data   Data that takes on values which are names or labels, such as sweet colour or dog breed.

census   A study that obtains data from every member of a population – often not practical, because of the cost and/or time required.

centre   The point inside a circle equidistant (an equal distance) from all points on the circumference.

chord   A line joining two points on the circumference of a circle.

circle   A set of points equidistant (an equal distance) from a single point.

circumference   The perimeter of a circle.

cluster sampling   Technique of dividing the population into unordered groups or clusters and then selecting individual subjects from each cluster by either simple random or systematic random sampling.

coefficient   A number written in front of a variable in an algebraic term; for example, in $8x$, 8 is the coefficient of x.

collinear   Three or more points which lie on a single straight line are described as collinear.

column vector   A vector written in the form $\begin{bmatrix} a \\ b \end{bmatrix}$ where a and b are the coefficients of $\mathbf{i}$ and $\mathbf{j}$.

combination   The number of choices of r objects from a total of n objects, when the order in which the objects are arranged does not matter. This is denoted by ${}^nC_r$ and given by the formula ${}^nC_r = \begin{pmatrix} n \\ r \end{pmatrix} = \dfrac{n!}{(n-r)!r!}$

complement   The probability of an event *not* happening.

completing the square   Rewriting expressions of the form $x^2 + bx + c$ as $\left(x + \frac{b}{2}\right)^2 - \left(\frac{b}{2}\right)^2 + c$ and $ax^2 + bx + c$ as $a\left(x + \frac{b}{2a}\right)^2 - \frac{b^2}{4a^2} + \frac{c}{a}$

congruent   Two shapes are said to be congruent if they are identical (all their sides and angles are the same and in corresponding positions).

consecutive   One immediately after another.

constant of integration   An indefinite integral will include an arbitrary constant c, known as the constant of integration.

constant of proportionality   A multiplier linking two variables.

continuous data   Variables that can take on any value between their minimum and maximum limits.

continuous random variable   A random variable that can take any value in the given interval.

correlation   A positive correlation means that if one variable gets bigger, the other variable tends to get bigger. A negative correlation means that if one variable gets bigger, the other variable tends to get smaller.

cosine   The cosine of a (or $\cos a$) is the x-coordinate of a point on the circle of radius 1 unit. The variable a can take any value.

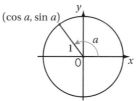

critical region   The area of the sampling distribution of a statistic that will lead to the rejection of the hypothesis tested when that hypothesis is true.

cubic   An expression or similar containing terms in one variable where all the indices are greater than or equal to zero and the highest index is 3.

cumulative distribution function   A running total of probabilities, which gives the probability that the random variable X will be less than or equal to a particular value. It is defined as $F(x_n) = P(X \leqslant x_n)$.

cumulative frequency   In a data set, the cumulative frequency for a value x is the total number of scores that are less than or equal to x.

data   Information collected to find out answers to problems.

data cleaning   A procedure to identify and remove invalid data points from a data set.

deceleration   A negative acceleration, i.e. where the final velocity is less than the initial velocity.

definite integral   An integral between two limits that results in a numerical value. It is written as $\int_a^b y \, dx$ where a and b are the limits.

denominator   The number or expression below the line in a fraction. Also known as the divisor.

dependent (response) variable   If there is a relationship between two variables and the value of one variable, y say, is dependent on another variable, x say, then y is said to be the dependent (response) variable.

derivative   The result of differentiating a function. It is written as $\frac{dy}{dx}$ or $f'(x)$.

deviation   The difference between a raw score and the mean.

diameter   A line joining two points on the circumference of a circle that passes through the centre of the circle.

differentiation   The process of finding an expression for the gradient of a curve.

direct proportion   A relationship in which one variable increases or decreases at the same rate as another. In the formula $y = 3x$, x and y are in direct proportion.

discontinuity   Where the value of x or y is not defined.

discrete data   Data with exact values.

discrete random variable   A random variable that can only take specific values, for example whole number values, in the given interval.

discriminant   The expression $b^2 - 4ac$ based on the general equation $ax^2 + bx + c = 0$. If the discriminant of a quadratic equation is negative, then the equation has no real roots.

displacement   The distance in a certain direction from an object's starting point.

disproof   A mathematical operation to show that a mathematical statement is not true.

distance   How far an object travels.

elimination   Given a pair of simultaneous equations with two variables, you can manipulate one or both equations to remove or eliminate one of the variables by a process of substitution, addition or subtraction.

equal vectors   Two vectors are equal if they have the same magnitude and the same direction.

equation A relation in which two expressions are separated by an equals sign with one or more variables. An equation can be solved to find one or more solutions, but it may not be true for all values of x.

event A set of possible outcomes from an experiment, for example rolling an even number on a die.

expansion of brackets A mathematical operation to remove the brackets from an expression by multiplication.

exponent See **index**.

exponential growth or decay Change over time represented by an exponential function. The rate of change of the function is proportional to the value of the function.

expression A group of numbers, symbols or operators. For example, $2x \times 5y^2$.

extrapolation Estimating a value outside the range of the data that you have. As it is outside the data you have, extrapolated values can be unreliable.

factor A divisor of a number with a zero remainder.

factor theorem A method to identify a factor of a polynomial expression, $f(x)$. If $f\left(\dfrac{m}{n}\right) = 0$, then $(mx - n)$ is a factor of $f(x)$.

factorial The product of the whole number n and all the whole numbers less than n down to 1. It is written as $x!$. For example, $5! = 5 \times 4 \times 3 \times 2 \times 1 = 120$.

factorisation The arrangement of a given number or expression into a product of its factors (from the verb: to factorise).

force An influence which causes an object to accelerate.

formula A mathematical rule, using numbers and letters, that shows a relationship between variables. For example, the conversion formula from temperatures in Fahrenheit to temperatures in Celsius is: $C = \dfrac{5}{9}(F - 32)$.

frequency The number of times an event occurs.

function An algebraic expression in which there is only one variable, often x.

gradient The slope of a line between two or more points, calculated as the vertical difference between the coordinates divided by the horizontal difference.

gravity The attractive force between bodies due to their mass.

highest common factor (HCF) The largest number that is a factor common to two or more other numbers.

histogram A diagram consisting of rectangles with area proportional to the frequency of a variable and width equal to the class interval.

hypothesis A theory or prediction that a statistical experiment is designed to test.

hypothesis test A formal process to determine whether to reject a null hypothesis, based on sample data.

i j notation A way of representing a vector in two dimensions, where $\mathbf{i}$ is one unit east and $\mathbf{j}$ is one unit north.

inclusive inequalities Inclusive inequalities contain an 'equals' element, specifically, 'less than or equal to' ($\leq$) and 'greater than or equal to' ($\geq$).

indefinite integral The result of integrating a function. when no limits are defined. An indefinite integral is written in the form $\int y \, dx$ and requires the addition of an unknown constant, usually written as $+c$.

independent events Two events, A and B, are independent if the fact that A occurs does *not* affect the probability of B occurring.

independent (explanatory) variable If there is a relationship between two variables and the value of one variable, y say, is dependent on another variable, x say, then x is said to be the independent (explanatory) variable.

index The power to which a base number is raised. For example, in 3^4, 4 is the index and 3 is the base number.

inequality A statement that one expression is greater or less than another, written with the symbol > (greater than), < (less than), $\geq$ (greater than or equal to) or $\leq$ (less than or equal to) instead of = (equals). See **strict inequalities** and **inclusive inequalities**.

integer Any positive or negative whole number, including zero.

integration The opposite to differentiation. Finding a function that, when differentiated, results in a specified function.

intercept The point where a line or curve cuts or crosses the axis.

interpolation The process of finding a value between two points.

interquartile range A measure of variability, based on dividing a data set into quartiles, equal to Q_3 minus Q_1.

irrational number A number that cannot be expressed as the ratio (fraction) of two integers. For example, $\sqrt{2}$.

linear An expression or similar containing terms in one variable where all the indices are greater than or equal to zero and the highest index is 1.

linear coding A change to a variable characterised by one or more of the following operations: adding a constant to the variable, subtracting a constant from the variable, multiplying the variable by a constant, dividing the variable by a constant.

logarithm The logarithm of a number is the power of the base that will give that number. If $a^x = b$ then $x = \log_a b$.

magnitude The size (or length) of a vector.

maximum point A point on a graph where the gradient is zero, which is higher than the points either side of it. Can be the global maximum where it is the highest point anywhere on the graph or a local maximum where it is the highest point in that part of the graph.

mean Average – the sum of individual items divided by the number of items.

median A simple measure of central tendency. If there is an odd number of observations, the median is the middle value. If there is an even number of observations, the median is the average of the two middle values.

member of a set An item contained in a set. See **set notation**.

midpoint The point halfway between two other points.

midpoint rule For position vectors $\mathbf{a}$ and $\mathbf{b}$ from the same point, the position vector of $\mathbf{c}$ which divides AB in the ratio 1 : 1 is given by $\frac{1}{2}(\mathbf{a} + \mathbf{b})$.

minimum point A point on a graph where the gradient is zero, which is lower than the points on either side of it. Can be the global minimum where it is the lowest point anywhere on the graph or a local minimum where it is the lowest point in that part of the graph.

mode The most frequently appearing value in a population or sample.

mutually exclusive events Two or more events which cannot occur simultaneously.

natural logarithm A logarithm to base e, where $e = 2.718281...$ If $e^x = a$ then $\ln a = x$.

Newton's first law A particle remains at rest or continues to move at a constant velocity in a straight line unless acted upon by an external force.

Newton's second law The force, F, acting upon a particle is proportional to the particle's mass, m, and acceleration, a.

Newton's third law Every action has an equal and opposite reaction.

normal to a curve A normal to a curve at a given point is a straight line through that point which is perpendicular to the tangent at that point.

null hypothesis A hypothesis which the researcher tries to disprove, reject or nullify in a statistical experiment.

numerator The number or expression above the line in a fraction.

numerical data Data representing a measurable quantity.

opportunity sample A technique that uses people from a target population available at the time and willing to take part. It is based on convenience.

opposite vectors Two vectors are opposite if they have the same magnitude but point in opposite directions.

origin The point O (0, 0) on Cartesian coordinate axes.

outcome One possible result from an experiment, for example rolling a 6 on a die.

outlier A data point that diverges greatly from the overall pattern of data.

Pascal's triangle A number pattern named after the French mathematician Blaise Pascal (1623–62). It is a triangular pattern of binomial coefficients.

parabola The shape achieved when a cone is cut by a plane that is at the same incline as the side of the cone.

parallel Two or more lines are parallel if their gradients, m, are the same.

parallelogram law If the vectors **a** and **b** are drawn in that order and then the vectors **a** and **b** are also drawn in the opposite order starting from the same point, a parallelogram is produced.

percentiles The values that divide a rank-ordered set of elements into 100 equal parts.

perpendicular Two straight lines with gradients m_1 and m_2 are perpendicular when $m_1 m_2 = -1$.

point of inflection Where the gradient of the curve is zero and it is not a turning point.

point of intersection A point where lines or curves intersect.

polynomial An algebraic expression containing two or more terms such as $2x^3 - 4x + 3$. Possible operations include addition, subtraction and multiplication.

population The underlying set from which samples are taken.

position vector The relative displacement of a point from the origin (e.g. a point with coordinates $(-2, 3)$ has the position vector $(-2\mathbf{i} + 3\mathbf{j})$.

power See **index**.

probability The measure of the chance that an event will occur.

probability distribution The set of all possible values a random variable can take, with the associated probability of each value occurring.

probability distribution function A function that defines how the probabilities in a probability distribution are to be determined.

proof A mathematical operation to show that a mathematical statement is true.

pulley A wheel which a string passes over, often for the purpose of moving an object.

quadrant One of the four sections between a pair of axes, crossing at the origin, with positive and/or negative values.

quadratic An expression or similar containing terms in one variable where all the indices are greater than or equal to zero and the highest index is 2.

qualitative data Data that can only be listed, such as make, model or colour.

quantitative data Data that is numerical, such as shoe size, temperature or height.

quartic An expression or similar containing terms in one variable where all the indices are greater than or equal to zero and the highest index is 4.

quartiles Divisions of a rank-ordered data set into four equal parts.

quota sample A sample that is simply selected with a predefined category. This is quick and easy to do but may not represent the whole population.

quotient A number or expression obtained after dividing a number or expression by a divisor.

radius A line joining the centre of a circle to its circumference.

random sampling A procedure for sampling from a population in which the selection of a sample unit is based on chance and every member of the population has the same probability of being selected.

range The difference between the largest value and the smallest value in a set of data.

rate of change The rate of change of y with respect to x is $\frac{dy}{dx}$.

ratio theorem Given position vectors **a** and **b** from the same point, the position vector of **c** which divides AB in the ratio $m : n$ is given by $\frac{1}{m+n}(n\mathbf{a} + m\mathbf{b})$.

rational number A number that can be expressed as the ratio (fraction) of two integers. For example, $\frac{1}{3}$.

rationalise To remove a surd from a denominator (by multiplying the numerator and denominator by that surd).

reaction force The force an object experiences when it is in contact with a surface – always acts perpendicular to the surface.

reciprocal The result of dividing a number into 1, so 1 divided by the number is its reciprocal.

regression line A line of best fit for a given set of values, using the equation of a straight line, $y = a + bx$.

relative displacement The displacement of one object compared with another object.

resistive force A force which opposes motion, reducing acceleration or causing deceleration.

roots The points on a graph where it crosses the x-axis.

rough If a surface is rough, it will produce a resistive frictional force.

sample space The range of values of a random variable.

sample survey A study that obtains data from a subset of a population, in order to estimate population attributes.

sampling frame A list of the items or people forming a population from which a sample is to be taken.

sampling unit A single section selected to research and gather statistics of the whole population.

second derivative The result of differentiating a function twice. It is written as $\frac{d^2 y}{dx^2}$ or $f''(x)$.

set A collection of objects or elements.

set notation A way of listing the members of a set.

SI units An internationally recognised system of units from which all other units can be derived.

significance level The null hypothesis is rejected if the p-value is less than the significance level.

simplify To make an equation or expression easier to work with or understand by combining like terms or cancelling.

simultaneous Two or more equations that are true for the same set of values for their variables.

sine The sine of a (or $\sin a$) is the y-coordinate of a point on the circle of radius 1 unit. The variable a can take any value.

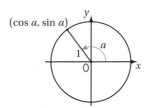

skew When data are displayed graphically, some distributions have many more observations on one side of the graph than on the other. Distributions with fewer observations on the right are positively skewed; distributions with fewer observations on the left are negatively skewed.

smooth If a surface is smooth, it will not produce any resistive frictional force.

speed The magnitude of the velocity.

standard deviation A numerical value used to indicate how widely individuals in a group vary, equal to the square root of the variance.

stationary point A point on a curve where $\frac{dy}{dx} = 0$.

stratified sampling Technique of dividing the population into separate groups, called strata, then drawing a sample from each group.

strict inequalities Do not contain an equals sign; specifically, 'less than' ($<$) and 'greater than' ($>$).

substitution Replace a variable in an expression with a number and evaluate it.

surd An irrational number found by taking a root of a number.

systematic sampling From a list of every member of a population, every nth element on the list is selected.

tangent (1) A straight line through a point on a curve that has the same gradient as the curve at that point. (2) The tangent of a (or $\tan a$) is defined by the formula $\tan a = \frac{\sin a}{\cos a}$.

target population The total group of individuals from which the sample might be drawn.

taut Pulled tight.

tension The force in a string when it is taut.

term   A part of an expression, equation or formula.

test statistic   A statistic used to test whether the null hypothesis should be accepted or rejected in a statistical experiment.

thrust   A force which pushes an object.

transformation   A change to a function, such as a translation, rotation, reflection or enlargement.

triangle law   Vectors can be added using the triangle law. If the vector $\overrightarrow{AB}$ (**a**) is followed by the vector $\overrightarrow{BC}$ (**b**), then the result is the vector $\overrightarrow{AC}$ (**a** + **b**).

turning point   A peak (maximum) or a trough (minimum) on a curve.

uniform distribution   A probability distribution where each value is equally likely, for example the rolling of a fair die.

unit vector   A vector with a magnitude of 1.

variance   A numerical value used to indicate how widely individuals in a group vary, equal to the square of the standard deviation.

vector   A quantity with a size and a direction.

velocity   The rate of change of displacement.

velocity vector   A vector for which the magnitude is the speed and the direction is the direction in which the object is moving.

weight   The force caused by the gravitational acceleration experienced by the object – weight always acts vertically downwards.

INDEX

Acknowledgements

The publishers wish to thank the following for permission to reproduce photographs. Every effort has been made to trace copyright holders and to obtain their permission for the use of copyright materials. The publishers will gladly receive any information enabling them to rectify any error or omission at the first opportunity.

Cover & p1 Paladin12/Shutterstock

p1 INTERFOTO/Alamy, p31 Amy Myers/Shutterstock, p58 stocker1970/Shutterstock, p100 zhykova/Shutterstock, p126 Media Services Asia Pacific/Alamy, p154 3Dsculptor/Shutterstock, p179 hxdyl/Shutterstock, p212 supergenijalac/Shutterstock, p243 Radu Bercan/Shutterstock, p260 Ian Shaw/Alamy, p285 BlueSkyImage/Shutterstock, p299 ibreakstock/Shutterstock, p342 ktsdesign/Shutterstock, p368 Christian Bertran/Shutterstock, p391 NASA, p425 OLOS/Shutterstock.